Interaction Between
Ions and Molecules

NATO ADVANCED STUDY INSTITUTES SERIES

A series of edited volumes comprising multifaceted studies of contemporary scientific issues by some of the best scientific minds in the world, assembled in cooperation with NATO Scientific Affairs Division.

Series B: Physics

Volume 1 – Superconducting Machines and Devices
edited by S. Foner and B. B. Schwartz

Volume 2 – Elementary Excitations in Solids, Molecules, and Atoms
(Parts A and B)
edited by J. Devreese, A. B. Kunz, and T. C. Collins

Volume 3 – Photon Correlation and Light Beating Spectroscopy
edited by H. Z. Cummins and E. R. Pike

Volume 4 – Particle Interactions at Very High Energies
(Parts A and B)
edited by David Speiser, Francis Halzen, and Jacques Weyers

Volume 5 – Renormalization and Invariance in Quantum Field Theory
edited by Eduardo R. Caianiello

Volume 6 – Interaction between Ions and Molecules
edited by Pierre Ausloos

Volume 7 – Low-Dimensional Cooperative Phenomena
edited by H. J. Keller

The series is published by an international board of publishers in conjunction with NATO Scientific Affairs Division

A	Life Sciences	Plenum Publishing Corporation
B	Physics	New York and London
C	Mathematical and Physical Sciences	D. Reidel Publishing Company Dordrecht and Boston
D	Behavioral and Social Sciences	Sijthoff International Publishing Company Leiden
E	Applied Sciences	Noordhoff International Publishing Leiden

Interaction Between Ions and Molecules

Edited by

Pierre Ausloos

Chief, Radiation Chemistry Section
National Bureau of Standards
U. S. Department of Commerce
Washington, D. C.

PLENUM PRESS • NEW YORK AND LONDON
Published in cooperation with NATO Scientific Affairs Division

Library of Congress Cataloging in Publication Data

NATO Advanced Study Institute on Ion-Molecule Interactions, Biarritz, 1974.
  Interaction between ions and molecules.

  (NATO advanced study institutes series: Series B: Physics; v. 6)
  Includes bibliographies and index.
  1. Chemical reaction, conditions and laws of—Congresses. 2. Ions—Congresses. 3.
Molecules—Congresses. I. Ausloos, Pierre J., ed. II. Title. III. Series.
QD501.N36 1974   539.7'54   74-31389

ISBN-13: 978-1-4613-4457-5   e-ISBN-13: 978-1-4613-4455-1
DOI: 10.1007/ 978-1-4613-4455-1

Lectures presented at the 1974 NATO Advanced Study Institute on Ion—Molecule
Interactions, held in Biarritz, France, June 24—July 6, 1974

© 1975 Plenum Press, New York

Softcover reprint of the hardcover 1st edition 1975

A Division of Plenum Publishing Corporation
227 West 17th Street, New York, N.Y. 10011

United Kingdom edition published by Plenum Press, London
A Division of Plenum Publishing Company, Ltd.
4a Lower John Street, London W1R 3PD, England

Preface

In the years since the early 1950's when the first quantitative observations of elementary reactions between ions and molecules were reported, a community has grown up in laboratories throughout the world of scientists whose primary interest is the exploration of various aspects of ion-molecule interactions. Nevertheless, until the summer of 1974, the working physicists and chemists who are the citizens of this "empire of ion-molecule interactions" (as it is christened in this volume by Michael Henchman) had never come together as an integrated group just for the purpose of exploring their common area of interest. The format of a North Atlantic Treaty Organization Advanced Study Institute seemed ideal for bringing about such a gathering, since in the intended atmosphere of a summer school, not only would the leading experts on the subject come together to learn from one another, but also many of the talented younger scientists who were just entering the field would be able to actively participate. The response from numerous laboratories to the initial suggestion for the organization of such an Institute was immediate and enthusiastic. And so it was that with the blessing of the Scientific Affairs Division of NATO a hundred or so scientists gathered in Biarritz, France, from June 24 to July 6, 1974, to discuss ion-molecule interactions.

The scope of the program of this Advanced Study Institute was broad, covering nearly every aspect of the field of ion-molecule interactions including the dynamics of the collisions of charged species with polar and non-polar molecules, the mechanisms, thermochemistry, and theoretical descriptions of ion-molecule reactions, as well as discussions of ion-molecule reactions in planetary ionospheres, interstellar space, flames, and systems under high energy irradiation. Few, if any of the participants came to Biarritz well versed in all of these areas. Thus, we all became students as well as teachers, and the most common complaint heard during the two week Institute was that there was such a vast amount of material to be assimilated. Nevertheless most of the participants courageously followed the various lectures and panels, and despite the lure of the Biarritz beach just outside the doors of the conference room, there were few drop-outs.

I would like to take this opportunity to express my sincere appreciation to all those who helped make this undertaking a success. Heading the list of those who deserve special credit is Rose Marx of the Universite de Paris-Sud, who as the member of the Organizing Committee from the host country graciously gave of her time and the facilities of her office to arrange all of the practical details of our stay in France. Special thanks also to Keith Jennings of the University of Warwick, whose attention to organizational details before and during the Institute helped to keep things running smoothly, and to Sharon Lias of my laboratory, who with the help of Carole Lamb, handled most of the voluminous correspondence involved in organizing the Institute. Thanks also to the other members of the Organizing Committee, Eldon Ferguson of NOAA Environmental Research Laboratory, Arnim Henglein of the Hahn-Meitner Institute, and Paul Kebarle of the University of Alberta, whose advice and collaboration in planning the program of the Institute was essential to its success.

We were especially fortunate in having as the secretary for the Institute Denise Duranthon. Not only was she outstandingly efficient, she anticipated our needs before they occurred, and more important, she remained cheerful and cooperative above and beyond the call of duty, in spite of the numerous demands made on her time and patience.

I believe that I can speak for all of the participants to use this occasion to acknowledge our appreciation to the staff of the Miramar Hotel, Biarritz, France. Our relationship was not always an easy one. The Miramar is an old hotel, with an atmosphere of gentility, formality, and elegance. The staff hardly knew what to think of us--we picnicked in our rooms, took over their lobbies for our conversations, and congregated in their dining room for breakfast when they were prepared to serve us in our rooms, then reversed ourselves and ate in our rooms when they expected us to come to the dining room. Yet, in spite of these and other cultural differences, the members of the staff remained friendly and cooperative, and even indicated at the end of the Institute that they would welcome a return visit by such an unpredictable group of scientists.

Thanks also to all the speakers and panel leaders whose efforts made up the real substance of the Institute, to all the participants, whose hard work and lively interest made this a memorable and stimulating summer school.

P. Ausloos

September 10, 1974

Contents

PANEL REPORTS

A HISTORY OF THE STUDY OF ION MOLECULE REACTIONS

J. L. Franklin

Department of Chemistry

William Marsh Rice University, Houston, Texas   77001

Most of us are inclined to think of ion-molecule reactions as a recent discovery. However, literature on the study of canal rays refers frequently to particles forming masses that were difficult to interpret; that is, the masses did not occur at positions corresponding to the expected sum of the atomic masses thought to be present. For example, Thompson (1913) refers to his observation of a particle at mass 3 in his study of the parabolas of ions formed in his equipment. He speculated that this might be either triply charged carbon or triatomic hydrogen and thought it probably was the latter. However, the mechanism of its formation was not established. Dempster (1916) using a very early approximation to a mass spectrograph observed an ion at M/e = 3 and concluded that it probably resulted from the ionization of a temporary combination of H_2 and H atoms. Hogness and Lunn (1925) again studied the ionization of hydrogen and concluded that the ion at M/e = 3 was indeed H_3^+. They observed that its intensity increased with pressure and concluded that the probable mechanism was the collisional dissociation of H_2^+ followed by the attachment of the proton to either its collision partner or to another molecule of hydrogen. They also observed a greater abundance of H^+ and very little H_2^+ or H_3^+ in mixtures of H_2 with He and concluded that H_2^+ is much more readily dissociated by collision with helium than with H_2. (Of course, this probably was the result of dissociative charge exchange with a helium ion.) However, the authors also reported the formation of HeH^+ as well as an ion at M/e = 6 which they thought might be HeH_2^+. Smyth (1925) again investigated hydrogen and showed that H_3^+ increased with pressure. He thought that the H_3^+ might be formed by reaction of H_2^+ with hydrogen atoms; of H^+ with H_2; or by the formation of an H_3

neutral which was subsequently ionized by electron impact. However, we find Smyth (1931) in his review "Products and Processes of Ionization by Low Speed Electrons" saying that the balance of evidence favors the reaction

$$H_2^+ + H_2 \rightarrow H_3^+ + H \tag{1}$$

Hogness and Harkness (1928) made the first really definitive study of ion-molecule reactions in a simple system. They studied both the positive and negative ions formed in iodine and observed charge transfer from both positive and negative atomic ions to form the corresponding molecular ions, and, in addition, observed the ion-molecule reactions

$$I_2^+ + I_2 \rightarrow I_3^+ + I \tag{2}$$

$$I_2^- + I_2 \rightarrow I_3^- + I \tag{3}$$

Since I_2^- is itself formed by charge exchange, the I_3^- is the result of a ternary reaction, probably the first such that had been determined. They also noted that the occurrence of charge exchange showed the electron affinity to I_2 to be greater than that of I. More recent results suggest that in fact the electron affinity of I_2 is slightly less than that of I, so the charge transfer was brought about, in the case of Hogness and Harkness (1928), by the relatively large collision energy.

After about 1931 there was very little activity in the study of ion-molecule reactions. At about the end of the 1920s the advent of good pumping systems and the development of relatively modern types of mass spectrometers encouraged physicists, the main users of this instrument, to emphasize investigations of ionization and appearance potentials and bond dissociation energies. Some studies of electric discharges were carried out, but little direct information resulted from them. It is of interest, however, that Eyring, Hirschfelder and Taylor (1936), in an effort to interpret the effects of radiation, developed a theoretical treatment of the rate constant for the reaction of H_2^+ with H_2 and obtained a result in almost exact agreement with experimental values measured some 20 years later. The treatment was based on absolute rate theory and assumed that the difference between rotational energy and the charge-induced dipole energy would result in a pseudo-activation energy. When this was incorporated into the absolute rate equations a quite satisfactory calculation of the rate constant resulted. It should be mentioned here that this treatment was in fact a calculation of the collision rate, but the rate of ion-molecule collisions is often the same as the reaction rate.

Except for a brief study by Mann, Hustrulid and Tate (1940) of the reaction

$$H_2O^+ + H_2O \rightarrow H_3O^+ + OH \tag{4}$$

made as part of a study of the appearance potentials of the various ions formed from water, no further work on ion-molecule reactions as such was carried out until the early to mid 1950s.

During a period from about 1952-57 several laboratories independently undertook studies of ion-molecule reactions employing the ionization chambers of mass spectrometers as reactors. Stevenson and Schissler (1955, 1956, 1958), Field, Franklin and Lampe (1957) and Meisels, Hamill and Williams (1956) in the US, and Tal'Roze and Lyubimova (1952) in Russia studied methane, ethylene, acetylene and other compounds. All observed the principal reactions occurring in methane, namely:

$$CH_4^+ + CH_4 \rightarrow CH_5^+ + CH_3 \tag{5}$$

$$CH_3^+ + CH_4 \rightarrow C_2H_5^+ + H_2 \tag{6}$$

Tal'Roze seems to deserve credit for being the first to observe the formation of CH_5^+ in methane, but the availability of Russian literature in the United States was very limited and the Russian results did not become available to American investigators until after they had begun publication.

Apparently, ion-molecule reactions in this period was a subject whose time had come. Although important work on the appearance potentials and bond association energies continued, several investigators, including those mentioned above, became keenly interested in other applications of mass spectrometry, and the quite surprising results obtained with methane encouraged other investigators to undertake these studies also. Since that time interest in ion-molecule reactions and the number of publications in the field have increased almost exponentially until at this time there are probably more articles published per year on this subject alone than had been published in all time on all aspects of mass spectrometry prior to about 1955. An indication of this is the fact that Field and Franklin (1957) in the brief book published in that year included about three or four pages on the then very limited amount of information available in the literature of ion-molecule reactions. Lampe, Franklin and Field (1961) published a brief review of the kinetics of ion-molecule reactions which included over 50 references to studies of the subject. A book by McDaniel et al. (1970) and a two-volume set edited by Franklin (1972), both devoted entirely to ion-molecule reactions,

were published. There is no indication that interest in the field
is declining.

The early "modern" studies of ion-molecule reactions were
largely carried out in the conventional sector-field type mass
spectrometer such as one normally used for measuring appearance
potentials or for chemical analysis. Although the range of source
pressures which could be employed was very limited, reasonably
reproducible rate constants could be measured by varying the
source pressure and by calculating the time on the assumption of
the free fall of the ion through the repeller field in the ionization chamber. A considerable number of systems were studied in
this way. The method and equipment, while primitive by current
standards, gave in many instances surprisingly good results. One
of the very startling discoveries was the fact that the rate constant was in many instances considerably greater than the classical
collision rate constant that would be calculated on the assumption
that Van der Waals forces prevailed. It was further evident that
these very large reaction rates could not occur if activation
energy were involved, and studies by Stevenson and Schissler (1958)
showed that for several reactions increasing the temperature did
not affect the rate constant. Similarly, in an early study, Field,
Franklin and Lampe (1957) showed that some rate constants decreased
markedly as the repeller field was increased from about one volt
per cm to about 10 volts per cm. This again showed that there was
no energy barrier to reaction.

In spite of these earlier successes it soon became apparent
to many investigators that the single continuous extraction source
typical of the conventional sector-field mass spectrometer was not
an ideal device for the study of ion-molecule reactions. Among
the difficulties and deficiencies of this instrument are the
following:

(1) It is extremely difficult, and in most instances it is
impossible to detect a secondary ion having the same mass as a
primary. Although in some instances it has been possible to circumvent this condition, the instrument is not in general well
suited to such studies. Thus it is very difficult in the continuous extraction source to study charge exchange reactions, to
detect products of hydride ion transfer and in many instances to
identify the precursors of a certain product ion.

(2) Rate constants determined in the continuous extraction
source represent an average over energy as the primary ion
accelerates over the flight path in the source. So long as the
field strength in the source is sufficiently small and the reaction
is exothermic, this will make a relatively small difference, but
in many cases the rate constants determined are not satisfactory.

(3) A single source, especially with continuous extraction, is not well suited to investigations of the effects of collision energy upon reaction and is especially poorly suited to the measurement of reaction rate constants for ions having thermal energy.

(4) The usual electron impact source suffers from the fact that the energy spread in the electron beam is quite large, normally in the order of 1/2 to 1 electron volts. This is so great that reaction rates for individual vibrational and electronic states cannot be determined.

(5) The conventional source cannot be operated at elevated pressures without narrowing the slits and, in most instances, installing differential pumping.

Naturally, investigators had many reasons for wishing to circumvent these problems and various approaches designed to accomplish certain specific results were undertaken. One of the first of such changes was made by Field (1961) to enable him to operate at pressures up to 0.5 to one torr instead of pressures in the order of a micron or less that had been the upper limit of conventional instruments. As would be expected, many studies at pressures of 0.1 to 1.0 torr resulted in products and reactions that would not have been suspected from studies at the lower pressures. Thus Field's (1961) first paper on ethylene detected some reactions that were at least 6th order kinetically.

Subsequent studies by Field and Munson (1965) showed that when extremely pure methane was employed the principal secondary ions, CH_5^+ and $C_2H_5^+$, did not react further. They showed that the addition of very small amounts of ethane or propane resulted in rapid depletion of one or both of the reagent ions. (Munson and Field 1965). They deduced that these ions might well be used as reagent ions to react with small amounts of polyatomic compounds and thus serve as a means of qualitative identification of pure compounds and perhaps of quantitative analysis of mixtures. They accordingly undertook to determine the spectra of a large number of compounds of various classes when these were added to the methane reagent gas. (Field et al., 1966) They found that the spectra were usually simpler than electron impact spectra and often included peaks one mass unit above or below that of the parent for compounds that gave no detectable ions near the parent under electron impact. From this they developed the chemical ionization process that is now used widely to identify very small amounts of complex molecules.

At sufficiently high pressures and long retention times it is possible to achieve equilibrium in certain ion-molecule reactions

and to determine the equilibrium constant. If this can be done at
several temperatures it is then possible to compute the free energy,
heat and entropy of reaction in the same fashion that chemists
employ in condensed systems. Equilibrium is readily determined in
proton transfer reactions and several such studies have been
reported. This has been an important means of measuring proton
affinities and has yielded values for water, (Chong et al., 1972)
benzene (Chong and Franklin 1972) and many others (Beauchamp 1971).
Similarly, relative values of gas phase acidities can be determined
by finding the direction of proton transfer in the process
$B^-(HA,HB)A^-$. By studying an homologous series in this way an
understanding of the effect of structure upon acid strength in the
absence of solvent can be determined. Thus Brauman and Blair
(1968, 1970) determined the relative acidities of the normal
alcohols and found the acidity to increase with chain length. This
is just opposite to the behavior in solution, where the acidity of
normal alcohols decreases with increasing chain length.

It has been observed by many workers that NH_4^+ and H_3O^+ add
molecules of ammonia or water stepwise as the pressure is increased.
Kebarle has pioneered in measuring the equilibrium constants at
various temperatures for the stepwise addition of solvent molecules
to ions. He has studied the ammonia and water systems and in addition has determined the thermodynamic constants for the stepwise
solvation of several positive and negative ions, including the
alkali positive ions and the halide and hydroxide negative ions.
This work is summarized in Kebarle's (1972) chapter.

The problem of evaluating rate constants, determined at
conditions in which the collision energy is uncertain, have been
resolved in a number of ways. Probably the earliest one was to
employ a pulsed ion source. (Tal'roze and Frankevich 1959, 1960;
Hand and von Weyssenhoff 1964; Lampe and Hess 1964) In such a
source ions are formed during a brief period, normally a quarter
of a micro second, while the electron beam is passed through the
ionization chamber. After the electron beam is cut off, the ions
may be retained in the ionization chamber from zero to several
micro seconds before being withdrawn and analyzed. Thus the rate
of the reaction is determined as chemists normally measure reaction
rates; namely, by measuring the variation of concentration or
intensity with time. Further, the ions retain their initial translational energy throughout their lifetime in the ionization
chamber. Thus, some of the uncertainties concerning collision
energy inherent in the continuous extraction source were removed.
Unfortunately, primary ions formed in fragmentation processes often
are formed with considerable translational and internal energy.
These have seldom been measured, so that their influence on
reaction rate was not established. Thus the pulsed source
represented an improvement but was not a complete solution to the
problem of control of the collision energy.

Several more sophisticated approaches to the problem have been quite successful. One of these is the use of crossed molecular beams. Probably the first "modern" study of reactions of negative ions using a beam apparatus was by Muschlitz (1957) in which he studied the reaction

$$H^- + H_2O \rightarrow OH^- + H_2 \tag{7}$$

Subsequent investigations by Henglein,(Henglein et al., 1965; Henglein 1966) Wolfgang (1972),and others have made valuable contributions, but in general have been directed toward reaction dynamics and will be discussed later. A completely different approach was devised by Ferguson and his associates (Ferguson et al., 1969) for the purpose of investigating ion molecule reactions at thermal conditions. For this they devised the extremely productive flowing afterglow method. In this method ions are formed in an electric discharge or by passage of a stream of electrons into the gas. Usually one of the rare gases is used. The ions leaving this primary source flow rapidly down a flight tube in which the time is determined by the rate of flow of the gas. A short distance downstream the rare gas ions encounter molecules injected into the stream and charge exchange with them to produce the reactant ions of interest. These are allowed to thermalize by flowing downstream a few centimeters,after which the neutral partner of the ion-molecule reaction is introduced into the stream. The reaction proceeds during the remaining time in the flight tube and the ions are sampled at the end of the tube into a quadrupole mass filter. This method has been extremely productive, and is the method of choice for the study of ion-molecule reactions at thermal conditions. Obviously, it is possible to vary the temperature of the system and thus to study the effect of temperature on reaction rate. The method has the extremely valuable property of making it possible to introduce normally unstable neutrals, such as atoms or free radicals, into the flowing afterglow and to observe their reactions with various ions in the system. This is the only method that has been successful in studying ion-molecule reactions involving unstable compounds, atoms and free radicals.

Of a somewhat similar nature, although probably not as completely reliable,is the study of ion-molecule reactions in a drift tube, in which ions drift through the gas under the influence of a small electric field through a rather long path. If the pressure in the system is sufficiently large and the field strength sufficiently small the mean energy of collision of ions with neutrals will approach quite close to the thermal energy. In this way rate constants determined from drift tube measurements can be looked upon as representing nearly thermal conditions. However, in order to obtain a completely valid measurement of the rates, it is necessary to measure the time distribution of ions in the drift tube and this is seldom done. The use of the drift tube to study the mobility of ions is an old technique, but unfortunately most of this work was done without the benefit of mass analysis and thus the results are often erroneous because the ions flowing were not the ones thought to be present. It is possible to estimate the mobility and thus the retention time if facilities are not available for measuring the time distribution. McDaniel and his associates (McDaniel et al., 1970) have pioneered studies of ion-molecule reactions in the drift tube with mass analysis, and several others, including Parkes (1972) and Long and Franklin (1974) have also carried out reactions in the drift tube to determine both reaction rates and equilibrium constants.

The ion cyclotron resonance mass spectrometer (Anders et al., 1966; Beauchamp 1971) has been developed primarily to study ion-molecule reactions and has proved gratifyingly successful. It offers the great advantage of giving absolute concentration measurements so that no assumptions need be made as to the relation of observed ion intensity to concentration. It offers the further advantage that the collision energy can be kept quite low and very close to thermal values. Thus,rate constants obtained by ICR approach closely to those for thermal systems. Further, ions may be retained in the instrument for seconds instead of micro seconds so that it is often possible to get equilibrium constants for exchange reactions.

The ion cyclotron resonance instrument is ideally suited to the identification of the precursor to certain secondary ions. This is accomplished by focusing on the secondary and increasing the energy input to the primary ion to indicate whether the secondary is affected. Thus, it is possible to detect the amount of reaction even though several primaries result in a common secondary or vice versa. The instrument has been used to great advantage for the purpose of sorting out various complex reaction systems and indeed this was the first major use to which the instrument was put.

In the conventional ion source it is almost always impossible to study ion-molecule reactions that result in secondary ions having the same mass as those of primary. About the only way that this can be accomplished in the continuous withdrawal source is by the method devised by Cermak and Herman (1961).   In this method electrons pass through the ionization chamber at energies below the ionization potential of the molecule under investigation. They are then accelerated in the trap region where they cause ionization. The positive ions formed are repelled into the ionization chamber where they can undergo charge exchange and/or ion-molecule reaction. In this method the product of the charge exchange process will be collectible since normally there is little momentum exchange in a charge exchange collision. Thus, the

charge exchange products can be detected. In some instances it
is also possible to observe the products of ion-molecule reactions.
Thus, one might expect to see products of a reaction involving a
long lived ion-neutral complex, or certain reactions in which a
positive fragment is transferred from the colliding ion to the
neutral.   Reactions of the latter kind often involve little or no
momentum change and thus the secondary can be collected. On the
other hand, if a neutral fragment is stripped from a molecule by an
ion, the resulting ion will not be collected.   Thus the method is
useful for certain kinds of processes and indeed can sometimes be
used to give insight into reaction mechanism. This method was
extended for pulsed sources by Hierl, Sen Sharma and Franklin
(1971) and several very good cross sections for charge exchange
have been determined in this way. (Sen Sharma, Hierl and Franklin
1972).

        In a molecular beam that has been subject to mass selection
before entering a collision chamber or a collision region there
should be no question about the nature of the precursor to a
given secondary. Thus, the reaction mechanism will be readily
established. This is also true of tandem mass spectrometer
systems. In such systems the primary ion is segregated by passing
through a mass spectrometer and introduced into a collision
chamber. The product ions are extracted and analyzed through a
secondary mass spectrometer system. For charge exchange studies
it is desirable to extract the secondary ions at right angles to
the path of the entering primaries. Lindholm (1954) has pioneered
in such studies. For studies of ion-molecule reactions it is
usually preferable to extract the secondary ions co-linearly with
the entering primary ion beam. Paulson (1970) has made several
studies of negative ions using this method. The tandem instrument,
when equipped with suitable lens, can be used to provide a wide
range of collision energies and further, the secondary ions can be
passed through an electro-static energy selector as well as the
mass selector to give a measure of the energy distribution of the
product ions. (Futrell and Miller, 1966; Franklin, 1972; Lifschitz
et al., 1971, 1973) The ultimate instrument allows the secondary
analyzer to be rotated around a point in the center of the collision zone so as to determine the spatial distribution of the
products as well. From this information it is possible to deduce
whether the reaction involves a long-lived complex or whether it
proceeds by a stripping or direct mechanism and thus provide
valuable information for theoretical study. Studies by Wolfgang
(1972) and others have provided valuable insights into the nature
of the chemical reactions and the manner in which energy is
distributed in the products. Thus the tandem mass spectrometer
and the crossed beam systems are generally fine instruments for
the study of ion-molecule reactions, but they suffer from one
disadvantage. It is usually impossible to stop the primary beam

down to energies much below about 3/10 volts and thus these
devices are not suitable for studies of thermal ionization.

It is often of considerable interest to study the effect of
collision energy upon ion-molecule reactions. As mentioned earlier
the cross sections for exothermic reactions do not change much or
tend to decline as the collision energy increases. For endoergic
processes, however, the reverse is true. There will be an onset
energy below which no reaction occurs and above which the cross
section will increase rather rapidly with increasing collision
energy. It is the energy in the center of mass that is effective
and this is related to the energy in the laboratory system by the
relation

$$E_{cm} = \frac{m_t}{m_t + m_i} E_{lab} \tag{8}$$

where m_i and m_t are respectively the masses of the ion and target
neutral. Thus by proper choice of ion and neutral it is often
possible to obtain quite sensitive control of the collision
energy. Obviously, where the attacking ion is massive and the
neutral is relatively light, it will be possible to change the
laboratory energy over rather a wide range while making relatively
small changes in the effective energy in the center of mass. This
method has been employed in several investigations as a device for
measuring thermochemical properties. Thus, Chupka et al. (1971)
using negative halide ions and especially I^-, have measured the
onset energy for the endothermic charge transfer reaction

$$y^- + X_2 \rightarrow X_2^- + y \tag{9}$$

By this method they have gotten quite precise values of the electron
affinities of the various halogen molecules. Recently Tiernan
and his associates (Lifschitz et al., 1971, 1973) have published
studies of a similar nature in which they measured the electron
affinities of several molecules. Obviously, the method is not
limited to charge transfer processes. In fact, one of the earliest
such studies was by Giese and Maier (1963) who impacted rare gas
ions on carbon monoxide and observed the onset for the formation
of O^+ with very good precision.

While several methods are available for measuring the effect
of collision energy, only one method, to my knowledge, is available for studying the effects of vibrational energy on ion-molecule
reactions. This involves the use of photoionization to prepare
the primary ion in the desired vibrational state. The ion may
then be allowed to react either in the ionization chamber or in a
collision chamber nearby where it may be possible to control the

collision energy. Chupka and Russell (1968) have studied the reactions of NH_3^+ with NH_3 and with H_2O this way. They observed a decrease in reaction cross section with increasing vibrational energy for the reaction with ammonia but found little change for the reaction with water. Chupka et al. (1968) studied the reaction $H_2^+(H_2,H)H_3^+$ and again found the cross section to decrease with increasing vibrational energy.

A few studies of the reactivity of electronically excited ions have been made using the ionization chamber as a reactor, and the rates of a few such reactions have been determined.

The reaction

$$N_2^+ \, (^4\Sigma) \; + \; N_2(^1\Sigma_g^+) \; \rightarrow \; N_3^+(^3\Sigma_g^-) \; + \; N(^4S) \tag{10}$$

has been observed by several investigators and the states involved have been established by Cermak and Herman (1962) and the reaction rate determined by Munson et al. (1962) and by Cress et al. (1966). The reaction of excited N_2^+ with Ar, Kr and Xe to form ArN^+, etc., has been reported by Kaul and Fuchs (1960) and by Munson et al. (1962). A few other reactions of excited ions are also reported by Curran, (1963) Franklin and Munson (1965) and Henglein (1962).

Almost unique in the investigation of ion-molecule reactions is the series of investigations by Ausloos and Lias. (Ausloos and Lias, 1972; Ausloos et al., 1966). In these studies a mixture of interest is subjected to ionizing radiation, either gamma rays or far ultra violet light and the total ion concentration determined from the saturation current between two parallel electrodes in the mixture. After the radiation is completed the product is analyzed. Ausloos and Lias (1972) have studied a number of systems in this way and have been able to draw some very interesting conclusions concerning the ions present and the effect of structure upon reactivity.

This paper has dealt with a few of the studies of ion-molecule reactions conducted for the interest inherent in these processes. There is, in addition, a great literature on the importance of ion-molecule reactions in various systems. Thus the ultimate purpose for developing the flowing afterglow method was for the light that ion-molecule reactions throw upon the properties of the upper atmosphere. Ferguson and Fehsenfeld (1967, 1968, 1969) have discussed this in several publications. The importance of ion-molecule reactions in radiation chemistry has been discussed by Lampe et al. (1961) and by several authors in a book "Fundamental Processes in Radiation Chemistry" edited by Ausloos (1968). Ion-molecule reactions in flames have been

studied by several workers and this field has been surveyed by
Calcote (1972). Electric discharges have been the subject of many
investigations but only a limited number of ion-molecule reactions
have been identified. Studniarz (1972) has summarized the
literature on this aspect of the subject.

In reviewing the development of our knowledge of ion-
molecule reactions one is impressed with the tremendous growth of
the subject and with the breadth and depth of understanding of
chemical processes that have resulted. My review barely outlines
the subject and omits important aspects of it. It is both
comforting and exciting to realize that we are engaged in very
important research and that the prospects for its continuing
development are excellent.

REFERENCES

Anders, L. R., Beauchamp, J. L., Dunbar, R. C., and Baldeschwieler,
J. D. (1966). J. Chem. Phys., 45, 1062.
Ausloos, P., Lias, S. G. and Scala, A. A. (1966). "ION-MOLECULE
REACTIONS IN THE GAS PHASE", Chap. 15, (Ed. P. Ausloos). Advances
in Chemistry Series 58, American Chemical Society, Washington,D.C.
Ausloos, P., ed. (1968). "FUNDAMENTAL PROCESSES IN RADIATION
CHEMISTRY". Interscience Publishers, New York.
Ausloos, P. and Lias, S. G. (1972). "ION-MOLECULE REACTIONS",
Chap. 16, (Ed. J. L. Franklin). Plenum Press, New York.
Beauchamp, J. L. (1971). "Annual Review of Physical Chemistry",
Vol. 22, see p 527 for a discussion of proton affinities,
especially those obtained by ion cyclotron resonance methods.
Brauman, J. L. and Blair, L. K. (1968). J. Am. Chem. Soc., 90,
6562.
Brauman, J. L. and Blair, L. K. (1970). J. Am. Chem. Soc., 92,
5986.
Calcote, H. F. (1972). "Ions in Flames", Chap. 15 in "ION-
MOLECULE REACTIONS", (Ed. J. L. Franklin). Plenum Press, New York.
Cermak, V. and Herman, Z. (1961). Nucleonics, 19, 106.
Cermak, V. and Herman, Z. (1962). Collection Czech. Chem. Commun.,
27, 1493
Chong, Shuang-Ling, and Franklin, J. L. (1972). J. Am. Chem.
Soc., 94, 6630
Chong, Shuang-Ling, Myers, R. A. and Franklin, J. L. (1972). J.
Chem. Phys., 56, 2427
Chupka, W. A. and Russell, M. E. (1968). J. Chem. Phys., 48, 1527
Chupka, W. A., Russell, M. E. and Refaey, K. (1968). J. Chem. Phys.,
48, 1518
Chupka, W. A. Berkowitz, J. and Gutman, D. (1971). J. Chem.
Phys., 55, 2724
Cress, M. C., Becker, P. M. and Lampe, F. W. (1966). J. Chem.

Phys., 44, 2212
Curran, R. K. (1963). J. Chem. Phys., 38, 2974
Dempster, A. J. (1916). Phil. Mag., 31, 438
Eyring, H., Hirschfelder, J. O. and Taylor, H. S. (1936). J. Chem. Phys., 4, 479
Ferguson, E. E. and Fehsenfeld, F. C. (1967). Rev. Geophys., 5, 305
Ferguson, E. E. and Fehsenfeld, F. C. (1968). J. Geophys. Res. 73, 6215
Ferguson, E. E. and Fehsenfeld, F. C. (1969). J. Geophys. Res. 74, 2217, 5743
Ferguson, E. E. Fehsenfeld, F. C. and Schmeltekopf, A. L. (1969). "Flowing Afterglow Measurements of Ion-Neutral Reactions" in ADVANCES IN ATOMIC AND MOLECULAR PHYSICS, (Eds. D. R. Bates and J. Estermann) Vol. 5, pp 1-56, Academic Press, New York
Field, F. H. and Franklin, J. L. (1957) "ELECTRON IMPACT PHENOMENA" Academic Press, New York
Field, F. H., Franklin, J. L. and Lampe, F. W. (1957). J. Am. Chem. Soc., 79, 2419, 2665, 6129
Field, F. H. (1961). J. Am. Chem. Soc., 83, 1523
Field, F. H., and Nunson, M.S.B. (1965). J. Am. Chem. Soc., 87, 3289
Field, F. H., Munson, M. S. B. and Becker, D. A. (1966). "ION MOLECULE REACTIONS IN THE GAS PHASE", Advances in Chemistry Series, 58, (Ed. P. Ausloos), American Chemical Society, Washington, D.C.
Franklin, J. L. and Munson, M. S. B. (1965). "Tenth Symp. (Int). on Combustion", p 561, The Combustion Institute
Franklin, J. L. (Ed.)(1972). "ION-MOLECULE REACTIONS" Plenum Press, New York
Futrell, J. H. and Miller, L. D. (1966). Rev. Sci. Instr. 37, 1521
Giese, C. F. and Maier, W. B. (1963). J. Chem. Phys., 39, 117,739
Hand, C. W. and von Weyssenhoff, H. (1964). Com. J. Chem. 42, 195
Henglein, A. (1962). Z. Naturforsch, 11a, 37
Hengleim, A. Lacmann, K. and Jacobs, G. (1965). Ber Bunsenges, Phys. Chem., 69, 279
Henglein, A. (1966). Chap. 5 p 63 et sef. in "ADVANCES IN CHEMISTRY SERIES", 58, (Ed. P. Ausloos) American Chem. Soc., Washington, D.C.
Hierl, P. M. Sen Sharma, D. K. and Franklin, J. L. (1971). Int. J. Mass Spectrom. Ion Phys., 6, 239
Hogness, T. R. and Lunn, E. G. (1925). Phys. Rev., 26, 44
Hogness, T. R. and Harkness, R. W. (1928). Phys. Rev. 32, 784
Kebarle, P. (1972). "HIGHER ORDER REACTIONS - ION CLUSTERS AND ION SOLVATION", Chap. 7 Ion-Molecule Reactions, (Ed. J. L. Franklin) Plenum Press, New York
Kaul, W. and Fuchs, R. (1960). Z. Naturforsch. 15a, 326
Lampe, F. W., Franklin, J. L. and Field, F. H. (1961). "KINETICS OF THE REACTIONS OF IONS WITH MOLECULES", Progress in Reaction

Kinetics, 1, 69, Permagon Press, London

Lampe, F. W. and Hess, G. G. (1964) J. Am. Chem. Soc., 86, 2952

Lifschitz, C., Tiernan, T. O. and Hughes, B. M. (1971). J. Chem. Phys., 55, 5692

Lifschitz, C., Tiernan, T. O. and Hughes, B. M. (1973). J. Chem. Phys., 59, 3162, 3182

Lindholm, E. (1954). Z. Naturforsch. 9a, 535

Long, J. W. and Franklin, J. L. (1974). J. Am. Chem. Soc., 96, 2320

Mann, M. M., Hustrulid, A. and Tate, J. T. (1940). Phys. Rev., 58, 340

McDaniel, E. W., Cermak, V. Delgarno, A. Ferguson, E. E. and Friedman, L. F. (1970). "ION-MOLECULE REACTIONS", Wiley Interscience, New York

Meisels, G. G. Hamill, W. H. and Williams, R. R. (1956). J. Chem. Phys., 25, 790

Munson, M. S. B., Field, F. H. and Franklin, J. L. (1962). J. Chem. Phys., 37, 1790

Munson, M. S. B. and Field, F. H. (1965). J. Am. Chem. Soc., 87, 3294

Muschlitz, E. E. (1957). J. Appl. Phys., 28, 1414

Parkes, D. A. (1972). J. Chem. Soc. Faraday Trans. I, 68, 613

Paulson, J. F. (1970). J. Chem. Phys., 52, 959, 963

Sen Sharma, D. K. Hierl, P. M. and Franklin. J. L. (1972). J. Chem. Phys., 56, 1095

Smyth, H. D. (1925). Phys. Rev., 25, 452

Smyth, H. D. (1931). Revs. Mod. Phys., 3, 347

Stevenson, D. P. and Schissler, D. O. (1955). J. Chem. Phys., 23, 1353

Stevenson, D. P. and Schissler, D. O. (1956). J. Chem. Phys., 24, 926

Stevenson, D. P. and Schissler, D. O. (1958). J. Chem. Phys., 29, 282

Studniarz, S. K. (1972). "Electrical Discharges", Chap. 14 in "ION-MOLECULE REACTIONS" (Ed. J. L. Franklin) Plenum Press, New York

Tal'roze, V. L. and Lyubimova, A. K. (1952). Dokl. Akad. Nauk. SSSR, 86, 909

Tal'roze, V. L. and Frankevich, E. L. (1959). Izv. Akad. Nauk. SSSR; Otd. Khim, Nauk, No. 7, p 1351

Tal'roze, V. L. and Frankevich, E. L. (1960). Zhur. Fiz. Khim. 34, 2709

Thompson, J. J. (1913). "RAYS OF POSITIVE ELECTRICITY". Longmans, Green and Co., London.

Wolfgang, R. and Herman, Z. (1972). Chap. 12 in "ION-MOLECULE REACTIONS" (Ed. J. L. Franklin) Plenum Press, New York

ION-MOLECULE REACTIONS: A PERSONAL VIEW OF PRESENT PROSPECTS AND PROBLEMS

Michael Henchman

Department of Chemistry

Brandeis University, Waltham, Mass. 02154

King George V is reputed to have uttered as his last remark on his death bed the question: "How's the Empire?" History does not record the answer which he was given but it is clear that the question is not susceptible to an answer which is both comprehensive and succinct. The same difficulties attend anyone attempting to assess the present status of the empire of ion-molecule reactions, difficulties compounded in the case of the present author by his own limitations for the task. The enduring legacy of the NATO Study Institute is the published proceedings, assembling a unique collection of data, concepts, models and theories, which represent the achievements within this field at this time. This compilation allows the reader to examine the pertinence of each individual contribution, experimental and theoretical, to a present understanding of the subject as a whole and to exploit this for its active development in the future. In preparing this lecture, I have sought to give a limited overview of certain areas and a review of some issues, whose relevance may extend beyond the confines of the particular topics of the individual conference sessions. In general the limitations of space, alas, allow me neither to illustrate these with suitable examples nor to present a sufficient citation of the literature to accord credit equitably wherever it is due.

We should first perhaps ask ourselves why the subject of ion-molecule reactions should be of general
interest and consequently indicate its contributions to
chemistry as a whole. In large part this has already
been accomplished most carefully by the Organizing Committee in its definition of topics for the individual
sessions. We can collate these in terms of three cate-
gories.

A). It is a major purpose to improve our understanding of those experimental situations where ions
are to be found. In the gas phase, this requires that
it be a "high-energy" environment, as illustrated in
the sessions devoted to planetary atmospheres, radiation chemistry, flames, plasmas, etc. In the liquid
phase, it relates principally to solutions in polar
solvents at laboratory temperatures. Many recent developments are of particular interest here and a representative sampling (Dubrin and Henchman, 1972) includes: the thermodynamic properties and structure of
solvated ions; the influence of selective solvation on
ionic reactivity; relative acidities; nucleophilic displacement reactions.

B). The subject of ion-molecule reactions has
particular contributions to make to chemical kinetics,
illustrated here by four representative examples: (1)
Unusual control of the energy of the reactants can be
exercised, both with respect to definition and range,
for both relative translational energy and the preparation of reactants in specific states. (2) Most exoergic ion-molecule reactions are not subject to energy
barriers and both reactivity and dynamics may be examined in the absence of the otherwise dominating influence of such barriers. (3) Even at the lowest energies, non-adiabatic processes may compete effectively
with adiabatic ones for ion-molecule reactions, since
the hypersurfaces of the ground and adjacent excited
states of the system are often comparatively close.
(4) As a consequence of (1) above, the influence of
both the excitation function of the reaction and the
relative velocity distribution, characteristic of a
particular technique, may be explored with respect to
their determination of the rate constant.

C). Through the development of chemical-ionization techniques, it is now clear that ion-molecule reactions have important applications in chemical analysis. Ironically they were first discovered as inconveniencies to be avoided in mass spectrometry in its major application to chemical analysis: today they are being exploited for the same purpose.

It is apparent therefore that the empire of ion-molecule reactions is today not only vigorous and expanding but serving valuable purposes of wider interest. A particularly noteworthy feature of the vigor and expansion is the continuing introduction of productive new techniques. The empire is clearly defined but there are disadvantages in it being too self-contained. It has attracted less attention from theoretical chemists than its counterpart, where bimolecular reactions of uncharged reactants are studied under single-collision conditions. Each constitutes a part of a new area of chemical kinetics, widely characterized as chemical dynamics, whose growth is facilitated by continuing integration of results from both areas and by the rapid deployment of developments in one to the other. Sometimes the cross-fertilization is rapid, e.g. the importance of molecular-orbital correlation diagrams (Mahan, 1971). Sometimes it is not, e.g. the importance of thermal target motion to beam/collision-chamber experiments (Bernstein, 1973). This type of communication problem is inherently simpler to solve than the integrative one. In a recent monograph on the chemical applications of molecular beam scattering, ion-molecule reactions receive little attention (Fluendy and Lawley, 1974) and in a forthcoming review of chemical dynamics (Farrar and Lee, 1974), the authors state that "the field of ion-molecule reactions is of sufficient importance that it deserves a separate review." The need for such valuable contributions to the literature is a continuing one; so too is the need for integrative reviews and it becomes an increasingly impossible task.

Granted the existence of the empire of ion-molecule reactions, one may ask what are the common features of the kinetic results which make the classification useful. Where certain reactions do not exhibit

these, other comparisons, i.e. different classifications, may be more fruitful. (Again this emphasizes
the integrative need, discussed above.) Where there
are common features, naturally one seeks to account
for these in terms of the initial condition, the interaction of an ion with a molecule, a situation characterized instinctively by the operation of powerful long-
range forces. For example it is a familiar result that
exoergic ion-molecule reactions exhibit excitation
functions with no energy threshold -- the cross section
increasing monotonically with decreasing collision
energy. (The equivalent statement is that the temperature dependence of the rate constant reveals the absence of an activation energy.) This finding is often
associated with the presence of long-range attractive
forces: instead it is a general feature of systems which
manifest attractive basins or wells in the corresponding potential-energy hypersurface. Indeed the presence
of an energy barrier requires that the hypersurface be
repulsive rather than attractive in this region of close
reactant proximity.

A case in point concerns the nominally thermoneutral reactions for which $\Delta E = -0.04$ eV. The excitation

$$D^+ + H_2 \rightarrow HD + H^+ \qquad (1)$$
$$D + H_2 \rightarrow HD + H \qquad (2)$$
$$D^- + H_2 \rightarrow HD + H^- \qquad (3)$$

function of (1) exhibits no threshold (Krenos et al.,
1974); that of (2) exhibits both a threshold and a
maximum (Rowland, 1970); while that of (3) shows a
striking similarity to that of (2) (Paulson, 1972).
The hypersurface of (1) is characterized by a deep
basin; that of (2) is repulsive with no basin; and
that of (3) is even more strongly repulsive -- this
progression resulting, in the language of molecular-
orbital theory, from the succession addition of non-
bonding electrons. To the extent that the long-range
forces operating in (1) and (3) may be expressed in
simple electrostatic terms (ion-induced dipole and
ion-quadrupole), they must be identical for both cases.
Yet the respective excitation functions could not be

more different. Useful comparisons should not be conducted here in terms of the classification of ion-molecule reactions, i.e. (1) and (3), but rather in terms
of the shape of the hypersurface, i.e. (2) and (3).
This is a simple example to illustrate why like behavior for positive and negative ion-molecule reactions
should not necessarily be expected, merely because in
each case one reactant is charged. An even more significant source of difference follows from the competition, for the negative-ion reactions, from associative
and collisional detachment at low energies (see the
lecture by Fehsenfeld). The obvious lesson is to preserve an open mind about the behavior to be expected
for ion-molecule reactions and to base one's expectations on a variety of considerations, which may be still
unclear in some cases. For example, we expect positive
ion-molecule systems to exhibit basins in their hypersurfaces -- an expectation rationalized simplistically
in terms of the electrophilic nature of the ionic reactant -- yet certain positive-ion reactions, which are
strongly exoergic, still exhibit thresholds in their
excitation functions (Wyatt et al., 1974).

With the conclusion of these general remarks, attention is now devoted to some specific topics. My
intention is to give some indication of the present
state of achievement by reviewing some of the developments which have occurred during the past decade, and
some present problem areas and needs.

ELASTIC AND INELASTIC SCATTERING

Throughout the past decade, dramatic advances in
the study of the elastic scattering of neutral particles have revealed exceedingly detailed knowledge of
intermolecular potential functions, both from the
measurement of differential cross sections and the velocity dependence of integral cross sections (Toennies,
1973). In contrast, progress for the case of elastic
scattering of ions off neutral particles has been
slower.

Experiments of the second type yield knowledge
of the potential at comparatively large inter-particle
separations. We still have no established experimental

means of studying these for ion-molecule interactions.
ICR line-width measurements and mobility data from
drift-tube measurements offer relevant data but their
"inversion" to yield detailed knowledge of the potential would appear to be a formidable task.

As Ding's lecture in this volume demonstrates, impressive progress has been achieved recently in the
first category. Observation of both supernumerary
rainbows and the rapid oscillations in the differential
scattering of protons off atomic targets allows the
potential well to be characterized -- to the extent
that a recent evaluation states that the data are now
as accurate as the procedures used for their inversion
(Buck, 1973). An obvious application of these results
is to yield important and hitherto elusive thermochemical information, e.g. $\Delta H_f^0(KrH^+)$. Such absolute determinations of proton affinities allow the calibration of
scales of relative proton affinities obtained in equilibrium measurements using the flowing afterglow (see the
lecture by Bohme) and ICR.

Two comments may be made with regard to the thermochemical application of these measurements (Henglein,
1972). Where the target atom has a lower ionization potential than the hydrogen atom, e.g. in the H^+ + Xe
experiments, it may not be possible to obtain reliable
proton affinities from the elastic scattering measurements. Reasons for this may include the competition of
charge transfer or the inability of the elastic scattering measurements to sample the ground-state potential
curve, e.g. of XeH^+. The second consideration alone
might suggest that the elastic-scattering measurements
would provide a lower bound to the proton affinity; yet
for the example of xenon, the scattering value of ~ 6.5
eV (Henglein, 1972) is higher than the afterglow value
of ~ 5.0 eV (Bohme's lecture, this volume).

Similar considerations may apply when either or
both reactants consist of more than one atom: at the
collision energies used, the region of configuration
space corresponding to the well in the hypersurface may
not be sampled in the elastic scattering. Such crude
arguments suggest that the elastic scattering of H_2^+
off Kr, for example, would yield only an upper bound to

$\Delta H_f^0(KrH_2^+)$ and that the value obtained would not require the presence of an energy barrier for the exoergic reaction $Kr^+(^2P_{3/2})(H_2,H)KrH^+$ (Henglein, 1972).
Indeed the activation energy inferred experimentally is
negligible (Fennelly et al., 1974). It is not clear
to what extent these elastic-scattering measurements
using polyatomic reactants can yield information on
potential-energy hypersurfaces.

In concluding this section, I would emphasize without discussion the important and continuing contributions of ion-molecule systems to the study of inelastic
processes, as outlined in recent surveys (Doering,
1973; Faubel, 1973) and further illustrated by the contributions of Mahan and Harrison to this volume. At
higher energies, this leads to collision-induced dissociation (Los, 1973).

RATE CONSTANTS

The development of experimental techniques during
the past decade has been particularly spectacular, as
reviewed elsewhere (Henchman, 1972). Rate constants in
the thermal energy range may now be measured by a wide
variety of techniques: stationary afterglow, flowing
afterglow, high-pressure mass spectrometry, drift tube,
and ICR. (When performed under conditions of zero field
and with time resolution, the high-pressure mass spectrometer is equivalent to the stationary afterglow and
is now employed in many laboratories e.g. Field,
Kebarle, Meisels.) The first three techniques yield
rate constants under truly isothermal conditions and
the temperature range accessible to the flowing afterglow is now 80-900°K (see the lecture by Ferguson). The
last two yield rate constants also throughout an energy
range, extending to several eV for the drift tube and
even higher for ICR. The Panel Discussion on Rate Constants in this volume provides recent references and
fuller discussion; of particular interest are the recent
developments in the ion-trapping technique (Ryan) and
the flow-drift technique (Ferguson).

These techniques have recently provided a wealth
of data, with regard to which two questions are of current interest. First, do the measurements of the rate

constants of a particular reaction using different techniques yield compatible results i.e. ultimately the
same excitation function? Second, the availability of
rate constant data <u>in the thermal-energy range</u> allows a
proper comparison to be made with theoretical predictions, based on simple electrostatic models for the ion-
molecule potential.

The first question cannot be answered in the general case. The "inversion" of rate-constant data to
yield the excitation function is a difficult task even
when the relative-velocity distribution function $f(v_r)$
is known i.e. under isothermal conditions. Where conditions are not isothermal, our lack of knowledge about
the distribution function makes it even harder. In
practice, one is forced to compare the magnitudes of
<u>rate constants</u>, measured by different techniques, even
though such a comparison is formally invalid where the
distribution functions differ. (Such a comparison involves the characterization of a non-isothermal distribution in terms of what is essentially only a single
moment of the distribution, expressed as a mean relative
kinetic energy or an "equivalent temperature".)

For example, if we compare a rate constant, measured in a drift tube, where the distribution function is
characterized by a certain equivalent temperature T,
with that, measured for the same reaction in the flowing
afterglow at the same (but now real) temperature T, we
cannot necessarily expect that the two measured values
will be identical -- simply because the two distribution functions differ. This follows from the familiar
statistical relationship

$$k = \int_0^\infty f(v_r) \cdot \quad \sigma(v_r) \cdot v_r \cdot dv_r.$$

The magnitude of the difference between the two measured
values will depend on the functional form of the excitation function. If $\sigma(v_r) \propto v_r^n$, this difference is always zero when n = -1. As n departs from this value,
the magnitude of the difference will in general increase.

Analysis of the significance of this consideration
is frustrated by our ignorance of both excitation functions at low relative velocities and distribution

functions for non-isothermal techniques. Accordingly
there is a need to extend our knowledge of both these
matters. Much useful discussion of the latter occurred
at the Rate Constant Panel Discussion, reported in this
volume, including the reassuring preliminary result of
Su and Bowers that, at thermal energies, the distribution function for the ICR techniques is close to Maxwellian. Other interest in this includes an attempt to
measure $f(v_r)$ experimentally for the drift tube (Moruzzi
and Harrison, 1974) and theoretical analysis for the
ICR technique (Bloom and Riggin, 1974).

We have made a crude, exploratory examination of
this problem for the drift-tube case, using, in the
absence of other tractable alternatives, the distribution function proposed by Woo and Wong (1971). For the
case of exoergic reactions, inversion of $k(T)$ data
measured on the flowing afterglow yields an approximate
form for the excitation function, in the functional
form indicated above, and reveals that $0 > n > -3$.
Examination of the problem for extreme values of $n = -3$
and $T = 900°K$ suggests that the difference in the rate
constants should be observable, the drift tube value
being $\sim 30\%$ lower. The situation is different for endoergic reactions, exhibiting an energy threshold. Our
comparison of rate constants, calculated by Woo and
Wong (1971) for $O^-(O_2, O)O_2^-$, at the same equivalent
temperature, suggests that drift-tube and isothermal
values could differ by as much as two orders of magnitude: this result is plausible since the rate constant
is determined by the form of the high velocity-tail of
the distribution function. Obviously when comparisons
are made with rate constants measured by beam techniques, where the distribution functions approach delta
functions, the difference could be even more marked.
This has possible relevance to drift-tube measurements
of rate constants where the mean relative energies E_r
lie in a range where $dk/dE_r \gg 0$: a priori one would not
expect those rate constants to be identical to values
obtained by beam techniques, unless the distributions
happen to be similar (McFarland et al., 1973).

We consider now the second question of the comparison of experimental rate constants at thermal energies
with values predicted by the Langevin theory for

non-polar molecules and its development for polar molecules, the ADO theory. The new data now available show
that the agreement is often impressively good (see the
lectures by Bohme and Bowers), demonstrating the general
predictive capacity of these theories for thermal rate
constants and justifying their general application for
that purpose. What are the implications of this result,
with respect to reservations expressed previously about
the assumptions of such models (Henchman, 1972)? It
would be useful to establish the consequences of the
failure of such models to satisfy the requirements of
detailed balancing. More significantly, the work of
Hyatt and Stanton (1970, 1971 and 1973) emphasizes that
rate constants are highly averaged quantities; agreement
between theory and experiment for rate constants does
not require that the molecular dynamics envisaged in
the models are a reliable representation of the actual
molecular dynamics; nor does it require that excitation
functions will be predicted correctly; nor does the
agreement necessarily establish the form assumed in the
model for the ion-molecule potential. In the systems
considered by Hyatt and Stanton, involving neutrals
both linear and symmetric top, they conclude that if
the potential is indeed given by the simple electrostatic potential, the Stark effect will operate. For
the polar molecules which they considered, the dipole
may not oscillate in the field induced by the approaching ion, as assumed in the ADO theory, but will rather
adopt a _fixed_ orientation, determined by its rotational
state. These interesting questions deserve further
study, including hopefully further experimental measurements of excitation functions where the neutral reactant
is in selected rotational states, e.g. $Ar^+ + H_2$ (Hyatt
and Stanton, 1970).

CROSS SECTIONS

It is the excitation function of a chemical reaction that summarizes the information about the intrinsic
reaction probability. One obvious insight gained during
the past decade is that the excitation function may depend significantly on the internal energy states of the
reactants and this has established the need to determine
excitation functions for state-selected reactants (see

in particular the lectures by Chupka and Koski). Additionally there is the need to extend the range of collision energy down to the thermal energy range. This latter need can be met by the merged-beams technique but often only at the cost of producing the neutral reactant in excited states, since it must be formed by charge exchange. In certain extreme cases, such as $O^+(N_2,N)NO^+$, this can be a severe limitation (Neynaber and Magnuson, 1973). Yet a particularly dramatic result was reported by Buttrill at the meeting that relative energies as low as 0.005 eV had been achieved for the reaction $D_2^+(N,D)ND^+$ in Gentry's laboratory. This technique clearly has immense potential if a wide variety of neutral reactants can be prepared in controlled internal-energy states.

The traditional approach presents the neutral target as a thermal gas in a collision chamber. As exploited extensively in tandem mass spectrometers, two complications arise -- the uncertain collection efficiency of the product ions from the collision chamber and the need to deconvolute the effect of thermal target motion. This latter problem can be solved for measurements on endoergic reactions involving monoenergetic ion beams (Chantry, 1971) but not for exoergic reactions at the lowest energies, where the energy spread in the ion beam can no longer be ignored and where a functional form must be assumed for excitation functions down to the limit of zero relative velocity. In contrast, the guided-ion beam technique (Teloy and Gerlich, 1974) offers many improvements on the standard tandem technique, with respect to both control of the states and energy of the ionic reactant and particularly the collection efficiency of the product ions. For the exoergic reaction studied, $Ar^+(D_2,D)ArD^+$, the thermal-target-motion problem led the authors to present their results in the form of rate constants. The development of this technique is of particular interest, as indeed are the new results on the above, much-studied reaction, since they disagree with earlier conclusions at the lowest energies (Henchman, 1972).

In summary the present need is for excitation functions for state-selected reactants down to the lowest energies. Moreover it may be noted that, even for the

most-studied reactions, the information to hand may be
subject to revision as new techniques are evolved.

REACTIVE SCATTERING

The major development here of the past decade has
been the measurement of intensity contour maps which
summarize information concerning the angular and velocity distributions of product ions. Again a wealth of
important new information has been obtained, revealing
knowledge of collision mechanism, energy disposal, product states, etc. Here too it has been demonstrated
that the internal energy states of the ionic reactant
can exercise a profound influence on the collision
dynamics. Major developments have occurred in the
laboratories of Henglein, Mahan and Wolfgang. Recent
surveys have been published (Herman and Wolfgang, 1972;
Dubrin and Henchman, 1972; Herman and Birkinshaw, 1973)
and also an interesting discussion of several topics of
current interest (Durup, 1974).

For the most part, investigation has been limited
to collision energies above 1-2 eV. (There are exceptions, e.g. $Ar^+(H_2,H)ArH^+$, where the inherent kinematic
leverage has reduced this by an order of magnitude.)
Chemists are much interested in acquiring mechanistic
information in the energy range approaching thermal
energy. Indeed there would appear to be interesting
mechanistic changes in this range for several reactions,
e.g. $O^+(N_2,N)NO^+$ and certain simple charge-transfer reactions (McFarland et al., 1973). These inferences come
from measurements of rate constants and excitation functions: however the mechanistic interpretation of such
results, including isotope effects in particular, is
fraught with ambiguity (Henchman, 1972). We really
need the intensity-contour maps in an energy range which
seems perhaps technically inaccessible at this time.
However the recent, remarkable development of the technique at Herman's laboratory at Prague reveals its
limits have not been reached. Both Birkinshaw and
Futrell reported at the meeting the detection of thermal
product ions at collision energies of < 1 eV -- an
achievement of great relevance to the present interest
in the mechanism of charge-transfer reactions at low
energies (see the lecture by Birkinshaw and the Charge

Transfer Panel Report).

We also need information on the distribution of internal states of the products. The achievement of this through the observation of chemiluminescence, exploited so effectively for neutral-neutral reactions (Carrington and Polanyi, 1972), is now beginning to be undertaken in several laboratories, e.g. Brandt et al., (1973); similarly the technique of laser-induced fluorescence holds promise for the future (Farrar and Lee, 1974). In view of the fact that many reactions appear to proceed via long-lived complexes, there is much interest in the structure and the distribution of the lifetimes of such complexes. Further insight into the behavior of complexes comes from measurement of product velocity distributions (Lee et al., 1972) and, in this regard, it is of interest to note the recent finding that the energy is not randomized within long-lived complexes of certain neutral-neutral reactions (Parson et al., 1973).

As experimental intensity contour maps have become available, there has been increased interest in providing theoretical descriptions in terms of trajectories over potential-energy hypersurfaces (Tully, 1973). The study of the $H^+ + H_2$ system (Krenos et al., 1974) will surely rank as a classic contribution in this area: in particular, it has revealed a detailed description of what has been considered traditionally as competition between ion-molecule reaction and charge transfer. A theoretical analysis of the system $Ar^+ + H_2$ has been conducted at a single energy (Chapman and Preston, 1974): at the next level of refinement, we have essentially no information, theoretical or experimental, as to how the state of the argon ion, $^2P_{3/2}$ or $^2P_{1/2}$, might affect these results. But the major activity has concentrated on the reaction $O^+ (N_2, N) NO^+$ and several proposals have been advanced as to the nature of the hypersurfaces and crossings involved (O'Malley, 1970; Pipano and Kaufman, 1972; Smith and Cross, 1974; Hopper and Tiernan, 1974). These possibilities raise certain general questions. On theoretical grounds alone, can the relative contribution of these to the reaction

be assessed? Are the experimental data available sufficient to distinguish between possible alternatives?
(In certain cases the answer is clearly affirmative
(Cohen, 1972).) Is, as Kaufman proposed at the meeting,
the limiting factor the lack.of quantitative knowledge
of the hypersurfaces and, if so, can this be rectified
in the near future -- both for economic and technical
reasons? Science has been called "the art of the
soluble": will the application of this approach to
successively more complicated systems result in clear
progress? Obviously, at the present time we are far
from reaching the stage where increasing investment
yields diminishing returns.

The trajectory approach considers each system as
a particular case, since the hypersurfaces are particular to the system. Traditionally chemists often seek
simpler theories or models, which attempt to correlate
the behavior of many systems. Though often comparatively naive, these can provide an overview, offering
instructive insight and predictive capacity. To what
extent can insight into ion-molecule reactions be
achieved in such terms, as distinct from the invaluable
detailed insight into specific systems gained from the
powerful trajectory calculations?

It is in this context that the impulsive model,
developed by Mahan in his lecture, is so appealing.
Simple, both conceptually and in application, it can
demonstrate, at high energies, how the general features
of contour maps change with energy. It also shows how
the maximum intensity in the contour map can be found
at the "spectator-stripping" position as a consequence
of a complex sequence of two-body interactions, i.e.
how two entirely different mechanisms can yield the
same experimental results. (This illustrates again the
difficulties inherent in the inverse procedure of in-
ferring mechanism from experimental data; there is much
discussion in the literature of spectator-stripping
mechanisms; to what extent are these established?) Two
recent papers by Mahan (1974) demonstrate again the insight which may be gained from another simple model,
involving collinear collisions on the simplest model
surfaces.

As a further example, both Henglein (1970) and
Wolfgang (1970) have used a model in the same spirit
to account for the mechanistic behavior of several reactions in terms of the thermochemistry of reactants,
products and the collision intermediate, and an "effective" number of degrees of freedom. The underlying
argument that the mechanism should be affected by the
height of the exit barrier (intermediate to products),
the energy available to surmount this and the molecular
complexity of the system is both attractive, plausible
and useful. Yet it is not clear with what confidence
the model may be used. The quantitative formulation
depends on the RRK theory, which is known to be invalid.
Second, Mahan, (1971) has shown for several systems
that the reaction does not proceed on the hypersurface
of lowest energy: there may be deep wells which are inaccessible, resulting in the mechanism being direct.

Given the complexity of these systems and of the
detailed experimental information which is being obtained, there is a continuing need for means of summarizing this information. One notes for example· the
successful application of an information-theoretic
approach to the problem of energy disposal and consump-
tion in neutral-neutral reactions (Levine and Bernstein,
1973), whereby, for the former, the specificity of
product-state distributions can be simply characterized.
One may hope that the extensive information available,
in the latter case, for ion-molecule reactions using
photo-ionization techniques, may be similarly systema-
tized. To date only one such application has been made
-- to a study of collision-induced dissociation (Rebick
and Levine, 1973).

CONCLUSION

In conclusion, it is appropriate to return from
flights of speculation to the solid ground of the
achievements of the present. Ten years ago, essentially
nothing was known about elastic and inelastic scattering;
likewise the influence of reactant states on rate or
mechanism; only the Talrose pulse technique was available to measure thermal rate constants; tandem measurements of cross sections were just appearing but relative
energies of 0.005 eV were unthinkable; and Henglein's

mechanistic inferences (of great significance in the
evolution of chemical dynamics) relied on velocity
spectra. This provides one context in which the contents of the present volume may be viewed.

REFERENCES

Bernstein, R. B. (1973). Comments Atom. Mol. Phys. $\underline{4}$,
    43.
Bloom, M. and Riggin, M. (1974). Can. J. Phys. $\underline{52}$, 436.
Brandt, D., Ottinger, Ch. and Simonis, J. (1973). Ber.
    Bunsenges. Physik. Chem. $\underline{77}$, 648.
Buck, U. (1973). "The Physics of Electronic and Atomic
    Collisions: Invited Lectures and Progress Reports:
    VIII ICPEAC". (Ed. B. C. Cobic and M. V. Kurepa).
    Institute of Physics, Beograd. p. 481.
Carrington, T. and Polanyi, J. C. (1972). "MTP Inter-
    national Reveiw of Science. Physical Chemistry.
    Series One", Vol. 9, "Chemical Kinetics". (Ed.
    J. C. Polanyi). University Park Press, Baltimore.
    p. 135.
Chantry, P. J. (1971). J. Chem. Phys. $\underline{55}$, 2746.
Chapman, S. and Preston, R. K. (1974). J. Chem. Phys.
    $\underline{60}$, 650.
Cohen, R. B. (1972). J. Chem. Phys. $\underline{57}$, 676.
Doering, J. P. (1973). Ber. Bunsenges. Physik. Chem.
    $\underline{77}$, 593.
Dubrin, J. and Henchman, M. J. (1972). "MTP Interna-
    tional Review of Science. Physical Chemistry.
    Series One", Vol. 9, "Chemical Kinetics". (Ed.
    J. C. Polanyi). University Park Press, Baltimore.
    p. 213.
Durup, J. (1974). Adv. Mass Spectrom. $\underline{6}$, 691.
Farrar, J. M. and Lee, Y. T. (1974). Ann. Rev. Phys.
    Chem. $\underline{25}$, in press.
Faubel, M. (1973). "The Physics of Electronic and
    Atomic Collisions: Invited Lectures and Progress
    Reports: VIII ICPEAC". (Ed. M. Cobic and M. V.
    Kurepa). Institute of Physics, Beograd. p. 543.
Fennelly, P. F., Payzant, J. D., Hemsworth, R. S. and
    Bohme, D. K. (1974). J. Chem. Phys. $\underline{60}$, 5115.
Fluendy, M. A. and Lawley, K. P. (1974). "Chemical
    Applications of Molecular Beam Scattering". Hal-
    sted Press, New York.

Henchman, M. J. (1972). "Ion-Molecule Reactions". (Ed. J. L. Franklin). Vol. I. Plenum Press, New York. p. 101.

Henglein, A. (1970). J. Chem. Phys. 53, 458.

Henglein, A. (1972). J. Phys. Chem. 76, 3883.

Herman, Z. and Birkinshaw, K. (1973). Ber. Bunsenges. Physik. Chem. 77, 566.

Herman, Z. and Wolfgang, R. (1972). "Ion-Molecule Reactions". (Ed. J. L. Franklin). Vol. II. Plenum Press, New York. p. 553.

Hopper, D. G. and Tiernan, T. O. (1974). Presented at the Twenty Second Annual Conference on Mass Spectrometry and Allied Topics, Philadelphia, Pennsylvania, May 19-24.

Hyatt, D. and Stanton, L. (1970). Proc. Roy. Soc. A 318, 107.

Hyatt, D. and Stanton, L. (1971). Chem. Phys. Letters 10, 10.

Hyatt, D. and Stanton, L. (1973). J. Chem. Soc. Faraday Trans. II. 69, 340.

Krenos, J., Preston, R. K., Wolfgang, R. and Tully, J. C. (1974). J. Chem. Phys. 60, 1634.

Lee, A., LeRoy, R. L., Herman, Z., Wolfgang, R. and Tully, J. C. (1972). Chem. Phys. Letters 12, 569.

Levine, R. D. and Bernstein, R. B. (1973). Faraday Discussions Chem. Soc. 55, 100.

Los, J. (1973). Ber. Bunsenges. Physik. Chem. 77, 640.

Mahan, B. H. (1971). J. Chem. Phys. 55, 1436.

Mahan, B. H. (1974). J. Chem. Educ. 51, 308, 377.

McFarland, M., Albritton, D. L., Fehsenfeld, F. C., Ferguson, E. E. and Schmeltekopf, A. L. (1973). J. Chem. Phys. 59, 6620.

Moruzzi, J. L. and Harrison, L. (1974). Int. J. Mass Spectrom. Ion Phys. 13, 163.

Neynaber, R. H. and Magnuson, G. D. (1973). J. Chem. Phys. 58, 4586.

O'Malley, T. F. (1970). J. Chem. Phys. 52, 3269.

Parson, J. M., Shobatake, K., Lee, Y. T. and Rice, S. A. (1973). Faraday Discussions Chem. Soc. 55, 344.

Paulson, J. F. (1972). Bull. Am. Phys. Soc. Ser. II 17, 401.

Pipano, A. and Kaufman, J. J. (1972). J. Chem. Phys. 56, 5258.

Rebick, C. and Levine, R. D. (1973). J. Chem. Phys.
    58, 3942.
Rowland, F. S. "Proceedings of the International School
    of Physics: Enrico Fermi. Course XLIV. Molecular
    Beams and Reaction Kinetics". (Ed. Ch. Schlier).
    Academic Press, New York. p. 293.
Smith, G. P. K. and Cross, R. J., Jr. (1974). J. Chem.
    Phys. 60, 2125.
Teloy, E. and Gerlich, D. (1974). Chem. Phys. 4, 417.
Toennies, J. P. (1973). Faraday Discussions Chem. Soc.
    55, 129.
Tully, J. C. (1973). Ber. Bunsenges. Physik. Chem. 77,
    557.
Wolfgang, R. (1970). Acc. Chem. Res. 3, 48.
Woo, S. B. and Wong, S. F. (1971). J. Chem. Phys. 55,
    3531.
Wyatt, J. R., Strattan, L. W., Snyder, S. C. and Hierl,
    P. M. (1974). J. Chem. Phys. 60, 3702.

ELASTIC SCATTERING OF IONS

Adalbert Ding

Hahn-Meitner-Institut für Kernforschung Berlin GmbH

1 Berlin 39

1. INTRODUCTION

Elastic scattering experiments have provided valuable insight into the understanding of intermolecular forces (for reviews see e.g. Pauly and Toennies (1968a, 1968b), Bernstein (1966), Bernstein and Muckermann (1967), Toennies (1973a, 1973b) and for ion-neutral scattering, Weise (1973). This is particularly true for interactions involving ions, as optical methods, which are a valuable source of information in the case of neutral-neutral interactions, have hardly been applied to molecular ions because of their instability.

Here we shall restrict ourselves to the elastic scattering of ions in the energy-range of approximately 1 eV to 1000 eV. In all cases we shall deal with the interaction of two particles A and B, which may be atoms, ions, or molecules with the kinetic energies T_A, T_B, the internal - i.e. electronic, vibrational or rotational - energies $E_{int, A}$, $E_{int, B}$ and the interaction potential V_{AB}. Without narrowing the scope of this discussion we shall use particle A as the experimentally investigated particle (generally an ion). The conservation of energy yields the total energy E as

$$E = T_A + T_B + E_{int, A} + E_{int, B} + V_{AB} = const. \qquad (1.1)$$

Long before and long after the collision event the interparticle potential V_{AB} will be negligible, so that

33

$$E = T_A^{(i)} + T_B^{(i)} + E_{int, A}^{(i)} + E_{int, B}^{(i)} \qquad \text{i: initial state} \quad (1.2)$$

$$= T_A^{(f)} + T_B^{(f)} + E_{int, A}^{(f)} + E_{int, B}^{(f)} \qquad \text{f: final state}$$

Elastic collisions are defined as the ones with constant internal energies $E_{int, A}^{(i)} = E_{int, A}^{(f)}; \quad E_{int, B}^{(i)} = E_{int, B}^{(f)}$

We have therefore for elastic collisions

$$E = T_A^{(i)} + T_B^{(i)} = T_A^{(f)} + T_B^{(f)} \qquad (1.3)$$

implying the kinetic energies T_A, T_B are changed in the course of the collision.

The kinetic energies depend on the choice of the coordinate system. Only for the centre of mass system defined as the one where the sum of the momenta is zero does the relation

$$T_A^{(i)} = T_A^{(f)}; \qquad T_B^{(i)} = T_B^{(f)} \qquad (1.4)$$

hold. In the laboratory system this is usually not the case; particles elastically scattered into different angles will have different speeds. In the absence of inelastic processes there is no need for an energy analysis; in the presence of inelastic processes an angle dependent energy discriminator has to be used. This applies as well to purely elastic scattering for certain mass ratios ($m_A > m_B$), where there will be contributions from two different cm-angles to one lab-angle, as shown in Fig. 1. Here again an energy analysis of the detected particles is necessary.

2. CLASSICAL TREATMENT OF ELASTIC SCATTERING

Scattering probability and cross section

The differential cross section $d\sigma/d\Omega\,(E, \Omega)$ is defined as the number of particles $dn^{(f)}$ being scattered into angle Ω of size $d\Omega$ per unit time, unit primary current j^i, unit density of the target beam n_T, and unit scattering volume V_T

$$\frac{d\sigma}{d\Omega}\,(E, \Omega) = \frac{dn^{(f)}}{dt}\;\frac{1}{n_T \cdot V_T \cdot j^i} \qquad (2.1)$$

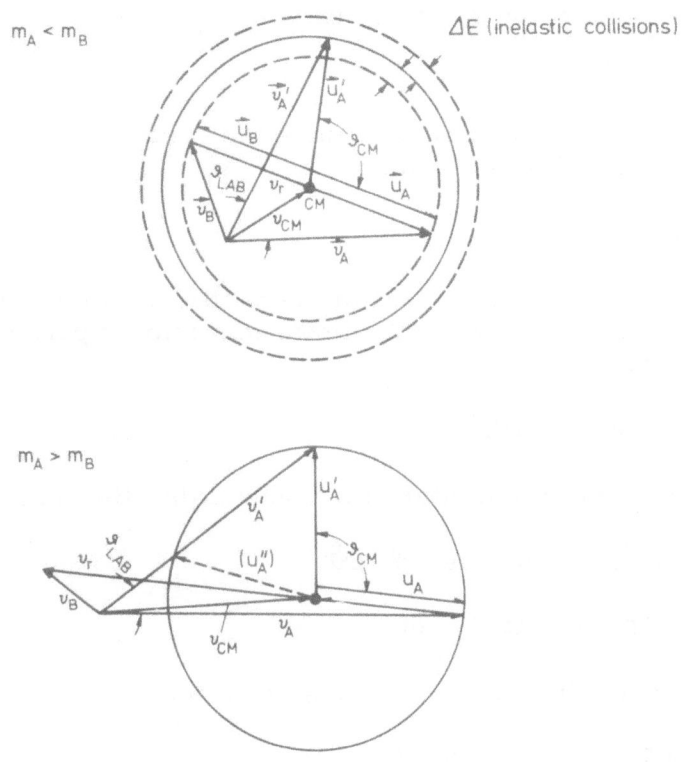

Fig. 1 - Newton diagrams for different mass combinations m_a, m_B v and u designate the velocities - the index denoting particle A or B - in the laboratory and the centre of mass frame, the primed quantities are the velocities after the collision. For a given energy change ΔE the vector u'_A lies on a circle around the centre of mass.

The integral cross section is the integral over all angles

$$\sigma(E) = \int_\Omega \frac{d\sigma}{d\Omega}(E,\Omega)\,d\Omega \tag{2.2}$$

The conversion between centre of mass (CM-) system and laboratory (LAB-) system is discussed elsewhere (Toennies, 1973a). The transformation formulas can be extracted from Fig. 1

$$\tan\vartheta_{LAB} = \frac{\sin\vartheta_{CM} \cdot u_A^f}{v_{CM} + u_A^f \cdot \cos\vartheta_{CM}} \tag{2.3}$$

$$\vartheta_{LAB} \approx \frac{m_B}{m_A + m_B}\vartheta_{CM} \quad \text{for small angles} \tag{2.4}$$

$$(2.3 \text{ and } 2.4 \text{ valid only for } v_B = 0)$$

In the following we will always assume the variables to be expressed in the centre of mass system.

The deflection function

Classically the interaction of the particles is described by the deflection function $\vartheta(b)$ which connects impact parameter b with deflection angle

$$\vartheta = \vartheta(b)$$

for spherical symmetric interaction potentials. Because

$$d\Omega = d\varphi \cdot \sin\vartheta \, d\vartheta$$

and $\qquad d\sigma = 2\pi \cdot b \cdot db$

we obtain for the differential cross section (cf. Fig. 2)

$$\frac{d\sigma}{d\Omega} = \frac{2\pi \cdot b \cdot db}{\int_{\varphi=0}^{2\pi} d\varphi \cdot \sin\vartheta \cdot d\vartheta}$$

$$= \frac{b}{\sin\vartheta} \cdot \frac{db}{d\vartheta} \qquad\qquad (2.5)$$

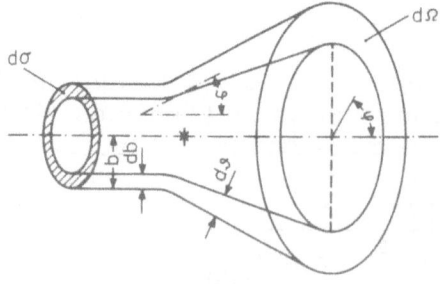

Fig.2 - Definition of the differential cross section: particles with impact parameters between b and db are scattered into angles between ϑ and $\vartheta + d\vartheta$

As the inverse deflection function may be a multivalued function the summation has to be performed over all possible branches

$$\frac{d\sigma}{d\Omega} = \frac{1}{\sin\vartheta} \sum_i b_i \frac{db_i}{d\vartheta} \qquad (2.6)$$

The conservation of energy and angular momentum yields in polar coordinates (Hasted, 1972)

$$E = \frac{1}{2}\mu\,(\dot{r}^2 + r^2\,\dot{\vartheta}^2) + V(r) = \frac{p_i^2}{2\mu} \qquad (2.7)$$

$$L = \mu \cdot r^2 \cdot \dot{\vartheta} = p_i \cdot b \qquad (2.8)$$

$$(\frac{dr}{d\vartheta})^2 \equiv \frac{\dot{r}^2}{\dot{\vartheta}^2} = \frac{2(E - V(r))}{\mu \cdot \dot{\vartheta}^2} - r^2$$

$$\frac{d\vartheta}{dr} = \frac{b}{r^2}\left\{1 - \frac{V(r)}{E} - \frac{b^2}{r^2}\right\}^{-1/2}$$

$$\vartheta = \pi - b \cdot \int_{r_o}^{\infty} \frac{dr}{r^2 \cdot \sqrt{1 - V(r)/E - b^2/r^2}} \qquad (2.9)$$

The term $\dfrac{V(r)}{E} - \dfrac{b^2}{r^2}$ can be rewritten using relation 2.7, 2.8

$$\frac{V(r)}{E} + \frac{b^2}{r^2} = \frac{1}{E} \cdot \left\{V(r) + \frac{L^2}{2\mu r^2}\right\} \equiv \frac{V_{eff}(r)}{E} \qquad (2.10)$$

and can be regarded as a reduced effective potential, consisting of the real and a centrifugal potential.

It is often useful to use reduced variables like

reduced energy K $\qquad = \dfrac{E}{\varepsilon}$

reduced potential U $\qquad = \dfrac{V}{\varepsilon}$

reduced impact parameter ß $= \dfrac{b}{r_m}$

reduced radius $\qquad \rho = \dfrac{r}{r_m} \; ; \quad \rho_o = \dfrac{r_o}{r_m}$

so that follows from 2.9

$$\vartheta = \pi - \beta \int_{\beta_0}^{\infty} \frac{d\varrho}{\varrho^2 \left(1 - \frac{U}{K} - \frac{\beta^2}{\varrho^2}\right)} \qquad (2.11)$$

Evidently the deflection function does not change if the reduced variables remain the same .

    If the potential has an attractive minimum, as it is usually the case with chemically important potentials, the deflection function will show a minimum, too. Fig. 3 shows trajectories in such a

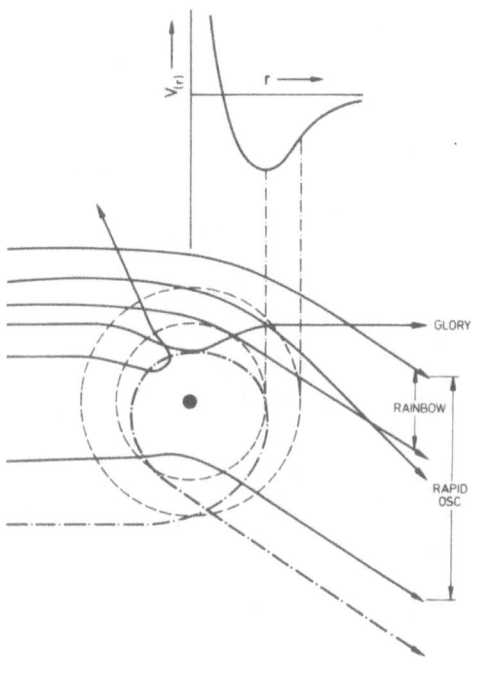

Fig. 3 -
Classical trajectories for different impact parameters: trajectories leading to the same angle give rise to interference effects.

potential: At large impact parameters the particle will experience only weak forces which will generate a slightly inward bent trajectory. Reducing the impact parameters the deflection angle will increase as the probed potential becomes stronger till it reaches a maximum approximately at the point where the forces are strongest (maximum slope of the potential). Further decreasing the impact parameter will cause a smaller deflection angle till finally the trajectory will come so close to the centre of the potential that a net repulsion will occur. It should be mentioned that there is one impact parameter b_o, where attraction and repulsion effects cancel each other, so that there will be no deflection at all. b_o approximately coincides with the position of the potential minimum r_m.

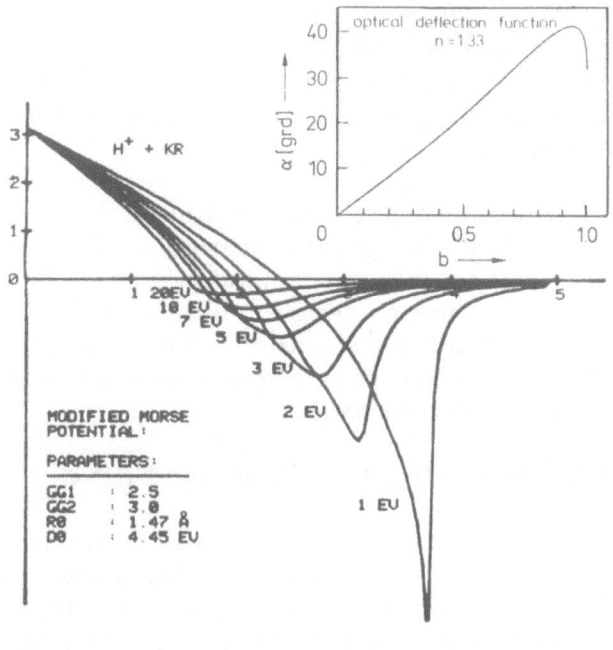

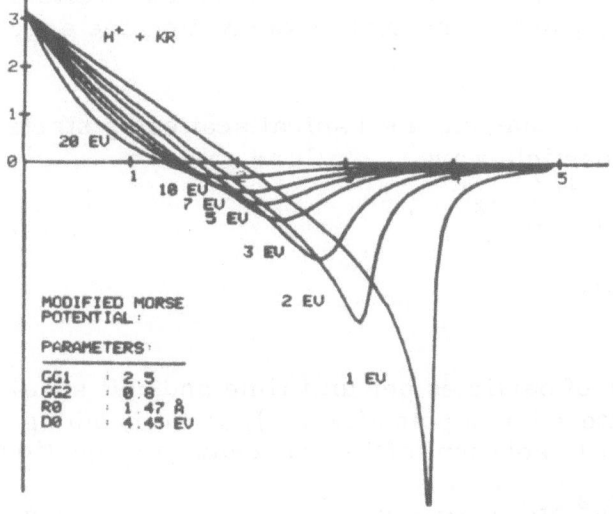

Fig. 4 – Deflection functions for several energies and different potential parameters. The potentials used have identical attractive parts but differ in the repulsive branch (bottom: soft core potential, top: hard core potential).

Fig.4 shows a typical set of deflection functions for a soft and a hard core potential. The potential parameters have been taken from experiments (H^+-Kr) (Weise, Mittmann, Ding and Henglein, 1971). According to 2.6 the cross section becomes singular for this angle as well as for $\vartheta = 0$. In analogy to the optical phenomenon this is called rainbow scattering, and the angle where this occurs, rainbow angle ϑ_r (the light deflection function of a water droplet is shown in the right upper part of Fig.4, which has a minimum at approximately 42°, depending only on the refractive index, not on the size of the droplet).

For deflection angles smaller than the rainbow angle three impact parameters b_1, b_2, b_3 contribute to the scattering into one angle, for greater deflection angles only one impact parameter makes a contribution. For very low kinetic energies deflection angles of more than 360° do occur, describing trajectories surrounding the scattering centre several times. In that case more than 3 trajectories contribute to the scattering into one angle.

3. QUANTUM MECHANICAL DESCRIPTION OF SCATTERING

The classical treatment of scattering is only valid if the de Broglie wavelength $\lambda = \dfrac{h}{m \cdot v}$ is small compared with the dimension of the scattering centre. Using normal atom masses this is not the case in the chemically interesting energy range (1...200 eV). Therefore interference effects are to be expected and the scattering has to be dealt with using quantum theory (for derivation see e.g. Hasted, 1972a).

The rigid treatment of quantum mechanical scattering starts from the idea of a plane particle wave

$$\psi_i = (\frac{A_i}{v})^{1/2} \cdot e^{ikz} \qquad (3.1)$$

z: direction of propagation
i: initial state

of flux density A (number of particles per unit time and unit area) and velocity v being scattered by a potential $V(r)$. Both incoming and outgoing waves have to be solutions of the Schroedinger-equation

$$\nabla^2 \psi + [k^2 - V(r)] \cdot \psi = 0 \qquad (3.2)$$

$k = \dfrac{2\pi}{\lambda}$ wave number

The outgoing wave will consist of the attenuated plane wave plus a spherical wave describing the scattered particles as shown in Fig. 5.

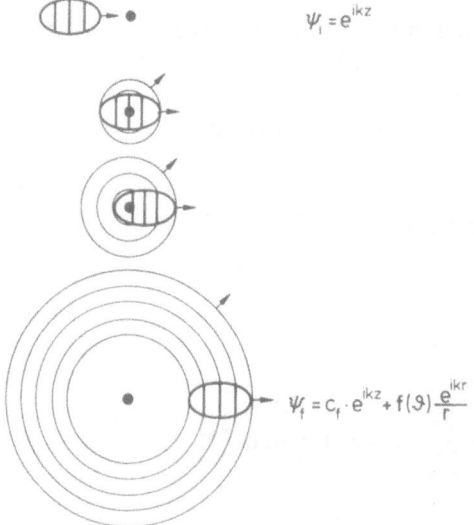

$\psi_i = e^{ikz}$

$\psi_f = c_f \cdot e^{ikz} + f(\vartheta) \frac{e^{ikr}}{r}$

Fig. 5 -
Schematic representation of a wave packet scattered by spherical potentials

$$\Psi_f = (\frac{A_f}{v})^{1/2} e^{ikz} + f(\vartheta) \cdot \frac{e^{ikr}}{r} \tag{3.3}$$

The scattering amplitude $f(\vartheta)$ may be derived by expanding in- and outgoing wave functions in the following form:

$$\psi(r,\vartheta,\varphi) = \sum_{l=0}^{\infty} \sum_{m=-1}^{+1} c_{lm} \cdot \frac{u_l(r)}{r} Y_{lm}(\vartheta,\varphi) \tag{3.4}$$

The radial functions $\frac{u_l(r)}{r}$ are solutions of the radial Schroedinger equation

$$[-\frac{\hbar}{2\mu} \frac{d^2}{dr^2} + \frac{l(l+1)\hbar^2}{2\mu r^2} + V(r)] \cdot u_l \equiv [-\frac{\hbar}{2\mu} \frac{d^2}{dr^2} + V_{eff}(l,r)] \cdot u_l$$

$$= E \cdot u_l \tag{3.5}$$

and have the asymptotic form

$$\frac{u_1(r)}{r} = \frac{\sin(k \cdot r - \frac{1 \cdot \pi}{2} + \eta_1)}{k \cdot r} \tag{3.6}$$

Inserting 3.4 and 3.6 into 3.3 yields an expression for the scattering amplitude (Faxen and Holtsmark, 1927)

$$f(\vartheta) = \frac{1}{2 \, ik} \sum_{l=0}^{\infty} (2\,l+1)\,[e^{2i\eta_l} - 1]\,P_l(\cos\vartheta) \tag{3.7}$$

The differential cross section is given by

$$\frac{d\sigma}{d\Omega}(E, \vartheta) = \left| f(\vartheta) \right|^2 \tag{3.8}$$

This yields for the total cross section

$$\sigma(E) = \int_{\Omega} \frac{d\sigma}{d\Omega} \, d\Omega = \frac{4\pi}{k^2} \sum_{l=0}^{\infty} (2\,l+1) \cdot \sin^2 \eta_l \tag{3.9}$$

The information about the scattering potential is contained in the phase shift η_l. The phase shifts have to be computed using the numerical solution of the radial Schroedinger-equation and comparing it with its asymptotic form. As this requires an immense computational effort, several computationally simpler approximations for the phase shift are used. For normal ion-atom (as well as for atom-atom-) scattering, the wavelength is small compared with the extension of the scattering centre so that the JWKB-approximation can be applied

$$\eta_1^{JWKB} = k \cdot r_m [\int_0^{\infty} (1 - \frac{U(\varrho)}{K} - \frac{\beta^2}{\varrho^2})^{1/2} \, d\varrho - \int_{\beta_0}^{\infty} (1 - \frac{\beta^2}{\varrho^2})^{1/2} \, d\varrho] \tag{3.10}$$

This expression is closely related to the classical action (F.T. Smith, 1964)

$$A = S - S_0 = \int p\,dq - \int p_0\,dq_0 \tag{3.11a}$$

$$= \int p_r\,dr - \int p_{r_0}\,dr + L \int_0^{\vartheta} d\vartheta \tag{3.11b}$$

$$= \tilde{\eta}(E, L) - L \cdot \vartheta \tag{3.11c}$$

$\tilde{\eta}$ (E, L) is called the classical phase and represents the limit of
tue quantal phase shift

$$\tilde{\eta}\,(E,\,L) = \lim\, 2\hbar \cdot \eta_1(E); \qquad (1 + \tfrac{1}{2}) \cdot \hbar \;\rightarrow\; L = v \cdot b \qquad (3.12)$$

It should be noted that in the JWKB-approximation the centrifugal
potential $\dfrac{\beta^2}{\beta^2}$ contains $\beta = (1 + \tfrac{1}{2})^2$ instead of $\beta^2 = 1\,(1+1)$ (Langer
modification) (Langer, 1937). The JWKB-phase shift η^{JWKB} can
be related to the deflection function $\vartheta\,(\beta)$ as follows

$$\frac{d\eta^{JWKB}}{d\beta} = \frac{k \cdot r_m}{2} \cdot \vartheta(\beta) \qquad (3.13)$$

For large angular momenta a further simplification is due to Jeff-
reys and Born.

$$\eta_1^{JB} = \frac{k \cdot r_m}{2} \int\limits_{\beta_1}^{\infty} \frac{U(\rho)\,\dfrac{\varepsilon}{E}}{[1 - \dfrac{\beta_1^2}{2}]}\, d\rho \qquad (3.14)$$

Several authors have compared the exact solutions of the Schroedinger equation with the JWKB solutions and have found good agreement.

Fig. 6 shows a comparison between interaction potential deflection function and phase shift. The filled circles in the potential
curve show the points of closest approach and have corresponding
circles in the deflection function. Trajectories with too large impact parameters $b > b_{max}$ will not experience any significant potential. This gives an upper limit for the angular momentum l using
relation (3.12):

$$1 < \frac{b_{max}}{k}; \qquad V(r \gtrless b_{max}) = 0 \qquad (3.15)$$

Usually a few hundred to a few thousand angular momenta are sufficient for accurate calculations.

The semiclassical approximations

Even using JWKB-phase shifts the numerical computations
are time-consuming. Approximately 80 % of the computing time is
used for calculating phase shifts; the rest is needed for the sum-

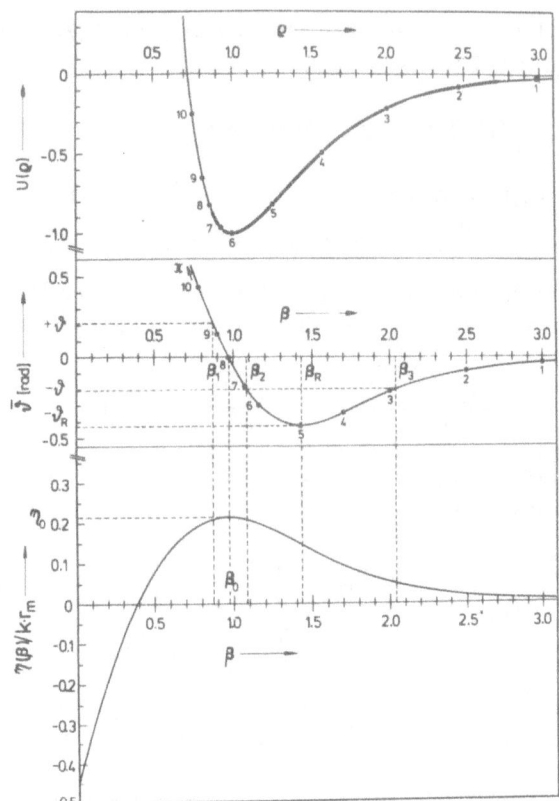

Fig. 6
Potential U(ϱ), deflection function ϑ(ß)
and phase η (ß) (in reduced units, E_c = 3,
$k \cdot r_m$ = 100). The filled circles denote the
turning points on the
potential for different
impact parameters.

mation of the partial waves and for data input and output. Therefore, further approximations have been used; they have the additional advantage of allowing an interpretation in terms of simple
physical pictures.

The semiclassical approximation (Berry, 1966; Berry and
Mount, 1972) consists of three stages. In the first stage, the summation over the partial waves (3.7) is replaced by an integration
using the Poisson summation formula.

$$\sum_{l=0}^{\infty} g(l) = \frac{1}{\hbar} \sum_{M=-\infty}^{+\infty} e^{-iM\pi} \cdot \int_{0}^{\infty} dL \, g \left\{ \left(\frac{L}{\hbar} \right) - \frac{1}{2} \right\} e^{2\pi iML/\hbar} \quad (3.16)$$

The second step replaces the quantities in the summands g(l) by
their asymptotic form for large L (using 3.12)

$$\eta\left\{\frac{L}{\hbar} - \frac{1}{2}\right\} \approx \frac{\tilde{\eta}(L)}{\hbar} \tag{3.17}$$

The spherical harmonics are expressed in terms of exponential, trigonometric, or Bessel functions, depending on the angular range. This yields for the scattering amplitude, using 3.11c, 3.12

$$f(\vartheta) = \frac{-i}{2(\pi\hbar\, mE\, \sin\vartheta)^{1/2}} \sum_{m=-\infty}^{+\infty} e^{-iM} \left(e^{-i\pi/4} \cdot I_m^+ + e^{+i\pi/4} \cdot I_m^- \right) \tag{3.18}$$

with

$$I_m^{\mp} = \int_{\hbar/2}^{\infty} dL\, L^{1/2} \exp\left(\frac{i}{\hbar}\left\{2\tilde{\eta}(L) + L\cdot\left(\mp\vartheta + 2M\pi\right)\right\}\right)$$

In the third step, the integral $I_m^{\pm}$ is evaluated using the method of stationary phase. The stationary points of the exponent occur only at those values $L_i(\vartheta)$ of the classical angular momentum which satisfy

$$2\frac{d\tilde{\eta}}{dL}(L_i(\vartheta)) = \mp\vartheta - 2M\pi \tag{3.19}$$

thus showing that only the classical paths leading to an observation angle contribute to $f(\vartheta)$ in the semiclassical limit. We obtain, therefore, the semiclassical scattering amplitude

$$f(E,\vartheta) \approx \sum_i \delta_i \left\{ \frac{L_i(\vartheta)}{2m\cdot E\cdot\sin\vartheta\, \frac{d}{dL}|(L_i(\vartheta))|} \right\}^{1/2} \exp\left(\frac{i}{\hbar} A_i(\vartheta)\right) \tag{3.20}$$

the δ_i denoting that only those angular momenta have to be summed which contribute classically to the trajectories with net deflection angle ϑ. We obtain for the cross section

$$\frac{d\sigma}{d\Omega}(E,\vartheta) \approx \sum_i \sum_j \frac{\delta_i\,\delta_j}{2m\cdot E\cdot\sin\vartheta} \cdot \frac{L_i(\vartheta)\cdot L_j(\vartheta)}{\left|\frac{d\vartheta}{dL}(L_i(\vartheta))\right| \cdot \left|\frac{d\vartheta}{dL}(L_j(\vartheta))\right|}$$

$$e^{\frac{i}{\hbar}(A_i(\vartheta) - A_j(\vartheta))} \tag{3.21}$$

The method of stationary phase, however, breaks down if the stationary points L_i, L_j are too close together. A different approximation, which overcomes this difficulty, has been developed by

Berry, and is discussed later in this chapter.

Eq. 3.21 can be used to explain the different interference effects in elastic scattering. Before going into the details, there are some general statements to be made. As shown in Fig. 3, there are several different trajectories leading to the same deflection angle . Besides the two attractive and the one repulsive trajectory, there may be more trajectories, which circle around the scattering centre M-times, and end up with the same deflection angle ϑ. Therefore, equations 3.16 - 3.19 contain a summation over all possible M. In the semi-classical limit different trajectories leading to the same net deflection angle can interfere with each other. If there are n different trajectories having the same net deflection angle ϑ, there will be n-1 distinct interference frequencies (i.e. in the described case 2). The frequency is determined by the difference in classical action

$$\exp(-\frac{i}{\hbar}\left\{A_i - A_j\right\})$$

In the classical limit, the cross terms (i, j) will be averaged out, and the classical expression 2.6 is obtained.

Eq. 3.21 can be used to explain the different interference effects in classic scattering, as shown in Fig. 7.

Total Cross Section

a) Glory Undulation

Interference of trajectories without net deflection - which may have encircled the scattering centre many times - with the unscattered beam will cause energy dependent undulations in the total cross section.

b) Symmetry (charge exchange) undulations will be produced if both ion and neutral target have the same nucleus. Back-scattered particles which have undergone a charge exchange will interfere with forward scattered ions.

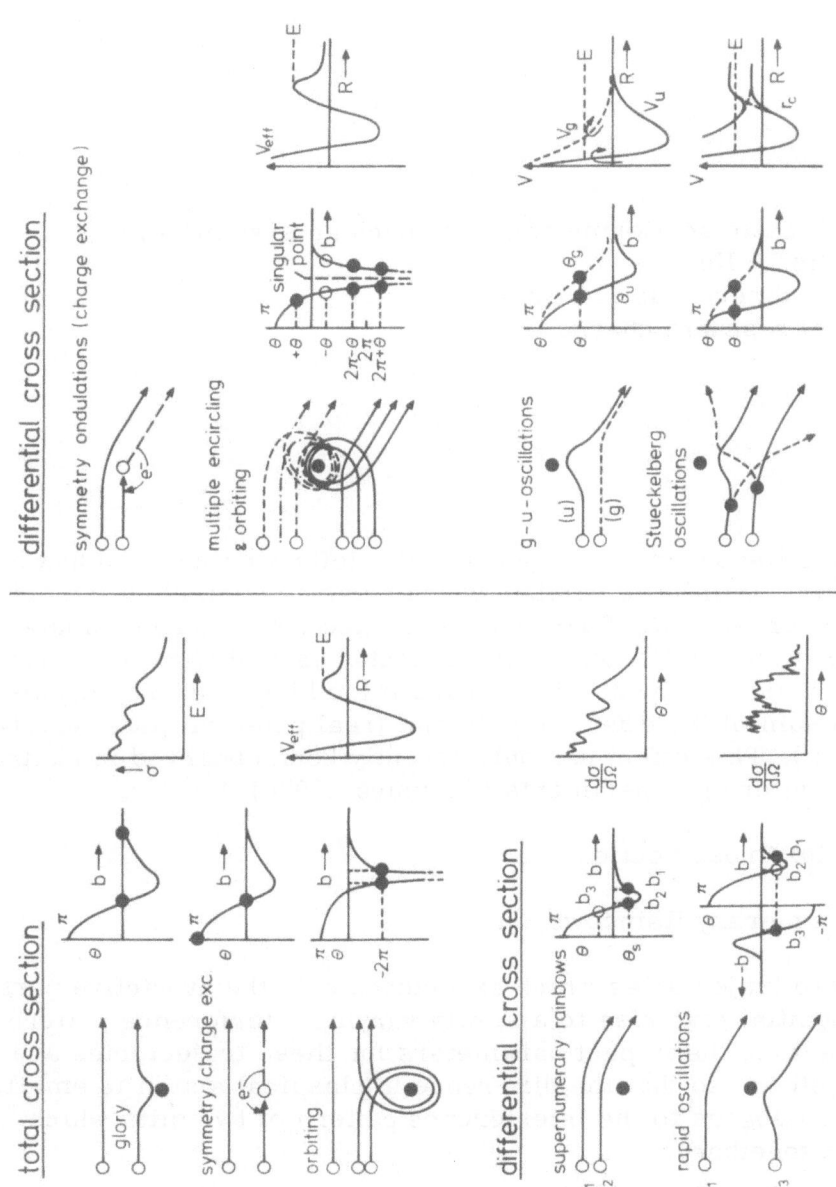

Fig. 7: Schematic representation of the different interference effects

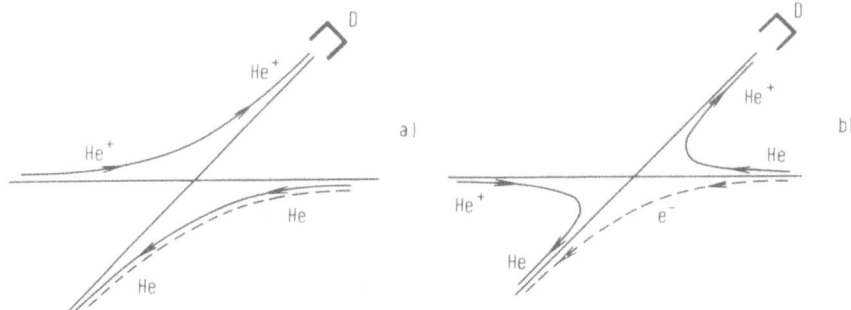

Fig. 8 - Elastic scattering of a homonuclear system, e.g.
    He^+ + He:
    a) direct elastic scattering
    b) resonant charge transfer

c) Orbiting Resonances will occur if the deflection function has a
singularity. An infinite number of trajectories encircling the scattering center will interfere. The difference from glory undulations is that an infinite number of trajectories of different impact
parameter are involved. This occurs if the kinetic energy equals
the maximum of the effective potential (real potential plus centrifugal term). This effect has only recently been observed in neutral-
neutral scattering experiments (Toennies, 1974)

Differential Cross Section

d) Supernumerary Rainbows

    The two trajectories which experience only the attractive part
of the potential give rise to a slowly varying interference pattern;
this is because the impact parameters for these trajectories are
close together, so that the difference in classical action is small.
(This is analogous to the interference pattern of two slits which
are close together.)

e) Rapid Oscillations in the interference pattern can be caused by
one repulsive trajectory and one attractive trajectory. The high

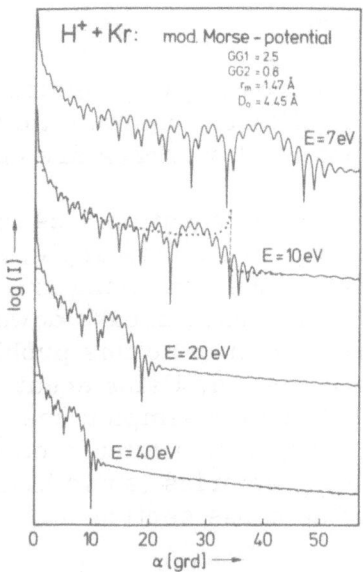

Fig. 9 - Differential cross-section for the system H$^+$ + Kr calculated using the partial wave expansion and WKB phases.
The dotted line is the appropriate classical cross section

frequency of these oscillations is caused by the large difference in
impact parameters (analogous to the interference of two slits which
are wide apart). Fig. 9 shows a calculated cross section which
shows clearly rainbow and rapid oscillations. With decreasing
energy the rainbow is shifted to higher angles.

f) Symmetric (charge exchange) oscillations
see b)

g) Orbiting
see c)

h) g-u-Oscillations:

In systems with potentials of different parity V_g, V_u which are
degenerate for r → ∞, interference between trajectories experiencing the two different potentials occur.

i) Stueckelberg Oscillations

In the case of pseudocrossing of potentials, interferences due to particles experiencing different parts of the potential by cross-ing over into different potential branches are observed.

As pointed out earlier the equation 3.24 cannot describe the scattering around the classical rainbow angle, as the impact para-meters $b_{i,j}$ contributing to the scattering are so close together that the method of stationary phase breaks down. Berry (Berry, 1966) has given a different solution to this problem by applying the method of uniform approximation. Using equations 3.16 - 3.18 he maps the integrand in 3.18 onto a simpler one with the same sta-tionary point structure. Neglecting interference effects between attractive and repulsive trajectories (smoothing out the rapid os-cillations), he finds for the cross section

$$\frac{d\sigma}{d\Omega}^{uniform}(\vartheta) = \frac{\pi}{2mE\sin\vartheta} \left[\left\{ \left(\frac{L_1}{|\vartheta'(L_1)|} \right)^{1/2} + \left(\frac{L_2}{|\vartheta'(L_2)|} \right)^{1/2} \right\}^2 \frac{(-\zeta)^{1/2}}{\hbar^{1/3}} Ai^2 \left(\frac{\zeta}{\hbar^{2/3}} \right) \right.$$

$$\left. + \left\{ \left(\frac{L_1}{|\vartheta'(L_1)|} \right)^{1/2} - \left(\frac{L_2}{|\vartheta'(L_2)|} \right)^{1/2} \right\}^2 \frac{\hbar^{1/3}}{(-\zeta)^{1/2}} Ai'^2 \left(\frac{\zeta}{\hbar^{2/3}} \right) \right] \quad (3.22)$$

with

$$\zeta = \left\{ -\frac{3i}{4} (A_2(\vartheta) - A_1(\vartheta)) \right\}^{2/3}$$

and Ai being the Airy function.

For angles well below the rainbow angle, this reduces to equation 3.21. For angles around the rainbow angle, this can be approximated by

$$\frac{d\sigma}{d\Omega}^{trans} (E,\vartheta) = \frac{L_r \cdot \pi}{\hbar^{1/3} \cdot m \cdot E \cdot \sin\vartheta \cdot \{\tilde{\eta}(L_r)\}^{2/3}}$$

$$Ai^2 \left(\frac{\vartheta - \vartheta_r}{\hbar^2 \cdot \tilde{\eta}(L_r)^{1/3}} \right) \quad (3.23)$$

This is the "transitional" approximation originally derived by Ford and Wheeler (Ford and Wheeler, 1959a, 1959b). Fig.10 shows a comparison of the three approximations. The uniform approximation describes the scattering well over the whole angular range, while expression 3.21 gives a good fit only to the supernumerary rainbows, equation 3.23 fits only the rainbow itself.

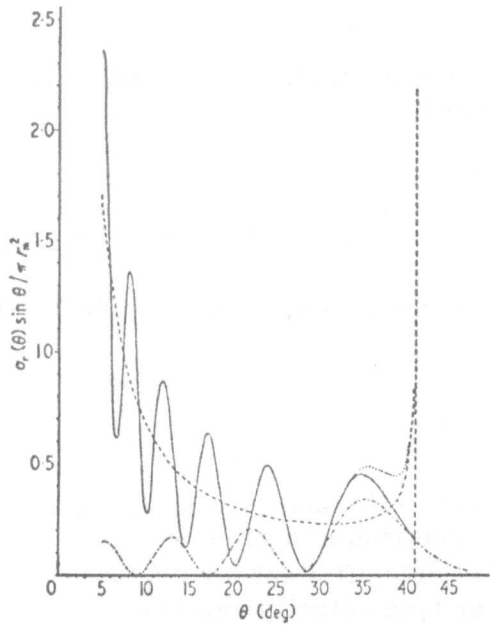

Fig. 10 -
Semiclassical approximation: Comparison of
— the uniform approximation (3.22),
-.- the transitional approximation (3.23),
.... the stationary phase approximation (3.21)
--- the classical cross section

(potential: L-J (12, 6); K = 3, $\frac{K}{r_m}$ = 200). The quantum mechanical result averaged over the rapid oscillations is within 2% of the uniform approximation (after Berry and Mount, 1972)

From 3.21 the distance between the supernumerary rainbows can be calculated using 3.11c

$$2\pi \cdot \hbar = A_1(\vartheta) - A_2(\vartheta) = \int_{b_{i,1}}^{b_{u,1}} db \cdot \frac{dA_1(\vartheta)}{db} - \int_{b_{i,2}}^{b_{u,2}} db \frac{dA_2(\vartheta)}{db}$$

$$\approx \hbar \cdot k \cdot (b_2 - b_1) \cdot \Delta\vartheta_s$$

$$\Delta\vartheta_s \approx \frac{2\pi}{k(b_2 - b_1)} \tag{3.24}$$

Similarly, one obtains for the period of the rapid oscillations $\Delta \vartheta_r$

$$\Delta \vartheta_r \approx \frac{2\pi}{k(b_2+b_3)} \approx \frac{\pi}{k \cdot b_o} \quad \text{as} \quad 2 b_o \approx b_2+b_3 \tag{3.25}$$

for the g-u-oscillations

$$\Delta \vartheta_{gu} \approx \frac{2\pi}{k(b_g+b_u)} \tag{3.26}$$

b_g, b_u are the impact parameters of trajectories experiencing the g-potential and the u-potential respectively,

and for the Stueckelberg oscillations,

$$\Delta \vartheta_{St} \approx \frac{2\pi}{k (b_I - b_{II})} \tag{3.27}$$

b_I, b_{II} are impact parameters of trajectories experiencing different branches of the potential.

4. EXPERIMENTAL RESULTS

A typical experimental set-up is (Mittmann, Weise, Ding, and Henglein, 1971) shown in Fig. 11. The ions are generated in a plasma ion source, which is particularly efficient for protons, then mass analyzed by a Wien-filter type velocity selector. The

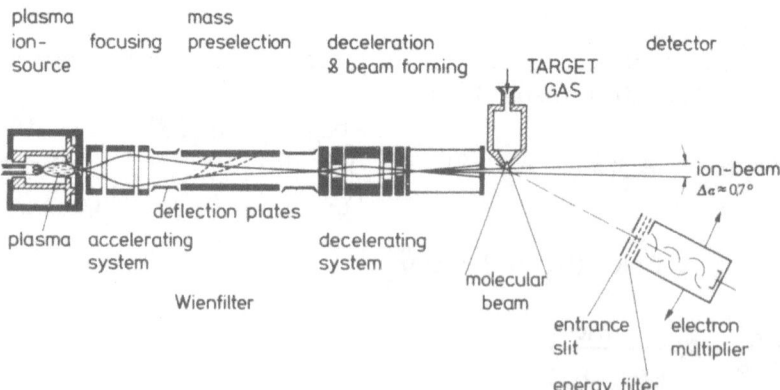

Fig. 11 - Schematic diagram of a typical elastic scattering experiment for measuring the differential cross section of ions

ions are decelerated down to the desired energy by a system of
lenses, pass a double slit system to improve the angular width
of the beam, and, subsequently, cross a molecular beam. The
scattered particles are detected by an open electron multiplier,
inelastically scattered particles being removed by a retarding
field. The scattered ion intensity is recorded as a function of the
scattering angle. There have been experiments performed using
different types of ion sources, mass analysis and deceleration
systems. Some authors use a scattering chamber instead of the
molecular beam.

The angular resolution in an experiment of this type is of
the order $0.5\ldots1^{\circ}$, the energy spread of the beam 0.1-0.3 eV.

As one can see from equation 3.24, 3.25, quantum mechanical interference effects are more pronounced if the deBroglie
wavelength is large. This is the case if the reduced mass of the
system and the CM-energy are small. Fig. 12 shows the effect
of different reduced masses in the same potential. The bottom
of the diagram shows the differential cross section for the system
H^{+}-Ar and D^{+}-Ar. As the deflection function only depends on the
intermolecular potential and not on the wavenumber, the rainbow
angle does not depend on the choice of the isotope. The distance
between the supernumerary rainbows changes the ratio by appro-
ximately $\dfrac{k_1}{k_2} = \left(\dfrac{\mu_1 \cdot E_1}{\mu_2 E_2}\right)^{\frac{1}{2}} \approx \left(\dfrac{m_1}{m_2}\right)^{\frac{1}{2}} = 1.4$. A similar behaviour is
observed for the rapid oscillations. In the top part of Fig. 12
the differential cross section of the system H^{+}-He, D^{+}-He is displayed. Again the oscillations become faster going from the lighter to the heavier isotope.

Experiments have been performed for a series of systems,
and the scattering of protons is best understood at the present
date. Table 1 gives a survey of data available today.

It would be beyond the scope of this talk to give an extensive
survey of all available data. Therefore only a few examples for
the different observed effects will be given.

Proton - Rare Gas. Rainbow Scattering

Different groups have investigated the elastic scattering of
protons (3 eV to a few 100 eV). In most of the cases the obtained

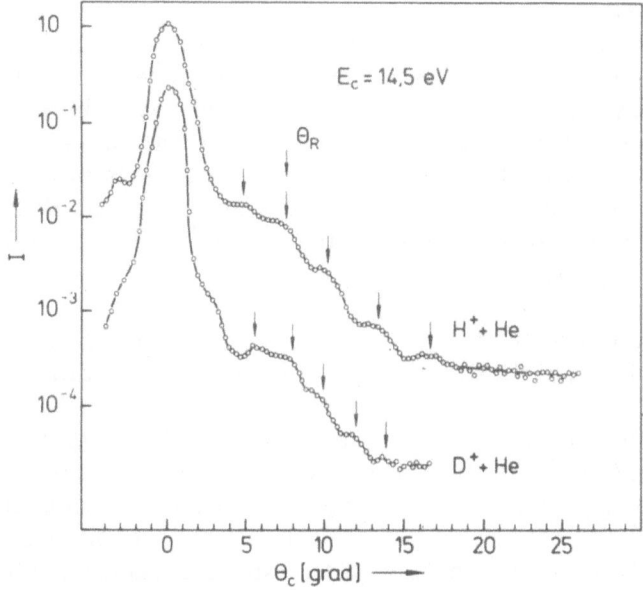

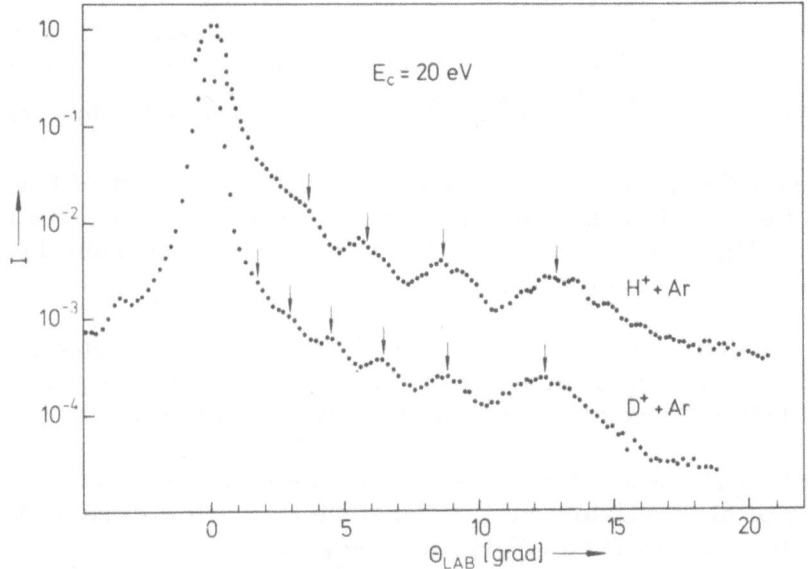

Fig. 12 - Isotope effects at the elastic scattering of protons
          by He and Ar

Table 1: Experimental work in elastic ion-scattering*

H^+ + rare gas	Mittmann, Weise, Ding and Henglein (1971a)	V
	Weise, Mittmann, Ding and Henglein (1971)	V
	Rich, Bobbio, Champion and Doverspike (1971)	V
	Usdeth, Giese and Gentry (1971)	H
	Herrero, Nemeth and Bailey (1969)	M
H^+ + molecules	Mittmann, Weise, Ding and Henglein (1971b)	V
H^+ + rare gas	Smith, Marchi, Aberth, and Lorentz (1967)	H
	Weise and Mittmann (1973)	V
Symmetric rare gas ion-atom	Lorents and Aberth (1965 , 1966) (1965, 1966)	H
	Aberth and Lorents (1965)	H
	Mittmann and Weise (1974)	V
Li^+ + rare gas	Böttner, Dimpel, Ross and Toennies (1974)	V
N^+ + rare gas	Weise, H.P., and Mittmann, H.U. (1974)	V
Ar^+ + CH_4	Arikawa, Hirakawa, Nishikawa and Watanabe (1973)	H
Cs^+ + Ar, Kr, N_2	Menendez and Datz (1965)	M
K^+ + rare gas (integral cross sections)	Menendez and Redmon (1969) Boerboom, Dop and Los (1970) Inonye and Kita (1972)	

*) Resolution: V: very high, H: high, M: medium

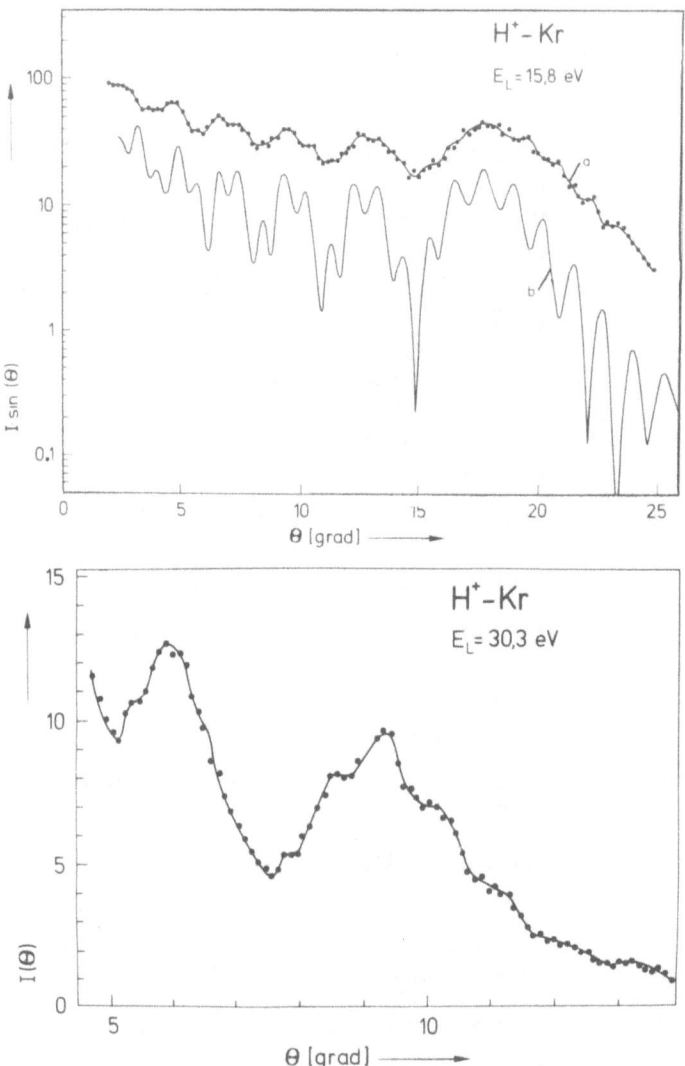

Fig. 13 - Typical experimental results for the system H^+-Kr:
top: differential elastic cross section (upper curve) show-
            ing supernumerary rainbows, and calculated cross
            section (lower curve) parameters mod. Morse-poten-
            tial   = 4.45 eV; r_m = 1.47 Å, G_1 = 2.5, G_2 = 0.8
bottom: high resolution measurement to determine the rapid
            oscillations

potentials can be compared with ab initio-calculations. Fig. 13 shows a typical experimental example together with a calculated cross section based on a fitted potential. The rainbow structure and the supernumerary oscillations, which could be even better separated in a high resolution experiment (right part of Fig. 13), are easily recognized. Fig. 14 shows the fitted potentials for the whole series, together with ab initio calculations for the systems H^+-He, H^+-Ne, H^+-Ar, which are in good agreement with the experimental results.

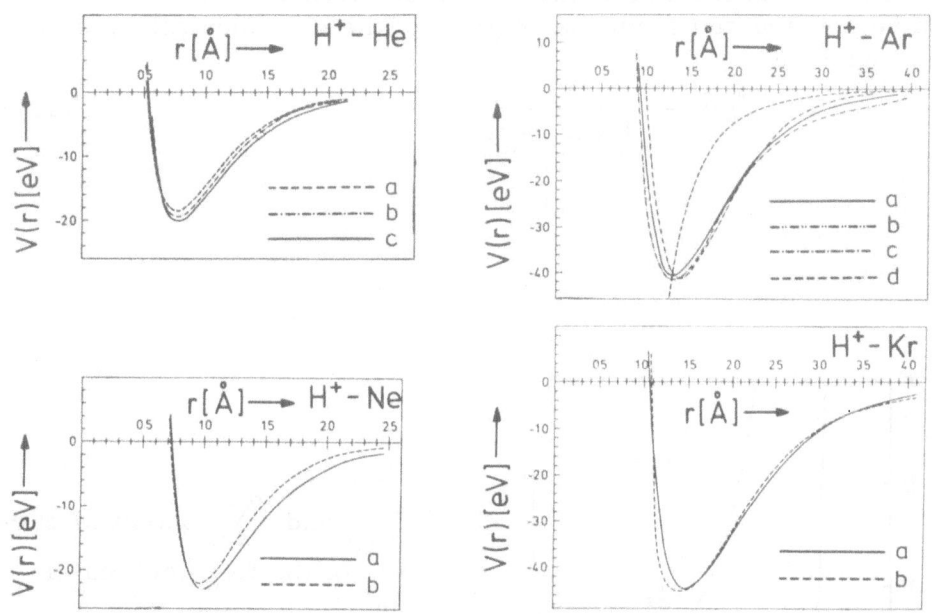

Fig. 14 - Results for the proton-rare gas systems:

      H^+-He: a - ab initio calculation by Michels (1966)
                b - ab initio calculation by Peyerimhoff (1965)
                c - best fit modified Morse-potential
      H^+-Ne: a - best fit modified Morse-potential
                b - ab initio calculations by Peyerimhoff (1965)
      H^+-Ar: a - best fit modified Morse-potential
                b - best fit modified L-J-potential
                c - ab initio calculations by Roach and Kuntz (1970)
                d - ion-induced dipole potential
      H^+-Kr: a - best fit modified Morse potential
                b - best fit modified L-J-potential

g-u-oscillations

     As pointed out in chapter 3, interference effects are to be expected if there are two potentials which are degenerate for infinite particle distance. One of the best investigated examples is the system $He^+ + He$. Fig. 15 shows the attractive $^2\Sigma_u^+$ and the repulsive $^2\Sigma_g^+$ potential curve for this system. At low energies the attractive curve gives rise to the ordinary rainbow scattering. At higher energies an interference structure due to the scattering on the gerade and ungerade branches of the potential is observed. The difference between both deflection functions can be constructed from the position of the g-u-undulations using relation 3.26.

$$\Delta \vartheta = \frac{2\pi}{k \mid b_g - b_u \mid}$$
(4.1)

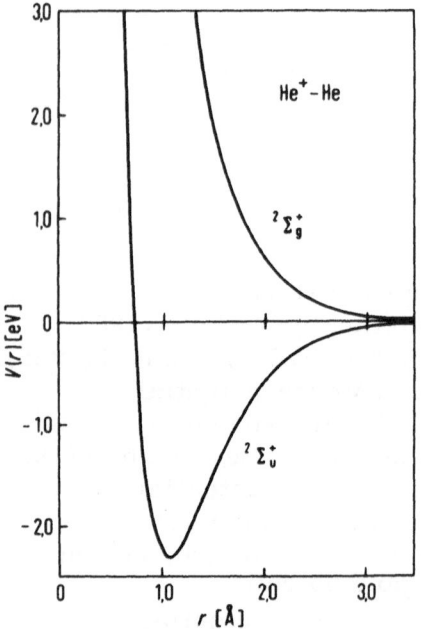

He$^+$-He

$^2\Sigma_g^+$

$^2\Sigma_u^+$

$V(r)$ [eV]

r [Å]

Fig. 15 –
$^2\Sigma_g^+$ and $^2\Sigma_u^+$ potential curves of the He$_2^+$-molecular ion

The quantum mechanical treatment leads to the expression

$$\frac{d\sigma}{d\Omega}(E,\vartheta) = \frac{1}{4}\cdot\left| f_g(E,\vartheta) + f_u(E,\vartheta)\right|^2 \tag{4.2}$$

f_g, f_u being the scattering amplitudes calculated from the gerade and ungerade potential using equation 3.10.

At high energies further modulation is to be seen superimposed onto the g-u-undulations. This is caused by resonant charge transfer between He-ion and He-atom. As a He-ion scattered into angle ϑ is indistinguishable from an ion being scattered into the opposite direction $(\pi - \vartheta)$ and having its charge transferred to the atom, which is scattered into the same angle ϑ, the scattering cross section has to be written as

$$\frac{d\sigma}{d\Omega}(E,\vartheta) = \frac{1}{4}\left| f_g(E,\vartheta) + f_u(E,\vartheta) + f_g(E,\pi-\vartheta) - f_u(E,\pi-\vartheta)\right|^2 \tag{4.3}$$

This interference effect will be more pronounced at higher energies as the ratio $f(E,\pi-\vartheta)/f(E,\vartheta)$ increases with energy and angle. Fig.16 shows experiments with $^4He^+$-^4He, $^4He^+$-^3He by Lorents and coworkers (Lorents and Aberth, 1965 and 1966 ; Lorents, Aberth and Hestermann, 1966; Aberth, Lorents, Marchi and Smith, 1965). The two different interference patterns can easily be seen, the undulations due to nuclear symmetry not being present using ^3He as a target.

Stueckelberg Oscillations

The potential curves we have dealt with so far have always been calculated under the assumption of adiabaticity. At medium and high energies this assumption breaks down if there are neighbouring potential curves of equal symmetry (Landau, 1921; Zener, 1932; Stueckelberg, 1932).

Landau, Zener and Stueckelberg have treated a system of two potential curves, where there is one point of closest (adiabatic) approach, which is called the critical radius r_c or the point of pseudocrossing. At the critical radius they find a limited probability, P, for a transition from one curve to the other; depending on the relative velocity, v_r, of the particles, and on the properties of the potential:

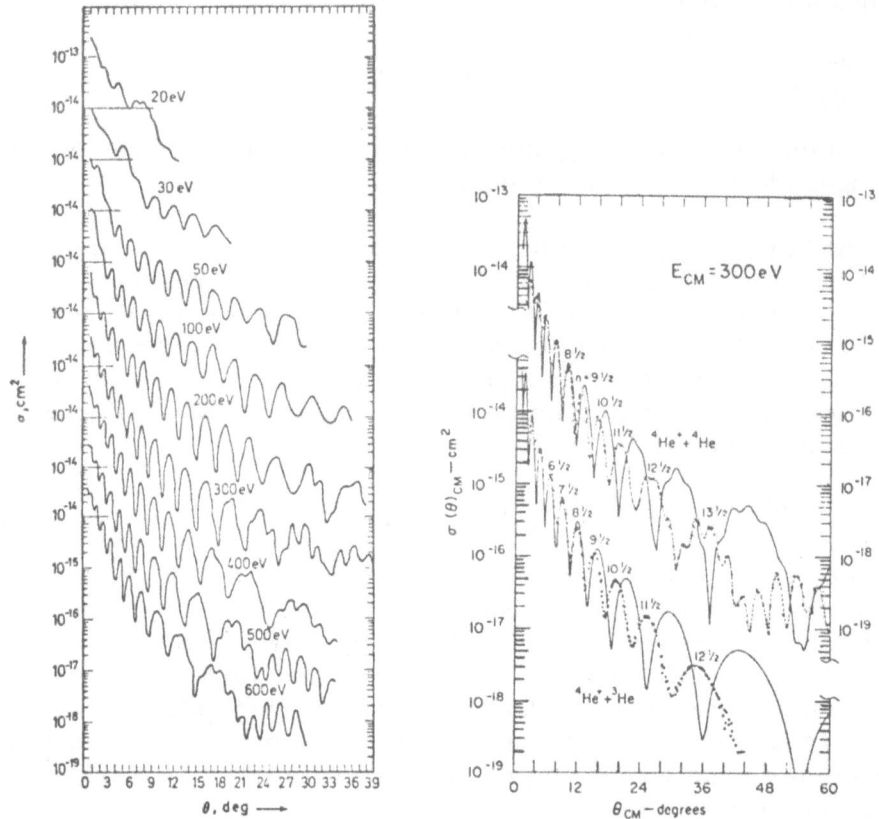

Fig. 16 - Elastic scattering of He⁺ on He after Lorents and co-workers (Lorents and Aberth, 1965; Aberth and Lorents, 1965)
left: differential elastic cross section for the scattering of ^{4}He⁺ on ^{4}He at various energies.
right: comparison of the elastic scattering of ^{4}He⁺ on ^{4}He and ^{4}He⁺ on ^{3}He. The nuclear symmetry oscillations cannot be detected in the latter case (o-o-o: experimental points, --- theory)

$$P = \exp \left(\frac{-2 \pi \cdot H_{12}^2}{\hbar \cdot v_r \left| \frac{d}{dr} (H_{11} - H_{22}) \right|_{r=r_c}} \right) \qquad (4.4)$$

H_{12} is the energy gap at the pseudocrossing, $\frac{d}{dr} (H_{11}-H_{22})$ the

difference in slope of both potential curves (cf. Fig. 15). A particle coming in on the lower potential curve can undergo four possible "trajectories", two of which will end at the same level.
Stueckelberg showed that four phases η_1 to η_4 are sufficient to
describe the interference effects due to superposition of trajectories experiencing different parts of the potential. These phases
are combinations of partial phases δ_1 to δ_4, which describe the
phase difference on four different branches of the potential, as
shown in Fig. 17. The scattering amplitudes are as follows:

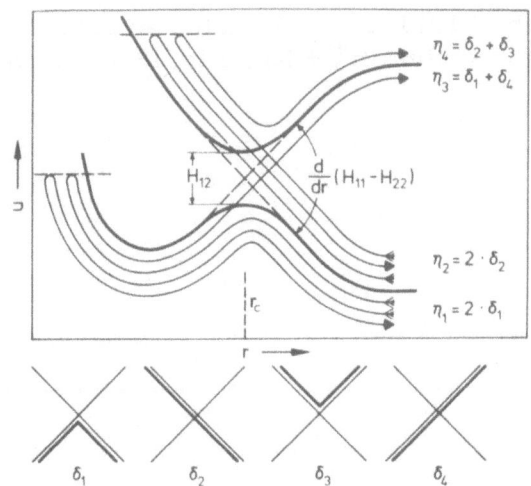

Fig. 17 - Schematic representation of the interference due to curve
crossing giving rise to Stueckelberg oscillations. The
scattering can be described by four partial phase-shifts
δ_1 to δ_4, the experimentally relevant phases η_1 to η_4
being combinations of these

elastic part: $\quad f_a(\vartheta) = \dfrac{1}{2ik_a} \quad (2l+1)\ (S_a(l)-1)\ P_l(\cos\vartheta)$ $\quad$ (4.5)

inelastic part: $f_b(\vartheta) = \dfrac{1}{2ik_b} \quad (2l+1)\ S_b(l)\ P_l(\cos\vartheta)$ $\quad\quad$ (4.6)

with

$$S_a(l) = (1-P)^2\ e^{i\cdot\eta_1} + P^2\cdot e^{i\cdot\eta_2} \quad\quad\quad (4.7)$$

$$S_b(l) = P(1-P)\left\{ e^{i\cdot\eta_3} + e^{i\cdot\eta_4} \right\} \quad\quad\quad (4.8)$$

$$\eta_1 = 2\cdot\delta_1 \quad\quad\quad\quad\quad\quad\quad\quad (4.9)$$

$$\eta_2 = 2 \cdot \delta_2$$
$$\eta_3 = \delta_1 + \delta_4$$
$$\eta_4 = \delta_2 + \delta_3$$

One should note that only in the case of elastic scattering interference with the incident plane wave occurs (indicated by the -1 in 4.5). Furthermore, an interference structure will also occur for the inelastic channel - this is generally true for trajectories with equal energy loss and equal net deflection angle - and this interference will be strongly modulated. The oscillatory structure gives information about the difference of the potential inside the pseudo crossing.

Raabe (Raabe, 1971) has calculated an example of Stueckelberg oscillations for the system $H_2^+ + Kr$. Fig. 18 shows the two poten-

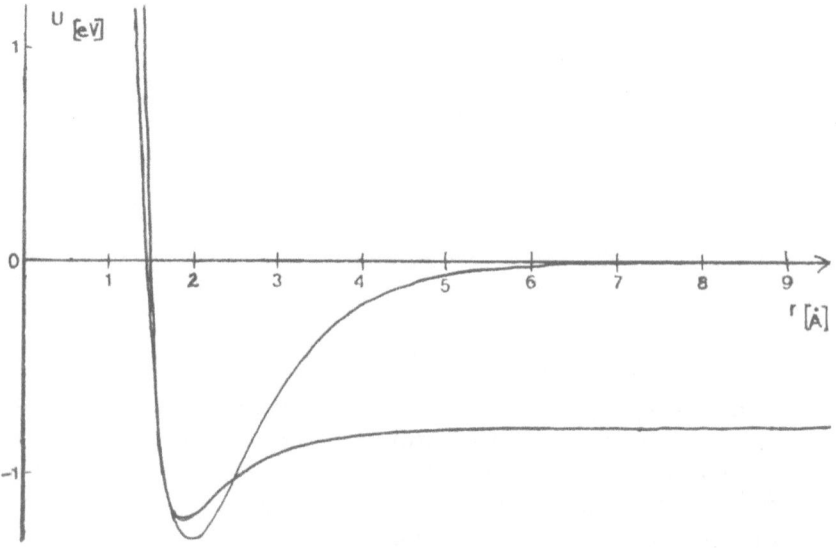

Fig. 18 - The potential $H_2^+ + Kr$ and the charge exchanged potential $H_2 + Kr^+$ (after Raabe, 1971)

tials involved, the upper one being the "diabatic" H_2^+-Kr potential, the lower curve representing the one for H_2-Kr^+. Fig. 19 shows calculated differential cross sections compared with experimental ones, the arrows indicating the rainbow positions. The oscillatory structure on the dark side of the rainbow is due to Stueckelberg interferences.

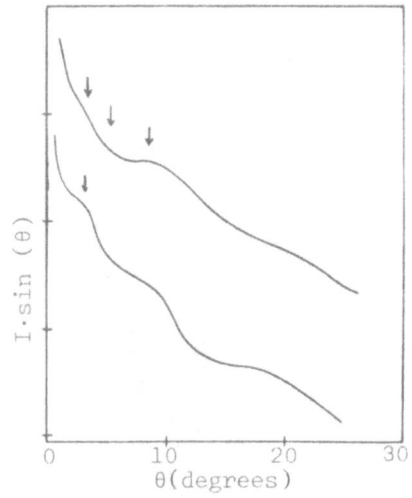

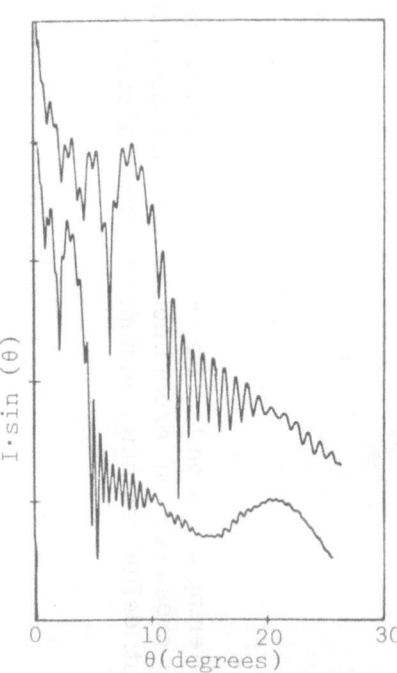

Fig. 19 -
Experimentally determined (upper
frame) and calculated (lower frame)
cross sections for the elastic scattering of H_2^+ on Kr. The upper curves in each frame correspond to
9.77 eV, the lower ones to 25.39 eV
ion-energy

As pointed out previously the inelastic channel will show similar interference effects. Delvigne and Los (1973) have given a
beautiful example by investigating the system $Na + I \rightarrow Na^+ + I^-$
and measuring the angular distribution of the sodium ion. Fig. 20
shows the experimental results, the deflection function, and semi-
classical calculations of the cross section (neglecting rapid oscil-

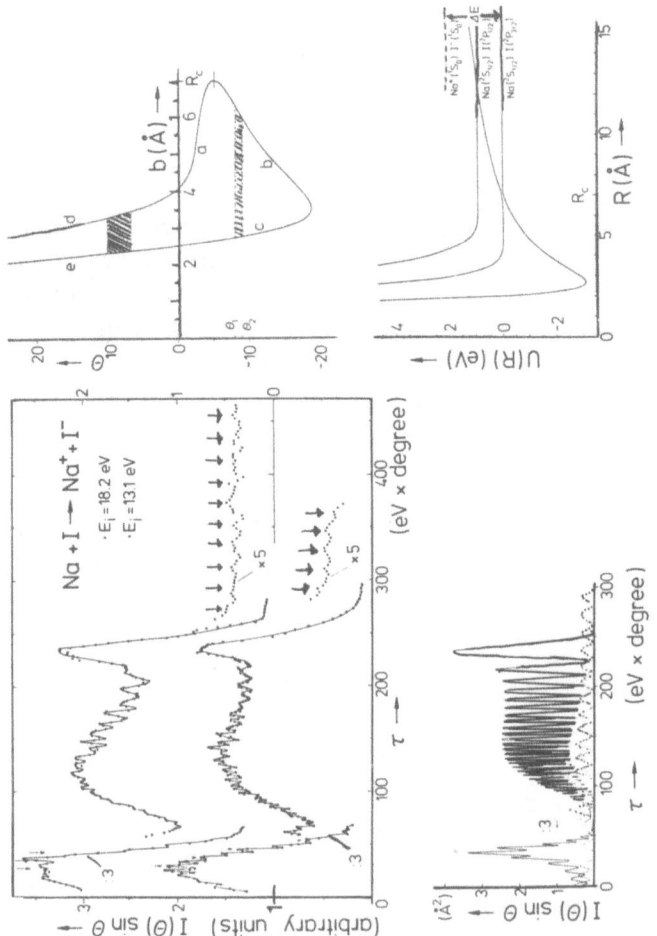

Fig. 20: Scattering of Na on I   Na$^+$ + I$^-$ (inelastic channel; Na$^+$ detected) left: cross section, experimental (top) and calculated (bottom); right: deflection function (top) and potential curves (bottom) (after Delvigne and Los, 1973)

lations), using an analytical expression for the interaction potential. The inelastic deflection function is not defined for impact parameters greater than the critical radius, as the inelastic channel is not accessible. The structure is due to interference between different parts of the deflection function: b + c yields rainbow scattering, a + c and d + e (dotted line) Stueckelberg oscillations. All other interference effects are neglected.

There has been considerable interest in this field even for systems with more than 2 potential curves: Kubach, Sidis and Durup (1974) have calculated elastic cross sections for the H^+-Xe system using a 3-state model. They compare their calculations with experimental results and find agreement except for large angles, which they explain by inelastic processes due to crossing with higher excited states.

5. INVERSION TECHNIQUES

A serious problem in elastic scattering is the extraction of the potential from the experimental data. The classical method is to use an analytic trial potential with several free parameters and compare the calculated angular distributions with the experimental one. The parameters are changed till experiment and calculations agree sufficiently well. The starting values for potential parameters are obtained from semiclassical expressions 3.24 - 3.27. A great number of analytic potentials have been described. The best known are the 3-parameter Morse-potential and the 4-parameter Lennard-Jones potential. Table 2 shows a list of frequently used potentials together with some of their properties.

In recent years, methods for directly obtaining potentials from the experimental data have been reported. These methods have the advantage of not requiring an analytical form for the potential and are faster than the above described fitting procedures. The latest development in this field has been reviewed by Buck (Buck 1974).

The inversion procedure is divided into two steps: The determination of the phase shifts from the cross section and the extraction of the potential from the phase shifts:

Table 2:　Semiempirical Potential Functions
(in brackets: number of parameters)

1.) Lennard-Jones n-m:　$U_{LJ} = \dfrac{n \cdot m}{n-m} \left(\dfrac{\varrho^{-n}}{n} - \dfrac{\varrho^{-m}}{m} \right)$
　　　(2)
　　　　　　　　　　　Usually m = 4 for ion-atom
　　　　　　　　　　　　　　　m = 6 for atom-atom

2.) Modified LJ-I: $U_I = U_{LJ} - (U_{LJ}+1) \cdot \Gamma_1 \exp\left(-\dfrac{-1}{\gamma_2} \right)$
　　　(4)

3.) Modified LJ-II: $U_{II} = \dfrac{W(GG_1 + W(GG_2 + WGG_3))}{GG_1 + GG_2 + GG_3} \cdot \dfrac{L}{2_{+1}}$; $W = U_{LJ}^{-1}$

4.) Kihara:　　　$U = \begin{cases} \infty \\[4pt] \dfrac{n \cdot m}{n-m} \left(\dfrac{\tau^{-n}}{n} \right) - \dfrac{\tau^{-m}}{m} \end{cases}$; $\tau = \dfrac{\varrho - \varrho_c}{1 - \varrho_c}$
　　　(3)

5.) 2-Piece LJ:　$U = U_{LJ} \begin{cases} m=m_1, & n=n_1 & \varrho < 1 \\ m=m_2, & n=n_2 & \varrho \geqslant 1 \end{cases}$
　　　(4)

6.) Morse:　　$U_M = \exp(2 \cdot G(1-\varrho)) - 2 \cdot \exp(G(1-\varrho))$
　　　(1)

7.) Buckingham: $U_B = \dfrac{\alpha}{\alpha - s} \left[\dfrac{s}{\alpha} \exp(\alpha(1-\varrho)) - \varrho^{-s} \right]$
　　　(2)

　　　　　　　　　　s = 4 for ion-atom
　　　　　　　　　　s = 6 for atom-atom

8.) 2-Piece Morse: $U_{M2} = \exp(2GG_1 GG_2 (1-\varrho)) - 2 \cdot \exp(GG_1 \cdot GG_2(1-\varrho))$
　　　(2)
　　　　　　　　　　$GG_2 = 1$　　　　　$\varrho > 1$
　　　　　　　　　　$GG_2 \neq 1$　　　　　$\varrho \leqslant 1$

9.) Lippincott:　　　$U_L = [1 - e^{-x}][1 - a \cdot b \cdot \sqrt{x} \cdot \exp(-b \cdot \left(\dfrac{x}{\varrho} \right)^{1/2})]$
　　　(3)

　　　　　　　$x = \dfrac{h \cdot c \cdot \omega_e^2 (\varrho - 1)^2}{4 B_e \cdot D \cdot \varrho \cdot r_m^2}$

---

for 1.),　4.),　6.),　7.), and 9.) see e.g. Hasted (1972b)
for 2.)　see Düren, Raabe and Schlier (1968)
for 3.),　5.),　and 8) see Mittmann, Weise, Ding and Henglein (1971a)

a) Extraction of the potential from the phase shifts

Using the substitution

$$s^2(r) = r^2(1 - V(r)/E)$$ (5.1)

equation 3.10 can be changed to

$$\eta(b) = k \cdot \int_b^\infty ds \, (s^2-b^2)^{1/2} \cdot \left(\frac{r'(s)}{r(s)} - \frac{1}{s}\right)$$ (5.2)

$$= k \cdot \int_b^\infty \ln[r(s)/s] \cdot s(s^2-b^2)^{-1/2} \, ds$$ (5.3)

This is an Abelian integral equation

$$\eta(b) = k \cdot \int_{\sqrt{x}}^\infty I(s) \, (x-y)^{-1/2} dx \qquad \begin{array}{l} x = s^2 \\ y = b^2 \\ I(s) = \ln\left(\frac{r(s)}{s}\right) \end{array}$$ (5.4)

with the solution

$$I(s) = \frac{2}{\pi \cdot k} \cdot \int_s^\infty db \, \frac{d}{db} \, \frac{1}{(b^2-s^2)^{3/2}}$$ (5.5)

Inverting 5.4 yields

$$r(s) = s \cdot e^{I(s)} \qquad \text{and}$$ (5.6)

$$V(r) = E\left(\frac{1-s^2}{r^2(s)}\right) = E\left\{1 - e^{-2I(s)}\right\}$$ (5.7)

It should be noted that using the semiclassical relation $\frac{2}{k} \frac{d\eta}{db} = \vartheta(b)$ can be rewritten as:

$$I(s) = \pi^{-1} \int_s^\infty db \, \vartheta(b) \cdot (b^2-s^2)^{-1/2}$$ (5.8)

thus establishing a direct connection between deflection function and potential.

b) Construction of the phase shift or deflection function
   from experimental data

   As it is possible to obtain the interaction potential by inverting either phase shift or deflection function, several solutions have
been reported.

   (aa) Several authors parametrize the phase shift as $\eta = c \cdot b^{-(n-1)}$
for potentials of the form

$$V = -c \cdot r^{-n} \tag{5.9}$$

or

$$\eta = c_1 e^{c_2(b-c_2)^3} + c_4 \cdot e^{c_5 \cdot (b-c_6)} + c_7 \cdot c_8(b-c_9)^2 \tag{5.10}$$

   The parameters c_i are evaluated by a least square fit of the
calculated cross sections - by partial wave summation or a semi-
classical method - with the experimental ones.

   (bb) Buck (1971) employs a method of constructing a piecewise
analytic deflection function. For angles beyond the rainbow maximum, he is able to invert the cross section unambiguously. He
approximates the deflection function by a straight line around ß= 1.0
by a parabola in the minimum, and by an inverse power function
for large impact parameters, making use of the known long range
potential using eq. 5.9. He appliesBerry's equation 3.22 to get an
expression for the position of maxima and minima, which he fits
with a non-linear least square method, using known molecular constants and some properties of the deflection function. He inverts
the so-constructed deflection function with equ. 5.8.

   (cc) Klingbeil (1972) uses an analytic phase shift expression for
large impact parameters and treats all other phase shifts as free
parameters. Using a good set of starting values, he again mini-
mizes the difference between the computed and experimental cross
sections, using a least square analysis. This method does not require
any functional form for the phase shifts, but needs initial phases
of good quality.

   (dd) Pritchard (1972) starts from the differences in classical
action (cf. eq. 3.21) responsible for supernumerary rainbows α_s
and rapid oscillations α_r

$$\alpha_s(\vartheta) = h^{-1} \left\{ A(\vartheta, b_2) - A(\vartheta, b_1) \right\} \tag{5.11}$$

$$\alpha_r(\vartheta) = h^{-1} \cdot \left\{ A(\vartheta, b_3) - A(\vartheta, b_1) \right\} \tag{5.12}$$

Using a parabolic approximation for the minimum of the deflection function,

$$\vartheta(b) = \vartheta_R - q \cdot \left\{ k(b - b_R) \right\}^2 \tag{5.13}$$

he obtains

$$\alpha_s(\vartheta) = \frac{4}{3} q^{-1/2} (\vartheta_R - \vartheta)^{3/2} \qquad 0.75 \leq N_s \leq 1.5 \tag{5.14}$$

and

$$\alpha_s(\vartheta) = 2\pi \cdot (N_s - \frac{3}{4}) \qquad\qquad N_s > 1.5 \tag{5.15}$$

and

$$\alpha_r(\vartheta) \approx \alpha_s(\vartheta) - 2k \cdot b_o \vartheta \tag{5.16}$$

As α_s represents the width of the deflection function, he is able to construct such a function using information on the absolute differential cross-section, or on the ratio of the upper and lower envelopes of the differential cross section.

(ee) Remler (1971) uses a special analytical representation for the S-matrix, $S = e^{2i\eta_l}$, which he splits up into a part S_A, describing the attractive part of the deflection function, and S_R, describing the repulsive part

$$S = S_A \cdot S_R = e^{2i(\eta_A + \eta_R)} \tag{5.17}$$

Inserting 5.17 into 3.7 he gets

$$f(\vartheta) = f_A(\vartheta) + f_R(\vartheta) \tag{5.18}$$

with

$$f_A(\vartheta) = \frac{1}{2ik} \cdot \sum (2l+1) \cdot P_l(\cos\vartheta) \cdot [S_a - 1] \tag{5-19}$$

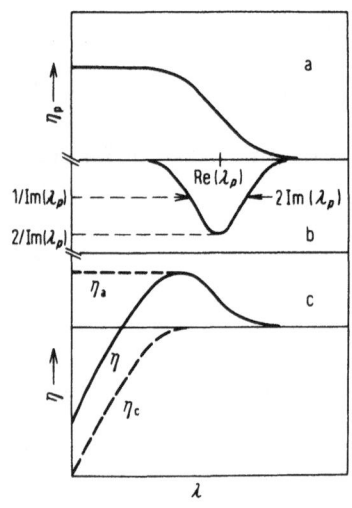

Fig. 21 -
Remlers method: a: attractive
phase shifts η_p.
b: deflection function, and
c: total phase shift $(\eta = \eta_p + \eta_c)$
is the core phase shift

$$f_R(\vartheta) = \frac{1}{2ik} \cdot \sum (2e+1) \cdot P_1(\cos\vartheta) \cdot [S_a(S_r-1)] \qquad (5\text{-}20)$$

He parametrizes S_a by

$$S_a = \prod_{p=1}^{N} \left\{ \frac{\lambda^2 - \lambda_p^{*2}}{\lambda^2 - \lambda_p^2} \right\} \qquad (5.21)$$

which contains N poles in the complex plane, and computes the scattering amplitudes, using the Regge-Watson-Sommerfeld transform. He obtains

$$f_A(\vartheta) = \frac{1}{2ik} \sum_{p=1}^{N} \frac{\pi}{\cos\pi\cdot\lambda} P_{\lambda-\frac{1}{2}}(\cos\vartheta) (\lambda_p^2 - \lambda_p^{*2}) \prod_{\substack{i=1 \\ p \neq i}}^{N} \left\{ \frac{\lambda_p^2 - \lambda_i^{*2}}{\lambda_p^2 - \lambda_i^2} \right\}$$

$$(5.22)$$

The computation of $f_R(\vartheta)$ is done by partial wave summation.

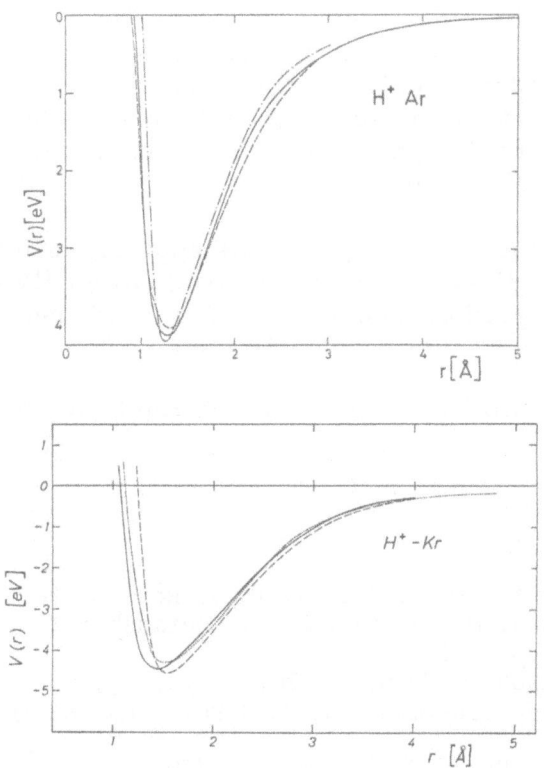

Fig. 22 - Comparison of different inversion procedures (after
          Buck, 1973):
          top: potential of ArH$^+$ obtained by inversion of the
          data by Rick, Bobbio, Champion and Doverspike
          (1971) using Remlers method (-.-), of the same data
          using Klingbeil's method (-). The dashed line is the
          result of Mittmann, Weise, Ding and Henglein (1971)
          using a fit procedure
          bottom: potential of KrH$^+$ obtained by inversion of the
          data of Rick, Bobbio, Champion and Doverspike (1971)
          using Remlers method (--) by inversion of the data of
          Weise, Mittmann, Ding and Henglein (1971) using
          Bucks method (...) by a fit procedure using the same
          data (-)

Thus an expression is obtained which requires only the summation over a few poles and over a few repulsive phase shifts. The position of real and imaginary part of the poles determine the position and the width of the deflection function as shown in Fig. 21. With this fast procedure, he generates an optimized set of phase shifts which he inverts using equ. 5.5. With his method, Remler does not get the correct asymptotic behavior of the potential, which can be a disadvantage.

Fig. 22 shows a comparison of potentials applying different inversion procedures for H^+-Kr (top) and H^+-Ar (bottom (Buck, 1973). They show a fairly good agreement; the small differences might be due to the different experimental data used.

The author would like to thank Dr. P. Kuntz for helpful discussions.

REFERENCES

Aberth, W. and Lorents, D. C. (1965). Phys. Rev. Letters 14, 776.
Arikawa, T., Hirakawa, H., Nishikawa, S., Watanabe, T. (1973). VIII ICPEAC, Beograd, p. 77.
Bernstein, R. B. (1966). Advan. in Chem. Phys. 10, 75.
Bernstein, R. B. and Muckermann, J. T. (1967). Advan. in Chem. Phys. 12, 389.
Berry, M. V. (1966). Proc. Phys. Soc. 89, 479.
Berry, M. V. and Mount, K. E. (1972). Rep. Prog. Phys. 35, 315.
Boerboom, A.J.H., van Dop, H., and Los, J. (1970). Physica 46, 458.
Boettner, R., Dimpel, W., Ross, U. and Toennies, J. P. (1974). Ann. Spring Meeting German Physical Society on Atomic and Molecular Coll., Stuttgart.
Buck, U. (1974). Rev. Mod. Phys. 46, 369.
Delvigne, C. A. L. and Los, J. (1973). Physica 67, 166.
Dueren, R., Raabe, G. P. and Schlier, Ch. (1968). Z. Phys. 214, 410.
Faxen, Von H. and Holtsmark, J. (1927). Z. Phys. 45, 307.
Ford, K. W. and Wheeler, J. A. (1959a). Ann. Phys. 7, 259.
Ford, K. W. and Wheeler, J. A. (1959b). Ann. Phys. 7, 287.
Hasted, J. B. (1972a) in Physics of Atomic Collisions, p. 85 ff., (Butterworths, London).
Hasted, J. B. (1972b). Physics of Atomic Collisions, p. 67 ff., (Butterworths, London).
Inonye, H., and Kita, S. (1972). J. Chem. Phys. 56, 4877.
Klingbeil, R. (1972). J. Chem. Phys. 56, 132.
Kubach, C., Sidis, V., and Durup, J. (1974). to be published.
Landau, L. D. (1932). Physik. Z. Sovjetunion 2, 46.
Langer, R. E. (1937). Phys. Rev. 51, 669.

Lorents, D. C. and Aberth, W. (1965). Phys. Rev. 139, A 1017.
Lorents, D. C. and Aberth, W. (1966). Phys. Rev. 144, 109.
Menendez, M. G. and Datz, S. (1965). 6th Int. Conf. on Phys. and
    Atomic Collisions, Quebec, Canada, p. 264.
Menendez, M. G., Redmon, M. J. and Aebischer, J. F. (1969). Phys.
    Rev. 180, 69.
Michels, H. H. (1966). J. Chem. Phys. 44, 3834.
Mittmann, H. U. and Weise, H. P. (1974a). Z. Naturforsch. 29a, 400.
Mittmann, H. U., Weise, H. P., Ding, A. and Henglein, A. (1971a).
    Z. Naturforsch. 26a, 1112.
Pauly, H. and Toennies, J. P. (1968a). Advan. Atmos. Mol. Phys. 1, 195.
Pauly, H. and Toennies, J. P. (1968b) in Methods of Experimental
    Physics, Vol. 7a, p. 227 (Academic Press, New York).
Peyerimhoff, S. (1965). J. Chem. Phys. 43, 998.
Pritchard, D. E. (1972). J. Chem. Phys. 56, 4206.
Raabe, G.-P. (1971). Ph.D. Thesis, University of Gottingen.
Remler, E. A. (1971). Phys. Rev. A3, 1949.
Roach, A. C. and Kuntz, P. J. (1970). Chem. Comm. 1336.
Smith, F. T. (1964). J. Chem. Phys. 42, 2419.
Smith, F. T., Marchi, R. P., Aberth, W., Lorents, D. C. and Heinz,
    O. (1967). Phys. Rev. 161, 31.
Stueckelberg, E. C. G. (1932). Helv. Phys. Acta 5, 369.
Toennies, J. P. (1973a). Phy. Chem. an Advan. Treatise, Vol 4,
    Chapter 6 (Academic Press, New York).
Toennies, J. P. (1973b). Faraday Disc. Chem. Soc. 55, 129.
Toennies, J. P. (1974). private communication.
Weise, H. P. (1973). Ber. Bunsenges. 77, 578.
Weise, H. P. and Mittmann, H. U. (1973). Z. Naturforsch. 28a, 714.
Weise, H. P. and Mittmann, H. U. (1974). Z. Naturforsch., in press.
Weise, H. P., Mittmann, H. U., Ding, A. and Henglein, A. (1971).
    Z. Naturforsch. 26a, 1122.
Zener, C. (1932). Proc. Roy. Soc. A137, 696.

AN ANALYSIS OF DIRECT ION-MOLECULE REACTIONS

Bruce H. Mahan

Department of Chemistry, and Inorganic Materials Research
Division of the Lawrence Berkeley Laboratory, University
of California, Berkeley, California 94720

One of the major goals of the study of molecular collision
phenomena is to learn how to analyze or anticipate the dynamics of
an elementary reaction without engaging in extensive numerical
calculations. This is of particular importance in ion-molecule
chemistry, where often the reaction dynamics are affected by more
than one potential energy surface. The accurate calculation of
these surfaces, and their use to investigate the exact classical
collision dynamics, while highly edifying, can be quite expensive
and time consuming. It is of interest, therefore, to explore the
efficacy with which simple models for the reaction process can be
used to understand and predict the energy and angular distributions
of products, isotope effects, and total reaction cross sections.

It has proved convenient to describe the dynamic mechanism of
an elementary bimolecular chemical reaction as involving either a
short-lived, direct interaction of collision partners, or a long-
lived collision complex. In the former case, the collision partners are close (within approximately an equilibrium bond distance)
for a time comparable to a vibrational period, but less than a
full rotational period. In the latter case, the partners are
close and strongly interacting for several rotational periods. The
dividing line between the two classifications can be hazy, and it
is also unrealistic to believe that a reaction can proceed <u>exclu-
sively</u> via a long-lived collision complex. Examples of ion-molecule
reactions which fall in each extreme classification are now known
(for reviews and references to the original literature, see Dubrin
and Henchman, 1972 and Mahan, 1974). Examples of intermediate behavior have also appeared (Chiang, Gislason, Mahan, and Werner,
1971, and Mahan and Sloane, 1973).

Unfortunately, there has been some rather careless usage of the term "collision complex" which has led to confusion concerning its meaning. Some workers have interpreted any scattered product which appears at the center-of-mass velocity in a product velocity vector intensity map as indicative of "complex formation". This is incorrect, since at the threshold for an endoergic process, the product will appear at the center-of-mass velocity whether the event proceeds by a direct interaction or by a long-lived complex. Other workers have used the term "complex" to describe any intimate interaction of collision partners, with the unsupported assumption or implication that the lifetime of the interaction is long compared to molecular rotation periods. Use of the term "complex" in this manner would be acceptable, as long as care is taken to remove unsupported implications concerning the lifetime of the complex. Finally, some workers seem to confuse the properties of a long-lived collision complex with the "activated complex" of absolute rate theory. The latter occurs in both direct and long-lived complex collision processes as the reactants pass through a surface in phase space that divides them from products. Perhaps the best way to see the distinction between activated complexes and long-lived complexes is to study the unimolecular reaction problem (Robinson and Holbrook, 1972).

For reactions which proceed through a long-lived collision complex, the interaction between all atoms may be strong enough so that the accessible phase space of the complex is explored fairly uniformly. In these cases, we can hope that the statistical or phase space theories of chemical reaction can reproduce and predict such things as the relative yields of products, isotope effects, and energy partitioning. The effectiveness of the present forms of statistical theory is still an open question, however.

For reactions which proceed by a direct interaction mechanism, there is also a relatively simple model available: the classical trajectory calculation with Monte Carlo sampling of a properly weighted set of initial conditions. As mentioned above, this approach can be expensive, and can produce more information than can be readily assimilated. In this paper we shall use a simple sequential impulse model to analyze the dynamics of direct ion-molecule reactions. The experimental studies which have prompted this analysis have been largely concerned with exoergic or thermoneutral hydrogen atom transfer reactions. To illustrate the nature of these findings we shall summarize some of the recent results obtained for the $O^+(H_2,H)OH^+$ reaction (Gillen, Mahan, and Winn, 1973 abc).

DIRECT HYDROGEN ATOM TRANSFER PROCESSES

Figure 1 shows the velocity vector distribution of OH^+ from the $O^+(H_2,H)OH^+$ reaction as measured in ion beam scattering experiments. This distribution has features which are quite characteristic of the results obtained for a number of exoergic hydrogen atom transfer reactions. The results are displayed by plotting contours of constant intensity in a polar coordinate system which has an origin which moves at the velocity of the center-of-mass of the collision partners. Thus the radial coordinate gives the speed of OH^+ relative to the centroid of the O^+-H_2 system. Small values of the radial coordinate correspond to small values of the final relative translational energy of the products, and therefore, by energy conservation, to large product internal excitation. The large labeled circles give the locations of two values of Q, the translational exoergicity. By energy conservation, Q can be

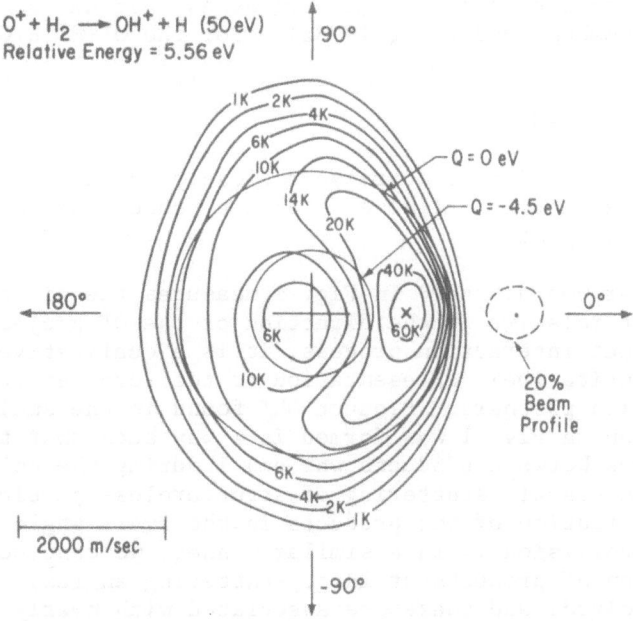

Fig. 1. A contour map of the specific intensity of OH^+ formed by the $O^+(H_2,H)OH^+$ reaction at an initial relative energy of 5.56 eV. The radial coordinate is the speed of OH^+ relative to the center-of-mass of the entire system. The angular coordinate measures the deflection in the center-of-mass system, of the OH^+ from the original direction of the O^+ projectile. The spectator stripping velocity is indicated by a small cross.

written as

$$Q \equiv \frac{\mu'(g')^2}{2} - \frac{\mu g^2}{2} = \Delta E_o^o - U. \tag{1}$$

Here μ is the reduced mass and g is the relative speed of the products (primed) and reactants (unprimed), E_o^o is the internal energy change for the reaction, and U is the internal excitation energy of the products.

For reactions in which the products are an atom and a molecule in their ground electronic states, Q is bounded by the situations in which U is zero or to D, the dissociation energy of the molecule:

$$-\Delta E_o^o - D \le Q \le -\Delta E_o^o. \tag{2}$$

The lower limit can be violated (apparently) if either product is in an excited electronic state whose dissociation limit lies above that of the ground state. The simplest way to take this into account is to recognize that for such processes, ΔE_o^o has a different value (Gillen, Mahan, and Winn, 1973a). For the $O^+(H_2,H)OH^+$ reaction,

$$-4.5 \le Q \le +0.43 \text{ eV},$$

and the Q circles in Fig. 1 correspond closely to these limits. in effect, these circles define a "stability zone" for OH^+ in its ground electronic state.

The angular coordinate θ in Fig. 1 measures the direction of the OH^+ product relative to the direction of the O^+ projectile. Thus for a direct interaction process, it is a qualitative (and eventually quantitative) representation of the force exerted between collision partners. Product OH^+ found in the small angle ($\theta \le 45°$) region in Fig. 1 was formed in a way such that the net integrated force between products was small during the collision. By analogy with elastic scattering of structureless particles, this implies formation of the products in the small angle region is by grazing collisions. In a similar manner, we conclude that in the formation of products at large scattering angles, large forces are involved, and these are associated with nearly head-on collisions between reactants. By using the impulse model of direct reactions, we hope to delineate what is meant by grazing and head-on collisions more clearly.

Figure 1 shows that there is a strong maximum in the intensity of OH^+ at the spectator stripping velocity: a scattering angle of zero degrees and a speed relative to the centroid consistent with the general expression

$$u = u_o \left(\frac{A}{A+B}\right) \left(\frac{C}{B+C}\right) \tag{3}$$

which applies to the reaction A(βC,C)AB. In Eq. (3) the letters represent the masses of the atoms, and u and u_o are respectively the product and projectile speeds relative to the centroid. Appearance of OH^+ at the spectator stripping velocity implies that the reaction occurred with no net integrated force on the freed hydrogen atom. With one exception, all exoergic hydrogen transfer reactions so far investigated have displayed a very prominent intensity maximum at or very near the spectator stripping velocity. The exception is apparently the ground state reaction $Kr^+(H_2,H)KrH^+$ (Henglein, 1972). Unfortunately, the stripping peak is frequently so prominent that reactions have often been rather carelessly described as "stripping processes", and the large angle scattering ignored or dismissed as unimportant. Without question, the idea that atom B can be transferred to A with no force being exerted on C is quite remarkable. It is therefore of considerable interest to determine in detail how this can occur, and how important it is to the overall chemical reaction cross section.

The experimental determinations of the final relative energy distributions of reaction products have been somewhat limited by the low velocity resolution employed so far. However, in most of the cases investigated, it is qualitatively clear that in the intermediate to high range of initial relative energies (>3 eV), the products in the small angle region are somewhat more excited internally than the products scattered through large angles. In this energy regime, much or most of the internal excitation of the products is supplied by the initial translational energy of the reactants. In the nearly head-on collisions which lead to large angle scattering, the large forces that occur provide the mechanism for disposing of some of this incipient product excitation as relative translational energy. There is less possibility for this disposal in the grazing collisions which produce the very small angle scattering. A more quantitative expression of these ideas is possible in terms of the sequential impulse model, as we shall see.

In the regime of high initial translational energy, the total reaction cross section is greatly influenced by the problem of stabilizing the product molecule against dissociation. This can be illustrated most clearly for the product formed at the spectator stripping velocity. The Q-value for this product is

$$Q_{ss} = -\frac{B}{A+B} E \tag{4}$$

where E is the laboratory energy of the projectile ion A. If E is made large enough, Q_{ss} will become more negative than the lower limit given by Eq. (2), and the molecular product in its ground electronic state will be unstable. In the early work which demonstrated the importance of spectator stripping in the reactions of Ar^+, N_2^+, and CO^+ with H_2, it was anticipated (Henglein, 1966) that the intensity peak at small angles would be lost entirely when E reached the critical value

$$E_C = (\frac{A+B}{B}) \ (\Delta E_o^o + D) \tag{5}$$

at which the internal energy of the stripped molecular product exceeds its dissociation energy. However, it was observed that for these systems, the forward scattered peak is not lost at high initial relative energies, but instead decreases in intensity and moves to speeds greater than the spectator stripping value. That is, some of the forward scattered molecules are stabilized by recoil which can evidently occur in grazing collision in these systems.

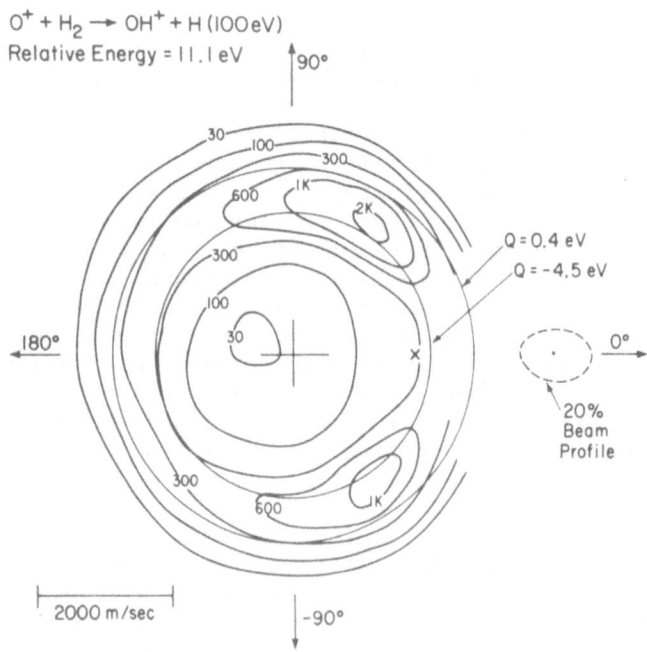

Fig. 2. A contour map of the specific intensity of OH^+ formed from collisions at 11.1 eV initial relative energy. Note the absence of an intensity peak at 0° and the appearance of peaks at ±60°. The spectator stripping velocity, marked by a small cross, lies inside the Q = -4.5 eV circle, where OH^+ in its ground state is unstable.

One reaction has been found that displays the loss of forward scattered products at initial energies above the critical value for spectator stripping. Figure 2 shows the velocity vector distribution of OH^+ from the $O^+(H_2,H)OH^+$ reaction at an initial relative energy of 11.1 eV. At this energy, the spectator stripping velocity (indicated by a small cross) lies in the zone where OH^+ in its electronic ground state is unstable. Indeed, the intensity peak so evident at lower initial relative energies (cf. Fig. 1) has been lost. Thus, the potential energy surface for the $O^+(H_2,H)$ OH^+ reaction lacks the features which allow stabilization by product recoil in the small angle region. It would be valuable to know what these critical features are, and in addition, to be able to understand the occurrence of the intensity peaks located at approximately 45 in Fig. 2. We shall find that the sequential impulse model illuminates this problem considerably.

THE SEQUENTIAL IMPULSE MODEL

A number of simple models for the atom transfer process have been proposed, and at least partially tested against molecular beam scattering data (Bates, Cook, and Smith, 1964; Light and Horrocks, 1964; Suplinskas, 1968; Kuntz, 1970; Chang and Light, 1970; Hierl, Herman, and Wolfgang, 1970; George and Suplinskas, 1971; Grice and Hardin, 1971; Marron, 1973). Even allowing for the necessity of using extremely simple approximations to potential energy surfaces and mechanical behavior, most of these models are lacking in generality or rigor, and some have not been particularly illuminating. The sequential impulse model proposed by Bates, Cook, and Smith (1964) is conceptually simple, and has the capacity for considerable refinement. In brief, the reaction A(BC,C)AB is viewed as an event in which A hits B impulsively and elastically, B then hits C in a like manner, and A then combines with B if their energy of relative motion is less than the dissociation energy of the product molecule. Suplinskas (1968) and George and Suplinskas (1971) have elaborated the model, and have shown that it can reproduce the major features of the Ar^+-D_2 reactive scattering. Gillen, Mahan, and Winn (1973c) found that a version of the model in which the atoms interact via hard sphere potentials is consistent with the distributions of the products of the reaction of O^+ with D_2 and HD in the regime of high relative energies. These two sets of applications involved calculation of the final product velocities from sampled initial conditions using large digital computers. However, to better discern and analyze the nature of the collisions which give products at various scattering angles and speeds, it would be valuable if the product distributions could be expressed analytically and evaluated with a small calculator. This proves to be possible, and the results will be reported in detail elsewhere. In what follows we shall demonstrate that a number of conclusions can be drawn from the model merely by using velocity vector diagrams.

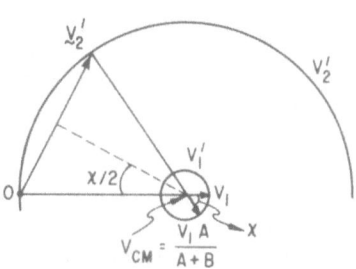

Fig. 3. A velocity vector diagram for the elastic collision of atom A with atom B. The circles marked V_1 and V_2' are, respectively, the loci of all possible final laboratory velocity vectors for atoms A and B. The scattering angle in the center-of-mass system is designated by χ. Note that the perpendicular bisector of any V_2' vector bisects χ and passes through the A-B centroid velocity.

First, let us review some fundamental features of elastic collisons which are essential to the development and understanding of the sequential impulse model. Consider atom A moving with an initial laboratory velocity V_1 toward atom B, which is initially stationary in the laboratory. The initial relative velocity g is equal to V_1, and the velocity of the center-of-mass of the A-B system is $V_1 A/(A+B)$. Regardless of the nature of the two body collision, the center-of-mass velocity is unchanged. Since the collision is assumed to be elastic, the final and initial relative velocity vectors have the same magnitude, but different direction. The final relative velocity vector is obtained by rotating the initial vector about the fixed center-of-mass velocity. The result, as is shown in Fig. 3, is that the final laboratory velocity V_1' of particle A is a vector which terminates on a sphere of radius $V_1 B/(A+B)$ centered at the centroid velocity. Similarly, V_2', the final laboratory velocity of B, lies on a concentric sphere of radius $V_1 A/(A+B)$.

The scattering angle χ_1, measured in the center-of-mass system of A and B, is also shown in Fig. 3. From the geometry, it is clear that the bisector of V_2' passes through the centroid velocity, and bisects the angle χ_1. As a result, we can write

$$V_2' = 2 \frac{A}{A+B} V_1 \sin(\frac{\chi_1}{2}) \tag{6}$$

for the magnitude of V_2'. This relation and the construction used to find it will be particularly useful later.

The vector relations just discussed give the possible values of the particle velocities after an elastic collision. The distribution of intensity is also important, and is expressed most compactly by the classical differential scattering cross section $I(\chi)$, where for a monotonic potential

$$I(\chi) = \frac{b}{\sin\chi \left|\frac{d\chi}{db}\right|} . \tag{7}$$

Here b is the aiming error or impact parameter. To evaluate $I(\chi)$, the relation between b and χ must be found from the intermolecular potential function. For hard spheres, the result is particularly simple:

$$I(\chi) = \frac{d^2}{4} \tag{8}$$

where d is the mutual collision diameter. Thus for this model, the scattered intensity is independent of χ. For more realistic potentials, $I(\chi)$ is large at small angles and drops rapidly as increases. In the range of angles from 60-180°, $I(\chi)$ decreases rather slowly, and in the large angle region, is pretty well represented by a constant term characteristic of hard sphere scattering. The hard sphere differential cross section is therefore a good first approximation to the intensity distribution, particularly for high energies and large scattering angles.

There is another feature of high energy collisions that is of importance. Such collisions, particularly those that produce large angle scattering, are impulsive. That is, the time during which a large force is exerted between a pair of atoms is small compared to the natural frequencies for nuclear motion in molecules. For example, if atoms repel each other according to the potential

$$\phi = \phi_o \, e^{-r/L},$$

where L is a range parameter, then the force is greater than 10% of its maximum value for a period of 3τ, where τ is a characteristic collision time defined by

$$\tau = 2L/g.$$

During this time, the relative velocity changes from approximately 90% of its initial value to 90% of its final value. For typical values of L and energies in the electron-volt range, τ is of the order of 2×10^{-15} sec. This is shorter than the vibrational period, and much shorter than the rotational period of H_2. Thus

for the case of a high energy atom A hitting a diatomic molecule
BC, it often may be quite reasonable to describe the process as an
elastic collision between A and B, followed by an independent
elastic collision between B and C. The initial condition for the
second condition is, of course, the final state of the first
collision.

The primary object of a model for the reaction process is to
calculate the intensity of scattered product AB as a function of
the scattering angle θ and speed relative to the center-of-mass of
the ABC system. Evaluating the intensity as a function of V_3'', the
final velocity in the laboratory system of the free atom C, is
completely equivalent to this, since by momentum conservation,
each value of V_3'' corresponds to a definite value of θ and the
final relative speed. Finding the magnitude of V_3'' is a simple
matter if one knows χ_1 and χ_2, the scattering angles for the A-B
and B-C collisions in their individual center-of-mass coordinate
systems. As indicated above, the magnitude of the laboratory
velocity of atom B after the A-B collision is

$$V_2' = 2 \ \frac{A}{A+B} \ \sin(\frac{\chi_1}{2}).$$

Now we simply regard V_2' as the initial velocity for the B-C
collision, and apply the analogous formula to get

$$V_3'' = 4(\frac{A}{A+B})(\frac{B}{B+C}) \ \sin(\frac{\chi_1}{2}) \ \sin(\frac{\chi_2}{2}). \tag{9}$$

Having found the laboratory velocity of atom C after a particular sequence of impulses, we must ask whether or not this constitutes a reactive collision. Our criterion for reaction is the
simplest possible: the value of V_3'' must lie in the stability zone
which corresponds to the internal energy of AB being less than its
dissociation energy. This is an important approximation, since it
allows us to disregard details of the trajectories such as the
possibility of additional collisions between C and B or A. However,
it is probably a good approximation, since for high energy collisions, the size of the cross section is governed largely by product
stability considerations. Moreover, hard sphere trajectory calculations (Gillen, Mahan, and Winn, 1973b) have demonstrated the
relative unimportance of additional impulses and other details of
the trajectories, and also the effectiveness of this reaction
criterion in reproducing experimental data. However, the approximation does restrict application of the model to reactions where
the potential energy surface has very simple properties: thermoneutral or nearly so, and no substantial wells or barriers.

Equation (9) suggests that a variety of impulse sequences can contribute to the product intensity at V_3''. The angle χ_1 may be large or small, as long as χ_2 has the appropriate small or large value consistent with the selected value of V_3''. However, there are limits to the range of χ_1 and χ_2 values that can be involved, and these limits are connected with the direction of V_3'', a property which we have not yet used.

To see how this limitation comes about, consider Fig. 4. Here we treat only those values of V_2' which lie in the plane defined by the vectors V_1 and V_3''. As indicated earlier, the possible values of V_2' lie on a circle of radius $V_1 A/(A+B)$ centered on V_1 at this distance from the origin of the laboratory coordinate system. The locus of all B-C center-of-mass velocities in this plane plays a very important role. It can be found by multiplying all possible V_2' vectors by the factor $B/(B+C)$, and plotting the points. The result is a circle of radius

$$R = (\frac{A}{A+B})(\frac{B}{B+C}) \, V_1 \tag{10}$$

centered on V_1 at a distance R from the laboratory origin. Let us call this the centroid circle.

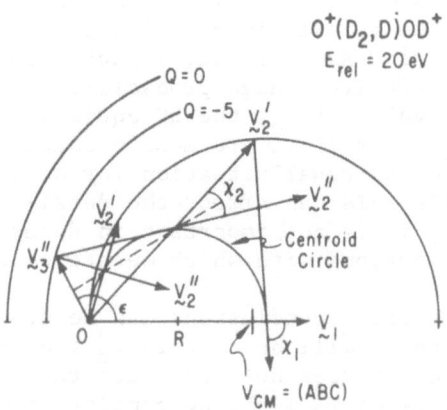

Fig. 4. A velocity vector diagram for the sequential impulse model in the plane of the initial projectile velocity V_1, and the final velocity of atom C, V_3''. The large Q circles indicate a part of the stability zone for the reaction: the velocity of atom C must lie in this zone if AB is to be stable to dissociation.

Now consider an arbitrary centroid velocity for the B-C
system just before (and after) their collision. These centroids
must be on the centroid circle, and must also lie on the perpendicular bisector of V_3''. As Fig. 4 shows, there are just two
centroids which satisfy both these conditions for any given V_3''
vector. One of these corresponds to a large χ_2 (and small χ_1),
the other to the values of χ_1 and χ_2 being interchanged. These
two angles are the extreme values of χ_1 and χ_2 that are consistent
with a selected V_3''.

The origin of the intermediate values of χ_1 and χ_2 becomes
obvious if we recognize that V_2' need not lie in the plane of V_1
and V_3''. Thus the centroid circle is really part of a centroid
sphere of radius R, and the perpendicular bisector of V_3'' is a
plane. The intersection of this bisecting plane with the centroid
sphere is a circle - the "magic circle" - perpendicular to the
V_1-V_3'' plane. As one moves along the magic circle, all the χ_1-χ_2
pairs that can contribute to scattering at V_3'' are encountered.
Thus the product intensity at V_3'' can be found by summing the
properly weighted contributions of all allowed χ_1-χ_2 scattering
pairs.

For the present purposes, the details of this weighted summation are not needed, but it is useful to note that the distribution over the various χ_1-χ_2 pairs is nearly uniform. The departure
from uniformity comes about because the angle α between V_2 and the
BC internuclear axis is distributed with a weighting factor of
$\sin\alpha$. Consequently, the BC axis is more likely to lie perpendicular to V_2 than parallel. As a result, impact parameters for the
B-C collision have a relatively high probability of being near
their maximum allowed value of r_o, the BC equilibrium bond distance. Thus smaller values of χ_2 are more probable than larger
values, in contrast to the usual situation for hard sphere scattering. However, while this can affect the details of the product
velocity distribution, it is not important in determining the gross
features of the distributions with which we are concerned here.

A number of qualitative conclusions can be drawn directly
from Fig. 4. First, there will be certain V_3'' vectors for which
the perpendicular bisector does not intersect the centroid sphere.
Even though these values of V_3'' might be consistent with the total
energy and momentum conservation laws, they can not be produced by
a sequence of two elastic impulses. For example, events in which
V_3'' is directed at 180° in the laboratory coordinate system can not
occur. Thus, there can be no backward recoil of particle C, and
no corresponding forward recoil of the AB product.

A little reflection shows that this forward recoil could occur
if, just before the A-B impulse, the vector V_1 were increased

in magnitude with the center-of-mass velocity held fixed. This could occur in a real system if there were an attractive potential between reactants, and this is in fact the mechanism for forward product recoil proposed in the so-called modified stripping model (Herman, Kerstetter, Rose, and Wolfgang, 1967). In addition, one can see that forward recoil could occur if, just prior to the B-C collision, the vector $\underset{\sim}{V_2'}$ were increased in length, so that this collision would appear to be super-elastic. This could come about if there were a repulsive energy release between B and C as the products separate. This is the basic idea involved in the so-called direct interaction with product repulsion (DIPR) model for reaction dynamics (Kuntz, 1970; Marron, 1973). The sequential impulse model thus clarifies the validity of either reactant attraction or product repulsion as sources of forward recoil.

It is evident that V_3'' vectors directed at angles other than 180° are accessible only if the magnitude of V_3'' is small enough so that there is an intersection of the bisecting plane and the centroid sphere. The condition for such an intersection can be found readily from the analytic geometry of the vector construction. The maximum values of V_3'' lie on a curve given by

$$\frac{(V_3'')_{max}}{2R} = \cos\varepsilon + 1 \tag{11}$$

where ε is the angle between $\underset{\sim}{V_3''}$ and $\underset{\sim}{V_1}$. Equation (11) represents a cardioid which has a cusp at the origin of the laboratory velocity coordinate system. There is a corresponding cardioid which gives the maximum values of the velocity of the AB product in the center-of-mass system, and this is illustrated in Fig. 5. The minimum values of the AB product velocity are just those given by the requirement that the excitation energy of the product AB must be less than its dissociation energy. Thus the zone in velocity space that is allowed is bounded from the inside by the stability circle, and from the outside by the limiting cardioid.

The size of the limiting cardioid is proportional to R, and thus scales with V_1. However, the size of the stability circle is determined by the magnitude of Q_{min}, a fixed number independent of V_1. Thus the size of the kinematically allowed zone can be represented as a function of initial relative energy by one cardioid, if the units of the diagram are changed as the energy changes. However, in this case there is a different sized stability circle for each initial relative energy, as indicated in Fig. 5. As the initial relative energy increases, the diameter of the stability circle increases, and eventually it intersects the limiting cardioid at the cusp. This corresponds to reaching the critical projectile energy above which products formed by spectator

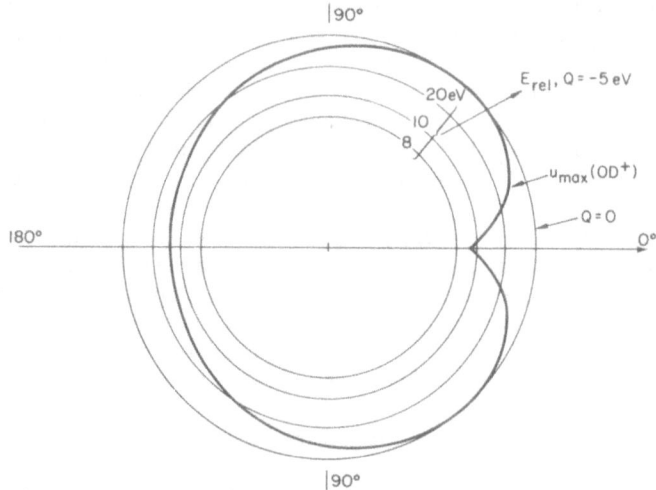

Fig. 5. The cardioid which gives the maximum velocities of OD$^+$ from the O$^+$(D$_2$,D)OD$^+$ reaction according to the sequential impulse model. The maximum velocity of OD$^+$ according to overall energy conservation is the Q = 0 circle. The three smaller circles give the minimum velocity of OD$^+$ consistent with product stability (i.e., the Q = -5 eV limit) for the three values of the initial relative energy indicated.

stripping are unstable. As the initial relative energy is increased still further, increasing amounts of the accessible small angle scattering region pass into the unstable zone, and the outline of the product distribution assumes a crescent-like shape. The experimentally observed distributions for the O$^+$-H$_2$ and O$^+$-D$_2$ reactions have just this shape when the initial relative energy is in the 11-30 eV range. Moreover, the observed decrease of the total reaction cross section with increasing energy can be in large measure attributed to the concomitant diminution of the size of the product stability zone.

The considerations just outlined provide an explanation of why the spectator stripping peak and all small angle scattering is lost at high energy in the O$^+$(H$_2$,H)OH$^+$ reaction, but is stabilized by forward recoil in the reactions of N$_2^+$, CO$^+$, and Ar$^+$ with H$_2$ and D$_2$. In the O$^+$-H$_2$ case, the reaction is only slightly exoergic (ΔH = -0.43 eV) and there is no obvious mechanism for producing large amounts of forward recoil. In contrast, the reactions of N$_2^+$, CO$^+$, and Ar$^+$ are notably more exoergic ($\Delta H \cong$ -1.4 eV). If all of this exoergicity were to be released as product repulsion in some of the grazing collisions, forward recoil and product stabilization could occur, as is observed.

Having delineated the general limits and energy dependence of
the product velocity vector distribution predicted by the sequential impulse model, we can now turn to some of the details of the
intensity variations. From Fig. 4 it is evident that $\underset{\sim}{V}_3$ vectors
of small magnitude directed approximately perpendicular to $\underset{\sim}{V}_1$ will
have bisecting planes which intersect the centroid sphere to
generate magic circles of large radii. There is, therefore, a
relatively large range of $\chi_1-\chi_2$ pairs which can produce these
events. The intensity in the small angle scattering region will
thus be large if the initial relative energy is low enough to place
the small angle scattering region in the stability zone. As V_3''
increases in magnitude, the size of the magic circle decreases,
and the product intensity goes down.

In order to see this effect develop systematically, in Fig. 6
we have plotted the $\chi_1-\chi_2$ pairs that produce scattering at various
fixed values of the product scattering angle θ. The calculations
apply to the $O^+(D_2,D)OD^+$ reaction at an initial relative energy of
20 eV, a situation in which the very small angle scattering does
not lie in a stable region of velocity space. The solid lines
refer to products formed with the minimum allowed Q value of

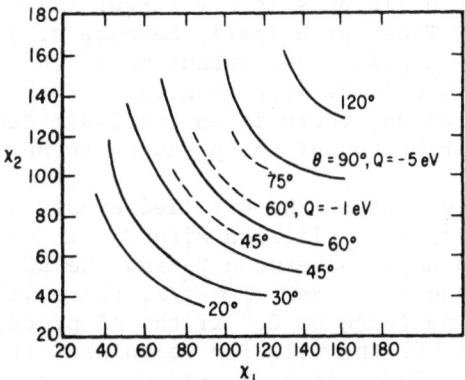

Fig. 6. The $\chi_1-\chi_2$ pairs that contribute to the intensity at
various values of the product scattering angle θ for the reaction
$O^+(D_2,D)OD^+$ at 20 eV initial relative energy. The solid lines
pertain to product at Q = -5 eV, the dashed lines to product at
Q = -1 eV.

-5 eV (the correct value, if the exoergicity is ignored), while
the dotted lines correspond to a Q value of -1 eV.

Figure 6 exposes the reason for the intensity maximum observed
experimentally near $45° \leq \theta \leq 60°$. For this region, the range of
$\chi_1-\chi_2$ pairs that can produce stable products reaches a maximum.
At smaller values of θ, the range of allowed $\chi_1-\chi_2$ pairs drops
abruptly, and the observed product intensity does also. At values
of θ greater than 90°, the allowed range of $\chi_1-\chi_2$ pairs again
diminishes, and the expected and observed product intensities
diminish.

Notice that the values of χ_1 and χ_2 which produce large
values of θ are themselves large. This is consistent with the
idea that backscattered products do come from nearly head-on collisions. In order to have both χ_1 and χ_2 large, A must hit B
nearly head-on, and B must hit C in a like manner. This implies a
nearly collinear ABC conformation at the beginning of the collision. Similarly, we can see that the values of χ_1 and χ_2 which
contribute to small values of θ are of modest magnitude ($\sim$35-90°).
Thus it is moderately accurate to associate the region of small
with "grazing" collisions, although in some of the events that
contribute, substantial deflections of A by B or of B by C do
occur. It is probably better to think of $\theta < 15°$ as the grazing
collision region.

Figure 6 also shows that the range of $\chi_1-\chi_2$ pairs that can
produce scattering at $Q = -1$ eV is smaller at any value of θ than
the corresponding range for $Q = -5$ eV. Moreover, the $\chi_1-\chi_2$ pairs
for a given value of θ lie at slightly larger values for $Q = -1$ eV
than for $Q = -5$ eV. Thus, principally because of a smaller
allowed range of $\chi_1-\chi_2$ pairs, the intensity of the lesser internally excited products will be less than that of the more excited
products. In other words, there is an intrinsic tendency for the
internal energy distribution of the products to be inverted.

So far we have discussed the detailed events in which A hits
B, and B has a hard sphere collision with C. In order for such
events to occur, the angle α between $\underset{\sim}{V_2}$ and the BC internuclear
axis must be less than $\pi/2$. For $\alpha > \pi/2$, there will be no B-C
collision, and thus no force on C. If the AB product of these
events is stable, it has the velocity calculated from the spectator stripping model. Thus, if A, B, and C are treated as hard
spheres, spectator stripping comes largely from events in which A
strikes and combines with the second atom it sees as it approaches
BC. Stripping processes are also possible for values of α somewhat smaller than $\pi/2$ if the mutual hard sphere diameter of the
B-C pair is less than the impact parameter of the second collision.
In the limit of vanishing hard sphere diameter for B and C, all
collisions will be spectator stripping processes.

These considerations help to make clear why spectator stripping is so prominent in the product velocity vector distributions of ion-molecule reactions. If the potential energy surfaces for these reactions have only a weak dependence on the ABC angle, then trajectories are possible in which the projectile A strikes and combines with the second atom without exerting force on the free or spectator atom. Moreover, if there are strong attractive forces between A and B, but not between B and C, there will be trajectories of the stripping type even when α is significantly less than $\pi/2$. Note that if spectator stripping is described as involving grazing collisions, it is the B-C interaction, and not necessarily the A-B collision which is of the grazing type.

Spectator stripping resembles both the rainbow and glory effects in atomic elastic scattering (Bernstein, 1966). Like rainbow scattering, it appears that there is in the reactive situation a range of initial conditions (in this case, the angle α) which gives product scattered at or very near to one point in velocity space. The fact that this point is at a scattering angle of zero degrees is also significant, since just as in glory scattering, there is an integration over all values of the azimuthal angle which is performed by the detector only when θ equals zero degrees. These two factors and the relatively low apparatus resolution employed so far combine to give spectator stripping a fame which it perhaps does not fully deserve. After considering realistic potential energy surfaces, it is very difficult to accept the fact that B and C can separate with a truly zero force between them. There is probably a partial or nearly total cancellation of forces on atom C: for example, an attraction toward A and B on the incoming leg of the trajectory followed by a repulsion between B and C on the outgoing leg. Such force compensation can give angular distributions strongly peaked near the spectator stripping velocity, and produce the illusion that no forces acted on particle C. In general, this compensation can not be expected to be perfect. In the future, when product distributions are examined with high resolution, some or all of the spectator peaks may be found not at $\theta = 0°$, but at small but finite scattering angles. Even now it should be realized that to go from the intensity contour maps of Figs. 1 and 2 to actual total reaction cross sections, one must apply a weighting factor of $\sin\theta$, and then integrate the intensity over angle and speed. Thus, product at $\theta = 0°$ is given zero weight, and that near $\theta = 90°$ contributes most heavily to the total reaction cross section. In other words, most of the chemistry is done by the type of events described at least approximately by the sequential impulse model.

Acknowledgement: This work was supported by the U. S. Atomic Energy Commission. Discussions with Dr. John S. Winn and Dr. Keith T. Gillen contributed significantly to the formulation of the ideas developed here.

REFERENCES

Bates, D. R., Cook, C. J. and Smith, F. J. (1964). Proc. Phys. Soc. 83, 49.

Bernstein, R. B. (1966). Advances in Chemical Physics, 10 (Ed. J. Ross). John Wiley and Sons, New York.

Chang, D. T. and Light, J. C. (1970). J. Chem. Phys. 52, 5687.

Chiang, M. H., Gislason, E. A., Mahan, B. H., Tsao, C. W. and Werner, A. S. (1971). J. Phys. Chem. 75, 1426.

Dubrin, J. and Henchman, M. J. (1972). MTP International Review of Science Series I, 9, 213 (Ed. J. C. Polanyi). Butterworth and Co., London.

George, T. F. and Suplinskas, R. J. (1971). J. Chem. Phys. 54, 1037.

Gillen, K. T., Mahan, B. H. and Winn, J. S. (1973a). J. Chem. Phys. 58, 5373.

Gillen, K. T., Mahan, B. H. and Winn, J. S. (1973b). Chem. Phys. Lett. 22, 344.

Gillen, K. T., Mahan, B. H. and Winn, J. S. (1973c). J. Chem. Phys. 59, 6380.

Grice, R. and Hardin, D. R. (1971). Molecular Physics, 21, 805.

Henglein, A. (1966). Advan. Chem. Ser. 58, 63.

Henglein, A. (1972). J. Phys. Chem. 76, 3883.

Herman, Z., Kerstetter, J., Rose, T. and Wolfgang, R. (1967). Disc. Faraday Soc. 44, 123.

Hierl, P. M., Herman, Z. and Wolfgang, R. (1970). J. Chem. Phys. 53, 660.

Kuntz, P. J. (1970). Trans. Faraday Soc. 66, 2980.

Light, J. C. and Horrocks, J. (1964). Proc. Phys. Soc. 84, 527.

Mahan, B. H. (1974) MTP International Review of Science, Series II, 9 (Ed. D. R. Herschbach). Butterworth and Co., London.

Mahan, B. H. and Sloane, T. M. (1973). J. Chem. Phys. 59, 5661.

Marron, Michael T. (1973). J. Chem. Phys. 58, 153.

Robinson, P. J. and Holbrook, K. A. (1972). Unimolecular
    Reactions, Wiley-Interscience, New York.

Suplinskas, R. J. (1968). J. Chem. Phys. 49, 5046.

ANGULAR DISTRIBUTION STUDIES OF ION-MOLECULE REACTIONS

K.Birkinshaw[+], V.Pacák[*], Z.Herman[*]

[+]Physics Dept., University College London,
Gower Street, London WC1E 6BT, England

[*]The Heyrovsky Institute of Physical Chemistry
and Electrochemistry, Czechoslovakian Academy
of Sciences, Máchova 7, Prague 2, Czechoslovakia

1)INTRODUCTION

In order to find out what happens during a collision
between two particles, we need to know all the initial
conditions and measure all the final conditions. Of course
it would be ideal to prepare the reactants in known single
states and to cross a beam of oriented ions with a beam of
oriented molecules, both with a precisely known velocity,
but at every stage of definition of the conditions a large
fraction of the intensity would be lost and at the present
time the intensity barrier is insurmountable. The best
that can be done in many cases is to work with a known
distribution of initial states and attempt to unfold the
data obtained, but often the shape of the distribution
is not well known.

Two properties of a collision which are now often
measured simultaneously are the angular and velocity
distributions of the scattered products. What are the uses
of angular distribution measurements? For a collision
between two atomic particles where the interaction
potential is spherically symmetrical, the only collisions
which will give scattering along the initial relative

velocity vector are those with impact parameter (b) equal
to zero or to other values given by equation 1 with χ=n,
where n=π,0,-π,-2π etc. All other impact parameters will

$$\chi = \pi - 2b \int_{r_m}^{\infty} \frac{dr}{r^2 \left(1-\frac{b^2}{r^2}-\frac{V(r)}{E}\right)^{\frac{1}{2}}} \tag{1}$$

result in scattering at an angle given by a graph of the
form of fig.1 which was calculated from equation 1
using a certain value of E (total energy) and a certain
form of V(r) (interaction potential). r is the internuclear separation and r_m is the distance of closest
approach. The form of fig.1 depends on the function V(r),
and it is clear that the majority of ions are scattered
at wide angles giving an angular distribution from which
can be extracted much useful information on the
interaction potential. Scattering from potentials which
have deflection functions such as that of fig.1 lead to
interference effects known as 'rainbow' and 'glory'

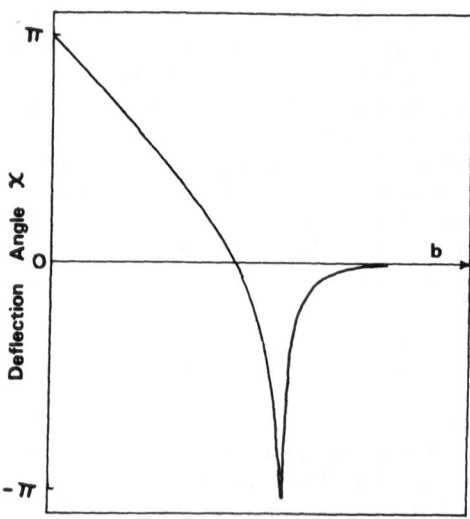

Fig.1   Deflection function for an elastic collision
         between two atoms calculated from equation 1.

scattering which are well understood, and by their effect on the scattered products yield information on $V(r)$. Angular distribution measurements are therefore important in the understanding of interactions between particles.

However, to chemists interested in collisions involving molecules, which do not give rise to a spherically symmetric potential field, and in which three or more atoms are involved, fig.1 is of limited use as the scattering angle also depends on the rotational and vibrational phase of the molecule and in general rearrangement of the nuclei occurs. When we also take into account potential energy surface crossings the situation becomes very complicated and it is not possible to give a unique scattering angle even when the initial conditions are classically well defined. Studies of classical trajectories over hypersurfaces where crossings are allowed show that there will in general be several possible angular deflections, and the best that can be done is to give the probability of ending up at each of these possible deflections. In some cases the difficulty also arises that when following trajectories over a surface, the crossings lead to trajectory branching, and each branch might lead to further branching with the result that if every trajectory is followed the number of branching points becomes prohibitively high or might even diverge. In this case a statistically sound criterion for ignoring or weighting against low probability branches must be used.

It is apparent that for collisions involving molecules there exists no simple deflection function nor analytical equation with which to obtain one. Unlike the simpler atomic case, the angular distribution does not give clear answers to questions concerning details of the interaction potential without more involved calculations. The key to the information locked up in intensity contour diagrams is trajectory studies on potential energy surfaces. General features of the potential energy surface, such as minima favouring sticky collisions or the positions of barriers in the entrance or exit channels (Kuntz,Nemeth and Polanyi, 1970) can often be obtained by examining the contour diagrams and their energy dependence, but the construction of a surface necessary for trajectory calculations requires more than general features and has been carried out by several workers (Kuntz,Nemeth,Polanyi,1970; Steiner,Certain, Kuntz,1973; Tully,1973).

Simple angular analysis of collision products is of

course of less value than simultaneous angular and velocity
analysis because chemists are very interested in the
effects of different types of energy on a reaction and
how it is distributed amongst the products. Velocity
analysis of products without simultaneous angular
analysis is more interesting as important conclusions
concerning the mechanism of a reaction or the presence of
different channels with different energy defects can be
drawn. In the latter case, product ions are usually
formed by passing a beam of ions into a collision
chamber and most of them are focussed into some type of
energy analyser. This means that product ions scattered
at large angles are dumped onto the 0°-180° line and the
observed velocity distribution is distorted. An example
of this is given in the very good work of Neynaber and
Magnuson (1973) where the authors fully realised the
problem and were able to make a correct qualitative
analysis of their data. This work will be examined later.
In some types of apparatus, for example a mass
spectrometer ion source, it is not possible to focus
widely scattered ions onto the 0°-180° line. When this
type of apparatus is used to measure rate constants for
ion-molecule reactions, it is possible, depending on the
geometry of the system, that there will be discrimination
against products scattered at wide angles. If this is the
case it is necessary to correct the results using
measured angular distributions where available.

Unlike in neutral-neutral studies where significant
data were first obtained through angular distribution
measurements alone, in ion-molecule reactions the
laboratory angular distribution is not usually very
informative. Because of the relatively large ion velocity
vector, the CM velocity vector usually lies at small
laboratory angles and often product ions are concentrated
around the direction of the ion beam unless pronounced
back scattering occurs. On the other hand, in neutral-
neutral crossed beam studies the CM velocity vector often
lies at large laboratory angles.

In this paper we will include a short section on ion
beam scattering carried out at energies higher than most
people at this summer school are familiar with. The
reason for this inclusion is that for the particular
case of charge exchange, results obtained at high (2Kev)
energy are applicable at low energy because of the slow
variation of cross section with energy. The results from
the type of work we will describe give insight into the
problem of the crossing of potential energy curves which

is of general interest in chemistry and is important in
trajectory calculations where more than one potential
energy surface is involved.

2)INTENSITY CONTOUR DIAGRAMS

a)Construction

      During the past five years, intensity contour
diagrams for a wide variety of ion-molecule reactions
have been measured. The raw experimental data must be
plotted in a framework which allows display of any
symmetry properties of the collision and which facilitates
a qualitative visual assessment of the results. Possible
procedures have already been described (Wolfgang and Cross
1969) and are in wide use but will be mentioned here for
the sake of completeness.

      For a collision at a given initial relative kinetic
energy, two types of data are taken - an angular
distribution of the scattered products, and energy
distributions at several strategically chosen angles. All
measurements are made in the plane of the colliding beams
in the case of crossed beam experiments or in any plane
containing the ion beam in the case of beam chamber
experiments. The energy distributions are then converted
to velocity distributions by multiplying the intensity at
a given energy by the velocity corresponding to that
energy. This is because when we take particular ion
intensities as we move along the energy distribution we
implicitly assume constant energy elements dE. However,
when dE is constant the element of velocity - dv - changes
as we have:

$$dE = m.v.dv \qquad\qquad (2)$$

In velocity space we require that dv is constant and so
in order to preserve the relationship:

$$I'(E).dE = I'(v).dv \qquad\qquad (3)$$

(where I' is intensity) we have $I'(E).m.v=I'(v)$.
Therefore if we multiply $I'(E)$ by v (neglecting the
constant m) we obtain the velocity distribution in
laboratory coordinates. Unfortunately in this coordinate
system the space is symmetrical about the laboratory
origin but not about the CM origin as shown in fig.2:

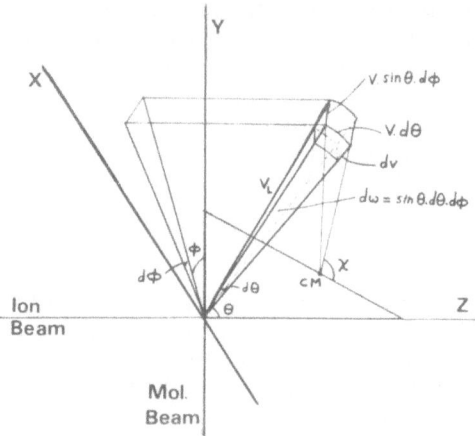

Fig.2   Velocity vector diagram with ions and molecules
        traveling in the Z and Y directions respectively.
        The X axis is also shown.

The CM coordinate system is more useful as in this
system the contour diagram should be symmetric about the
relative velocity vector, indicating the correct
functioning of the apparatus, and the division of this
system into two halves by a line passing through its
origin and perpendicular to the relative velocity vector
correctly defines the forward and backward scattering
directions with respect to the colliding beams. Symmetry
about this line is a necessary result if a long-lived
complex is formed (although such symmetry does not prove
that long-lived complex formation has occurred)

The coordinate system most used in presenting
experimental data is the cartesian system. In the lab.
system the volume elements are not of equal size but are
given by $\sin\theta.d\theta.d\phi.dv.v^2$ as can be seen from fig.2. We
have:

$$I(\theta,\phi,v)v^2.\sin\theta.d\theta.d\phi.dv = \text{fraction of incident}$$

$$\text{flux scattered into } d\omega$$
$$\text{and between v and v+}dv$$
$$= I(\theta,\phi,v)v^2.d\omega.dv$$

$$d\omega = \sin\theta.d\theta.d\phi$$

$$I = \text{differential cross section} \qquad\qquad (4)$$

If we divide our volume elements by v^2 or equivalently divide the scattered intensity by v^2 our space now has equal volume elements and is called cartesian space. For further discussion of coordinate systems the reader is referred to Wolfgang and Cross (1969) and to Hierl, Herman, Wolfgang(1970).

As a result of the above two transformations it can be seen that if the energy distribution of the intensity is divided by v - the lab. velocity of the product ions - we have in effect performed a transformation to a velocity space in which the volume elements have a constant size. These velocity distributions are now plotted along the directions in which they were measured and finally points of equal intensity are joined together to give the intensity contour diagram.

Now we can look at the types of contour diagram to be expected from various types of idealised mechanism.

b)Types of intensity contour diagram

Before looking at some experimental results we will look qualitatively at some kinematically possible contour diagrams. In order to illustrate the type of information obtainable from angular distribution studies, simplyfied schematic contour diagrams are presented in fig.3. The diagrams refer to a mass ratio of $m_A:m_B:m_C = 3:2:1$ with A^+ approaching from the left and BC from the right. The collision occurs at the CM. For this mass ratio the position of the CM divides the relative velocity vector into two equal and opposite parts.

Figs.3(a),(b) and (c) show three cases in which if complex formation occurs (symmetry about a line passing through the CM and perpendicular to the relative velocity vector is a necessary but not a sufficient condition for this) the complexes differ in the detailed role of the angular momentum. Important generalisations concerning the shapes of contour diagrams for products formed via a long-lived complex mechanism have been made by Miller, Safron and Herschbach (1967) which will be looked at in more detail later. In fig.3(a) the isotropic intensity distribution implies that the angular momentum of the intermediate complex is small and its temperature is large. The forward backward peaking of fig.3(b) indicates

$A^+ + BC \longrightarrow$

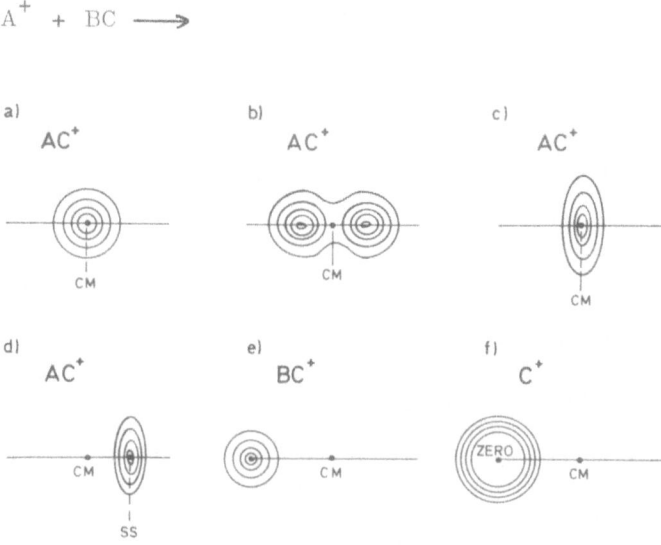

Fig.3   Simplyfied schematic contour diagrams resulting
         from different reaction mechanisms. The mass ratio
         is $m_A:m_B:m_C = 3:2:1$. A^+ approaches from the left
         and BC from the right.

that the orbital angular momentum of the species forming
the complex dominates as would be the case for a prolate
complex. Fig.3(c) would be favoured if the complex were
oblate.

    Figs.3(d),(e) and (f) show three cases in which the
reaction mechanism is direct. Whereas in complex formation
the contour diagram must be symmetrical about a line
passing through the CM and perpendicular to the relative
velocity vector, direct reactions normally give asymmetric
diagrams. Fig.3(d) suggests that a spectator stripping
mechanism may be operative. In this type of mechanism
first postulated by the Henglein group, the momentum of
atom B is unchanged by the collision and the product AC^+
falls at a position which can be calculated from the
momenta of A^+ and C alone. Fig.3(e) shows the type of
contour diagram to be obtained from charge transfer
without momentum transfer, and 3(f) shows the possible
outcome of a dissociative charge transfer reaction in
which no momentum is transferred in the initial charge
transfer step. The wheel like structure of the diagram

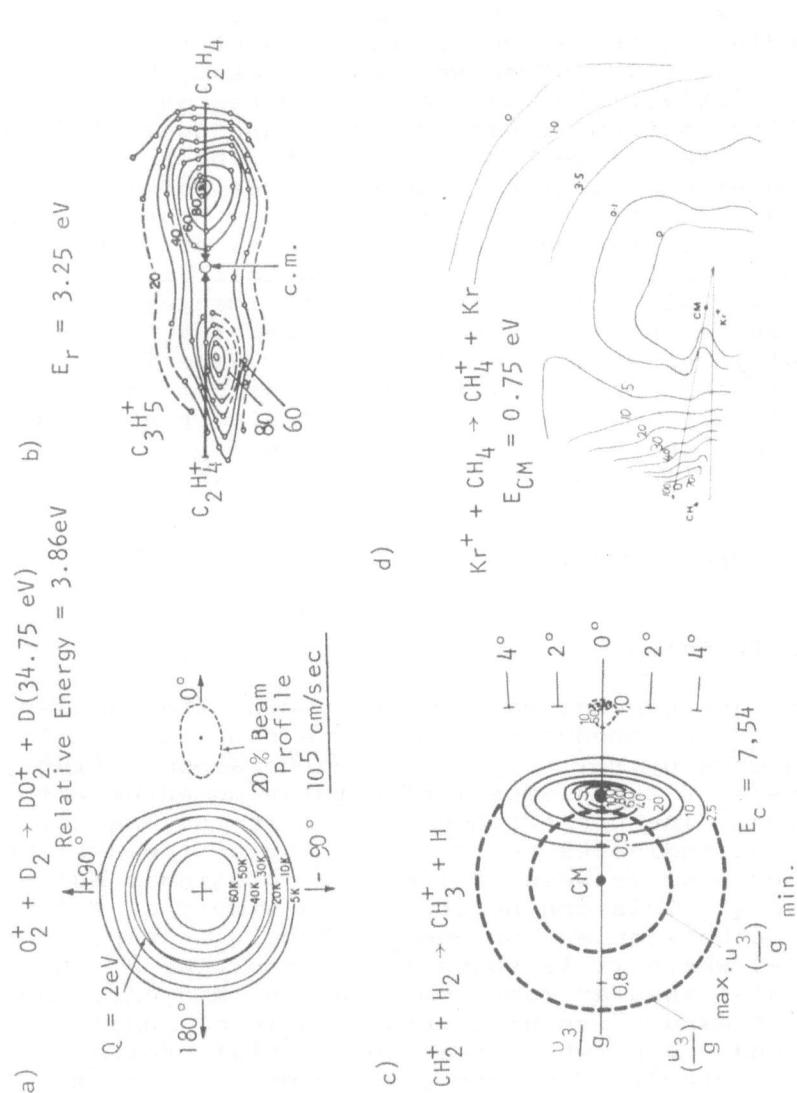

Fig.4 Experimental results demonstrating different types of reaction mechanism.

would be caused by the fact that the initial charge
transfer products are formed without change of velocity
but the distribution 'explodes' when the charge transfer
products break up. The latter is a possibility to be
considered when looking at dissociative charge transfer
reactions.

Examples of experimental contour diagrams
corresponding to the schematic diagrams of fig.3 are shown
in fig.4. Fig.4(a) is from work by Gislason, Mahan, Tsao,
and Werner (1969); 4(b) is from Herman,Lee and Wolfgang
(1969); 4(c) is from Eisele, Henglein and Bosse (1974);
4(d) is from recent work carried out by the present
authors in Prague. Charge transfer is not uncommon in
ion-molecule reactions but the low energy charge transfer
ions formed from thermal neutral beams in crossed beam
experiments, or from thermal gas in a collision chamber,
are extremely difficult to handle experimentally, and
fig.4(d) shows the first charge transfer reaction to be
successfully measured in a crossed beam experiment. From
the data the CM angular distribution was obtained
showing clearly a backward charge transfer peak with a
large amount (approximately 50%) of isotropic scattering.
An experimental example of fig.3(f) is unknown.

3)REACTION MECHANISMS

a)Complex formation

Important generalisations concerning the angular
distributions of products formed by a complex mechanism
have been made by Miller, Safron and Herschbach (1967)
on the basis of the disposal of angular momentum within
the complex. "The preferred direction in the angular
distribution indicates the partitioning of the available
angular momentum between orbital and rotational motion of
the products". This can be considered simply as follows.
A collision between an ion and a molecule is represented
in fig.5(a) where it is shown to be occurring at the CM
and the total angular momentum is J. In this case the
molecule is assumed to have zero rotational angular
momentum and J = L (where L is the initial orbital
angular momentum). If a long-lived complex is formed
(ie. an intermediate whose lifetime is longer than
several rotational periods) which loses its memory of the
way in which it was formed, then the products will
separate along any diameter of the circle shown in fig.5(a)

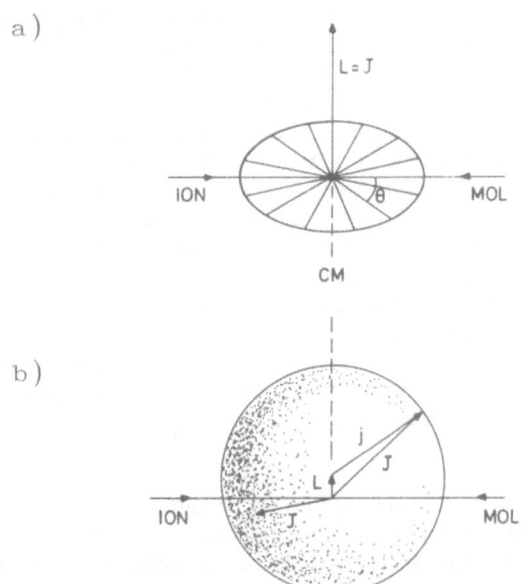

Fig.5   Angular momenta involved in an ion-molecule
        collision. a) No rotational angular momentum (j)
        in the reactants or products. The intensity
        distribution can be generated by rotating J about
        the initial relative velocity vector. This gives
        contour diagrams of the type shown in fig.4(b).
        b) The rotational angular momentum of the reactants
        (j) is large compared with the orbital angular
        momentum (L) and the total angular momenta (J) are
        distributed almost spherically symmetrically about
        the CM. This gives diagrams of the type 4(a).

with equal probability (assuming the products have zero
rotational angular momentum). The L vectors – and also
the J vectors in this particular case – are uniformly
distributed with all azimuthal angles equally likely, and
therefore the total angular distribution is obtained by
integrating the distribution over all azimuthal angles.
When this is done it can be seen that there will be a
build up of intensity along the relative velocity vector
thus giving rise to the forward-backward peaking of
fig.4(b). It is clear from fig.5(b) that if L is small
compared with the randomly oriented rotational angular
momenta of the reactant ion and molecule (j) then the Js
are distributed almost spherically symmetrically about the
CM and therefore so will be the scattered product ions.

This would result in a contour diagram of the type in
fig.4(a) as the experimental measurements are made in a
single plane containing the initial relative velocity
vector. Miller et al. also consider more complicated cases
in which the internal rotational angular momentum changes
from zero for the reactants to non-zero for the products.
In this case the final relative velocity vector is not
necessarily in the plane of the circle but will be
distributed over a range of out-of-plane angles. The total
angular distribution can be generated by a uniform
rotation of the product relative velocity vector
distribution about J and then rotating J about the
initial relative velocity vector. If there is also a
distibution of initial Js because of initial rotational
angular momentum of the reactants, this must also be
taken into account.

The concept of an intermediate complex is not well
defined. It is clear that the intermediate must be long-
lived to qualify for the name complex, and could be
defined according to its lifetime, but within this
definition the nature of the intermediate is not
specified. It is interesting to note that a long-lived
intermediate could occur in collisions between two atomic
particles. For example, if two particles approach along
a given potential energy curve and because of a curve
crossing attempt to separate along a curve with a higher
assymptotic energy than the total energy of the colliding
atoms, they will have insufficient energy to separate and
the intermediate would remain together until they
separate along a lower potential energy curve. Although
it is a gross oversimplification to apply this idea to
colliding molecules, it could help to visualise collisions
in which two colliding molecules form a long-lived
intermediate but each molecule retains its chemical
identity.

b)Direct reactions

Asymmetric angular distributions such as that shown
in fig.4(c) are interpreted in terms of a direct mechanism.
The lifetime of the intermediate must be less than a few
rotational periods, and the intermediate does not forget
the way in which it was formed. Examples of such reactions
are well known (Henglein, Lacmann(1964); Herman, Kerstetter,
Rose,Wolfgang(1967); Gentry, Gislason, Lee, Mahan, Tsao,
(1967)), for example the well known reaction:

$$Ar^+ + D_2 \longrightarrow ArD^+ + D$$

shows a strong forward scattered peak, presumably from grazing collisions, and an almost isotropic contribution at large scattering angles resembling hard-sphere scattering, presumably from near head-on collisions.

In cases where a complex is formed at low energy, there is normally a transition to a direct mechanism as the relative kinetic energy increases and the intermediate lifetime decreases (Robinson, Holbrook(1972); Herman, Birkinshaw(1973)) until it approaches the time of several rotational periods. At this time the lifetime becomes insufficient to ensure a uniform distribution of products around J (see fig.5) and the contour diagram becomes asymmetric.

Some qualitative generalisations can be made relating the shape of the angular distribution to other properties of the collision which however are not always borne out experimentally. Firstly, if a complex is formed which gives rise to an isotropic intensity distribution about the CM (fig.3(a)) as opposed to a distribution peaked symmetrically about the CM and situated along the relative velocity vector (fig.3(b)), this indicates that the angular momentum of the complex is small in comparison with its rotational temperature and it could be expected that only small impact parameters contribute to the reaction thus giving a small cross section (of the order of $1\overset{o}{A}{}^2$ or less). On the other hand, the symmetrically peaked distribution could result from large impact parameter collisions and hence the cross section could be large. Secondly, if the reaction proceeds by a direct mechanism, a rebound peak would result from small impact parameter collisions giving a low cross section, whereas stripping can occur at larger impact parameters with a larger cross section. So that the terms 'forward' and 'backward' are clear we have included fig.6 which is self explanatory.

In the light of trajectory studies it is clear, as noted by Kuntz (1972), that the products of a collision process could possibly form a peak at the stripping point after a complicated history in which the 'spectator' was by no means a spectator in the normal sense. Therefore the observation of a peak at the spectator stripping point does not necessarily mean that ideal spectator stripping has taken place.

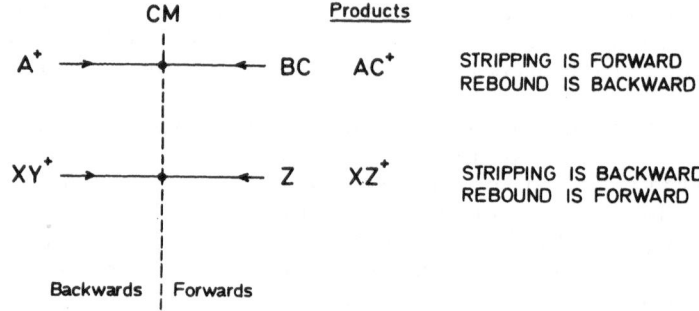

Fig.6   The explanation of the terms 'forward' and
        'backward' in relation to rebound and the ideal
        stripping mechanism.

c)Charge transfer

        The study of ion-molecule reactions in which charge
transfer is the only channel has been relatively neglected.
A good deal of work has been done by physicists on charge
transfer between an atomic ion and a neutral atom,
usually at high energies, but little work has been done
on charge transfer involving molecules (see however
Lindholm(1966); Champion,Doverspike(1968); Birkinshaw,
Hasted(1971)). This is partly due to the well known
problems of detecting low energy ions, and partly due to
the fact that physicists do not get excited about
molecules and chemists do not get excited about charge
transfer. However, because of the interest in trajectory
calculations which involve surface transitions, it is
important to study these transitions and probably the
best way to do this is to study reactions in which there
is only one channel and this channel involves a surface
crossing. Charge transfer reactions of the type:

$$A^+ + BC \longrightarrow A + BC^+$$

could fit these requirements if the energy is kept low
enough so that all other channels are closed. It has been
postulated that charge transfer is the first step in
some chemical reactions (Tully, Herman, Wolfgang(1971);
Masson, Birkinshaw, Henchman(1969)) and charge transfer
has been shown to be important in trajectory calculations.
With the added importance of being a method of forming

molecular ions in the upper atmosphere which can
subsequently recombine with electrons, these reactions
are worthy of more study at low energy.

Possible angular distributions resulting from simple
and dissociative charge transfer are shown in figs.3(e)
and (f) respectively. Measured distributions could differ
from these idealised cases as shown in fig.4(d) and as
indicated on the film shown by Wolf at this conference.
The locations of transition regions between potential
energy surfaces is an important factor influencing the
angular distribution of ions scattered in a charge
transfer reaction. Dealing with the simpler case of
potential energy curves, it should be noted that
transitions are possible between curves at a position
which is not obvious from the curves themselves (both the
diabatic and the adiabatic) as they do not cross. This
type of case has already been treated (Bates(1960);
Demkov(1963)) and we can look at it in the diabatic
basis set for which the coupled equations of the semi-
classical two-state approximation are given by:

a) $\qquad i.\hbar.\dfrac{db_2}{dr} = \dfrac{H_{12}.b_1}{v}.\exp\left(+\dfrac{i}{\hbar.v}\int\left(H_{22}-H_{11}\right) dr\right)$

$$(5)$$

b) $\qquad i.\hbar.\dfrac{db_1}{dr} = \dfrac{H_{12}.b_2}{v}.\exp\left(-\dfrac{i}{\hbar.v}\int\left(H_{22}-H_{11}\right) dr\right)$

where r is the internuclear separation; v is the relative
velocity; H_{aa} gives the potential energy of state a; H_{ab}
gives the interaction matrix element. The terms $|b_1|^2$
and $|b_2|^2$ give the probabilities of being on curves 1 and
2 respectively. In order to understand why there is a
transition region we need only examine the differential
equations. We first assume that the initial conditions
are $b_1(\infty) = 1$ and $b_2(\infty) = 0$, and we then plot the r
dependent terms of eq.5(a) individually as in fig.7,
assuming that b_2 remains very small with respect to b_1.
Only the real terms are shown. The imaginary terms will
behave similarly. If we examine the schematic diagram we
can see that H_{12} multiplied by the exponential term
gives the variation of b_2 with r (assuming b_2 and v are
constant) and if b_2 is integrated over r, as in fig.7(e),
its value could be non-zero. The probability of a

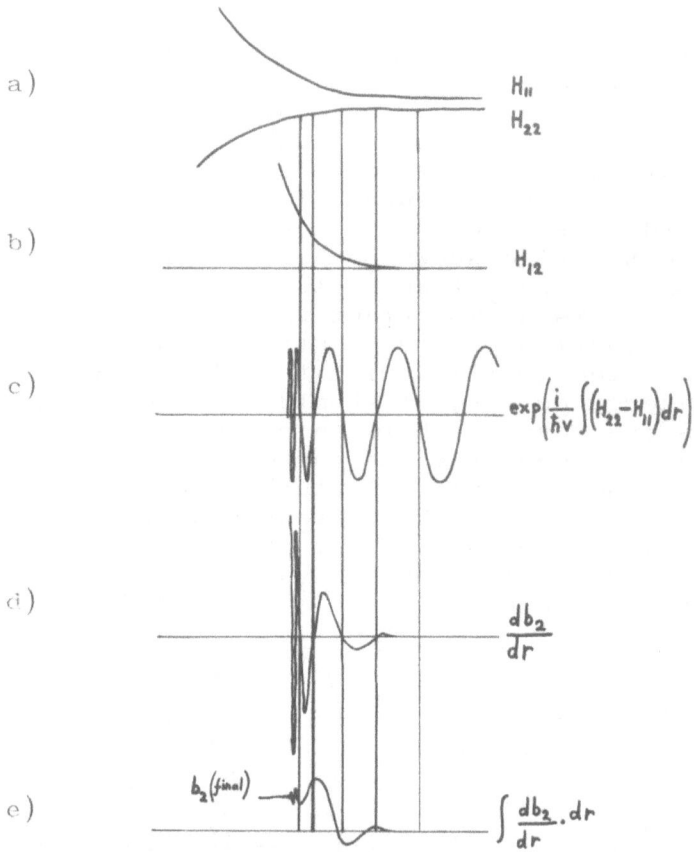

a) H_{11}
      H_{22}

b) H_{12}

c) $\exp\left(\dfrac{i}{\hbar v}\displaystyle\int (H_{22}-H_{11})dr\right)$

d) $\dfrac{db_2}{dr}$

e) $b_2(final)$ $\displaystyle\int \dfrac{db_2}{dr}.dr$

Fig.7   Individual terms from the coupled equations of the
         semi-classical two-state approximation.

transition is non-zero because the exponential term does
not oscillate unduly rapidly at large internuclear
separation, and when we pass through the region in which
the interaction matrix element begins to build up we do
not get a cancellation of the transfer probability as we
would if $H_{11}-H_{22}$ were large. If the diabatic curves
shown in fig.7(a) cross at low values of r instead of
diverging as shown in this case, we have the case of an
extended transition region instead of the transition point
often assumed in curve crossing calculations. Transitions
between close lying surfaces have in fact been seen by

Tully (private communication) during trajectory
calculations on the reaction between D^+ and H_2 in which
the coupled equations were numerically integrated
throughout the trajectories. These transitions 'occurred
only in nearly symmetrical configurations, and only if
there was a fairly significant component of the velocity
in the direction that would break the symmetry'.

4)EXPERIMENTAL WORK

     Intensity contour diagrams often contain a great
deal of information on the dynamics, energetics, surface
crossings etc. all mixed together. In this section we
will look at an ion beam scattering experiment by
Hasted, Iqbal, Yousaf(1971) which has been used to
concentrate on the study of curve crossings, and then at
recent work carried out in Prague by the Herman group.

a)High energy scattering.

     As already mentioned there is usually no simple
deflection function for the collisions of ions with
molecules and it would be experimentally difficult to get
some idea of how the outcome of a collision varies with
impact parameter. In the case of the scattering of atomic
ions from atoms however, there is a relation between the
scattering angle and the impact parameter given by
equation 1. It might be argued that angular distributions
from high energy (2Kev) ion beams gives information only
on the repulsive part of the potential which is of little
interest to chemists working at low energies, but on the
other hand it could be said that it is through the
repulsive potential that we are able to relate the
scattering angle to the impact parameter and subsequently
study the outcome of collisions as a function of impact
parameter.

     Reactions of the type:

$$A^{++} + B \longrightarrow A^+ + B^+$$

have been studied by Hasted, Iqbal, Yousaf(1971). This
type of reaction is particularly useful for the study of
curve crossing because the dominant forces are a coulomb
force and an ion-induced dipole force (if the crossing
occurs at a sufficiently large distance) and the factors
which are important in the theory of curve crossing such
as the position of the crossing point and the slopes of

the curves at the crossing point, are well defined.
Results were obtained for the collisions of C^{2+}, N^{2+},
and O^{2+}, with He, Ne and Ar in the energy (lab.) range
1–3Kev. Typical results are shown in fig.8 for the
collision between C^{2+} and Ar. On the basis of the Landau-
Zener theory no oscillations would be expected in the
curve as the calculated crossing probability is predicted
to decrease with increasing impact parameter so that the
total charge transfer probability is given by $2P(1-P)$ and
has at most one maximum. However, a more realistic
solution of the coupled equations of the semi-classical
two-state approximation carried out by Bates, Johnston,

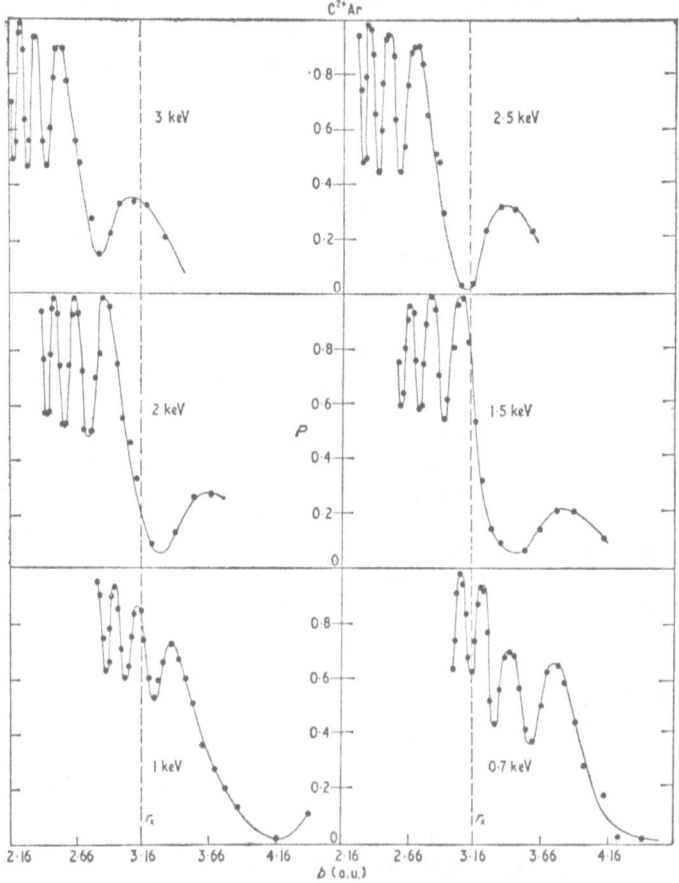

Fig.8   Probabilities of transition P as functions of impact
        parameter b for collisions of C^{2+} with argon.

Stewart(1964) does produce curves of the type shown in
fig.8, and a comparison between the results of Hasted
and the calculations of Bates is said to be encouraging.

Here then we have an example of a high energy angular
scattering experiment which gives information useful at
low energies. To carry out the experiment at low energy
would be more difficult and the curves (fig.8) would not
have such well defined peaks because of the increasing
importance of the attractive potential and the
correspondingly more complicated trajectories.

b)Crossed beam studies on H_2^+ + He $\longrightarrow$ HeH$^+$ + H

This reaction has been studied by the Herman group
in Prague over the relative energy range 0.35ev to 3.5ev
using a crossed beam apparatus. The reactant ions were
formed by bombardment with 100ev electrons and were
distributed among the vibrational energy levels in a
way determined by the Franck-Condon factors, which
predict that the v=2 state is the most highly
populated. The laboratory angular resolution was about
1° which gave in this case a CM angular resolution of
about 2-3°. The reaction is endothermic by 0.81ev using
the dissociation energy of 1.84ev for HeH$^+$ obtained by
Chupka and Russel(1968).

The reaction has been studied by Leventhal(1971)
using a beam-chamber method and by Neynaber and Magnuson
(1973) using their merged-beam apparatus. Leventhal
initially concluded that the product HeH$^+$ is predominantly
forward scattered, but later (Leventhal (1973)) re-
examined his results and showed that backward scattered
ions are also present. Neynaber and Magnuson showed the
existence of both forward and backward scattering in this
reaction and also measured its relative cross section
(averaged over all states in the H_2^+ beam) over the energy
range 0.05ev to 12ev. Their measurements were made without
angular analysis and consisted of energy profiles
measured along the relative velocity vector. As noted in
the paper, these energy profiles include intensity
scattered at large CM angles, and the shapes of the
profiles and positions of the peaks will therefore be
influenced.

The experiments of Herman et al. provide intensity
contour maps for this reaction, two of which are shown in
fig.9. At the lowest relative energy of these experiments

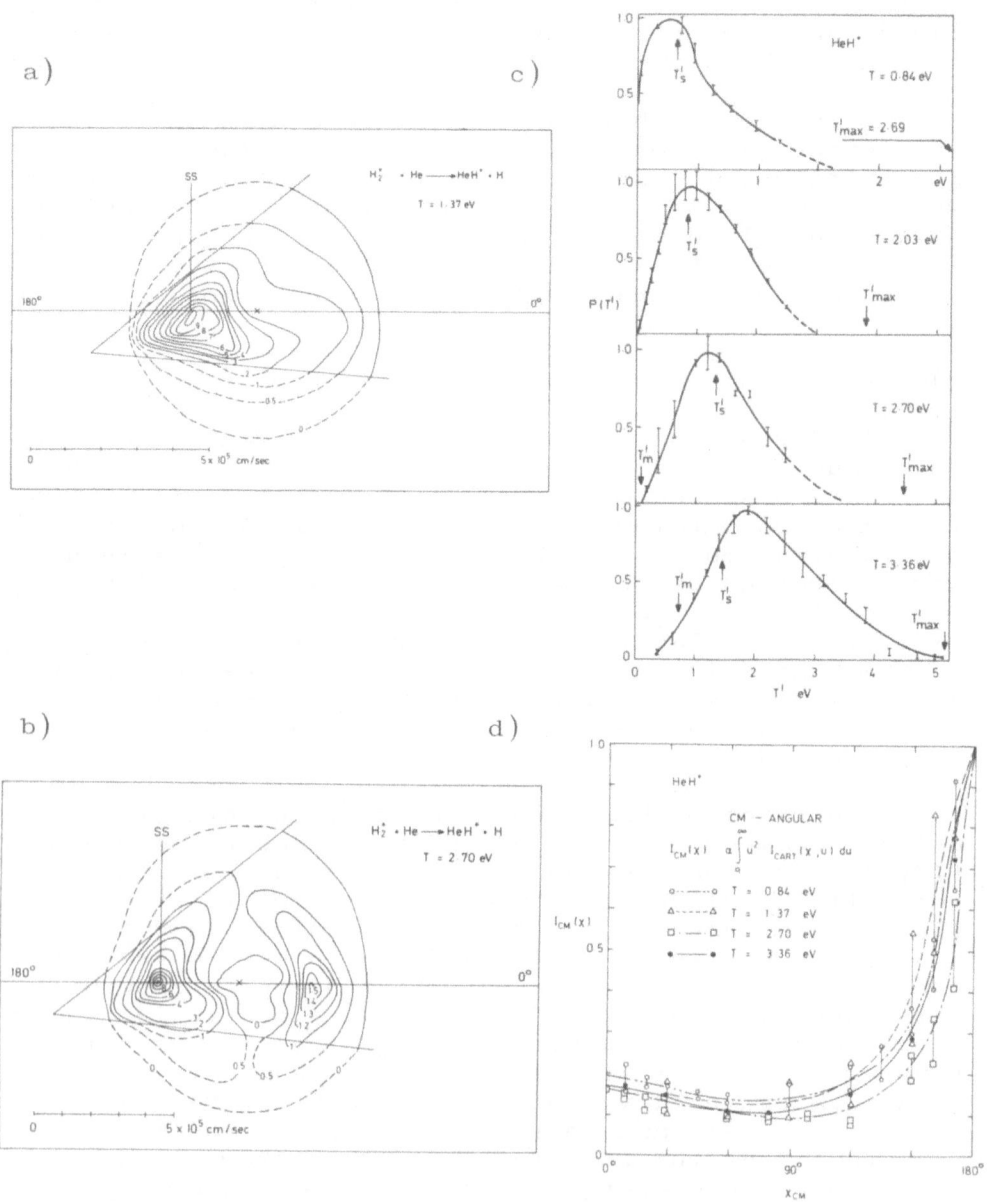

Fig.9 Experimental results from the study of
H₂⁺+He ➔ HeH⁺+H. (a) and (b) Contour diagrams at
relative energies 1.37ev and 2.70ev respectively.
(c) CM energy distributions of HeH⁺.
(d) CM angular distributions of HeH⁺.

(0.35ev) the energy available is not sufficient to overcome the endothermicity of the reaction and therefore the observed products originate from the vibrationally excited H_2^+ ions. The contour diagram shows a backward peak and as the collision energy increases, a hole in the intensity begins to appear around the CM as this is the position at which the products have the least kinetic energy and therefore the most internal energy. When the internal energy is sufficiently high the product HeH^+ begins to dissociate. A backward and forward peak appear with the backward peak always dominant, and at a realative energy of 3.36ev the height of the forward peak is about 30% of the height of the backward peak.

In order to examine the contour diagrams further, the CM kinetic energy distributions and angular distributions were found by integration of the contour diagrams and the results are shown in fig.9(c) and 9(d). The CM angular distributions show clearly the large backward peak and isotropic elastic like scattering at large angles. As the kinetic energy is decreased, the effect of the vibrational energy of H_2^+ would be expected to increase the width of the backward peak but there seems to be very little effect experimentally. The relative translational energy distributions of the products are given in fig.9(c) and show that within experimental error the peak sits at the ideal stripping point at all energies except the highest. The forward shift of the high energy peak is caused by the hole which eats away the low kinetic energy side of the peak.

The comparison between the conclusions of Herman et al. and those of both Neynaber et al. and Leventhal is instructive. In the case of the latter two workers, angular analysis of the products was not carried out and the intensity peaks did not lie at the ideal spectator stripping point as in the Herman experiments. As already mentioned this is probably because the energy profiles include intensity scattered at large CM angles and the peak positions will be influenced. This shows rather clearly the importance of angular measurements. It should be noted again however that an intensity peak at the ideal stripping point does not prove that ideal stripping has taken place.

c)Crossed beam studies on dissociative charge transfer

Preliminary experiments have been carried out in

Prague on the dissociative charge transfer reactions:

$$Ar^+ + CH_4 \longrightarrow CH_3^+ + H + Ar \qquad (6)$$

$$Ar^+ + CH_4 \longrightarrow CH_2^+ + H_2 + Ar \qquad (7)$$

$$Ar^+ + CD_4 \longrightarrow CD_3^+ + D + Ar \qquad (8)$$

$$Kr^+ + CH_4 \longrightarrow CH_3^+ + H + Kr \qquad (9)$$

reaction 6 having received the most attention. These
reactions have previously been studied in detail by
Henchman and Masson (Masson (1970)) over the CM energy
range 1.4ev to 6ev using a mass spectrometer double
chamber ion source. In the Prague experiments the energy
has ranged from 0.4ev to 4.2ev. Both the $^2P_{3/2}$ and $^2P_{1/2}$
states of Ar^+ were present in the ion beam which was
formed using 100ev electrons. For the $^2P_{3/2}$ state, simple
charge transfer between Ar^+ and CH_4 is exothermic by
3.1ev (using $IP(CH_4)=12.62ev$ from Chupka,Berkowitz(1971)),
reaction 6 is exothermic by 1.8ev, and reaction 7 is
exothermic by 0.6ev. For the $^2P_{1/2}$ state these figures
are higher by 0.2ev. In order to be able to study this
type of reaction in which a large proportion of the
products are formed with a very low laboratory energy,
careful shielding of the reaction region and careful
design and setting of the detection system is very
important. The Prague machine was capable of detecting
these low energy products. Work on these reactions is
proceeding at the moment and although the results
presented here are preliminary and require confirmation,
some qualitative observations can be made.

Fig.10 shows three intensity contour diagrams and
some of the relevant results of Henchman and Masson.
The latter results were obtained using a double chamber
mass spectrometer ion source without angular analysis
and therefore the velocity profile will be distorted to
some extent as already discussed. First figs.10(a) and
(c) can be compared with 10(d). For the product CH_3^+
ions the contour diagram and the velocity profile
of Henchman and Masson both show a minimum at the CM with
large backward and small forward scattering. The minimum
of the Henchman results will be filled in to some extent
by products scattered at large CM angles. The forward
Henchman peak lies at a somewhat smaller velocity than
the most forward Prague peak. There does not seem to be
two forward Henchman peaks. It is difficult to compare
the back scattered ions. A comparison between the CH_2^+
results also show an encouraging similarity. In both

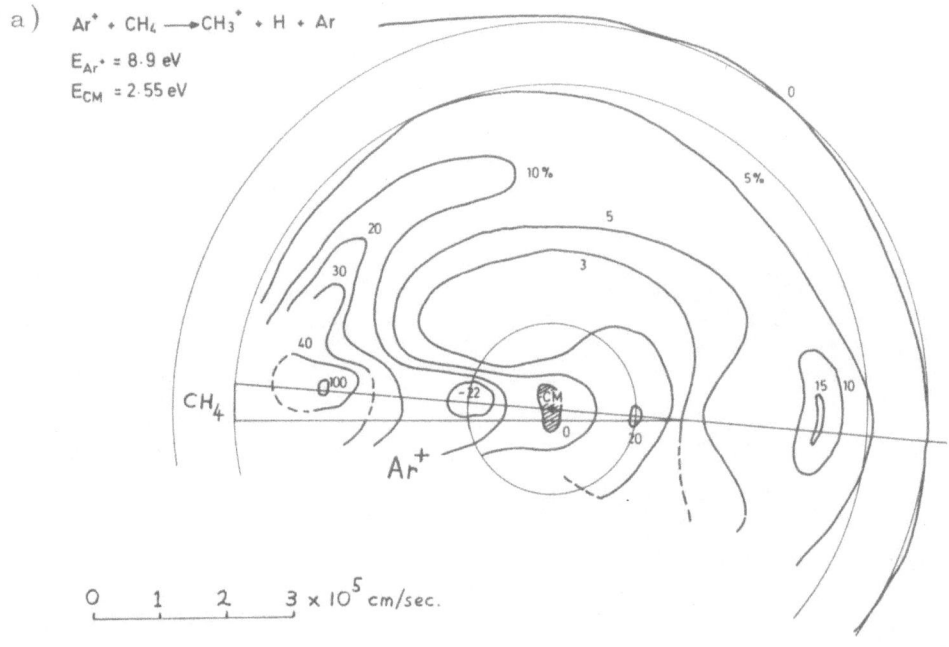

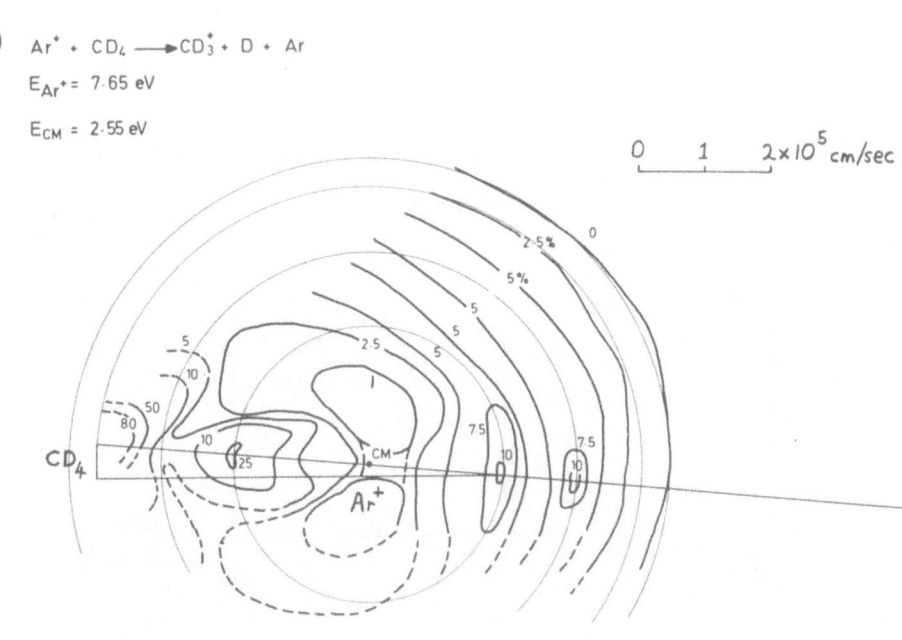

Fig.10   Intensity contour diagrams for the reactions
         shown. Dotted lines show hypothetical contours
         outside the angular range or areas of uncertainty.

c)

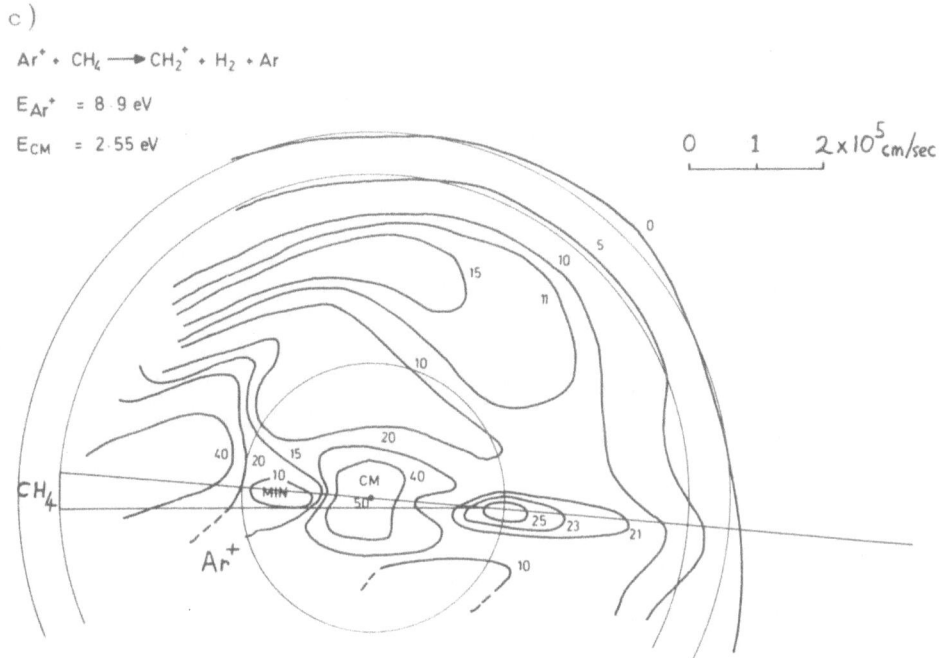

d)

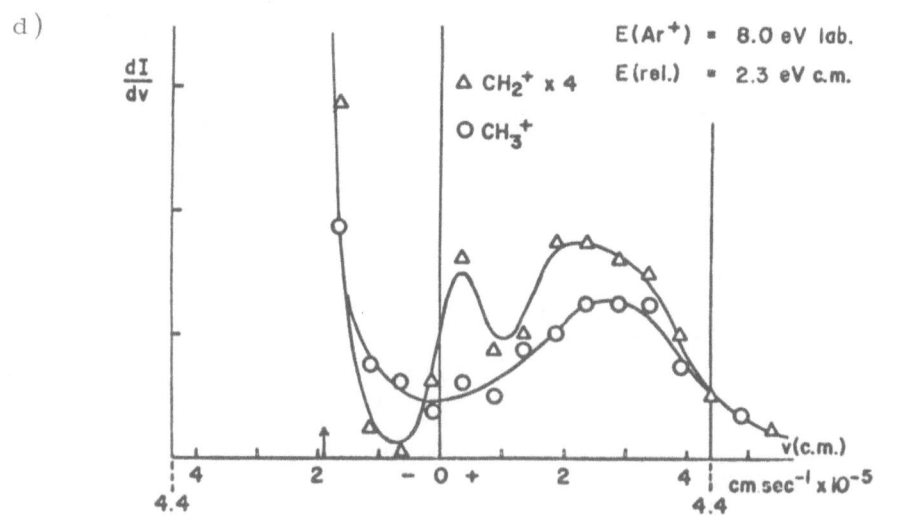

Fig.10   c)Intensity contour diagram for the reaction
            indicated.
          d)Velocity profiles for CH_3^+ and CH_2^+ formed by
            the reaction between Ar^+ and CH_4, from the
            work of Henchman and Masson.

cases there is a large backward peak followed by a
backward minimum, a peak near the CM , and an elongated
forward peak. The positions of the peaks are in reasonable
agreement considering the preliminary nature of the Prague
results and the possible distortion of the Henchman
profiles.

Interesting possibilities arise when attempting to
interpret the results. Points which must be considered
include:
i)There are several competing channels, the most
important leading to CH_3^+ and CH_2^+.
ii)If there is a transition region at large internuclear
separation which gives charge transfer, then there should
be a probability that some reactants will pass through
this region without undergoing charge transfer. We
therefore have two possible types of intimate collision,
one between Ar^+ and CH_4 and one between Ar and CH_4^+
(probably in an excited state).
iii)The stage of the collision at which H (or H_2) is lost
will have a possibly large effect on the reaction. For
example, if during a collision between Ar^+ and CH_4 an
H atom was ejected, then the intermediate could be
stabilised and result in some of the products appearing
at the CM. If an H atom were not ejected but instead
CH_4^+ were first formed, then backward and forward peaks
could result from large impact parameter and small
impact parameter collisions respectively. Subsequent
break-up of the CH_4^+ could lead to contours of the type
shown in fig.3(f) if sufficient energy were released.
iv)How is the energy divided among the kinetic and
internal energy of the products? In the case where three
particles are formed as products the intensity contour
diagram does not give an unambiguous answer to this
question.

In spite of these complications, some conclusions
can be attempted from the results at hand. Firstly,
considering reaction 6, the dominant reaction channel
appears to be charge transfer with little or no momentum
transfer. The precise location of the backward peak is
experimentally difficult, small variations of the energy
scale causing large peak shifts. Secondly, from fig.10(a)
and from experiments at lower energies, the symmetry of
the contour diagrams suggests the possibility of complex
formation of two types, one characteristic of a complex
formed with large angular momentum (ie. giving peaks along
the relative velocity vector and symmetrical about the
CM), and one giving a peak at the CM. The peak at the CM·

behaves rather strangely in that it is present at low energies, disappears at intermediate energies (2ev to 4ev approximately) and intensity at the CM begins to reappear at higher energies. This effect was also seen by Henchman and Masson. Thirdly, at the lowest energy of these experiments (0.4ev CM) the most forward scattered peak appears to fall on the elastic circle. As the collision energy is increased, this peak moves in from the elastic circle indicating that it is formed from low impact parameter rebound collisions which become more inelastic as the energy increases. In most experiments there seems to be an intensity bulge between the most forward peak and the elastic circle indicating a range of inelastic energy loss. If these suggestions are correct then the experiments with CD_4 indicate that the forward CD_3^+ rebound peak suffers a greater energy loss than CH_3^+.

In the case of CH_2^+ formation from Ar^+ and CH_4 at 2.55ev relative energy, long lived complex formation could be occurring giving a peak at the CM (as already mentioned, a peak at the CM does not prove complex formation), and an elongated forward peak can be seen indicating forward rebound with a range of inelastic energy loss. There is again a large amount of back scattering but it does not seem to be as dominant as in the case of CH_3^+ formation.

Work is proceeding on dissociative charge transfer reactions at the present time.

AKNOWLEDGMENTS

We would like to thank Prof.M.J.Henchman, Prof. F.A.Wolf, and Prof.J.B.Hasted for helpful comments, and we are grateful to the participants of this study institute for their constructive criticism.

REFERENCES

Bates,D.R.(1960). Proc.Roy.Soc.A257.22.
Bates,D.R.,Johnston.H.C.,Stewart.I.(1964). Proc.Phys.Soc. 84.517.
Birkinshaw.K.,Hasted.J.B.(1971). J.Phys.B:Atom molec. Phys. 4.1711.
Champion.R.L.,Doverspike.L.D.(1968). J.Chem.Phys.49.4321.

Chiang.M.,Gislason.E.A.,Mahan.B.H.,Tsao.C.W.,Werner.A.S.
      (1970). 52.2698.
Chupka.W.A.,Berkowitz.J.(1971). J.Chem.Phys.54.4256.
Chupka.W.A.,Russel.M.E.(1968). J.Chem.Phys.49.5426.
Demkov.Y.N.(1963). Atomic Collision Processes, Proc.Third
      Int.Conf.Electrn.Atom. Collns., London 1963
      (Amsterdam: North-Holland) Pp.831-8.
Eisele.G.,Henglein.A.,Bosse.G.(1974). Ber.Bunsenges.
      Physik.Chem.78.140.
Gentry.W.R.,Gislason.E.A.,Lee.Y.T.,Mahan.B.H.,Tsao.C.W.
      (1967). Disc.Faraday Soc.44.137.
Gislason.E.A.,Mahan.B.H.,Tsao.C.W.,Werner.A.S.(1969)
      50.5418.
Hasted.J.B.,Iqbal.S.M.,Yousaf.M.M.(1971). J.Phys.B:Atom
      molec.Phys.4.343.
Henglein.A.,Lacmann.K.(1964). Advan.Mass Spectry.3.331.
Herman.Z.,Birkinshaw.K.(1973). Ber.Bunsenges.Physik.
      Chem.77.566.
Herman.Z.,Kerstetter.J.,Rose.T.,Wolfgang.R.(1967). Disc.
      Faraday Soc.44.123.
Herman.Z.,Lee.A.,Wolfgang.R.(1969). J.Chem.Phys.51.453.
Hierl.P.M.,Herman.Z.,Wolfgang.R.(1970). J.Chem.Phys.
      53.660.
Kuntz.P.J.(1972). VII ICPEAC Invited Papers and Progress
      Reports, page 427.
Kuntz.P.J.,Nemeth.E.M.,Polanyi.J.C.,Wong.W.H.(1970)
      J.Chem.Phys.52.4654.
Leventhal.J.J.(1971). J.Chem.Phys.54.3279.
Leventhal.J.J.(1973). J.Chem.Phys.58.4711.
Lindholm.E.(1966). Ion-Molecule Reactions in the Gas
      Phase, (American Chemical Society, Washington D.C.
      1966); Adv.Chem.Series,58.1.
Masson.A.J.(1970). Ph.D. thesis, Brandeis University.
Masson.A.J.,Birkinshaw.K.,Henchman.M.J.(1969). J.Chem.
      Phys.50.4112.
Miller.W.B.,Safron.S.A.,Herschbach.D.R.(1967). Disc.
      Faraday Soc.44.108.
Neynaber.R.H.,Magnuson.G.D.(1973). J.Chem.Phys.59.825.
Robinson.R.J.,Holbrook.K.A.(1972). "Unimolecular
      Reactions", (Wiley-Interscience, New York 1972).
Steiner.E.,Certain.P.R.,Kuntz.P.J.(1973). J.Chem.Phys.
      59.47.
Tully.J.C.(1973). J.Chem.Phys.58.1396.
Tully.J.C.,Herman.Z.,Wolfgang.R.(1971). J.Chem.Phys.
      54.1730.
Wolfgang.R.,Cross.R.J.(1969). J.Chem.Phys.73.743.

THE CLASSICAL TRAJECTORY METHOD

P.J.KUNTZ

Hahn-Meitner-Institut für Kernforschung Berlin GmbH
1 Berlin 39, West Germany, and St.Mary's University,
Halifax, N.S. B3H 3C3, Canada

INTRODUCTION

Trajectory calculations have long been of practical value in neutral systems in linking fundamental theory with scattering experiments. Their popularity stems from their applicability to many-atom systems with arbitrary interaction. Such calculations are applicable to ion-molecule systems also, since these are essentially the same as neutral ones, except that the possibility of charge transfer increases the complexity of the calculations somewhat. Previous descriptions (Bunker and Blais, 1962, 1963, 1964; Karplus and Raff, 1964, 1966; Karplus, Porter, and Sharma, 1964, 1965, 1966; Bunker, 1971; Kuntz, 1972a) of this technique are therefore pertinent.

Experimental results are ultimately a reflection of quantum mechanical laws, so that a fundamental description of ion-molecule interactions begins with the Schrödinger equation. Invoking the Born-Oppenheimer approximation allows independent descriptions of the electronic and nuclear motions: the electronic equation, when solved, provides a number of potential energy functions (or surfaces) which determine the motion of the nuclei. In trajectory calculations this motion is treated classically. Newton's equations describe the time development of the motion, influenced by forces obtained from the derivatives of the potential function. The resulting trajectories, which connect initial and final scattering states, are used to compute cross sections; hence, they provide a path from theory to experiment which requires only two major dynamical approximations: (i) separation of electronic and nuclear motion, and (ii) classical treatment of the nuclear motion.

The most satisfying link between theory and experiment is the non-empirical one outlined above, in which the potential function is calculated from first principles. The trajectories then predict experimental results. Such a calculation could also assess the theoretical errors associated with the approximations, making them amenable to experimental test. This is rare, however, even though several non-empirical studies have been done (Csizmadia, 1969, Krenos, 1974, Preston, 1973, MacLaughlin, 1973).

A more frequent type of calculation involves the rationalisation of existing experimental results in terms of an empirical potential function, which can then be viewed as a model for the system. The search for a realistic model potential is not easy, especially if there are a variety of experimental results which must be represented (Bernstein and Rulis, 1973, Bunker and Goring-Simpson, 1973, LaBudde et al., 1973); however, this kind of rationalisation is preferable to the invention of new dynamical models for each new experimental result. The resulting potential can (i) elucidate the reaction mechanism, and (ii) predict results for different experimental conditions or for related systems.

SUGGESTED APPLICATIONS TO ION-MOLECULE SYSTEMS

Trajectory calculations find their greatest use in describing those processes which are most sensitive to variations in the potential function. If experiments can be adequately described by a theory which makes no reference to the interaction potential, the method is of little use. At the other extreme, if some simple dynamical model (e.g. hard spheres) can describe experiment, then, aside from validating the model, trajectory calculations become supererogatory.

For ion-molecule systems, the ability to control the relative collision energy allows probing of the whole range of mechanisms from statistical "complex" at low energy to hard-sphere dynamics at high energy. (The notion of "complex" is defined operationally. The observable need only behave as expected from statistical calculations; hence, the energy below which "complex" formation occurs will depend on the observable.) Even though complex formation often precludes obtaining information about the potential, it is important that trajectories be able to show the onset of complex formation at the correct energy. Similarly, the approach to idealised high-energy mechanisms must be correct, something difficult to judge, unless the scattering behaviour is adequately represented by some model which describes the approach to the mechanism. Such models should express the deviation from ideality in terms of physical quantities such as masses, collision energy, and impulses, thereby allowing a quantitative characterisation of the approach to ideality with increasing collision energy, and facilitating comparison of different systems. It is also desirable for such models to be verified by the trajectories in order to avoid interpretations in terms of impossible mechanisms.

In the intermediate energy range, there may also be qualitative mechanisms dominant over part of the range. These are often difficult to foresee without the aid of trajectories, which can uncover some mechanisms, and should thereby increase our qualitative understanding of collision processes.

Since "complex" trajectories are costly to compute and require special treatment (Brumer and Karplus, 1973), we shall restrict discussion to direct interactions. Also, we consider the case where all degrees of freedom are treated classically, although it is possible to combine classical and quantal descriptions of some processes (Raff, 1967). The method outlined below is directly applicable to reactive events, but with slight modification applies to inelastic or elastic processes as well.

THE POTENTIAL ENERGY SURFACE

The potential function is the most important ingredient of the calculation. It is usually difficult to obtain, since for large systems accurate theoretical potentials are not available. Whatever the source of the potential, it is essential that the derivatives can be computed accurately and rapidly at all coordinates accessible to the system.

Non-empirical potentials are obtained from approximate solutions to the electronic Schrödinger equation, which means that the function is known accurately at only a limited number of geometries. It is necessary, therefore, to represent the function conveniently. This can be done by fitting global (Csizmadia et al. 1969) or local (MacLaughlin and Thompson, 1973) functions to the points, or by employing a semi-empirical method which has been calibrated to be (or in fact is) a good representation of the non-empirical points (Kuntz, 1972b).

If a non-empirical potential is not available, an empirical form must be used. Most often this is chosen to be both physically realistic and simple to use.

Potential functions for ion-molecule systems are similar to those for neutral-neutral ones, although curve crossing occurs more frequently. The long-range interactions are often important at low energy, but at higher energy their influence on the collision dynamics is insignificant compared to other regions of the potential.

CROSS SECTIONS AND DISTRIBUTIONS

The end result of trajectory calculations is a distribution function in one or many of the variables describing the outcome of the collision process. We will see that the whole trajectory calculation can be viewed as a means of transforming a distribution function which is known at the beginning of the collisions to one which is only known at the end of the collisions; the latter is proportional to a differential cross section. It is essential that some of the dynamical variables which describe the initial state be specified only by a distribution if differential cross-sections are to be computed at all.

To further elaborate upon this, we consider how the initial and final states are specified. In quantum mechanics, the states in the asymptotic regions are specified by quantum numbers for the complete set of commuting observables which describe the separated systems. The dynamical variables conjugate to these can only be assigned a probability distribution function (wave function).

In classical mechanics, a state is specified by the coordinate and momenta at some time, t. In the asymptotic regions, a transformation can be made to new variables, some of which are constants of the motion, n, (e.g. internal energy of a molecule), and some of which vary with time, ß, (e.g. vibrational phase). The constants of the motion n are chosen to correspond to the observables describing the initial state, while the variables ß are assigned distributions which depend in part on the equations of motion, and in part on the conditions of the experiment which the calculation is intended to represent.

The schematic outline of the trajectory calculation is thus quite straightforward: For given (fixed) values of the initial state constants, n, the other variables, ß, are selected statistically from the assigned distribution function. Each microscopic classical state so selected is used to define the initial conditions for the equations of motion for a particular trajectory, which is followed through the interaction region until the final asymptotic state is reached. At this point, the final state constants of the motion, N, are calculated for each trajectory, the other variables for the final state, B, being irrelevant. In this way, the distribution function of the initial state variables ß is transformed into a distribution function of the final state constants of the motion N. The final state distribution depends parametrically upon the initial state constants n, and can be averaged over these if necessary.

From the foregoing we see that the initial distributions must be chosen with great care, since the final differential cross sections are merely transformations of these. There is an important restriction on these distributions which must be observed if the results are not to depend in a spurious fashion on the precise times in the asymptotic regions at which the calculation is begun and finished. The distributions of those initial variables which are not constants of the motion (e.g. phase angles) must not vary with time; otherwise, the resulting differential cross sections would also vary with time in the final asymptotic region. In general,there is only one distribution which is time independent, and this is determined by the classical equations of motion themselves. Once the potential energy of the molecules in the asymptotic region has been specified, the distribution function for certain of the initial variables is also specified (Bowman et al. 1973).

THE CALCULATION OF A SINGLE TRAJECTORY

We now outline the procedure used to calculate a single trajectory connecting a given initial state with a final state which is determined by the trajectory. The method is quite general, but for definiteness, many of the equations here are appropriate to the 3-body process $A + BC \longrightarrow AB + C$, $A + BC$, $AC + B$, or $A + B + C$. The masses of the particles are denoted by A, B, C or by m_1, m_2, m_3, respectively; the internuclear distances BC, CA, and AB are labelled 1, 2, and 3.

For an N-body system, the equations of motion are first expressed in terms of 3N coordinates, $\tilde{q}_i$, and 3N momenta, $\tilde{p}_i$ which it is convenient to group into N vectors q_i and N vectors p_i. The problem can always be reduced to one involving 6N-6 differential equations by removing the motion of the centre-of-mass by the following transformation (Whittaker, 1964):

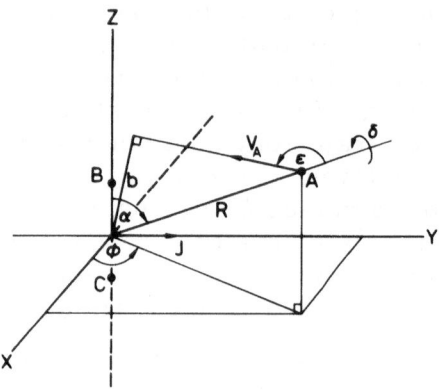

Figure 1. Coordinate System

$$Q_1 = q_1 \qquad P_1 = p_1 + \ldots + p_N$$

$$Q_i = q_i - q_N \quad P_i = p_i$$

for i = 1, N. The Hamiltonian function and equations of motion in the new coordinates are

$$H = \sum_{\substack{i=2}}^{N} \frac{P_i \cdot P_i}{2\,\mu_{i\,1}} + \frac{1}{m_1} \sum_{\substack{i=2}}^{N} \sum_{\substack{j \neq i}} P_i \cdot P_j + U(Q)$$

$$\dot{Q}_i = \frac{\partial H}{\partial P_i} \qquad\qquad \dot{P}_i = - \frac{\partial H}{\partial Q_i}$$

The trajectory is determined by specifying the values of Q_i and P_i appropriate to the initial state at some arbitrary time t_i and then integrating the above equations numerically for successive small time increments until a suitable later time t_f, when the system is deemed to lie in one of the possible asymptotic regions appropriate to the final state.

The initial values of Q_i and P_i are usually obtained from a set of coordinates more descriptive of the collision process (see fig. 1). The positions of B and C are fixed by requiring that the

atoms lie on the z-axis with their centre-of-mass at the origin.
Atom A is specified by the polar coordinates R, α, and $\emptyset$. The velocity of A is related to the polar variables V, ε , and δ . The
magnitude V is obtained from the translational energy of the whole
system, T.

$$V = [2T (B + C) / (AM)]^{1/2}$$

where M = A + B + C. The polar angle ε and impact parameter b
are related by

$$\varepsilon = \pi - Sin^{-1} (b/R)$$

If the angles α and ϕ were 0, the velocity of A would be $(V_x, V_y,
V_z) = V (\sin \varepsilon \cos \delta , \sin \varepsilon \sin \delta, \cos \varepsilon)$. This vector is rotated
through α and ϕ :

$$
\begin{pmatrix} V_{A_x} \\ V_{A_y} \\ V_{A_z} \end{pmatrix} = \begin{pmatrix} \cos\alpha \cos \phi & -\sin \phi & \sin\alpha \cos\phi \\ \cos\alpha \sin \phi & \cos \phi & \sin\alpha \sin \phi \\ -\sin\alpha & 0 & \cos\alpha \end{pmatrix} \begin{pmatrix} V_x \\ V_y \\ V_z \end{pmatrix}
$$

The velocities of B and C have a common component due to trans-
lation, $\vec{V}^T_B$, and an internal component, $\vec{V}^I_B, \vec{V}^I_C$.

$$\vec{V}^T_B = \vec{V}^T_C = - A \vec{V}_A / (B + C)$$

The internal component perpendicular to the BC bond is obtained
from the rotational angular momentum, $\vec{J}$, which we take to lie
along the y-axis. The magnitude of this component is

$$v_x = J / (\mu_{BC} R_{BC}),$$

and is divided between B and C. The internal component parallel
to the BC bond is obtained from v_x, the total internal energy E_{int},
and the potential energy $U_{BC} (R_{BC})$:

$$v_z = \overset{+}{_-} [2 (E_{int} - U_{BC}) / \mu_{BC} - v_x^2]^{1/2}$$

$$\vec{V}^I_B = (v_x, O, v_z) C/(B + C)$$

$$\vec{V}^I_C = - (v_x, O, v_z) B / (B + C)$$

$$\vec{V}_B = \vec{V}^T_B + \vec{V}^I_B \quad and \quad \vec{V}_C = \vec{V}^T_C + \vec{V}^I_C$$

For direct interaction trajectories, the progress of the trajectory with time is quite straightforward. After leaving the initial channel, it enters a region of phase-space where there is strong interaction between all particles, but after a short time in this region, it passes into one of the "product" channels, at which point the numerical integration is terminated and the final values of Q_i and P_i transformed to the constants of the motion and angle variables for the final state. (As the trajectory evolves, it is necessary to interrupt the integration procedure at regular intervals to test whether the product region has been reached.)

The final state constants of the motion of interest are the translational energy T', the molecular internal energy E_{int}, the rotational angular momentum of BC, $\vec{J}'$, the relative velocity vector $\vec{V}'_R$, the impact parameter b', and the scattering angle Θ_s. These are obtained from the Cartesian coordinates $\vec{R}'_i$ and velocities $\vec{V}'_i$. In the following equations, $\vec{v}'$ is the velocity of atom j relative to atom k (the atoms in the molecule), and $\vec{r}_i$ is the internuclear distance of molecule jk.

$$\vec{V}'_R = M \vec{V}'_i / (m_j + m_k)$$

$$T' = 0.5 \, m_i \, (m_j + m_k) \, (V'_R)^2 \quad M$$

$$E'_{int} = 0.5 \, (m_j \, m_k) \, (v')^2 / (m_j + m_k) + U_{jk} \, (r_i)$$

$$\vec{J}' = m_j \, m_k \, \vec{r}_i \times \vec{v}' / (m_j + m_k)$$

$$\Theta_s = \mathrm{Cos}^{-1} \, [-\vec{V}_R \cdot \vec{V}'_R / (V_R \, V'_R)]$$

$$\vec{K} = \sum_{i=1}^{3} m_i \, \vec{R}_i \times \vec{V}_i$$

$$\vec{L}' = \vec{K} - \vec{J}'$$

$$b' = ML' / [m_i \, (m_j + m_k) \, V'_R]$$

$$\vec{V}_R = M \vec{V}_A / (B + C)$$

PRODUCT DISTRIBUTIONS

The trajectories furnish a relation between the initial state constants and variables (n, ß) and the final state ones (N, B). We are interested in using the known distribution function of the variables ß for fixed n, f(n; ß), to obtain the distribution function of the variables N, $S(n; N)$. At fixed n, $f(n; \beta)$ can be regarded as a function of the final constants of the motion: $f(n; \beta(N))$. The formal relation for the product state distribution then involves a sum over all those values of β which yield the particular N sought, and is symbolized as

$$S(n; N) = \int d\beta\, f(n; \beta(\overline{N}))\ \left|\partial \overline{N} / \partial \beta\right|^{-1}\ \delta(\overline{N}(\beta)-N)$$

The function S may then be averaged over as many of the n and N as necessary to obtain a suitable differential cross-section.

The above development is useful in showing the connection between initial and final state distribution functions, but it is hardly ever used in practice, for computation of S involves a multidimensional search for all those values of the β which produce the desired N, and at each of these values the Jacobian factor $\left|\partial N / \partial \beta\right|^{-1}$ must be evaluated. This is too cumbersome and unnecessary for most applications.

A more direct procedure is to approximate S by an average over some small region of N, ΔN:

$$\overline{S}(n; N) = (\Delta N)^{-1} \int_{\Delta N} S(n; \overline{N})\, d\overline{N}$$

The Jacobian in the previous expression for S is removed by transforming variables from $\overline{N}$ to β:

$$\overline{S}(n; N) = (\Delta N)^{-1} \int P(\overline{N}(\beta)) \cdot f(n; \beta)\, d\beta$$

Here the integration is over the entire range of β, and the function P is 1 for all values of β which yield values of N within a range ΔN about N, and is zero otherwise. This integral is usually approximated by a sum over discrete points in β-space. M points are chosen, and for each one a trajectory is computed to ascertain the value of P for that point. The function $\overline{S}$ is then approximated by

$$\overline{S}(n; N) \simeq (\Delta N)^{-1} \sum_{j=1}^{M} w_j \cdot P_j(\overline{N}(\beta_j))$$

where w_j is the weight associated with the region $\Delta \beta_j$ about the point β_j:

$$w_j = \int_{\Delta \beta_j} f(n; \beta)\, d\beta$$

Simply stated, the range of N is divided into "bins" of equal size ΔN, and the amount w_j is added to the appropriate bin if the jth trajectory produces an N value lying within ΔN.

SELECTION OF INITIAL CONDITIONS

The initial variables ß must be selected from the distribution function f(n; ß). For the coordinate system in the figure this is

$$f(E, E_{int}, J; \alpha, \phi, b, \delta, R_{BC}) = const. b \sin\alpha \, p(R_{BC})$$

where E is the total energy and $p(R_{BC})$ is the weight function for the variable R_{BC}, and is proportional to the inverse of the relative velocity of C with respect to B.

The points $ß_j$ needed in the sum over j in the expression for $\overline{S}$ are chosen by a Monte Carlo technique (Hammersby and Handscomb, 1964): the variables ß are chosen randomly for each trajectory. If the $ß_j$ are chosen from a uniform distribution, then the weights w_j must be calculated from the above distribution. Alternatively, the $ß_j$ could be chosen randomly from the above distribution with $w_j = 1$. Another possibility is to choose the variables from some distribution which weights those trajectories which lead to the process under study; this is called importance sampling, and attempts to avoid excessive computation of trajectories which are not fruitful.

Recently, methods involving non-random selection of the $ß_j$ have been used (LaBudde and Bernstein, 1971). These involve choosing points according to a prescription such that errors associated with the integral approximation may decrease faster with the number of points chosen, at least for some types of integrands. A comparison between the usual Monte Carlo methods and the non-random ones for the calculation of cross sections for the $H + H_2$ system (Porter-Karplus potential surface) shows that the non-random method gives maximum probable errors which compare favourably with those expected for Monte Carlo calculations (Suzukawa et al., 1973). One drawback of the non-random methods is that they do not admit of a simple error analysis; furthermore, if a larger calculation is deemed necessary, it is sometimes not possible to add more points to already-existing ones.

CROSS SECTIONS

The distribution functions obtained from trajectory calculations will represent cross sections if the function f(n; ß) is chosen so that

$$\int f(n;\beta)\, d\beta = \pi b^2_{max}$$

The function $\bar{S}$ then approximates a differential cross section for the process $n \rightarrow N$.

In general, trajectory calculations approximate cross sections as follows: The process of interest must first be clearly defined; then, of the total number of trajectories, M, a certain number, M_S will lead to that process. The total cross section for the process is

$$\bar{\sigma} = \pi b^2_{max} \left(\sum_{j=1}^{M_S} w_j \right) / \left(\sum_{j=1}^{M} w_j \right)$$

Differential cross sections are computed in the same manner. For example, suppose we wish to find the differential cross section with respect to some constant of the motion N for the process S e.g. S could label all trajectories which react to form molecule AB, and N could be the internal energy of the AB molecule). We first divide the range of N into k equal intervals ΔN, and then count all those trajectories labelled by S which also fall into the range ΔN about a particular value of N. These trajectories, M_{SN} in number are a subset of the M_S trajectories: $M_S = \sum M_{SN}$. The differential cross section is then approximated by

$$\left(\frac{\Delta \sigma}{\Delta N} \right)_N = \frac{\pi b^2_{max}}{\Delta N} \left[\left(\sum_{j=1}^{M_{SN}} w_j \right) / \left(\sum_{j=1}^{M} w_j \right) \right]$$

The total cross section $\bar{\sigma}$ is related to this by

$$\bar{\sigma} = \sum_{N=1}^{k} \left(\frac{\Delta \sigma}{\Delta N} \right)_N \Delta N$$

SEMICLASSICAL METHODS

Recently, the results of classical trajectories have been used as input for classical-limit quantum mechanical formulations of cross sections for collision processes (Miller and Raczkowski, 1973). One method of obtaining these formulae has been to consistently apply the stationary-phase approximation to the relevant quantum formulae (Miller, 1974). This has the effect of selecting only the classical paths out of all the possible paths connecting the initial and final states. The scattering amplitude, whose squared modulus is related to the cross section, is in general complex, and in the classical limit its phase is determined by the classical action integral along the trajectory, and its amplitude by a mixed derivative of the action with respect to two of the dynamical variables. Both of these quantities can be determined from the trajectory by means of a simple quadrature to determine the action.

These methods may be important in two cases. The first arises when the cross-section for a state-to-state process is desired, and there is little averaging over initial conditions. Often, there is more than one path connecting the initial and final states, and in the semiclassical formulation, this gives rise to interferences which cause undulations in the cross sections. When the initial and final states are not well-defined, i.e. when there is averaging over several constants of the motion, there are too many classical trajectories connecting the states, and the interferences tend to cancel out, giving the same result as a normal Monte Carlo calculation. Even in this case, the semiclassical method has been of great use in pointing out how to do the classical calculation in a way which obtains the correct correspondence with quantum mechanics in the classical limit. In particular, the initial states should not be quantized, but rather averaged over small ranges, if one is to achieve cross sections which are "microscopically reversible" (Schreiber, 1973).

The other case involves processes which are energetically allowed, but classically forbidden dynamically. These include tunnelling through potential barriers, excitation of energy levels not reached classically, and transitions from one potential surface to another. In all of these cases information can be obtained from classical trajectories, provided they are allowed to take on complex values of all variables, including time.

Semiclassical methods may have many applications to ion-molecule systems because of the charge-transfer processes which can occur. These are essentially surface-crossing problems, and must be considered in any realistic computations.

TRAJECTORY CALCULATIONS IN ION-MOLECULE REACTIONS

There have been fewer trajectory calculations on ion-molecule systems than on neutral systems. The earlier studies are outgrowths of the Langevin theory of capture cross-sections resulting from the long-range interaction between an ion and an induced dipole (Dugan et al. 1967, 1968, 1969). Capture cross sections have been computed for the interactions of ion-dipoles and ion-quadrupoles. A basic drawback of these calculations is that they only describe the first half of a reactive collision; nevertheless interesting dynamical effects have been observed.

A more conventional study has been done for the model system $X^+ + H_2 \longrightarrow XH^+ + H$ (Kuntz and Roach, 1973). Product angular and energy distributions were computed for three different surfaces at three collision energies, for a potential function of the LEPS type with a superimposed r^{-4} term. Experimental data were qualitatively reproduced, particularly the behaviour of average translational exoergicity with relative collision energy. The exoergicity was demonstrated to be insensitive to the changes in potential surface. Trajectories were also compared with predictions of a direct interaction model of the near-stripping variety, and were shown to be quite different. It was also demonstrated that at intermediate energies a sharply forward-peaked angular distribution could possibly arise via a mechanism which is nearly direct, but quite far removed from spectator stripping, since it involves strong interactions of the X^+ ion with both ends of the H_2 molecule in a sequential fashion. Such a mechanism could always be possible in systems where atoms B and C are identical.

Interesting non-empirical calculations have been done on the H_3^+ system (Csizmadia et al. 1969) at low energies, and more recently a more complete non-empirical study on this system has been done using the surface-hopping technique (Preston and Cross, 1973; Krenos et al., 1974), since at higher energies there is a strong competition between charge-transfer and rearrangement. This method involves the use of two or more potential surfaces, and allows for transitions between them along "crossing seams", where there are avoided crossings. Trajectories are started on the surface appropriate to the reactants and then followed until a crossing is reached. At that point, a probability for making the transition is computed via the Landau-Zener approximation, and comparison of this probability with a random number between zero and one allows a decision to be made as to whether to remain on the old surface or cross to the new one. This procedure is followed at every crossing seam until the trajectory is out of the strong interaction region. Results of these calculations have been most promising, the most important one being that chemical reaction and charge transfer are very strongly coupled, and must be considered simultaneously. Similar empirical calculations have been done on the ArH_2^+ system at one energy (Chapman and Preston, 1974).

Non empirical studies have been done for $HeH^+ + H_2 \quad He + + H_3^+$, confining the system to C_{2v} geometry (MacLaughlin and Thompson, 1973). The calculations showed most of the energy available going into vibration of H_3^+.

Calculations for $He + H_2^+ \quad HeH^+ + H$ have also been done for the case that H_2^+ has no vibrational energy (North and Leventhal, 1973). This system may prove fruitful for future study since non-empirical surfaces are becoming available as well as more detailed experimental results.

REFERENCES

Bernstein,R.B. and Rulis,A.M. (1973), Faraday Discussions of the Chemical Society, 55, 293

Blais,N.C. and Bunker,D.L. (1962), J.Chem.Phys.37, 2713

Blais,N.C. and Bunker,D.L. (1963), J.Chem.Phys.39, 315

Bowman,J.M., Kuppermann,A., and Schatz,G.C. (1973), Chem. Phys.Letters, 19, 21

Brumer,P. and Karplus,M. (1973), Faraday Discussions of the Chemical Society 55, 80

Bunker,D.L. and Blais,N.C. (1964), J.Chem.Phys.41, 2377

Bunker,D.L. (1971), Methods Comput.Phys.10, 287

Bunker,D.L. and Goring-Simpson E.A. (1973), Faraday Discussions of the Chemical Society 55, 93

Chapman,S. and Preston,R.K. (1974), J.Chem.Phys. 60, 650

Csizmadia,I.G., Polanyi,J.C., Roach,A.C., and Wong,W.H. (1969), Can.J.Chem.47, 4097

Dugan,J.V.Jr. and Magee,J.L. (1967), J.Chem.Phys.47, 3103

Dugan,J.V.Jr., Rice,J.H., and Magee,J.L. (1968), Chem.Phys. Letters 2, 219; (1969), ibid. 3, 323

Hammersly,J.M. and Handscomb,D.C. (1964), "Monte Carlo Methods", (Methuen and Co., London)

Karplus,M. and Raff,L.M. (1964), J.Chem.Phys. 41, 1267

Karplus,M., Porter,R.N., and Sharma,R.D. (1964), J.Chem. Phys. 40, 2033; (1965), ibid. 43, 3259; (1966), ibid. 45,3871

Krenos,J.R., Preston,R.K., Wolfgang,R., and Tully,J.C. (1974) J.Chem.Phys. 60, 1634

Kuntz,P.J. (1972a), VII ICPEAC, ed.by T.R.Govers and F.J. DeHeer, (North-Holland, Amsterdam), p.427

Kuntz,P.J. (1972b), Chem.Phys.Letters, 16, 581

Kuntz,P.J. and Roach,A.C. (1973), J.Chem.Phys. 59, 6299

LaBudde, R.A. and Bernstein, R.B. (1971), J.Chem.Phys. 55, 5499
LaBudde, R.A., Bernstein, R.B., Kuntz, P.J., and Levine, R.D.
    (1973), J.Chem.Phys. 59, 6286
Miller, W.H., Raczkowski, A.W. (1973), Faraday Discussions of
    the Chem.Soc. 55, 45
Miller, W.H. (1974), Adv.Chem.Phys. 25, 63
North, G.R. and Leventhal, J.J. (1973), Chem.Phys.Letters 23,
    600
Preston, R.K. and Cross, R.J.Jr. (1973), J.Chem.Phys. 59,
    3616
Raff, L.M. (1967), J.Chem.Phys. 47, 4789
Raff, L.M. and Karplus, M. (1966), J.Chem.Phys. 44, 1212
Schreiber, J.L. (1973), Faraday Discussions of the Chemical So-
    ciety 55, 72
Suzukawa, H.H.Jr., Thompson, D.L., Cheng, V.B., and Wolfs-
    berg, M. (1973), J.Chem.Phys. 59, 4000
Whittaker, E.T. (1964), "A Treatise on the Analytical Dynamics
    of Particles and Rigid Bodies", 4th ed., (Cambridge Univer-
    sity Press, London)

PHASE SPACE THEORY AND ITS APPLICATION TO TRI-ATOMIC COLLISION EVENTS

P.F. Knewstubb

University of Cambridge, England
Department of Physical Chemistry, Lensfield Road

The theoretical description of reactive molecular encounters has been attempted from several standpoints, which may be set out in very brief summary as follows:

1) Quantum theory and treatment by wave mechanics; agreed to be the correct method, but almost impossibly unwieldy.

2) Semiclassical methods; S-matrix and R-matrix approaches with a "short wavelength approximation" and allowing representation of momentum as a complex quantity. A considerable amount of work has been done using this formalism.

3) Classical treatments; these represent momentum solely as a real quantity, and are usually basically acceptable as an approximation for heavy particles moving at other than very low velocities. A brief and probably over-simplified subdivision of the methods is

a) trajectory analysis; Monte Carlo methods, with each trial giving the single result of a single collision.

b) statistical or phase space theory; this may give the average result of a single collision.

c) thermodynamic equilibrium or absolute rate theory; this gives, in a sense, the average result of an average collision.

INTRODUCTION TO PHASE SPACE THEORY

An intermolecular collision event involving N atoms is in classical terms describable by specifying the variation of 3N coordinates of distance and of momentum as a function of time. A device which has been used for many years is to regard this description in terms of a single phase point moving in a 6N-dimensional phase space, and describing a certain trajectory in it. A reduction in the number of dimensions to 6N-6 is usually achieved by ignoring position and momentum of the centre of mass of the system. The trajectory then represents all motions of the system relative to the centre of mass.

A further development was prompted by the appearance of quantum theory and was incorporated as a minimal recognition of the wave nature of the event, while seeking to avoid a fully detailed calculation. It is recognised that in each distance q_i and conjugate momentum p_i there are essential uncertainties such that the values p_i and q_i are to be specified only within the limits of a "cell" of area $\delta p_i \times \delta q_i = h$. When applied to all coordinates it is seen that the complete volume of phase space is divided into (6N-6)-dimensional cells of individual volume h^{3N-3}. It is convenient to define different types of phase space as follows:

A μ-space or molecule space exhibits the 6N-6 coordinates of a molecule of N atoms, and a trajectory represents the internal motions of the molecule. An ensemble of μ-spaces identical in energy is an alternative representation to a study of the evolution of one μ-space as a function of time. If all points of the ensemble are mapped into one μ-space, there appears a density of points which represents the relative probability of different regions of μ-space. We expect for a random event that all accessible space will be equally populated.

If one now regards a system of ν such molecules, the total state of the system is described in $\nu \times 6N$ coordinates which form a γ-space or gas space. In a γ-space the subspaces of μ space need not be identical, but may be regarded as freely exchanging energy while the total energy of the γ space is constant. Thus one may model a system at a specified temperature. A trajectory in γ-space represents the motions of all ν molecules, plus their internal motions. Rather than trace the evolution of a γ-space as a function of time, it is easier to base arguments (as for μ-space) on the simultaneous state of a large number of replicas of the γ-space - a Gibbs ensemble. The representative points may all be mapped into one γ-space to give a density representing relative probability of various configurations, for a given temperature.

For discussion of reactions between species, there is some convenience in defining an r-space (reaction space) for two reactive species in close proximity. Such a space is created when two particles approach within an arbitrary limit and is destroyed when they separate. Assuming that the boundary condition for an r-space lies outside the range of mutual influence of the molecules, simple collision theory can be used to give the rate of creation (and presumably hence the rate of decay) of the r-spaces. For example, if we take three Cartesian coordinates (with their conjugate momenta) to describe the approach of the molecules, and stipulate that all three must be less than a value ℓ before the r-space is created, then the relative position vector must cross one of six faces of a cube of side 2ℓ and we get, for a γ-space in thermal equilibrium at temperature T,

$$\text{Rate of creation of r-spaces} = 6\ell^2 \left(\frac{8kT}{\pi\mu} \right)^{\frac{1}{2}}$$

During the existence of an r-space one may describe the co-ordinates of an individual event (a trajectory representation), or one may map out some average behaviour of a large number of such events, and represent this as a density of points in the r-space.

THE LIOUVILLE THEOREM

This fundamental theorem is proved in most books on statistical thermodynamics, often by the following argument. Consider a rectangular solid element of phase space having one dimension dq_1 and area 'perpendicular' to this of $A = (dp_1 dq_2 dp_2 \ldots dq_m dp_m)/h^m$. The number of cells in the element is then $V = A \, dq_1$. We have also the flux of phase points entering one face perpendicular to $q_1 = A \, D \, \dot{q}_1$. The flux of phase points out of the opposite face

$$= A\left(D + \frac{\partial D}{\partial q_1} \, dq_1 \right)\left(\dot{q}_1 + \frac{\partial \dot{q}_1}{\partial q_1} \, dq_1 \right)$$

Hence the rate of accumulation of points in the cell is

$$\frac{\partial N}{\partial t}(q_1) = - \, A \, dq_1 \left[\dot{q}_1 \, \frac{\partial D}{\partial q_1} + D \, \frac{\partial \dot{q}_1}{\partial q_1} \right]$$

and a similar argument can be applied to all the other coordinates.

$$\therefore \quad \frac{\partial N}{\partial t} = - \, V\left[\sum_i \left(\dot{q}_i \, \frac{\partial D}{\partial q_i} + \dot{p}_i \, \frac{\partial D}{\partial p_i} \right) + D \sum_i \left(\frac{\partial \dot{q}_i}{\partial q_i} + \frac{\partial \dot{p}_i}{\partial p_i} \right) \right]$$

Now since Hamilton's equations of motion will apply to the system, the last bracket is zero, the terms cancelling in pairs. Hence

$$\frac{\partial D}{\partial t} = - \sum_i \left(\dot{q}_i \frac{\partial D}{\partial q_i} + \dot{p}_i \frac{\partial D}{\partial p_i} \right) = - \sum_i \left(\frac{\partial H}{\partial p_i} \frac{\partial D}{\partial q_i} - \frac{\partial H}{\partial q_i} \frac{\partial D}{\partial p_i} \right) = [H\ D]$$

gives the rate of change of point density at a fixed position.

To study an equilibrium or steady state condition, we require $\partial D/\partial t = 0$, and hence the density D is a function only of the Hamiltonian H, or the total energy E of the system. Two useful choices of function are, firstly, D(E) = constant over a range E to E + dE and zero otherwise (a microcanonical ensemble). Alternatively one may choose D(E) = C exp(-sE) with C and s as constants (a canonical ensemble).

If the density of phase points is not, as is often assumed, constant throughout a region, a further consequence of the Liouville Theorem may be seen. The density of points around a selected moving phase point is governed by

$$\frac{dD}{dt} = \frac{\partial D}{\partial t} + \sum_i \left(\frac{\partial D}{\partial q_i} \dot{q}_i + \frac{\partial D}{\partial p_i} \dot{p}_i \right)$$

which is seen from the above equations to be zero. Hence the local density must remain constant. From this one derives the model of transport of phase points as being like the flow of an incompressible fluid within the boundaries of the space.

THE VIEWPOINT OF INFORMATION THEORY

If one has a series of cells of phase space with a known *a priori* probability P of containing the representative point of a single system, and $\sum_i P_i = 1$, then the information function of Shannon (1949) is defined by $I = -K \sum_i P_i \ln P_i$. Here K is an arbitrary constant often taken as $1/\ln 2$, and I then represents, in "bits", the information value of discovering in which cell a phase point actually lies. Thus as soon as the position of the phase point is known, the information value of further investigation is zero. The information value is a maximum for a completely random system in which all P_i values are the same.

For a microcanonical ensemble, a certain volume (a number of cells, say N will be available depending on its energy and boundary conditions. This can be the case for γ, μ or r-spaces. If we make the normal statistical assumption that nothing more is known about the system, all cells are equally likely to be occupied and

Information function = $K \ln N$

If we now suppose that one of the coordinates of the space is

known to be in the range dq_i at q_i, the system now occupies a
reduced number of cells of phase space $\mathscr{A}_i$, which may be regarded as
the area of a cross section of phase space at q_i.

Information now lacking $\qquad = K \ln \mathscr{A}_i$

Information value of knowing $q_i = K \ln \left(\dfrac{N}{\mathscr{A}_i} \right)$

This area function is of some interest in studying the features of
an r-space (Fig. 1). In general it is to be expected that $\mathscr{A}$,
which can be defined for any trajectory in r-space, will show a
minimum at some point. For example, this might occur near a
potential energy maximum. For this trajectory the minimum of $\mathscr{A}$
denotes the configuration for which there is least lack of
information if occupation of the phase space is random, i.e. if it
is uniformly filled with points. The configuration is identified
as that of the transition state for reaction along the chosen
trajectory. If one uses the model of fluid flow suggested by the
Liouville theorem, the minimum value of $\mathscr{A}$ also appears as a bottleneck to the flow of phase points in the r-space. For two non-
interacting particles, for example, one may derive for approach
along any one coordinate

$$\mathscr{A}_i \;=\; \delta q_i \;\times\; \frac{8\ell^2}{h^3} \, (2\pi\mu)^{3/2} \left(\frac{E}{\pi} \right)^{\frac{1}{2}} * \, \rho_{int} \quad \text{cells}$$

where E is the energy of the approach and the last factor is a
convolution integral with a density of states function for any
internal degrees of freedom. For any r-space the area function
should tend to the above value as ℓ is increased, and hence at
least one minimum must occur along coordinate i within the r-space.

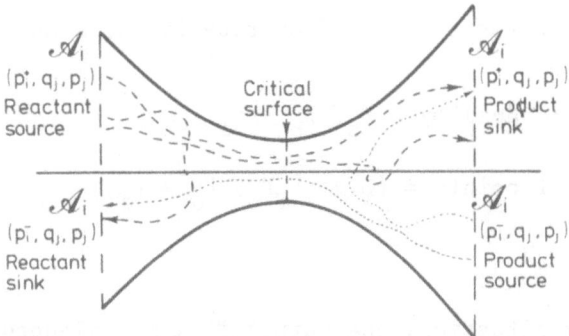

Fig. 1 Diagram of r-space with trajectories and critical surfaces

PHASE SPACE THEORY AND REACTION RATES

A theory of reaction rates has been proposed by J.C. Keck
(1960, 1967) in which a surface is defined which cuts through an
r-space so as to divide all reactant configurations from all product
configurations.   A perfectly-drawn surface would allow only one
crossing by any trajectory, but it is recognised that an imperfect
or "trial" surface may allow multiple crossings.   The theory
assesses the rate at which phase points cross the surface in one
direction; this gives a value for reaction rate which is an upper
limit to the true value.   Variation of the surface is attempted
so as to achieve a minimum value of the limit.   The general
expression for rate of flow of phase points used by Keck is

$$R_{fi} = \int_{S_{fi}} D\,(\underline{v}.\underline{n})\,ds$$

Where D = density of phase points, $\underline{n}$ is the normal to the surface
and $\underline{v}$ the transport velocity in phase space.   In terms of Fig. 1,
having defined some coordinate i leading to products, surfaces
normal to this may be tried and it seems intuitively likely that
one drawn at the bottleneck is the best choice, with the least
likelihood of re-crossing.

A viewpoint of reaction rates based on the fluid flow model
and statistical ideas was also advanced by Keck (1958).   This may
be discussed by reference to Fig. 2, which is similar to Fig. 1
except for showing the possibility of two minima in the $\mathcal{A}$ function,
between which lies a region of phase space which represents a
collision complex.   It is essential to the theory that one supposes
a uniform density of phase points in this region, and hence uniform
and equal surface densities on S_f and S_b.   The essential steps of
the theory may then be summarised as follows:

Input flux of points = R_f = $(D \times \Gamma_f)$ by the principle of
S_{f-}

detailed balancing, where Γ is a flow rate in the sense (volume/
unit time).

Transition probability = $\Gamma_b/(\Gamma_b + \Gamma_f)$

Output flux of points = $(D \times \Gamma_f\Gamma_b/(\Gamma_b + \Gamma_f))$
S_{b+}

In an equilibrium situation, the output flux is balanced by an
input flux of R_b.

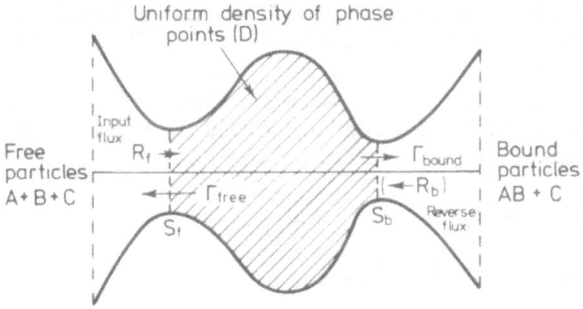

Fig. 2 Statistical model of J.C. Keck

A somewhat different statistical theory of reaction rates has been proposed and elaborated by J.C. Light and co-workers (1964, 1965(a,b), 1967).   This approach is set in the same context as Keck's theory in Fig. 3.   A reaction cross section for formation of a specific complex is defined, amounting to an expression of the area of phase space $\mathcal{A}_{in}$ which includes all classical trajectories leading through S_1 to the complex (E.J).   The decomposition of the complex is then discussed in terms of the relative probability for various states of reactants and products;   these are assessed after the particles have emerged from the complex and achieved definite states.   Again, the density of points in the phase space of the collision complex is assumed to be uniform.

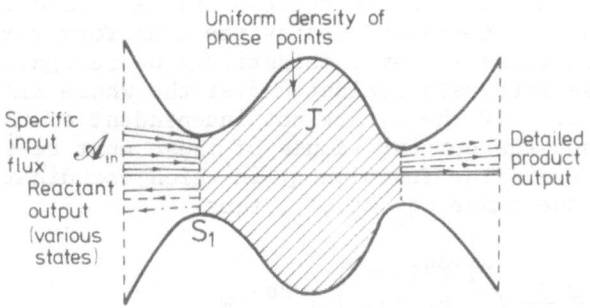

Fig. 3   Statistical model of J.C. Light

Light's theory has been applied to ion-molecule reactions since one may sieze upon the "Langevin radius" as providing a plausible limit for conditions leading to the randomised collision complex. The "Langevin cross section" (σ_L) is then employed as a normalising factor so that reaction cross sections can be expressed in absolute terms. Thus in Fig. 3 one has

Input flux, $R = \sigma_L \, u \, n_1 \, n_2 \, x \, P_{(J)}$

Yield of state $\beta = R \, x \, P_{(J \to \beta)}$

where u is an average relative velocity of approach, n_1, n_2 are number densities and P factors are probabilities. Whereas Keck's theory yields a very general and highly averaged formulation, the theory of Light offers very specific predictions about proportions of products in different energy states; averages can of course be taken as desired.

COMPUTATION OF THE AREA FUNCTION

The assessment of an available volume or area of phase space is simplest if the coordinates of the system can be made to refer to recognisable and non-interacting motions or "degrees of freedom"; unfortunately this is not in general possible. The volume would appear, for a microcanonical ensemble, as a total density of states function, obtainable from a repeated convolution of density functions for each coordinate. This operation could be treated by the convolution theorem for Laplace transforms, and at the intermediate stage of this process one would have in effect a product of partition functions thus

$$F \;=\; f_1 \; x \; f_2 \; x \; f_3 \, \ldots \ldots \; f_n$$

If $A \;=\; - \, (1/s) \, \ln F$

the function A is clearly representing a Helmholz free energy function for some temperature set by the transform variable s = $1/kT$. The functions f_i may alternatively be recognised as the classical phase integrals evaluated over the whole allowed region of coordinate i. If the f_i are not independent of the other co-ordinates j then the current values of these must be specified. We may define a similar function g_i in which coordinate i is restricted to the range δq_i at q'_i; thus

$$g_i = \int_0^E \left(\frac{2}{h} \frac{d}{dE} \int_{q'_i}^{q'_i + \delta q_i} p_i \, dq_i \right) e^{-Es} \, dE$$

Then just as $L(N_{(E)}) = F_{(s)}$

so $$L(\mathcal{A}_i) = g_i \times \prod_{j \neq i} f_j = g_i \, e^{-sA} / f_i$$

If δq is sufficiently small that we may take p_i as a constant in the interval we get

$$\dot{g}_i = (2\pi m/s)^{\frac{1}{2}} \, \delta q_i \, e^{-sV}/h$$

hence $L(\mathcal{A}_{i(E)}) = (2\pi m/s)^{\frac{1}{2}} \, \delta q_i \, e^{-s(A+V)}/h \, f_i$

from which one may write

$$\mathcal{A}_{i(E)} = \left[(2\pi m)^{\frac{1}{2}} \, \delta q_i / h \right] * H(V)$$

$$* L^{-1} \left[s^{\frac{1}{2}} \, e^{-sA} / f_i \right]$$

where asterisks denote convolution integrals and H is the Heaviside step function.

Computations of the $\mathcal{A}$ function in this form for trajectories in specific atomic systems would certainly be a considerable task. Interesting results can, however, be derived from the intermediate stage at which the A function is defined. The change of this quantity for a small displacement in a phase space is expressed as

$$dA = \frac{\partial A}{\partial q_1} dq_1 + \frac{\partial A}{\partial q_2} dq_2 + +$$

and $$\frac{\partial A}{\partial q_i} = -\frac{1}{s} \left(\frac{\partial \ln f_1}{\partial q_i} + \frac{\partial \ln f_2}{\partial q_i} + + \right)$$

Thus determination separately of the partial gradients of A at any point leads to a local value of the vector <u>grad A</u>, while integration can then yield a map of A relative to some reference level.

A computer programme has been devised to examine this type of calculation for arbitrary configurations of three atoms in an r-space. Recognising that coordinates showing minimum interaction would require continued adjustment, simplicity was sought in the use of six cartesian coordinates, which suffice also for the determination of the potential of the system, without reference to the conjugate momenta. Although independent in geometric space, these

coordinates are unsuitable for the description of large regions of
phase space; the form of the calculations, however, relies only
on comparisons over quite small displacements. The results should
therefore show the principal features, although some numerical
errors and approximations still remain. The example chosen for
illustration used a LEPS model potential surface for the reaction
I + HCl → HI + Cl and Fig. 4 shows results for this r-space at an
ensemble temperature of 1000 K with the restriction that the H – Cl
bond length is held at 1.25 Å. The arrows represent the vector
grad A as calculated at a regular network of points. They indicate
the type of surface which would represent the free energy function
itself.

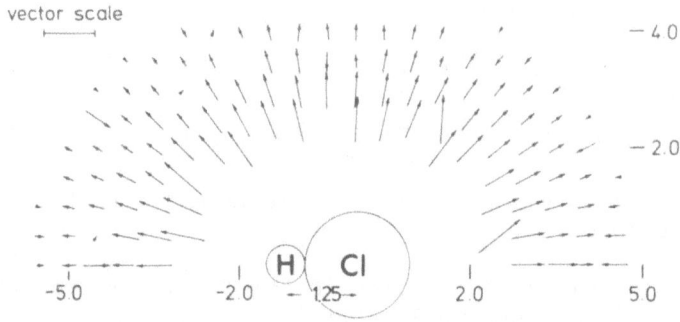

Fig. 4 Free Energy Gradients for I-H-Cl configurations with H-Cl
bond length 1.25 Å and LEPS potential. Rectangular coordinates (Å)
are referred to H-Cl centre of mass. Arrow lengths represent log
($|$grad A$|$) with scale shown for 10^5 cm^{-1}/Å.

The meaning of the map of the A function would be to represent
the ln(relative probability) of different configurations of the
atoms in an ensemble at the temperature of 1000 K, with the built-
in restriction on H-Cl bond length. The latter restriction could
be relaxed by producing a three-dimensional map in which one could
presumably trace a path of minimum free energy, and hence maximum
probability, which would be a thermodynamically averaged reaction
path. In the centre of the diagram, the repulsive potential is
clearly a dominant influence, but in the peripheral region, where
the interaction is considerably less than kT, a small lowering of
free energy relative to that at larger separations is apparent.
This arises from events with low initial momenta which, with even
small retardation, show an appreciable expansion of the phase space
in the momentum coordinates. Calculations of this type will
shortly be extended to selected ion-molecule interactions and to
the production of three-dimensional maps.

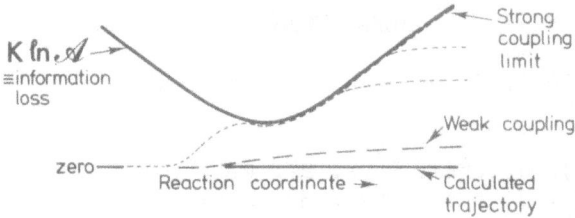

Fig. 5 Possible loss of information in a collision event.

INFORMATION LOSS IN ACTUAL COLLISION EVENTS

A final comment can be offered on the question of whether actual events are better described by trajectories or by a phase space theory. It will be recalled that the information function represents the value of characterising precisely an unknown or partially known situation. Thus, $I = K \ln A_i$ represents the information lacking if only one coordinate is known. The function also represents the maximum possible loss of information, starting from a known state which then evolves.

The fact that evolution of any real dynamic system will inevitably result in loss of information has been discussed by Brillouin (1962), and the idea is illustrated in Fig. 5. Starting from a defined condition (a known quantum state), a trajectory calculation assumes in effect that we retain complete information about the system. This is an unrealistic representation of an atomic event unless the interaction of the particles is zero, although such calculations are used in a statistical manner. If interactions are strong, we can make the 'strong coupling' approximation which will imply that the information loss rises rapidly to its limiting value as the particles approach, and breaks away from this limit only when the interaction has sufficiently decreased. A phase space theory is quite appropriate for discussion of an event of this nature. There will of course be all degrees of intermediate behaviour, such as the indication of 'weak coupling' in Fig. 5, and for such single events there is still no adequate description in classical terms.

REFERENCES

BRILLOUIN, L. (1962). Science and Information Theory. Academic
  Press, New York.

KECK, J.C. (1958). J. chem. Phys. 29, 410.
          (1960). J. chem. Phys. 32, 1035.
          (1967). Adv. chem. Phys. XIII, 85.

LIGHT, J.C. (1964). J. chem. Phys. 40, 3221.
LIGHT, J.C. and LIN, J. (1965). J. chem. Phys. 43, 3209.
LIGHT, J.C. (1967). Disc. Far. Soc. 44, 14.

PECHUKAS, P. and LIGHT, J.C. (1965). J. chem. Phys. 42, 3281.

SHANNON, C.E. and WEAVER, W. (1949). The Mathematical Theory of
  Communication. Univ. of Illinois Press, Urbana.

ION-MOLECULE CHARGE-TRANSFER COMPUTER-ANIMATED MOVIE-TRAJECTORIES

Fred Alan Wolf

Physics Dept., Birkbeck College
(Univ. of London)England and Collisions Ioniques,
Université Paris-Sud, France

Introduction

We shall describe these trajectories from the point
of view of potential energy curves, crossing in space at
a single point, using the Landau-Zener concept to
describe the charge transfer process.

These trajectories are computed using the classical
Lagrangian equations of motion, which we write down in
the next section. As each point of the trajectory is
computed numerically, giving both the position of the
reduced mass of the system and the orientation of the
dipole moment of the molecule with respect to the center-
of-mass of the molecule, a question is asked concerning
space-time coordinates of the system. This question
is: Does H_{11} equal H_{22}?* If the answer is yes we mark
the point on the actual trajectory seen in the film
save the information and then continue on the same path.
In this manner we obtain a sequence of "crossing points".
After completion of the primary trajectory we return to
each crossing point, assume that a charge transfer takes
place, and then follow in the same manner the secondary
trajectory. We also follow the same procedure of asking
the question $(H_{11} = H_{22}$?$)$, and continue to mark crossing
points when they occur on it. And so on. The purpose of
the film is not to compute probabilities and cross

---

* This is our criterion for a crossing from one curve
to another. Read on for more clarity.

sections for charge transfer although our computer
program enables us to do so, but it is to aid the
insight of the searcher (chemist or physicist or anyone
really) as to how a classically forbidden process
(electron transfer) changes the Newtonian deterministic
picture.

I - The Lagrangian Equations, Potentials and Matrix
    Elements Yielding Trajectories

        We view the collision between an argon ion A^+ and
a carbon monoxide molecule CO, at a low and therefore
interesting trajectory-producing energy. We formulate
the collision problem from a classical Lagrangian point
of view. In fig. 1. we see a coordinate layout for the
study indicating the classical coordinates of interest.

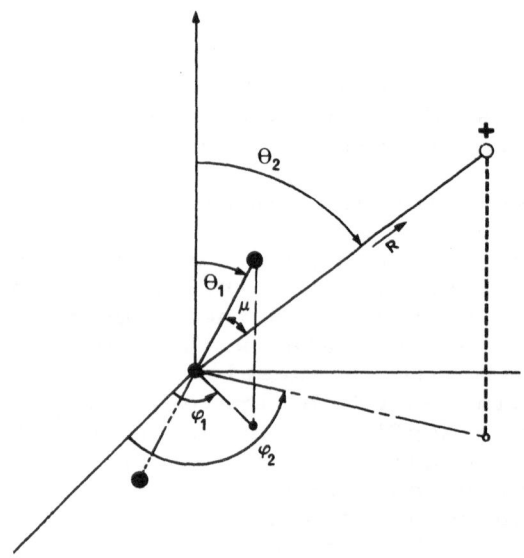

Fig. 1. Coordinate layout for trajectory study.

The dynamics of vibrational excitation are omitted - but
the process is allowed to take place as a result of the
charge transfer - that is, we can and do end up with the
vibrationally excited ionized molecule. Rotational

excitation does take place as a result of the classical coupling.

Our equations all follow from the classical Lagrangian, and its equation of motion. Terms have their usual meaning and can be understood when you look at fig.1.

$$\mathcal{L} = T - V \quad , \qquad \frac{d}{dt}\left(\frac{\partial \mathcal{L}}{\partial \dot{q}}\right) - \left(\partial \mathcal{L}/\partial q\right) = 0$$

motional   potentional
energy   energy

$$T = \frac{1}{2}mv^2 + L^2/2mR^2 + B_1 J^2$$

    kinetic   orbital   rotational
    énergy   energy   energy

Define $\longrightarrow \dot{R} \equiv v$

then $\longrightarrow \dot{v} = L^2/m^2R^3 - (1/m)\,\partial V/\partial R$

since $\longrightarrow \vec{J} + \vec{L} = $ constant (total angular momentum)

Define
$$\begin{cases} J^2 \equiv J_z^2/\sin^2\Theta_1 + J_p^2 \\[2mm] L^2 \equiv L_z^2/\sin^2\Theta_2 + L_p^2 \end{cases}$$

Then

$\longrightarrow \dot{L}_p = \cot\Theta_2\left(\dfrac{1}{mR^2}\right)\left(\dfrac{L_z}{\sin\Theta_2}\right)^2 - \left(\partial V/\partial\Theta_2\right)$

$\longrightarrow \dot{J}_p = \cot\Theta_1(2B_1)\left(\dfrac{J_z}{\sin\Theta_1}\right)^2 - \left(\partial V/\partial\Theta_1\right)$

$\longrightarrow \dot{L}_z = -\left(\partial V/\partial\varphi_2\right)$

$\longrightarrow \dot{J}_z = -\left(\partial V/\partial\varphi_1\right)$

$\longrightarrow \dot{\Theta}_2 = \left(L_p/mR^2\right)$

$\longrightarrow \dot{\Theta}_1 = 2B_1 J_p$

$\longrightarrow \dot{\varphi}_2 = L_z/(mR^2\sin^2\Theta_2)$

$\longrightarrow \dot{\varphi}_1 = 2B_1 J_z/\sin^2\Theta_1$

(1)

The potential for the system is also defined in terms of the angle μ.

$$V(R, \mu) = V(R, \Theta_1, \varphi_1, \Theta_2, \varphi_2)$$

$$= \underbrace{(A/R^{12})}_{\substack{\text{hard core} \\ \text{repulsion}}} - \underbrace{(\alpha_o e^2/2R^4) - (\alpha_1 e^2/2R^4) P_2(\cos \mu)}_{\text{induced dipole polarization}} \qquad (2)$$

$$+ \underbrace{(eD/R^2) P_1(\cos \mu)}_{\text{permanent dipole}} + \underbrace{(eQ/R^3) P_2(\cos \mu)}_{\text{quadrupole}} = H_{11}$$

where $\cos \mu = \cos \Theta_1 \cos \Theta_2 + \sin \Theta_1 \sin \Theta_2 \cos (\varphi_1 - \varphi_2)$

and P_1 and P_2 are Legendre polynomials.

We briefly discuss H_{11}. It consists of what we shall call long range separate ion-molecule interactions computed by using reactant ion/molecule wave functions. The electron in question is localized. The terms are in the order written: a hard core repulsion (R^{-12}), a charge induced dipole potential with tensor polarizability (two terms), a charge permanent dipole term, and a charge-quadrupole interaction.

Now the dynamics of a collision are perfectly capable of being studied using these equations. Such studies have been carried out and in fact several computer generated-animated film studies (Dugan et al. 1969) have indeed been made. Since V depends on μ, there is rotational energy transfer taking place and this has been shown in previous films (Dugan et al. 1969).

If we now ignore the problem of how dynamic charge transfer takes place and simply assume that one has occurred, the picture changes: we now view the + charge as residing on the CO molecule hence the potential V will no longer depend on the orientation but only upon the distance, R. This is of course due to the spherical symmetry of the charge distribution of the argon atom. Of course this is only roughly true. For during the time of the collision which is typically 100 to 300 collision units, where one collision unit is of the order of 10^{-15} seconds, during which the charge distribution of the atom fluctuates about 10 times, we also have some asymmetric dipole effects - but we can successfully ignore these by arguing that during one collision unit 10 fluctuations on the atomic scale occur - while nothing dynamic has happened to the nuclei. Hence the ionized molecule-atom interaction is averaged out to be a scalar quantity.

Our new potential is then written

$$W(R) = (A/R^{12}) - (\alpha_2 \ e^2/2R^4) = H_{22} + \Delta E \qquad (3)$$

When $\Delta E = \Delta E_{electronic} + \Delta E_{vibration} + \Delta E_{rotation}$.

In the above ΔE represents the energy defect for a particular set of reactant and product states. In the film we choose the transition

$$A^+ + CO(X, v = 0) \longrightarrow A + CO^+ (X, v = 6).$$

Hence $\Delta E_{electronic}$ and $\Delta E_{vibration}$ are known

(Bohme et al. 1968). Using the rotational constants B_1 and B_2 given by Bohme et al (with vibrational correction) we have $\Delta E_{rot} = (B_1 - B_2) \ J(R_x)^2$ where R_x is the crossing radius. Thus we point out that the energy defect depends on the history of the trajectory prior to reaching the crossing radius. We look at his point again in section III.

The equations describing the trajectory after charge transfer remain the same but now we replace V by W and, B_1 and B_2 and notice that W is free of dependence on any angular coordinates. Hence L_Z and J_Z are separable constants of motion.

Simplification: We shall restrict the motion to be planar - this is not a necessary restriction but it does reduce the complexity of the problem and enables us to make a simple computer-animated film. To do this we let $\theta_1 = \theta_2 = 90^o$. Whereby $\mu = \varphi_1 - \varphi_2$, $L_p = J_p = 0$, and $L = L_Z$, $J = J_Z$. It is this system which we will observe in the film. (See fig. 2 for a typical trajectory of the reduced mass ion).

Fig. 2. A typical reduced mass trajectory at 0.1 eV
kinetic energy with dipole orientation at coordinate
center. The distance from the dipole to the start of
the trajectory at the right border is ~ 10 bohr.

II - Landau-Zener Crossings

    We briefly examine the quantum mechanical equations
of motion. We do not solve these equations along with
the classical equations as our trajectory develops in
time. We only wish to point out that the condition
$H_{11} = H_{22}$ leads to the idea of a null point in the phase
of these equations. This null point is then labeled
as the crossing point, the point where the charge transfer
is most likely to take place. Let 1, 2 refer to reactants
and products respectively.

    The equations are well-known

$$i \hbar \dot{C}_1 = H_{11}C_1 \; + \; H_{12}C_2 \qquad (4)$$

$$i \hbar \dot{C}_2 = H_{21}C_1 \; + \; H_{22}C_2$$

By next writing $(i = 1, 2)$ (interaction picture)

$$C_i = a_i(t) \exp(-i \int^t H_{ii} \, dt'/ \hbar)$$

These equations become

$$i \hbar \, \dot{a}_1 = H_{12} \, a_2 \, e^{i\Phi(t)}$$

$$i \hbar \, \dot{a}_2 = H_{21} \, a_1 \, e^{-i\Phi(t)}$$

(5)

where $\qquad \Phi(t) = \int^t (H_{11} - H_{22}) \, dt'/\hbar$

Solutions to these equations are also discussed by Birkinshaw et al. in this volume. We only mention them here because it is important to see that the condition $H_{11} = H_{22}$ is the same as $d\Phi/dt = 0$, namely a point of stationary phase exists - or a kind of resonance takes place as the trajectory grows which allows the pre-set conditions at $t \longrightarrow -\infty$, $/a_1/ = 1$, $/a_2/ = 0$ to relax at the crossing pt. When $\Phi(t) \longrightarrow$ stationarity, a_2 can grow hence the time when $\Phi(t)$ is stationary is called the crossing time.

III - The Crossing Map

For our study, we treat the potential surface crossing as a curve-crossing in three dimensional space - that is, we imagine that - for sufficiently large separation of the ion and molecule, or atom and ionized molecule we can write at the crossing $H_{11} = H_{22}$, using equations (2) and (3)

$$aR^4 + bR^2 + cR + d = 0$$

(6)

where $a = \Delta E$, $b = eDP_1 (\cos \mu)$ $c = eQP_2 (\cos \mu)$, and

$$d = (\alpha_2 - \alpha_0 - \alpha_1 \, P_2 (\cos \mu)) \, e^2/2$$

Thus the condition $(H_{11} = H_{22})$ leads to a 4th order polynomial in $R(\mu)$ i.e. for each value of $\mu = \varphi_1 - \varphi_2$ there exists a real positive root R. Of course, there are in general four complex roots with positive and negative real parts. For positive ΔE there exists a real positive root for a range of values of μ. The locus of this root as a function of μ is shown in the crossing map in fig. 3.

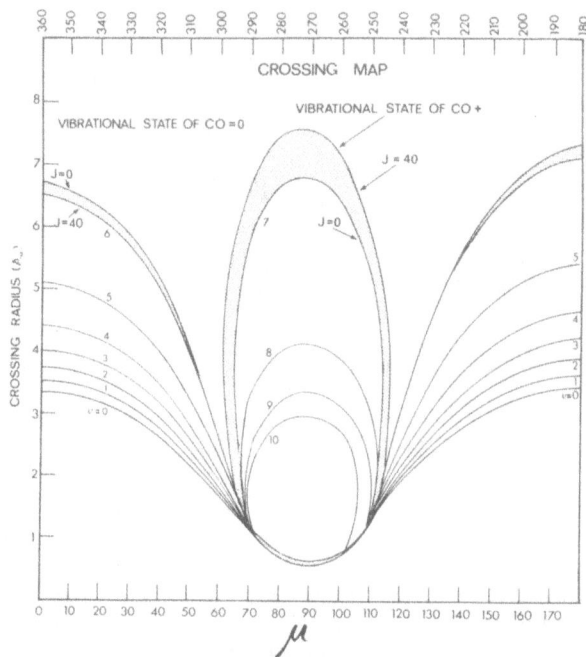

Fig. 3. The locus of charge transfer null points
$(H_{11} \rightleftharpoons H_{22})$ for 10 product vibrational states showing
rotational broadening for v=6 and v = 7.

Here we really show 10 crossing loci (crossing paths)
superimposed wherein each different product vibrational
quantum number gives rise to a new locus of crossing
points. We have chosen the reactant vibrational quantum
to be zero in each case. The effect of rotational history
is also shown for the product states v = 6,7. To interpret
this effect (i.e. of the term $\Delta E_{rotation}$) and to understand the use of the map let us imagine that a reactant
state collision is taking place. The position R and
dipole orientation angle $\mu = \Delta \varphi$ are given at each point
of the actual trajectory and can be placed on the crossing
map as a point. When this point crosses the desired
locus (crossing path) the null point mentioned in
section II is reached. But the crossing path is not at
rest, it moves depending on the rotational state of the
molecule, which in effect "broadens" the crossing path.
For the case v = 7, rotational quantum numbers in excess
of 40 allow crossings to take place at very large R for
$\mu = 90^{\circ}$ but for $\mu > 120^{\circ}$ or $\mu < 60^{\circ}$ no crossings occur
at all.

In fig. 4 we see some strips from our film showing indeed this development. View these sequences starting from the top left-hand corner down the page.

Here you see superimposed the crossing map and the collision picture in the upper central part of each frame.

The sequence of black dots show the R, μ, coordinates of the collision while the collision picture shows the actual trajectory. Thus we are able to show the distribution of the rotor as well as crossings at the same time.

IV - The Making of a "Movie"

The basic ingredients include four people, film, 16 mm in size, photo-sensitive to light phosphors from a cathode ray tube (C.R.T.), a cathode ray tube, an electron beam passing into the tube capable of being X-Y plotted intersectionally as it collides with the C.R.T's photon emitting phosphor surface. Except for the four people this is called a Calcomp 1670 micro-film plotter. You also need for a typical ten minute film, which at 25 frames per second programmed (for T.V. reproductability) is 15,000 frames, about 40 programs, each program sequence describing that movement that you wish to show. The idea is that you make a canonical program to give as output, information X-Y plotted. You then choose the information such that it moves the electron beam across the face of the cathode ray tube making traces in front of a 16 mm loaded camera with lens stop wide open so that all traces are film recorded.

Then you go to the next frame by computer programmed instruction for film advance and put on that frame whatever you wish. And so on.

The four people were Keith Birkinshaw, Ken Gray, David Clark and Fred Wolf. David Clark acted as a film editor, adviser, and sound track recorder, by courtesy of the University of London Audio Visual Centre. Keith Birkinshaw helped conceive of the film and also did much of the actual programming. Ken Gray performed as the main programmer. Fred Wolf directed, cooled fevered brows and kept sanity together for it seemed often if not daily that machines are breaking down for one reason or another. The last breakdown occurred

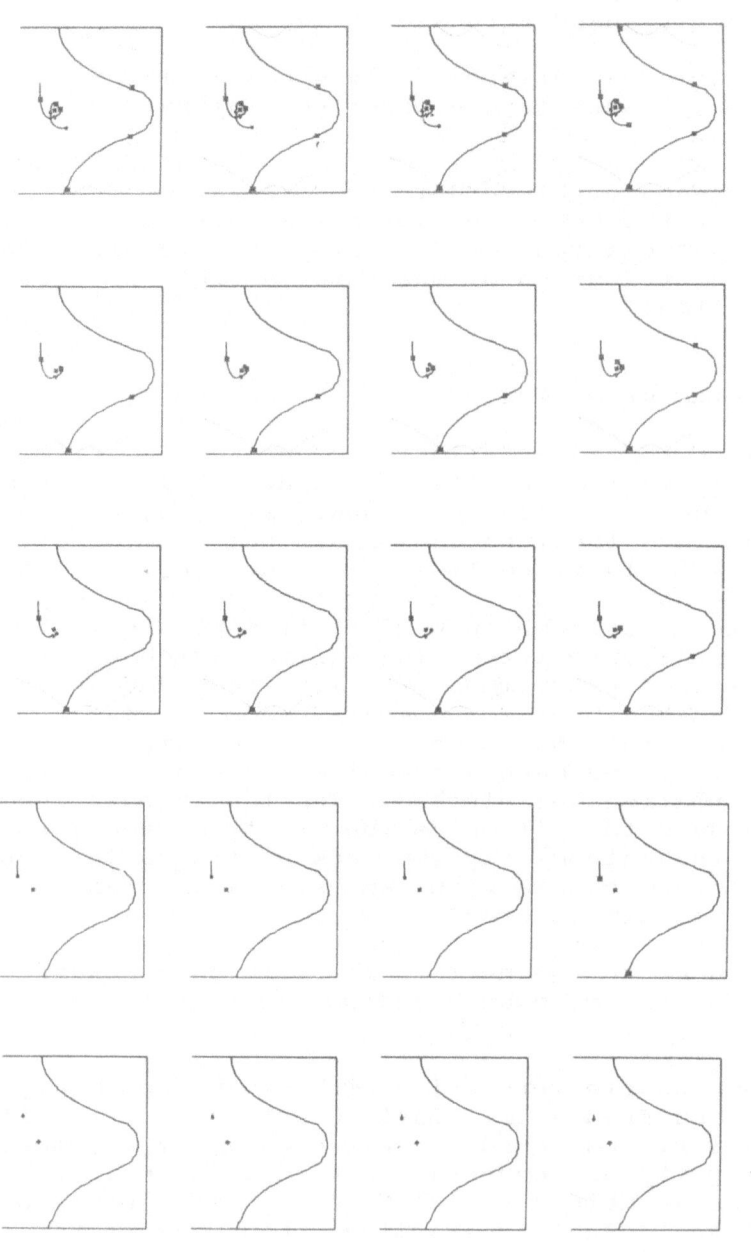

Fig. 4. Film strip sequence showing time development of crossing phenomenon, trajectory and rotational distribution. The ordinate and abscissa in each frame is R and μ respectively.

at sunset just as we were about to make the final film
sequences for editing and assembly. The machine's green
tuning eye blinks indicating normal operation. It only
stared. By performing a dance of wisdom around the
1670 plotter and inserting into its uncertain mouth a
delicious morsel called "confidence tape" (a brain
storm of Birkinshaw), the machine went from joyless
unblinking green lighted eye, non-willingless to perform,
into joyful acceptance of further programming, cheerily
blinking green lighted willingness. And a film was made.
There are others expected. Love.

V - Further Studies

One really needs to see the film several times in
order to comprehend all the ideas in it. One of the main
ideas shown is that the trajectory followed by the
system is capable of changing after passage through the
crossing point. That is, if a charge transfer occurs,
a sequence of multiple paths arising from a single set of
starting conditions is possible. In fig. 5, we see two
possible secondary trajectories as well as the primary
trajectory which is similar to that shown in fig. 2.
The primary shows the rather complicated hindered rotor
phenomenon shown in Dugan's films. An outer crossing
labeled 1, and an inner crossing 2, are also shown.

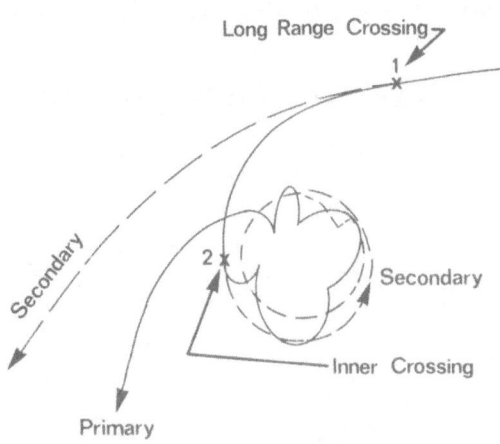

Fig. 5. Primary and secondary trajectories.

As a result of crossing 1 the secondary trajectory which develops is uncomplicated, but charge transfer at 2 leads to a complex formation with an enhanced lifetime and the complicated trajectory shown.

Because of this enhanced lifetimes of collision-formed complexes can take place. This process was unexpected when we began our calculations. What occurs is, the charge transfer takes place within the product centrifugal barrier. The three particle system remains bound until charge transfer takes place again.

Finally we wish to point out that we are continuing in our efforts to use the film media to gain better understanding of these complex phenomena. Quantum process resulting in the change of state of a system such as is imagined in charge transfer, happens without the passage of time (collapse of the wave-function). By using imaginary time (Miller and George, 1972) it is possible to"show"such phenomenon. We are presently looking into this in the hopes of using Feynman Path movies to describe ion-molecule reactions.

References

Bohme, D.K., Hasted, J.B. and Ong. P.P., (1968)
        J. Phys. B. 1, 879

Notably the many films of Jack Dugan. See for example
Dugan, J.V.,Jr., Canright, R.E.,Jr., and Palmer, R.W.
        (1969). Abstr. Sixth Int. Conf. on the Phys.
        Electr. and Atom. Colls. M.I.T. press, Cambridge,
        Mass. p 337

Miller, W.H. and George T.F. (1972) J. Chem. Phys.
        56, 5668

THEORY OF ION POLAR MOLECULE COLLISIONS

Michael T. Bowers and Timothy Su

Department of Chemistry
University of California
Santa Barbara, California, 93106

I.   INTRODUCTION

This paper is primarily concerned with low energy (less than a few eV) ion-polar molecule theories. A brief review of the Langevin theory for non-polar molecules is given for instructive reasons, however. Following this initial section a review of the various ion-dipole theories will be given, comparisons with experiments made, and conclusions drawn about the current utility of the various theories. Finally, a few remarks will be made about some very recent theoretical studies on quadrupolar and generally anisotropic molecules.

II.   PURE POLARIZATION THEORY

Theoretical expressions for the classical capture cross section, $\sigma_c(v)$, and capture rate constant k_c for a point charge colliding at relative velocity v with a point polarizable molecule were first derived by Gioumousis and Stevenson (1958). Subsequently the pure polarization theory has been reformulated and extensively discussed by a number of authors which include McDaniel (1964), Futrell and Tiernan (1968) and Henglein (1970). Hence the discussion given here will be brief and is primarily intended as a prelude to the more extensive description of the ion-polar molecule theories to be discussed next.

A schematic view of an ion of mass M^+ colliding at impact parameter b with a neutral with mass M is given in Fig. 1. For structureless point particles with no internal energy the relative energy of the system at infinite separation is given by Eq. 1

163

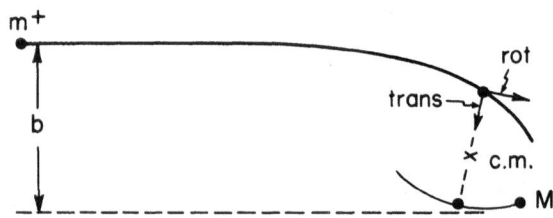

Fig. 1 :   Schematic ion-molecule collision

$$E_r = \frac{1}{2} \mu v^2 \tag{1}$$

where μ is the reduced mass of the system and v is the relative velocity.

As the particles approach, the classical potential in Eq. 2

$$V(r) = - \int_r^\infty F(r)dr = - \int_r^\infty \frac{2\alpha q^2}{r^5}\, dr = - \frac{\alpha q^2}{2r^4} \tag{2}$$

becomes increasingly important.   In Eq. 2, α is the polarizability of the neutral, q is the charge on the ion and r is the ion neutral separation.   Hence for $r < \infty$ the relative energy of the system is a sum of the instantaneous kinetic energy, $E_{kin}(r)$, and the potential energy

$$E_r = \frac{1}{2} \mu v^2 = E_{kin}(r) + V(r) \quad . \tag{3}$$

From Fig. 1 it is apparent that there are two components to the kinetic energy term,

$$E_{kin}(r) = E_{rot}(r) + E_{trans}(r) \tag{4}$$

where $E_{trans}(r)$ is the translational energy along the line of centers of the collision and $E_{rot}(r)$ is the energy of relative rotation of the particles.   The rotation energy of the system is given by

$$E_{rot}(r) = \frac{L^2}{2\mu r^2} = \frac{\mu^2 v^2 b^2}{2\mu r^2} = E_r \frac{b^2}{r^2} \tag{5}$$

where $L = \mu vb$ is the classical orbital angular momentum of the two particles. It is usual practice to couple $E_{rot}(r)$ with the system potential $V(r)$ to obtain an effective potential given in (6)

$$V_{eff}(r) = -\frac{q^2\alpha}{2r^4} + E_r\,\frac{b^2}{r^2} \qquad (6)$$

yielding

$$E_r = E_{trans}(r) + V_{eff}(r) \quad . \qquad (7)$$

A plot of $V_{eff}(r)$ <u>versus</u> r at constant E_r for several values of the impact parameter <u>is given</u> in Fig. 2. For b = 0 there is no

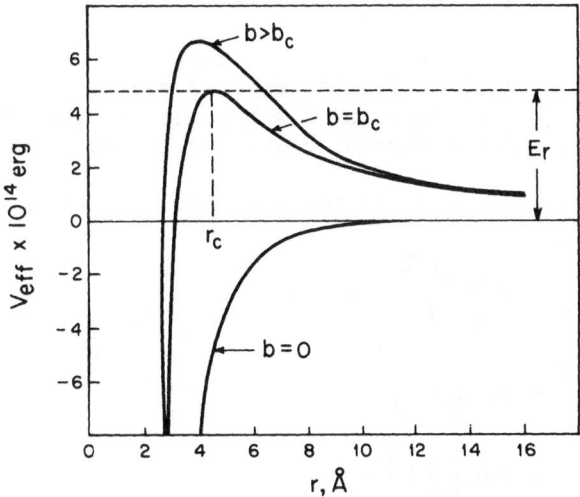

Fig. 2 : Plot of V_{eff} <u>vs</u> r from Eq.(6) for N_2^+ colliding with N_2

contribution from the repulsive rotational term (often called the centrifugal potential) and $V_{eff}(r)$ is attractive for all values of r. Hence, for point particles, the system passes through r = 0. When b > 0 the repulsive rotational term in $V_{eff}(r)$ creates a so-called "centrifugal barrier" to capture collisions. For a given initial relative velocity v there is always a maximum impact parameter, b_c, for which a capture collision can occur. (A

capture collision is defined as one in which the particles have
appropriate energy and impact parameters to pass through $r = 0$.
For real ions and molecules there would, of course, be some
minimum value of $r > 0$ due to coulomb repulsion and the Pauli
exclusion principle.)   This condition occurs when $E_{trans}(r) = 0$
at $r = r_c$.   Hence at $r = r_c$, $E_r = V_{eff}(r_c)$.   By setting
$\delta V_{eff}(r)/\delta r = 0$ it directly follows that

$$r_c = \frac{q}{b_c}\left(\frac{2\alpha}{E_r}\right)^{\frac{1}{2}} , \tag{8}$$

and

$$b_c = \left(\frac{2q^2\alpha}{E_r}\right)^{1/4} , \tag{9}$$

and

$$r_c = \frac{b_c}{\sqrt{2}} . \tag{10}$$

The capture limit impact parameter case is pictured as the middle
curve in Fig. 2.   For values of $b > b_c$, the centrifugal barrier
prevents capture and the particles are merely scattered at large r.

Finally, the capture cross section and rate constant are given
by

$$\sigma_c(v) = \pi b_c^2 = \pi q\left(\frac{2\alpha}{E_r}\right)^{\frac{1}{2}} , \tag{11}$$

$$k_c = v\sigma_c(v) = \pi q v \left(\frac{2\alpha}{E_r}\right)^{\frac{1}{2}}$$

$$= 2\pi q \left(\frac{\alpha}{\mu}\right)^{\frac{1}{2}} . \tag{12}$$

The predictions, then, of the pure polarization theory are that the
capture cross section is inversely proportional to v and that k_c is
independent of v.   The attempts to verify this velocity dependence
have met with only mixed success.   Friedman and Reuben (1971) have
pointed out that many of the reported conflicts between experiment
and theory arose due to the failure to recognize the importance of
accounting for all reaction channels of a given ion-neutral pair
as the energy was varied.   This accounting is particularly important when the absolute magnitude of an experimental rate constant is to be compared to theory.   Hamill and coworkers also
found it necessary to account for a hard-core gas kinetic cross-
section that gave rise to an inverse v^2 dependence of the cross

section at energies above a few eV (Boelnijk and Hamill, 1962; Pottie et al., 1962).   Finally, it was determined early on (Moran and Hamill, 1963) that contributions of the ion-dipole potential to the cross section could not be ignored for target molecules with significant permanent dipole moments.   The remainder of this chapter will be devoted to the recent theoretical developments on prediction of low energy ion-polar molecule cross sections and rate constants.

III.   ION-DIPOLE THEORY

     In this section the current state of the theory of ion-dipole collisions will be reviewed.   Emphasis will be on the "macroscopic" average dipole orientation (ADO) theory that predicts low energy rate constants and cross sections and their dependence on energy and temperature.   A brief summary of the elegant microscopic theory of Dugan and Magee (1967) will also be given.

A.   The Locked Dipole Approximation

     The classical potential for the ion point dipole interaction is given in (13)

$$V_D(r) = - \frac{\alpha q^2}{2r^4} - \frac{\mu_D q}{r^2} \cos\theta \qquad\qquad (13)$$

where μ_D is the permanent molecular dipole moment and θ is the angle between the line of centers of collision and the dipole axis. There are two limiting cases of (13):

(a) When $\theta = 90°$, $V_D(r)$ reduces to $V(r)$ given in Eq.(2) and the dipole has no effect on the ion motion.   Since most experimental data suggest the molecular dipole strongly affects the cross section and rate constant this approximation will give only a lower limit (eqs.(11) and (12));

(b) When $\theta = 0°$, the potential interaction is maximized and Eq.(14) results

$$V_{LD}(r) = - \frac{\alpha q^2}{2r^4} - \frac{\mu_D q}{r^2} \quad . \qquad\qquad (14)$$

Moran and Hamill (1963) and subsequently Harrison and coworkers (Gupta et al., 1967) have derived Eq.(15) for the locked dipole capture limit rate constant

$$k_{LD}(v) = \frac{2\pi q}{\mu^{\frac{1}{2}}} [\alpha^{\frac{1}{2}} + \frac{\mu_D}{v}] \qquad\qquad (15)$$

which for thermal ion distributions yields (16)

$$k_{LD}(\text{therm}) = \frac{2\pi q}{\mu^{\frac{1}{2}}} [\alpha^{\frac{1}{2}} + \mu_D \left(\frac{2}{\pi KT}\right)^{\frac{1}{2}}] \tag{16}$$

where K is Boltzmann's constant and T is the absolute temperature. Eqs.(15) and (16) provide useful upper limits to the ion-dipole capture rate constants. However, they seriously overestimate the magnitude of experimentally observed rate constants, as will be demonstrated shortly, because in reality the ion field does not completely quench the rotational angular momentum of the polar molecule and only partial "locking" can occur.

Bowers and Laudenslager (1972) made the first attempt to estimate quantitatively by experiment the degree of locking of the dipole.   They observed charge transfer reactions from rare gas ions to the set of geometric isomers of difluoroethylene, trans; 1,1; and cis which have dipole moments of 0.0, 1.38, and 2.42 Debye respectively.   Since all of these isomers have the same angle averaged polarizabilities and ionization potentials and similar photoelectron spectra (Bowers and Elleman, 1972; Bowers and Su, 1973) an incremental increase in rate constant with increasing polarity of the molecule presumably reflects the degree of locking of the dipole.   Bowers and Laudenslager assumed the rate constant can be approximated by the semiempirical form

$$k_{PLD} = \frac{2\pi q}{\mu^{\frac{1}{2}}} [\alpha^{\frac{1}{2}} + C\mu_D \left(\frac{2}{\pi KT}\right)^{\frac{1}{2}}] \tag{17}$$

where C is the dipole locking constant with values between 0 (no locking) and 1 (complete locking).   The constant C can be determined from the experimental data using Eq.(18)

$$C = \Delta k \left(\frac{KT\mu}{8\pi q^2 \mu_D^2}\right)^{\frac{1}{2}} \tag{18}$$

where Δk is the experimental incremental increase in rate constant for the polar isomer relative to the non-polar isomer.   The results are summarized in Table I.   The average values of the locking constants are $\overline{C}_{1,1} = 0.093 \pm 0.05$ and $\overline{C}_{cis} = 0.111 \pm 0.04$ which are values close to the zero dipole limit.   If these locking constants can be loosely thought to be related to a phenomenological average orientation angle of the dipole axis w.r.t. the line of centers of the collision (C = cos⟨θ⟩) then ⟨θ⟩$_{1,1}$ = 84°6' and ⟨θ⟩$_{cis}$ = 83°6'.   In this context it can be seen that the more polar cis isomer is oriented to a slightly greater extent than the 1,1 isomer although both isomers are only weakly oriented.   The

TABLE I: Locking Constants Derived from Charge Transfer Reactions
to Geometric Isomers of Difluoroethylene[a]

ion	1,1-$C_2F_2H_2$ Δk[b]	C	cis-$C_2H_2F_2$ Δk[b]	C
He^+	0.90	0.138	1.80	0.158
Ne^+	0.32	0.099	0.57	0.100
Ar^+	0.25	0.098	0.48	0.108
Kr^+	0.08	0.038	0.29	0.079

[a] Bowers and Laudenslager (1972).

[b] Δk in units of 10^{-9} cm^3/s. Δk is the experimental rate
constant of the polar isomer (1,1 or cis) minus the value
for the non-polar isomer (trans).

degree of locking of the dipole will be discussed in much greater
detail in the following sections, particularly the section on the
Average Dipole Orientation (ADO) Theory of Su and Bowers (1973a,b).

B.   Ion-Dipole Trajectory Calculations

In a classic paper on ion-polar molecule collisions, Dugan
and Magee (1967) have developed the formalism necessary for a
classical and semi-classical microscopic description of ion-polar
molecule collisions.   Their method used the Lagrangian formulation
of classical mechanics and entailed solving Eq.(19) for the time
dependence of the spatial variables

$$\frac{d}{dt} \frac{\partial L}{\partial \dot{q}} - \frac{\partial L}{\partial q} = 0 \tag{19}$$

where L is the Lagrangian, q a generalized coordinate and $\dot{q}$ the
time derivative of q.   The appropriate coordinate system for a
"diatomic" polar molecule and a positive ion is given in Fig.3,
and the Lagrangian in Eq.(20) for a fixed molecule and a moving
ion

$$L = \tfrac{1}{2}\mu(\dot{X}^2 + \dot{Y}^2 + \dot{Z}^2) + \tfrac{1}{2}I(\dot{x}_1^2 + \dot{x}_2^2 + \dot{x}_3^2) + \frac{\mu_D q}{r^2}\cos\theta + \frac{\alpha q^2}{2r^4} \tag{20}$$

where I is the moment of inertia of the polar molecule.   For a
diatomic rotor, the rotation about the axis can be ignored.
Hence x_3 can be expressed in terms of x_1 and x_2.   The Lagrangian

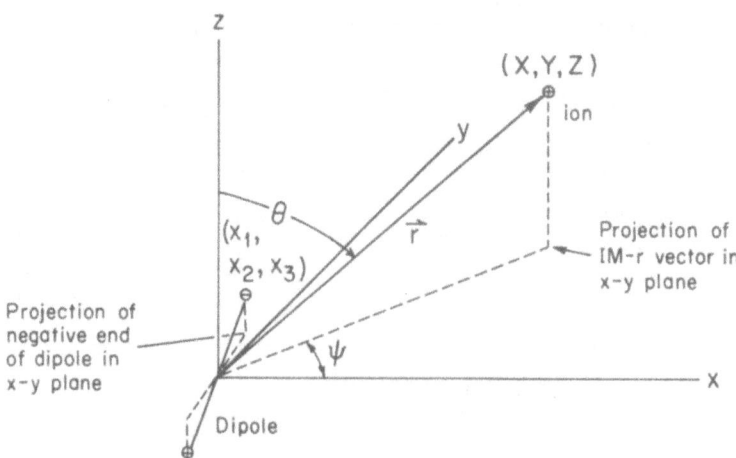

Fig. 3: Coordinate system for an ion-diatomic rotor collision

equations of motion are then solved numerically to obtain X(t),
Y(t), Z(t), x_1(t) and x_2(t).   The results are used in two ways:
(a) to trace out microscopic trajectories (time-history plots)
given a particular set of initial collision parameters, and (b)
statistically to generate the percentage of "capture" type
collisions by counting the percentage of collisions in which the
partners approach to a predetermined separation (2-3Å).

The time history plots allow a number of interesting phenomena
to be investigated.   For example, it has been demonstrated that
the time dependence of the centrifugal barrier in ion-polar
molecule collisions can sometimes lead to long lived ($\tau > 10^{-11}$
sec) complexes in the absence of chemical forces (Dugan et al.,
1969).   Further, the effect of vibrational energy on the detailed
time history plots has been made (Dugan and Canright, 1971) with
the tentative conclusion that vibrational energy can have a
dramatic effect on the capture ratio.   A third factor the trajectory calculations elucidate is the oscillatory nature of the
angular momentum transfer between the rotor and the system as the
particles approach.   A typical case is shown in Fig. 4.   It is
apparent that there is little net momentum transfer until small
collision radii are realized.   This fact will greatly simplify
the analysis in the average dipole orientation theory presented
next.

The statistical calculations compile the percentage of
collisions that reach a preset radii for a given velocity and for
a random selection of the initial orientation parameters.   The

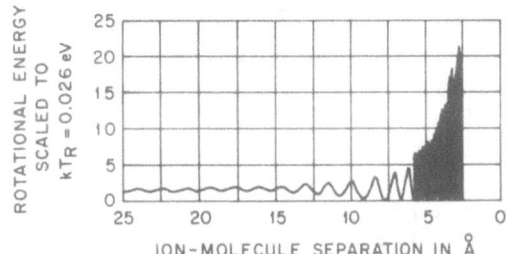

Fig. 4:   Variation in rotational energy of HCl as a function of
          ion-molecule separation for the collision of NO_2^+ on HCl
          (Dugan et al., 1969).

impact parameter is generally varied systematically and a plot of
the "capture ratio" versus impact parameter is made.   The
"capture" cross section is obtained from Eq.(21) by numerical
integration

$$\sigma_c(v) = \pi \int_0^\infty C_R(b)d(b^2) \tag{21}$$

of the plots and the corresponding rate constant from $k_c(v) =$
$v\sigma_c(v)$.   Velocity dependence is obtained by varying v, recal-
culating $C_R(b)$ versus b, and integrating.   Some comparisons with
experiment will be given in the following section on the ADO
theory.

     Dugan has estimated from a series of trajectory calculations
(Dugan and Canright, 1972) that the numerical capture cross
sections depend on relative translational energy, $\mathcal{E}$, approximately
as $\mathcal{E}^{-0.60}$ to $\mathcal{E}^{-0.75}$ at 300-500°K.   If Eq.(22) is assumed for the
cross section

$$\sigma_c(v) = (\sigma_o)_c \mathcal{E}^{-n} \tag{22}$$

then the Maxwell-Boltzmann averaged thermal rate constant is given
in (23) (Dugan, 1973)

$$k_c(\text{scaled}) = \frac{1.956 \times 10^{-4}}{T^{3/2} \mu^{\frac{1}{2}}}(KT)^{2-n}(\sigma_o)_c \Gamma(2-n) \tag{23}$$

where T is in °K, μ in a.m.u., KT in eV, $(\sigma_o)_c$ in $Å^2(eV)^n$.
Values of thermal energy rate constants predicted using Eq.(23)
will be compared with ADO predictions and experiment in a
following section.

C.   The Frozen Rotor Approximation

Dugan and Magee (1966) have determined the capture cross section for an ion colliding with a polar rotor molecule fixed at angle θ w.r.t. the line of centers of the collision with the result given in (24)

$$\sigma_\theta = \pi \ \frac{\mu_D q}{\varepsilon} \ \cos\theta + \left(\frac{2\alpha q^2}{\varepsilon}\right)^{\frac{1}{2}} \ . \tag{24}$$

This result can be spacially averaged over all orientations under the constraint that σ_θ is always positive.   The result is given in (25)

$$\overline{\sigma}_\theta = \pi \ \frac{\mu_D q}{4\varepsilon}(1 - \frac{\varepsilon}{\varepsilon_c}) + \left(\frac{\alpha q^2}{2\varepsilon}\right)^{\frac{1}{2}} + \frac{\alpha q}{\mu_D} \tag{25}$$

where $\varepsilon_c = \mu_D^2/2\alpha$. Dugan (1973) has obtained an expression for the thermally averaged rate constant from (25) given in (26)

$$\overline{k}_\theta = \left(\frac{8\pi}{\mu}\right)^{\frac{1}{2}}(KT)^{3/2} \ \frac{\mu_D q}{4\varepsilon}(KT) + \frac{\alpha q}{\mu} - \frac{\mu q}{4\varepsilon_c} \ (KT)^2$$

$$+ \frac{3}{2}\left(\frac{\alpha q^2}{2\varepsilon}\right)^{\frac{1}{2}}(KT)^{3/2} \ . \tag{26}$$

Comparison of the predictions of this frozen rotor approximation with predictions of the ADO theory and scaled numerical results of Dugan will be made in a following section.

D.   The Average Dipole Orientation Theory

The details of the average dipole orientation (ADO) theory of Su and Bowers (1973a, 1973b) have been published and will only be summarized here.   The basic intent of the theory is to provide a simple method of accurately estimating ion-polar molecule capture rate constants and their dependence on temperature and kinetic energy.   The theory is not intended to represent microscopic phenomena in any fashion.   It is a purely classical statistical theory relevant only for comparison with average, phenomenological rate data.

The primary assumptions of the theory are three:

1.   The average angle between the dipole axis and the line of centers of collision, θ(r), can be calculated as a function of r using statistical mechanics and the classical ion-polar rotor interaction potential.   This averaged angle can then

replace the microscopic angle in the expression for the effective potential;

2. There is no net angular momentum transfer between the rotating polar molecule and the system as a whole. Dugan and co-workers (see, for example, Dugan and Magee, 1967) have shown on the microscopic level there is an oscillatory momentum transfer between the rotor and the system. This occurs because the ion field induces changes in the angular velocity θ of the dipole as either the negative or positive end of the dipole approaches the ion. As the negative end of the dipole approaches a positive ion there is an acceleration of the dipole resulting in a net angular momentum transfer from the system to the rotor. As θ passes through zero, the negative end begins to rotate away from the positive charge resulting in a decrease in angular velocity of the rotor and a transfer of angular momentum from the rotor to the system. As the trajectory calculations confirm, this oscillating momentum transfer almost exactly cancels for collision separations equal to or greater than the capture radii for ion-polar molecule collisions;

3. The polar molecule can be approximated by a two dimensional rotor in the plane of collisions. This assumption is equivalent to assuming rotations about the dipole axis are unimportant. This assumption is rigorously true for diatomic molecules introducing an error of less than 5% in polyatomic molecules.

The method of approach is similar to the Langevin treatment discussed earlier and will only be outlined here. The effective potential is given in Eq. (27)

$$V_{eff}(r)_{ADO} = \frac{L^2}{2\mu r^2} - \frac{\alpha q^2}{2r^4} - \frac{q\mu_D}{r^2} \cos\overline{\theta}(r) \tag{27}$$

By differentiating $V_{eff}(r)_{ADO}$ w.r.t. r and setting it equal to zero it is straightforward to derive Eqs. (28) and (29) (Su and Bowers, 1973)

$$\langle \sigma(v) \rangle = \pi r_k^2 + \frac{\pi q^2 \alpha}{r_k^2 \mu v^2} + \frac{2\pi q \mu_D}{\mu v^2} \cos\overline{\theta}(r=r_k) \tag{28}$$

and

$$v^2 = \frac{q^2 \alpha}{\mu r_k^4} + \frac{q\mu_D}{r_k \mu} \sin\overline{\theta}(r=r_k) \left(\frac{\delta\overline{\theta}}{\delta r}\right)_{r_k} \tag{29}$$

where r_k is the value of r at the top of the centrifugal barrier. By specifying r_k, $\langle \sigma(v) \rangle$ can be plotted as a function of v. A typical plot is given in Fig. 5. For the important thermal case

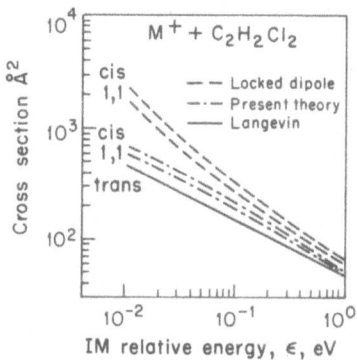

Fig. 5: Plot of $\langle \sigma(v) \rangle$ versus relative kinetic energy from Eqs. (28) and (29) for collisions of rare gas ions with dichloroethylene isomers.

the rate constant is given by Eq.(30)

$$k_{ADO}(\text{therm}) = \int_0^\infty v \langle \sigma(v) \rangle P(v) dv \tag{30}$$

where $P(v)$ is the normalized Maxwell-Boltzmann velocity distribution function.

The function $\bar{\theta}(r)$ is calculated using Eq.(31)

$$\bar{\theta}(r) = \frac{\int \theta P(\theta) d\theta}{\int P(\theta) d\theta} \tag{31}$$

where $P(\theta)$ is the distribution function of angle θ of thermal energy polar rotors in the ion field. The θ dependence is given in (32)

$$P(\theta) \propto \frac{\sin\theta}{\dot{\theta}} \tag{32}$$

where $\sin\theta$ is a spacial degeneracy factor and $\dot{\theta}$ is the θ dependent angular velocity of the ion. Using the ion-polar molecule interaction potential and the thermal energy rotational distribution function it is possible to solve (31).   Similar techniques are used for the derivative $\left(\dfrac{\partial\bar{\theta}(r)}{\partial r}\right)_{r_k}$.   Details are given elsewhere (Su and Bowers, 1973a,b).

A plot of $\bar{\theta}(r)$ versus r is given in Fig. 6.   At large r, $\bar{\theta}(r)$ is slightly larger than $\pi/2$ and approaches $\pi/2$ at infinity.

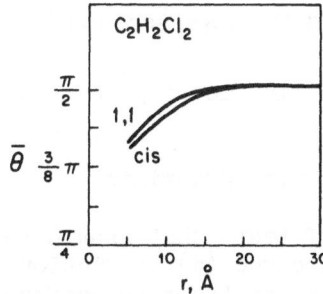

Fig. 6:   Plot of $\bar{\theta}(r)$ vs r for collisions of rare gas ions with dichloroethylene isomers.

The reason for the slightly repulsive orientation of the dipole at large r is the dominance of the θ term in $P(\theta)$.   In the repulsive phase the dipole is moving slowly (high probability) and in the attractive phase rapidly (low probability).   At shorter r this effect is swamped by the attractive coulomb interactions.

E.   Parameterization of the ADO Theory

Eq.(17) can be used to parameterize the ADO theory to permit a simple semiempirical calculation of ADO capture rate constants (Su and Bowers, 1973b).   It turns out that at constant tempera-ture the dipole locking constant C is a function of $\mu_D/\alpha^{\frac{1}{2}}$ only. Hence when k_{PLD} is calculated from the ADO theory a plot of C versus $\mu_D/\alpha^{\frac{1}{2}}$ can be made.   Such a plot is given in Fig. 7 for $T = 300^\circ K$.   Plots for values of T from $100^\circ K$ to $500^\circ K$ are being published (Su and Bowers, 1974a).   To calculate a capture rate constant, determine $\mu_D/\alpha^{\frac{1}{2}}$ for the polar molecule of interest, determine C from the published graph and then calculate $k_{PLD} \equiv k_{ADO}$(therm) from Eq.(17).

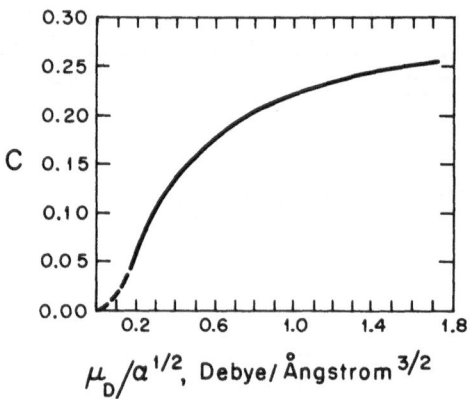

Fig. 7: A plot of the dipole "locking" constant, C, plotted
         against $\mu_D/\alpha^{\frac{1}{2}}$.

F.   Comparison of Theory with Experiment

     There are a number of tests that should be made of the
various theories. These include (a) prediction of the incremental
effect of the dipole moment on the rate constant; (b) prediction
of the absolute rate constant at thermal energies; (c) prediction
of the ion kinetic energy dependence of the rate constant; and
(d) prediction of the temperature dependence of the rate constant.
Reasonably extensive tests of the theories are available for (a)
and (b) and less extensive tests of (c).   The temperature
dependence of a suitable reaction over a suitable temperature
range is yet to be done.

     The incremental effect of the dipole on the rate constant can
be approximated experimentally by the technique described in the
locked dipole section.   The ratios of $\Delta k_{exp}/\Delta k_{ADO}$ for a number of
charge transfer reactions to the isomers of $C_2Cl_2H_2$ and $C_2F_2H_4$ are
given in Table II where Δk is the difference between the polar and
non-polar rate constants.   The numbers are clustered about 1.0
with a deviation of ca ± 30%.

     In Table III a comparison of $\Delta k_{exp}/k_{theory}$ is given for the
ADO theory, the Frozen Rotor approximation of Dugan and Magee
(Eq.26), and the scaled numerical results of Dugan (Eq.24).   All
the theoretical data were taken from the compilation of Dugan (1973).
The ADO theory on the average gives results that are incrementally
ca 10-40% larger than experiment for the geometric isomer substrates with a scatter of 50-60% positive and 30% negative.   The

TABLE II: Comparison of Experimental and Theoretical Incremental Effects of the Dipole Moment on Charge Transfer Rate Constants.

$$\Delta k_{exp}/\Delta k_{ADO}$$

	$M^+ + C_2Cl_2H_2$		$M^+ + C_6F_2H_4$	
	1,1	cis	Meta	Ortho
He^+	1.41	1.08	1.08	0.95
Ne^+	1.26	1.03	0.73	0.85
Ar^+	1.20	1.00	1.35	0.93
Kr^+	1.30	1.00	0.97	0.79
Dipole Moments	1.34	1.90	1.58	2.40

TABLE III: Average Values of $\Delta k_{exp}/\Delta k_{theory}$ for Various Theoretical Models.

No. of Expts.	ADO	Frozen Rotor[c]	Dugan scaled[c]
40^a	0.69	0.41	($\pm$ 5.0)
63^b	0.92	-	-

[a] Charge transfer and proton transfer to the geometric isomers of $C_6F_2H_4$, $C_2F_2H_2$, $C_2Cl_2H_2$. Experimental data from Su and Bowers (1973c) and Bowers and Laudenslager (1972).

[b] All UCSB experiments. References in (a) above plus Su and Bowers (1973d; 1974d).

[c] Dugan (1973).

Frozen Rotor approximation predicts incremental increases in the rate constant that are ca 150% larger than experimental while the scaled numerical results are very erratic giving no predictable pattern with large deviations from experiment observed. It is not surprising the scaled numerical method gives erratic results because it assumes the same scaling law for all systems ($\varepsilon^{-0.7}$ dependence). It is somewhat surprising, however, that the Frozen Rotor method gives incremental increases in rate constant much larger than experimentally observed since this method only spacially averages and seemingly does not benefit from the partial locking of the dipole. It does serve to demonstrate that "first glance" intuition can sometimes be grossly in error, however. Finally, it should be noted, although the data are not in Table III, the locked dipole approximation overestimates the incremental effect of the dipole on the rate constant by more than an order of magnitude.

Perhaps a better perspective of the utility of a collision theory can be had by comparison with total rate constants not just incremental dipole effects. For this purpose exothermic proton transfer reactions in cases where that is the only reaction channel appear to be best suited. More so than any other class of reaction, proton transfer reactions appear to occur on virtually

TABLE IV: Comparison of Theoretical Predictions with Experiment for Absolute Proton Transfer Rate Constants.[a]

Theoretical Average Percent Deviations[b]

No. of Expts.	ADO[c] (absolute)	ADO[d]	Langevin	Locked Dipole
47	9	-1.2	-33	+145

[a] All thermal energy positive ion proton transfer reactions to polar substrates that have been done at UCSB or communicated to us by D. K. Bohme at the University of York, Ontario, Canada.

[b] The percent deviation is obtained from $100(k_{theory}-k_{exp})/k_{exp}$.

[c] Average total deviation regardless of sign.

[d] Average deviation taking the sign of the deviation into account.

every capture collision.   A summary of the results is given in
Table IV.   Predictably, the Langevin theory gives results about
33% low while the Locked Dipole Theory is ca 150% high.   The ADO
theory, on the other hand, predicts capture rate constants that
are accurate within ± 10-15% for most cases, or about as well as
experiments measure them.   The ADO theory thus appears to be a
useful collision theory suitable for diagnostic purposes in ion-
polar molecule systems.   Additional comparisons of experiment to
theory are given by D. K. Bohme elsewhere in this volume.

A comparison between the ADO theory and experiment for the
kinetic energy dependence of two simple ion-molecule reactions
has been made.   Reaction (33) has been reported by Bowers,

$$CH_3OH^{+\cdot} + CH_3OH \rightarrow CH_3OH_2^{+\cdot} + CH_3O \tag{33}$$

Anicich and Su (1973) and reaction (34) by Bowers and Su (1973)

$$NH_3^{+\cdot} + NH_3 \rightarrow NH_4^{+} + NH_2\cdot \tag{34}$$

on data taken by Huntress et al. (1970).   In both cases the
agreement was excellent.   Dugan (1973) has recently compared the
ADO kinetic energy dependence of reactions (33) and (34) with
predictions from the trajectory calculations of Dugan and Magee
(1967), with the Frozen Rotor approximation of Dugan and Magee
(1966) and with the scaled numerical results of Dugan (1973).
The comparisons are given in Figs. 8 and 9.   The Frozen Rotor

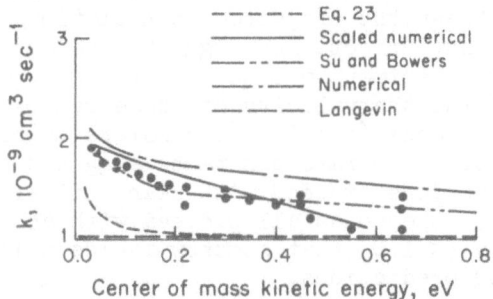

Fig. 8:   The rate constant versus relative kinetic energy for the
          reaction CH_3OH^{+} + $\overline{CH_3OH}$ → $CH_3OH_2^{+}$ + CH_3O. The ex-
          perimental data are from Bowers et al. (1973).   The
          numerical (— ·· —) prediction is from Dugan and Magee
          (1967) type trajectory calculations as reported by Dugan
          (1973).   The Scaled Numerical (———) and Frozen Rotor
          (------) predictions are from Dugan (1973).

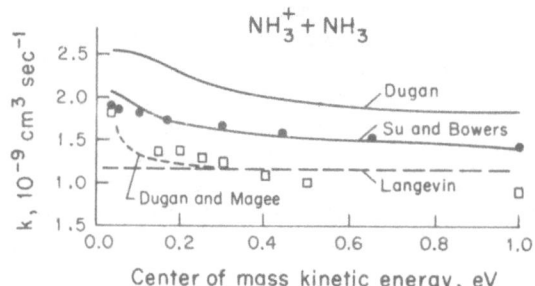

Fig. 9:   The rate constant versus relative kinetic energy for the
          reaction $NH_3^+ + NH_3 \rightarrow NH_4^+ + NH_2$.   The results of Dugan
          (————) are from numerical trajectory calculations of
          the Dugan and Magee (1967) type.   The results labeled
          Dugan and Magee (------) are the Frozen Rotor results of
          Eq.(26) as reported by Dugan (1973).   The data are from
          Huntress et al. (1970).

Model clearly has the wrong energy dependence, the scaled numerical
has about the right magnitude but the wrong shape, the "exact"
trajectory calculation results are about 30% high but have the
right shape, and the ADO results have both the right shape and
magnitude over the entire energy range.   Dugan (1973) ascribes
the discrepancy between the trajectory calculations and experiment
to reaction efficiency effects and suggests the agreement between
the ADO theory and experiment may be fortuitous.   While more data
is needed to determine the validity of these remarks, the fact that
the ADO theory reproduces (± 15%) the absolute magnitude of most
exothermic proton transfer rate constants suggests that perhaps
proton transfer reactions proceed with near unit efficiency.   If
this is not the case they must all proceed with about the same
reaction efficiency (< 1.0) that somehow accidentally nearly
correlates with ADO predictions.

      There have, as yet, been no temperature dependent studies of
ion-polar molecule reactions suitable for testing the various
collision theories.   The predictions of the ADO theory for the
reactions of CH_5^+ with CH_3CN and NH_3 are given in Fig. 10.   The
ADO theory predicts a substantial temperature dependence for
reactions such as these and hopefully data will soon be available
for testing the theory.

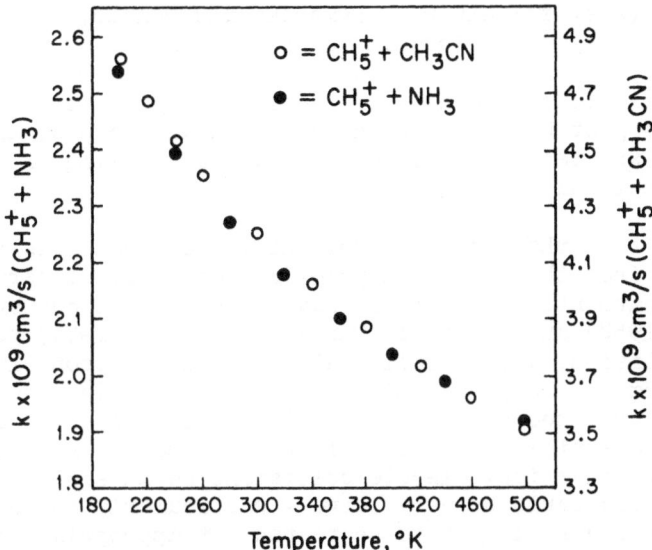

Fig. 10:   Predicted temperature dependence from the ADO theory for
the reactions of CH_5^+ with NH_3 and CH_3CN.

IV.   OTHER THEORIES

While the pure polarization and dipole terms are the two most
important terms in the ion-molecule potential function, a number
of other terms can substantially contribute to the capture rate
constant.   Su and Bowers (1974b) have developed a theory for the
ion-quadrupole interaction using a framework similar to the ADO
theory.   This average quadrupole orientation (AQO) theory indicates the quadrupole interaction can lead to contributions to
the rate constant between 0 and 20% of Langevin and gives good
agreement with experiment.   Dugan and Palmer (1972) had earlier
developed a "locked" Quadrupole Theory that indicated the quadrupole effect might be as large as Langevin.   The AQO calculations
as well as most experimental data indicate quadrupole locking does
not occur, however.

Another effect that has received attention is the anisotropy
in the polarizability.   Since most molecules are anisotropic,
the rotationally averaged polarizability usually used in Langevin
theory calculations is always smaller than the maximum component.

Some workers (see, for example, Dymerski and Dunbar, 1972) have suggested the ion might "lock in" on the maximum component of the polarizability thus explaining rate constants larger than Langevin predictions. A capture theory for anisotropic molecules has just been completed (Bass, Su and Bowers) and the results are expected to indicate only partial locking of the anisotropic molecule will occur.

Another term that may be important in some cases is the r^{-6} term arising from the induced dipole-induced dipole interaction. Empirical treatments of this term have been carried out by Mason and Shamps (1958) and Dymerski, Dunbar and Dugan (1974) for ion-molecule collisions and by Bell, Dalgarno and Kingston (1968) for neutral systems. Very recently Su and Bowers (1974c) have completed a capture theory of ion-molecule collisions using both the r^{-4} and r^{-6} potentials. It is felt to be a possibly important theory for negative ion-molecule collisions because negative ion polarizabilities could be very large - particularly for excited states. The theory is yet to be tested but appropriate model systems are being constructed.

Finally, Gentry (1974) has recently published an ion-dipole scattering theory suitable at high ion energies and low scattering angles. The theory compares qualitatively with experiment (Udseth, Giese and Gentry, 1974) for collisions of protons with CO, HCl and HF and been shown to have some diagnostic importance. It remains to be shown that the theory has general utility, however.

V.   ACKNOWLEDGEMENT

The support of the National Science Foundation under Grant GP-15628 for this work is gratefully acknowledged. Very useful conversations with Dr. J. V. Dugan during the development of much of this work are also acknowledged.

REFERENCES

Bass, L., Su, T., and Bowers, M. T., (1974) J. Chem. Phys. (to be. submitted for publication).
Bell, K. L., Dalgarno, A., and Kingston, A. E. (1968) J. Phys. B., 1, 18.
Boelrijk, N. and Hamill, W. H. (1962) J. Amer. Chem. Soc., 84, 529.
Bowers, M. T. and Elleman, D. D. (1972) Chem. Phys. Lett., 16, 482.
Bowers, M. T. and Laudenslager, J. B. (1972) J. Chem. Phys., 56, 4711.
Bowers, M. T. and Su, T. (1973) Advan. Electron. Electron Phys., 34, 223.

Bowers, M. T., Su, T., and Anicich, V. G. (1973) J. Chem. Phys.,
    58, 5175.
Dugan, J. V. and Magee, J. L. (1966), NASA TN D-3229
Dugan, J. V., Rice, J. H., and Magee, J. L. (1969) Chem. Phys.
    Lett., 3, 323.
Dugan, J. V., Canright, R. B., Palmer, R. W., and Magee, J. L.
    (1969) NASA TM X-52662.
Dugan, J. V. and Canright, R. B.(1971) Chem. Phys. Lett., 8, 253.
Dugan, J. V. and Canright, R. B. (1972) J. Chem. Phys., 56, 3623.
Dugan, J. V. and Palmer, R. W. (1972) Chem. Phys. Lett., 13, 144.
Dugan, J. V. (1973) Chem. Phys. Lett., 21, 476.
Dymerski, P. P. and Dunbar, R. C. (1972) J. Chem. Phys., 57, 4049.
Dymerski, P. P., Dunbar, R. C., and Dugan, J. V. (1974) J. Chem.
    Phys., 61, 298.
Friedman, L. and Reuben, B. G. (1971) Advan. Chem. Phys., 19, 33.
Futrell, J. H. and Tiernan, T. O. (1968). Ion Molecule Reactions,
    "Fundamental Processes in Radiation Chemistry " (Ed., P.
    Ausloos) Interscience Publishers, New York.
Gentry, W. R. (1974) J. Chem. Phys., 60, 2547.
Gioumousis, G. and Stevenson, D. P. (1959) J. Chem. Phys., 29, 294.
Gupta, S. K., Jones, E. G., Harrison, A. G., and Myher, J. J.
    (1967) Can. J. Chem., 45, 3107.
Henglien, A. (1970) Kinematics of Ion-Molecule Reactions.
    "Molecular Beams and Reaction Kinetics" (Ed., Ch. Schlier)
    Academic Press, New York.
Huntress, W. T., Mosesman, M. H., and Elleman, D. D. (1970),
    J. Chem. Phys., 52, 6009.
Mason, E. W. and Schamp, H. W. (1958) Ann. Phys. (N.Y.), 4, 233.
McDaniel, E. W. (1964), Collision Phenomena in Ionized Gases,
    Wiley, New York.
Moran, T. F. and Hamill, W. H. (1963) J. Chem. Phys., 39, 1413.
Pottie, R. F., Lorquet, A. J., and Hamill, W. H. (1962) J. Amer.
    Chem. Soc., 84, 529.
Su, T. and Bowers, M. T. (1973a) J. Chem. Phys., 58, 3027.
Su, T. and Bowers, M. T. (1973b) Int. J. Mass Spectrom. Ion Phys.,
    12, 347.
Su, T. and Bowers, M. T. (1973c) J. Amer. Chem. Soc., 95, 1370.
Su, T. and Bowers, M. T. (1973d) J. Amer. Chem. Soc., 95, 7609.
Su, T. and Bowers, M. T. (1974a) Int. J. Mass Spectrom. Ion Phys.,
    (in press).
Su, T. and Bowers, M. T. (1974b) Chem. Phys. (to be submitted).
Su, T. and Bowers, M. T. (1974c) J. Chem. Phys. (to be submitted).
Su, T. and Bowers, M. T. (1974d) J. Chem. Phys., 60, 4897.
Udseth, H., Giese, C. F. and Gentry, W. R. (1974) J. Chem. Phys.,
    60, 3051.

THEORETICAL CONSIDERATIONS OF POTENTIAL ENERGY SURFACES FOR ION-

MOLECULE REACTIONS

Joyce J. Kaufman

Department of Chemistry
The Johns Hopkins University
Baltimore, Maryland, 21218

INTRODUCTION

There are many more observed phenomena which need theoretical interpretation and guidance than will be possible in the forseeable future to compute accurately the corresponding potential energy surfaces. Extremely valuable insight can be gleaned from the application of the symmetry and spin restrictions on electronic states of an intermediate quasimolecule permissible from the symmetries and spins of the separated reactants or products or on electronic states of fragments permissible from a particular electronic state of a dissociating molecule. These restrictions are the three dimensional extensions of the original Wigner-Witmer (64) quantum mechanical derivations for diatomic molecules and are discussed in detail by Herzberg (23a). Recently, it was cogently pointed out by Hasted (21) that some experimentalists apparently do not grasp and, thus, do not apply even its most rudimentary concepts to their collisional problems. This is most noticeable in the discussions of spin permitted reactions. In these cases, it is not necessarily total spin which must be conserved from reactants to products, but only vector addition of total spin.

In the case of collisional phenomena, reactive or non-reactive, consideration of the appropriate surfaces is vital whether the intermediate complex is a stable bound entity or even a completely repulsive state since the reactants must approach and the products must recede along the potential energy curves of the quasimolecule. The symmetries and stabilities of these intermediates have a profound effect on the paths along which and the rates at which reactions proceed. We have been able to demonstrate the usefulness of this concept to ion-neutral reactions (32,33,37,48,49), neutral-neutral reactions and hot atom reactions (32,33,35).

The course of the reaction is determined by the shapes of these curves and by the kinds of curve crossings which do or do not take place. In this regard there are apparently two types of behavior followed: 1. adiabatic, where the reactants approach each other slowly and, at least for states of the same symmetry, the curves do not cross; 2. diabatic, where the reactants approach each other rapidly and diabatic transitions occur even between states of the same symmetry.

For the case of molecular dissociative phenomena we have noted an additional restriction which we have called a "uniqueness" criterion. For any specified electronic state of a molecule (or intermediate quasimolecule) along a particular dissociation coordinate there is one and only one set of dissociation products permitted to it by the symmetry and spin restrictions. The only way in which this specified molecular state can dissociate to products in different electronic states is by a curve crossing. This "uniqueness" restriction is of wide applicability and should be especially useful for determining the products in the case of dissociation from excited states vs. dissociation from the ground state. This is particularly pertinent to photodissociation and electron impact dissociation processes.

For states of different symmetries of diatomics the probability of crossing was derived by Landau and Zener. Even for diatomics the Landau-Zener formula has been shown to be inapplicable to certain instances and has been revised. For polyatomic molecules the Landau-Zener formula is not applicable as such and modifications have been made to permit its use for specified cases.

POTENTIAL ENERGY SURFACES

Due to space limitations only the important points on the correlation rules or on the various quantum chemical techniques available for calculating potential energy surfaces will be high-lighted here, but with sufficient references to enable the reader to go back to the original sources for the necessary background material to carry through such analyses.

A. Correlation Rules - General

1. Diatomics

There are three different schemes for building up the

electronic states of diatomic molecules (22a): a) from separated
atoms; b) from the united-atom and; c) from the molecular orbitals
of the diatomic molecule itself. The correlation between a) and c)
is most valuable for any general consideration of reactions and
excited states and is not limited solely to diatomic molecules but
also forms a valid approach for polyatomic molecular systems. The
correlation rules for diatomics are discussed in Herzberg (22a).
These correlation rules hold for the adiabatic potential curves of
the electronic states* when Russell-Saunders coupling is valid for
the separated atoms as well as the molecules.

2. Polyatomics

There are four different schemes for building up the electronic
states of polyatomic molecules (23a): a) from the united atom or
molecule; b) by starting from a geometrical conformation of
different symmetry (either higher or lower); c) from separated
atoms or molecular fragments; d) from the molecular orbitals of
the polyatomic molecule. In general, the most useful correlations
for elucidation of the mechanism of collisions involving polyatomic
intermediates are those between the states of the molecule formed
from separated atoms or molecular fragments and those built up
from the molecular orbitals of the polyatomic intermediate. For
the elucidation of molecular decomposition as well as formation,
in addition to these two correlations, the correlations by starting
from a geometrical conformation of different symmetry (either
higher or lower) is also extremely valuable. In all of the
correlations, as long as spin-orbit coupling is small, there must
be spin correlation. It should be emphasized that these are
spin correlations of the intermediate state and what must be conserved is vector addition of total spin.

a. From a geometrical conformation of different symmetry

For the question of collisions this correlation may be of
importance because during the collisional process the symmetry of
the intermediate often changes. An intermediate complex may only
be completely symmetrical when it is in its equilibrium conformation.
During the formation and breakup, the intermediate has had a conformation of lower symmetry. Herzberg Polyatomics gives the

---

*
In general, electronic motions are very fast compared to nuclear
motions. For slowly varying changes in the internuclear distance
the electronic motion adjusts itself to each change in nuclear
configuration without undergoing electronic transitions. This
behavior is called adiabatic.

correlation of species of different point groups corresponding to different conformations of a given molecule: $D_{2h} \to D_2 \to D_{2d}$; $D_{3d} \to D_3 \to D_{3h}$; $C_{2v} \to C_2 \to C_{2h}$ (23b). While the correlations of species of a point group of higher symmetry into those of a point group of lower symmetry are always unambiguous, the reverse correlations are not always unambiguous. The actual correlation depends on the path chosen. It is straightforward to decipher which states of a distorted molecule arise from certain states of a more symmetrical molecule; the reverse correlation is not always unambiguous.

In the case of molecular decompositions this correlation takes on even more importance. Oftentimes, the original molecule does possess a certain high or moderately high degree of symmetry. As the molecule decomposes, this symmetry distorts to that of a point group of lower symmetry but in this case it is always possible to trace the correlation uniquely.

b. From separated atoms or molecular fragments

The correlation between the electronic states of a polyatomic molecule and those of the separated atoms or molecular fragments can be derived from a generalization of the Wigner-Witmer correlation rules for diatomic molecules. (See references 32 and 33 for details which space does not permit here). Herzberg gives a useful table of the correlations permitted for an atom plus a diatomic or two diatomics to form linear intermediates (23c).

A table of molecular electronic states of linear molecules resulting from certain states of the separated atoms (23d) is much more difficult to use for conceptualizing collisional problems.

c. Symmetrical molecules $(D_{\infty h})$

If two like linear fragments (but in different electronic states) are brought together, the correlations mentioned in the previous section and listed in reference 23c are the ones used. Each state of that table now occurs twice since resolution of the resonance degeneracy leads to splitting into a g and a u state. If two identical linear fragments (in identical electronic states) are brought together, there is no resonance degeneracy and only the same total of states arise as for unequal groups. The symmetry character of the states alternates for different possible spin values and, therefore, some of the states are g and some are u. The electronic states of symmetrical linear molecules $(D_{\infty h})$ resulting from identical states of the separated equal groups are listed (23e).

3. Non-linear molecules

The building up of non-linear molecules from individual atoms leads to a very great number of molecular states. The procedure is outlined in Herzberg Polyatomics (23f) but will not be discussed here since it is not easily usable for the purposes of studying possible collisional intermediate complexes. Much more useful for this purpose is the building up of the intermediate molecules from separated fragments outlined below.

(a) Unlike groups

If neither of the two parts which are to be combined has a symmetry lower than that of the final molecule and if the full symmetry is retained during the approach of the two parts, then the correlation is the same as if the molecule were built up from separated atoms.

If the symmetry of at least one of the parts is lower than that of the complete molecule and if during the approach of the parts the molecule has the geometrical conformation of the part with the lower symmetry, the resulting states are obtained in terms of the lower symmetry. If in the final molecule the symmetry is now higher, the correlations rules between deformed conformations are used to find the relation of the states of the intermediate to the states of the final molecule.

A third case is where the symmetry of both parts is higher than that of the final molecule being built up but where the symmetry of the intermediate during the formation process is lower than that of the final molecule. The states of both parts must be resolved into the species of the final molecule.

The resulting molecular states of the final molecule are obtained by multiplication of the states of the two parts. For symmetrical final molecules this resolution is not completely unambiguous.

(b) Like groups

If two like fragments are brought together (even if the symmetry of the parts is lower than that of the final molecule) it is usually possible to determine unambiguously the resulting molecular states. This is possible because even at large separation of the two fragments the full symmetry of the molecule may exist.

If the two fragments are in different electronic states there

is a resonance degeneracy which leads to a splitting into a
symmetrical and an anti-symmetrical state. If the two fragments
are identical (in the identical electronic states) there is no
resonance degeneracy. The symmetries of the resulting states are
either g or u or else ' and ". These symmetries are determined by
using the correlation with the corresponding diatomic or linear
polyatomic molecules.

(Several useful general references on the above topic are
39, 40 and 58).

B. Correlations With Molecular Orbitals

It is not the intent of this chapter to present a detailed
review of quantum chemical theory. The few following portions on
the theoretical background serve merely to put the discussion of
the quantum chemical computations into the proper perspective.

1. Theoretical background

a. Separation of electronic and nuclear motion

Because, in general, electrons move with much greater velocities
than nuclei, to a first approximation electron and nuclear motions
can be separated, (Born-Oppenheimer theorem (3)). The validity of
this separation of electronic and nuclear motions provides the
only real justification for the idea of a potential energy curve of
a molecule. The eigenfunction Ψ for the entire system of nuclei
and electrons can be expressed as a product of two functions Ψ_e and
Ψ_n where Ψ_e is an eigenfunction of the electronic coordinates found
by solving Schroedinger's equation with the assumption that the
nuclei are held fixed in space and Ψ_n involves only the coordinates
of the nuclei (14a). The exact Hamiltonian operator may be written
as:

$$\mathcal{H} = - \sum_A \frac{\hbar^2}{2M_A} \nabla_A^2 - \sum_i \frac{\hbar^2}{2m} \nabla_i^2 + V_{nn} + V_{ne} + V_{ee}$$

where the first term represents the kinetic energy of the nuclei,
the second represents the kinetic energy of the electrons and V_{nn},
V_{ne} and V_{ee} are the contributions to the potential energy
arising from nuclear, nuclear-electronic and electronic interactions
respectively. If the nuclei were assumed to be fixed in space, the
Hamiltonian for the electrons would be:

$$\mathcal{H}_e = - \sum_i \frac{\hbar^2}{2m} \nabla_i^2 + V_{ne} + V_{ee} \quad .$$

The remaining terms are represented by $\mathcal{H}_n$

$$\mathcal{H}_n = - \sum_A \frac{\hbar^2}{2M_A} \nabla_A^2 + V_{nn} \qquad \text{and} \qquad \mathcal{H} = \mathcal{H}_e + \mathcal{H}_n$$

Ψ_e is defined as the function which satisfies the equation

$\mathcal{H}_e \Psi_e = E_e \Psi_e$ where E_e is the electronic energy. For the total system $\mathcal{H} \Psi_e \Psi_n = E \Psi_e \Psi_n$. (This holds only in a "loose" perturbation sense for $\mathcal{H}\Psi = E\Psi$. Ψ should be expanded in a complete set.) In the Born-Oppenheimer approximation $(\mathcal{H}_n + E_e)\Psi_n = E\Psi_n$ since small terms coupling the electronic and nuclear motion have been neglected. Thus the molecular energy as a first approximation can be considered as the sum of two terms: the energy that the electrons would have if the nuclei were at rest plus the repulsive mutual energies of these fixed nuclei. The neglected terms may be treated as a perturbation and will give rise to energy terms representing the interaction of electronic and nuclear motions. These perturbations prove of importance later when discussing probabilities for crossing from one potential curve to another.

b. Group theory and quantum mechanics (14a)

For a symmetrical atomic or molecular system, group theoretical considerations place a severe restriction on the possible eigenfunctions of the system since these must form bases for some irreducible representation of the group of symmetry operations. Their form is also determined to a large extent since they must transform in a quite definite way under the operations of the group.

c. Various quantum chemical computational methods

In order to render tractable the problem of determining the molecular electronic eigenfunction, Ψ_e, it is customary to assume the individual molecular orbitals to be functions of the atomic electron eigenfunctions, χ_r, centered on each atom. The molecular orbitals (MO's), ϕ_i, are taken to be linear combinations of the atomic orbitals (LCAO's), χ_r, thus $\phi_i = \sum_r \chi_r c_{ri}$. A configuration wave function Φ (where here Φ denotes the molecular electronic eigenfunction Ψ_e) is represented by antisymmetrized product wave function (59), since only states can occur whose eigenfunctions are antisymmetric with respect to exchange of any two electrons.

The total wave function Φ is normalized $(\Phi|\Phi) = \int \Phi^* \Phi d\tau = 1$

and $E_{el} = \dfrac{\int \Phi^* \mathcal{H}_{el} \Phi d\tau}{\int \Phi^* \Phi d\tau}$. The Rayleigh-Ritz variational method

is used to determine the coefficients, c_{ri}, corresponding to the best approximation to the minimum energy, E_{el}, for the system. It is convenient to make linear combinations of the atomic orbitals which form symmetry orbitals of the molecule in question. The allowed combinations of atomic eigenfunctions to form molecular eigenfunctions of the various symmetry types are determined (for a molecule of any given symmetry) directly from the character table of the irreducible representations of this group: (The various symmetry operations, the classification of molecular symmetry types according to the behavior of molecular electronic eigenfunctions with respect to the symmetry operations and character tables for all the groups of diatomic and polyatomic molecules are listed in Herzberg (22a, 23a).

Since one of the most useful correlations is with the molecular orbitals for the intermediate species involved in collisions or in molecular decompositions, it seems appropriate now to indicate the various quantum chemical calculational methods in most common usage.

(1) Non-empirical

These methods under the category of non-empirical fall into two subclasses. The first is the well-known Hartree-Fock-Roothaan (54) LCAO-MO-SCF (self-consistent field) method. By determining the spin-orbitals according to the Hartree-Fock equations, one gets the best approximate solution in the form of a single determinant. The second is an even more rigorous technique which includes either the effect of configuration interaction on the eigenfunction obtained in a non-empirical calculation or the multiconfiguration SCF method. Both these methods correct for the deficiency in the Hartree-Fock method which is necessary for calculating correlation energy corrections to the molecular energies.

To obtain a better solution, one must set up many determinantal functions, formed from different spin-orbitals, and must use an approximate wave function which is a linear combination of these determinantal functions, with coefficients to be determined by minimizing the energy. This process is called configuration inter-action (CI) or superposition of configurations. The determinantal functions must be linearly independent, and be eigenfunctions of the spin operators S^2 and S_z, and preferably belong to a specified row of a specified irreducible representation of the symmetry group of the molecule (16,50). Definite spin states can be obtained by applying a spin-projection operator to the spin-orbital product defining a configuration (20). From functions of the same symmetry as Φ_0, (the solution of the Hartree-Fock equation) one can build a wave function $\Phi = a_0\Phi_0 + a_1\Phi_1 + a_2\Phi_2 + \text{---}$. By optimizing the orbitals in each function and by variationally selecting the CI coefficients a_0, a_1, a_2 ---, one gets an eigenfunction as good or better than Φ_0. If the series is sufficiently long one can reach an exact solution. The trouble is that the necessary series is too long and usually converges slowly. In practice a truncated series is used. [A recent useful book by Schaefer includes examples of configuration interaction calculations (56).]

If when Φ_0 is constructed, Φ_1, Φ_2, etc. are constructed at the same time from a common orthonormal set of orbitals and one solves, not for the best possible Φ_0, but for the best Φ; then the variational principle, used simultaneously on both the a's (the CI coefficients) and the c's (the atomic orbital coefficients), will insure that the Φ_i will overlap as much as possible.

Complete multiconfiguration-self-consistent field (CMC-SCF) technique designates the method where a given occupied molecular orbital of the set is excited to all unoccupied molecular orbitals. If an occupied orbital is excited to one or more, but not all, of the unoccupied orbitals, the technique is described as incomplete MC-SCF (IMC-SCF) (8,24). The CMC-SCF formalism differs from most many-body techniques presented to date insofar as the Hartree-Fock energy is not assumed to be the zero-order energy.

(2) Semirigorous

Semirigorous LCAO-MO-SCF methods start with the complete many-electron Hamiltonian and make certain approximations for the integrals and for the form of the matrices to be solved. Some years ago we derived such a method starting with the correct many-electron Hamiltonian (in which interelectronic interactions are included explicitly) and the LCAO-MO-SCF equations of Roothaan and then making a consistent series of systematic approximations for the integrals involved (31).

The two most restrictive approximations A and B of our semi-rigorous scheme are similar conceptually to the CNDO method published by Pople at the same time and his later revision, the NDDO and INDO methods published subsequently (52). Our simplest, Method A, resembles Pople's CNDO (complete neglect of differential overlap) method. Method B resembles Pople's NDDO (neglect of diatomic differential overlap) or INDO (intermediate neglect of diatomic differential overlap) methods.

(There are several additional useful review articles in all-

valence electron semi-rigorous self-consistent field calculations
(28,30,47).

(3) Semi-empirical

The semi-empirical extended Hückel method, which takes into
account all valence electrons in a molecule, was introduced by
Wolfsberg and Helmholz (65), has been used over the years extensively
by Lipscomb and co-workers especially Hoffman (13,25,26), and has
been applied by a large number of investigators.

From a molecular orbital $\Phi_i = \sum_r \chi_r c_{ri}$ and by application of
the variation principle for the variation of energy, the following
set of equations for the expansion coefficients is obtained:

$(\alpha_r - ES_{rr}) c_r + \sum_{r \neq s} (\beta_{rs} - ES_{rs}) c_s = 0$; $S_{rs} = \int \chi_r^* \chi_s \, dv$ = overlap

integral; $H_{rr} = \alpha_r = \int \chi_r^* \mathcal{H} \chi_r \, dv$ = Coulomb integral; $H_{rs} = \beta_{rs} =$
$\int \chi_r^* \mathcal{H} \chi_s \, dv$ = Resonance integral $(r \neq s)$; $\mathcal{H}$ is an effective one
electron Hamiltonian representing the kinetic energy, the field of
the nuclei and the smoothed out distribution of the other electrons.
Electron-electron repulsion is neglected in this method.

The diagonal elements are set equal to the effective valence
state ionization potentials of the orbitals in question.   The off-
diagonal elements, H_{rs}, can be evaluated in several ways.   The
two expressions in most common usage are the original Wolfsberg-
Helmholz expression $H_{rs} = 0.5 k (H_{rr} + H_{ss}) S_{rs}$ and the more
recent Cusachs' expression (10):

$$H_{rs} = \frac{(H_{rr} + H_{ss})}{2} S_{rs} (2 - |S_{rs}|).$$

The total electronic energy of a particular system is taken
to be the sum of the orbital energies, ε_i, times their occupation
numbers. For a closed shell system $E_{elect} = 2 \sum \varepsilon_i$. The total
molecular energy can be written as:

$E = 2 \sum \varepsilon_i + \sum_{n,n'} E_{nn'} - \sum_{e,e'} E_{ee'}$ where $E_{nn'}$ and $E_{ee'}$ are
nuclear-nuclear and electron-electron repulsion energies.   The
success of these extended Hückel calculations in predicting
preferred geometrical conformations from calculated minimum
energies lies in the fact that the method of selecting the H_{rs}
values must simulate, within the calculated electronic energies,
the contribution of nuclear repulsions to the total energy (25a).
The nuclear-nuclear and electron-electron repulsion energies cancel approximately (60) and thus the simple sum of one-electron

energy behaves similarly to the true molecular energy. Due to neglect of electron-electron repulsion, the calculated energy values are, unfortunately, equal for the identical molecular states with different multiplicities.

2. Electronic configuration

The electrons are fed into the molecular orbitals according to the auf bau principle. To derive the term type (species) from the electronic configuration Russell-Saunders coupling is assumed:

$\vec{\Lambda} = \Sigma\vec{\lambda} : \vec{S} = \Sigma\vec{S_i}$ where, for non-equivalent electrons, the quantities

$\vec{\lambda_i}$ and $\vec{S_i}$ are added vectorially in all possible combinations. If the electrons are equivalent, the Pauli principle must be taken into account when adding the $\vec{\lambda_i}$ and the $\vec{S_i}$; the electrons must differ in m_ℓ or m_s. The energy difference between corresponding states of different multiplicity is due to the electrostatic interaction of the electrons. The state with the greatest multiplicity almost invariably lies lower in energy.

If equivalent as well as non-equivalent electrons are present the resulting states are found by first forming the resulting states of each group of equivalent electrons and then forming the direct product of the species so obtained. Since closed shells always give a single totally symmetric singlet state they can be entirely neglected in the determination of the resulting states.

Herzberg gives tables of the terms arising from nonequivalent electrons [diatomics (22b), polyatomics (23g)],equivalent electrons [diatomics (22c), polyatomics (23b)], and equivalent as well as non-equivalent electrons [diatomics (22d)].

In addition, for diatomics consisting of like atoms (or certain symmetrical linear polyatomic molecules) the resulting states are also either even (g) or odd (u). They are even if the number of "odd" electrons (σu, πu, ---) is even, whereas they are odd if the number of "odd" electrons is odd.

These terms of electron configuration arising from different partially filled orbitals are very valuable in deciphering what excited electronic states of molecules and intermediates are possible and lie in an energy range accessible to the particular experiment.

The ordering of the molecular orbital is dependent on the relative positions of the nuclei. In analyzing a collisional problem care must be taken to feed the electrons into the appropriate molecular orbitals correlating the reactants or products with the proper intermediate state as permitted by the symmetry and spin restrictions outlined above.

CROSSING OR PSEUDO-CROSSING OF MOLECULAR POTENTIAL ENERGY SURFACES

The course of behavior followed in a collisional process is greatly influenced by the crossings or pseudo-crossings of relevant molecular potential energy surfaces. For diatomic molecules these surfaces are simply curves, crossing in a point. But for a general molecule with N atoms, the surfaces are in a configuration space of dimension 3N - 6 (for a linear molecule 3N - 5)(9). For the convenience of visualizing these energy surfaces are often referred to as curves, and that practice will be followed in the present section.

The well known Landau-Zener (41,61,66) formula relating to the probability of an electronic jump near the crossing point of two potential energy curves or surfaces has been seriously critiqued (9, 1b). New treatments of greater validity have been formulated (1,9,42).

A series of papers by Lichten on the general topic of molecular wave functions and inelastic atomic collisions (43) introduced a new descriptive terminology which has since been used extensively. If the atoms approach each other slowly in state Φ_1 an adiabatic transition from state Φ_1 to state Φ_2 of identical spin and symmetry will occur. Lichten coined the expression "diabatic" to describe the behavior if the two atoms approached each other rapidly and a transition from Φ_1 to Φ_1 occurred. This type of a transition is called a "diabatic transition. Recent results in low-energy ion-atom and ion-molecule collisions indicated that these processes can be highly effective in producing optical excitation in contradiction to the well-known "adiabatic criterion" (12). The concept of diabatic transitions has proven particularly useful in delineating the processes in resonant and quasi-resonant charge exchange processes (12,43c,d). Lichten (43e) discusses some of the effects which cause transitions among diabatic MO's at crossings; electron-penetration, electron correlation and electronic interaction with nuclear motion.

EXAMPLES OF THE APPLICATION OF THESE CONCEPTS

A.   Theoretical Explanation for the Apparently Anomalous Low
     Energy Behavior of the $O^+ + N_2 \rightarrow NO^+ + N$ Reaction

The utility of the concepts described earlier in this chapter as an aid in deciphering and predicting experimental results in molecular collisional processes may be illustrated by the following theoretical justification derived for the experimentally observed apparently anomalous low-energy behavior of the $O^+ + N_2$ ion-molecule reaction.

It is generally observed that cross sections for exothermic ion-molecule reactions decrease with increasing ion kinetic energies indicating a negligible activation energy. An important exception to this behavior is exhibited by the exothermic reaction of ground states $O^+ + N_2 \rightarrow NO^+ + N$, the cross section of which rises with energy, goes through a broad maximum at 10 eV and then decreases (17). This is contrary to the expected behavior of an exothermic ion-molecule reaction since on the basis of the simple considerations of the Gioumousis and Stevenson (19) picture the cross section is expected to decrease with increasing energy. A recent review had mentioned that there was no satisfactory explanation for this apparent anomalous behavior. More recently Schmeltkopf, et al., (57) studied this reaction in the thermal region and found it proceeded with no activation energy and furthermore the rate of reaction increased rapidly with vibrational excitation of the N_2. It was very desirable to have detailed knowledge of the potential energy surfaces of this system since it would permit a more definitive interpretation of the experimental results. When we initiated our theoretical investigation, no accurate quantum chemical calculations had yet been made on these surfaces. However, application merely of the concepts described earlier in this chapter combined with the available theoretical, optical spectroscopic and mass spectro-scopic data which existed at that time for this system permitted us to outline schematically some of the pertinent potential energy curves (37). From these curves it was possible to explain qualitatively the apparently anomalous features of this reaction. Our more recent extensive configuration interaction calculations (49) have validated the qualitative theoretical explanation based on the symmetry and spin analyses.

The intermediate of the $O^+ + N_2$ reaction, N_2O^+, has 15 valence electrons and, thus, is assumed to have a linear configuration ($N-N-O^+$) in accordance with Walsh's rules (62) for a 15 valence electron triatomic. Experimentally the ground state of N_2O^+ is known to be $^2\Pi_i$ and its first excited state, lying at 3.49 eV, to be $^2\Sigma^+$ (6).

Using Koopmans' Theorem (38) the ordering of the topmost filled levels for N_2O^+ ($X\ ^2\Pi_i$) (at its equilibrium geometry) is $(1\pi)^4(7\sigma)^2(2\pi)^3$ (45) and at the same geometry, for the first excited state $^2\Sigma^+$ $(1\pi)^4(7\sigma)^1(2\pi)^4$; and for the second excited state, a $^2\Pi_i$ state $(1\pi)^3(7\sigma)^2(2\pi)^4$. From recent photoelectron

spectroscopic measurements on N_2O (5,46) it is substantiated that
the ground and first excited states of N_2O^+ are $\tilde{X}\,^2\Pi_i$ and $^2\Sigma^+$
(3.49 eV above ground state N_2O^+) and that the second excited state,
$^2\Pi_i$, lies only about 1.3 eV above the $^2\Sigma^+$ state. This $^2\Pi_i$ state
is attractive since transitions to at least fourteen different
vibrational levels were reported. The third excited state, $^2\Sigma^+$,
corresponds to the earlier state reported at ~7.19 eV above the
ground state of N_2O^+ (23i) and is also attractive since transitions
are reported to five different vibrational levels. From these
levels for N_2O^+, the experimental data for the reactants $O^+ + N_2$
(or $O + N_2^+$ which must also be taken into account when drawing a
full potential curve), the products $NO^+ + N$ (or $NO + N^+$), the
experimental dissociation energy of N_2O^+ to $NO^+ + N$ (18) and MOST
IMPORTANTLY from the rules for the possible symmetry- and spin-
permitted combinations of these reactants, (Table 1) and products
(Table 2) into the intermediate states of N_2O^+, the following
schematic potential diagram was drawn for the system (profiles
through sections of a cube) (39b) (Figure 1). The energies
indicated are those relative to the separated neutral atoms.

It becomes very obvious that ground state $O^+ + N_2$ can combine
uniquely only along the repulsive $^2\Sigma^-$ N_2O^+ curve which is then cut
by the attractive ground state $\tilde{X}\,^2\Pi_i$ N_2O^+ curve. The $\tilde{X}\,^2\Pi_i$ N_2O^+
curve indicates that predissociation from the $\tilde{X}\,^2\Pi_i$ N_2O^+ to ground
state $NO^+ + N$ can take place since the repulsive $^4\Sigma^-$ lower curve,
which separates to ground state NO^+ ($^1\Sigma^+$) and $N(^4S_u)$, cuts the
attractive $\tilde{X}\,^2\Pi_i$ ground state solid curve which would separate to
NO^+ ($^1\Sigma^+$) and $N(^2D_u)$. This behavior is substantiated by the
experimental observation that (although spin-forbidden)

TABLE 1.   NNO$^+$ reactant energy levels and symmetry- and spin-
            permitted intermediates

Reactants		Energy scale (relative to separated neutral atoms)
N_2 $+O^+$ $\tilde{X}^1\Sigma_g^+$ ^4S_u	$\longrightarrow$ N–N–O$^+$ $^4\Sigma^-$	3.859 eV
N_2^+ $+O$ $\tilde{X}^2\Sigma_g^+$ ^3P_g	$\longrightarrow$ N–N–O$^+$ $^{2,4}\Sigma^-$, $^{2,4}\Pi$	5.821 eV
N_2^+ $+O$ $A^2\Pi_u$ ^3P_g	$\longrightarrow$ N–N–O$^+$ $^{2,4}\Sigma^+$, $^{2,4}\Sigma^-$, $^{2,4}\Pi$, $^{2,4}\Delta$	6.941 eV
N_2 $+O^+$ $\tilde{X}^1\Sigma_g^+$ ^2D_u	$\longrightarrow$ N–N–O$^+$ $^2\Sigma^-$, $^2\Pi$, $^2\Delta$	7.184 eV

TABLE 2. NNO⁺ product energy levels and symmetry- and spin-
permitted intermediates

Products		Energy scale (relative to separated neutral atoms)
$NO^+ + N$ $\tilde{X}^1\Sigma^+$ ^4S_u	⟶ N–N–O⁺ $^4\Sigma^-$	2.760 eV
$NO^+ + N$ $\tilde{X}^1\Sigma^+$ ^2D_u	⟶ N–N–O⁺ $_{\circ}^2\Sigma^-$, $^2\Pi$, $^2\Delta$	5.144 eV
$NO^+ + N$ $\tilde{X}^1\Sigma^+$ ^2P_u	⟶ N–N–O⁺ $^2\Sigma^+$, $^2\Pi$	6.336 eV
$NO + N^+$ $\tilde{X}^2\Pi_i$ ^3P_g	⟶ N–N–O⁺ $^{2,4}\Sigma^+$, $^{2,4}\Sigma^-$, $^{2,4}\Pi$, $^{2,4}\Delta$	8.025 eV

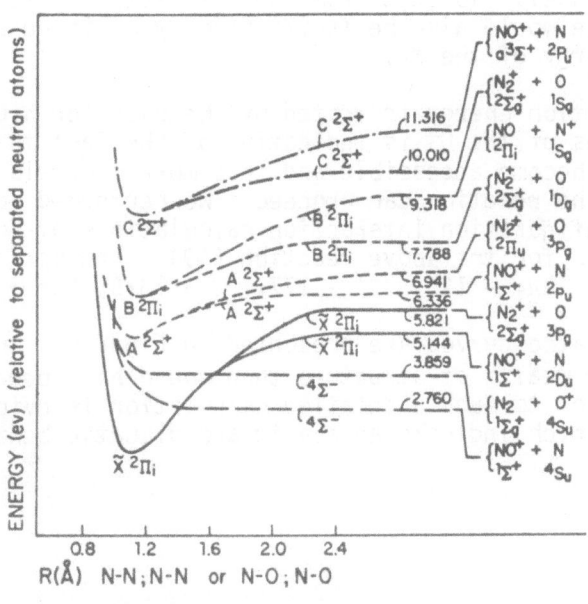

FIG. 1. N–N–O⁺ POTENTIAL ENERGY DIAGRAM (SCHEMATIC)

predissociation to ground state products NO^+ + N is noted to occur in the mass spectroscopic decomposition of ground state N_2O^+ (23j).

Since the ground state of O^+ and N_2 can combine uniquely only along the indicated $^4\Sigma^-$ curve, the lack of an activation energy for the production of NO^+ at thermal energies implies that this $^4\Sigma^-$ curve, even though repulsive, must come in with an almost flat slope until the O-N distance is in the region of the NNO^+ equilibrium distance. The very low reaction efficiency under these conditions when N_2 is in its ground vibrational state indicates that the reactants in the $^4\Sigma^-$ state must make a forbidden curve crossing into another state (in this case into the $\tilde{X}$ $^2\Pi_i$ state of NNO^+). This scheme permits the curve of the ground state reactants to cross into the attractive ground state of N_2O^+ with no activation energy and then to cross over into the $^4\Sigma^-$ state of the ground state products. The spin forbiddenness of the cross-overs then accounts for the observed low probability of reaction as suggested by Schmeltekopf, et al. (57).

The appearance potential for $N_2O \rightarrow NO^+$ + N of 15.01 eV (11) indicates the presence of some type of barrier in this exit channel (51) since the energy of the separated NO^+ + N is lower than 15.01 eV on the same relative energy scale. The detailed quantum chemical calculations for O^+ + $N_2 \rightarrow NO^+$ + N, (49) which will be described below, indicated an unsuspected channel, a low lying $^4\Pi$ state which cut through both the $\tilde{X}$ $^2\Pi_i$ and the $^4\Sigma^-$ states in the exit channel. To exit from the $^2\Pi_i$ state via the $^4\Pi$ state to the $^4\Sigma^-$ state would also be facilitated by an increase in the vibrational energy of the N_2.

The activation energy indicated by the data for higher kinetic energies of the O^+ is indicative of the fact that higher states of NNO^+ become accessible and thus more channels open up through which the reaction can proceed. We performed ab-initio large scale configuration interaction calculations of some of the pertinent states for the above reaction (49). (Number of configurations: $^2\Pi_i$ - 123, $^2\Sigma^+$ - 177, $^4\Pi$ - 1379, $^4\Sigma^-$ - 1310).

The calculated curves are presented in Figures 2 and 3. An even greater wealth of structure than would have been anticipated prior to such a detailed calculation is evident from these figures which indicate adiabatic and diabatic behavior.

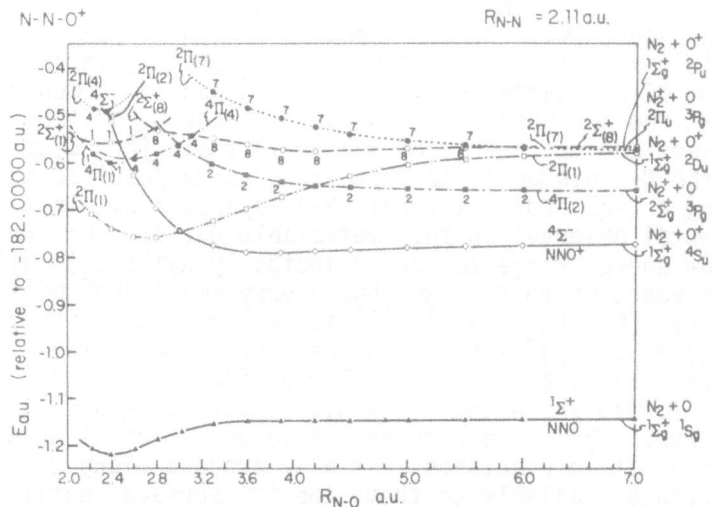

FIG. 2.   CI ENERGIES OF NNO$^+$ VS. R$_{N-O}$

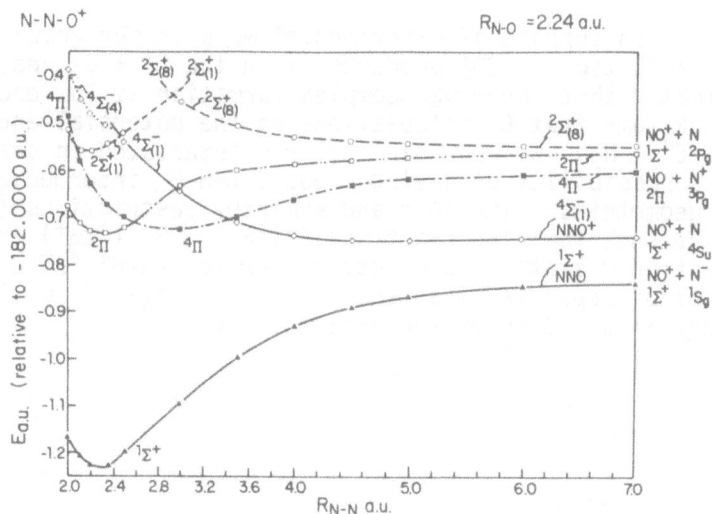

FIG. 3.   CI ENERGIES OF NNO$^+$ VS. R$_{N-N}$

The dominant contribution at each point is marked with the serial number in the configuration list corresponding to its MO filling. [For details and MO fillings, see reference (49)]. The energy distances between curves near the separation limit are in reasonable agreement with experiment. The apparent discrepancy in case of the two $^2\pi$ states of reactants [$^2\pi(1)$ and $^2\pi(7)$] can be accounted for by looking at the corresponding curves. The energy of the $^2\pi(7)$ state decreases and the energy of the $^2\pi(1)$ state increases while increasing the N-O distance, so that it is quite possible that the corresponding curves cross at larger N-O separations. This explains the experimental observation that metastable $O^+(^2D_u)$ ions react with N_2 in the low energy range to form principally $N_2^+(^2\pi_u)$ while the ion-molecule reaction to form NO^+ has a very small probability (55). The $^2\pi(1)$ curve [$O^+(^2D_u) + N_2(^1\Sigma_g^+)$] is near resonant with the $^2\pi(7)$ curve [$O(^3P_g) + N_2(^2\pi_u)$] for N_2^+ in the A $^2\pi_u$, v = 1 state. Thus charge exchange takes place at very large distances and the $O(^3P_g) + N_2^+(^2\pi_u)$ are formed where they then dissociate along a slightly repulsive curve. The CI results validated our original hypothesis that the ground state reactants initially proceed along the one potential surface available to them, the $^4\Sigma-$ surface, until they either curve cross into another surface or proceed far enough along that surface to enable reaction to take place. Likewise, the ground state products can only arise from the $^4\Sigma-$ surface.

B. $C^+ + H_2 \rightarrow CH^+ + H$

In 1971, in support of experimental work in our group on the angular distribution of CD^+ products, from the $C^+ + D_2$ reaction, which indicated that there was complex formation at low energies (27), we undertook some test CI calculations on the potential energy curves for $C^+ + H_2$. We were particularly interested in seeing if there was a possibility of having a bound C-H-H$^+$ intermediate even in linear geometries. The spin and symmetry restrictions (Tables 3 and 4) are such that the reactants $C^+(^2P_u) + H_2$ ($^1\Sigma_g^+$) in their ground electronic states could lead either to C-H-H$^+$ $^2\Sigma^+$ or $^2\pi$. The next set of separated fragments $C(^3P_g) + H_2^+(^2\Sigma_g^+)$ would lead to $^{2,4}\Sigma-$ and $^{2,4}\pi$ states of the intermediate.

TABLE 3. $C^+ + H_2$ or $C + H_2^+$ (Reactants)

(Energies relative to ground state neutral atoms, eV)

$C^+ (^2P_u) + H_2 (^1\Sigma_g^+) \rightarrow C\text{--}H\text{--}H^+$ 6.71 eV

$C_{\infty v}$	C_s	C_{2v}		$C_{\infty v}$	C_s	C_{2v}		$C_{\infty v}$	C_s	C_{2v}
$^2\Sigma_{(u)}^+$	$^2A'$	$^2B_2$		$^1\Sigma_g^+$	$^1A'$	$^1A_1$		$^2\Sigma^+$	$^2A'$	$^2B_2$
$^2\Pi_{(u)}$	$^2A'$	$^2A_1$						$^2\Pi$	$^2A'$	$^2A_1$
	$^2A''$	$^2B_1$							$^2A''$	$^2B_1$

$C (^3P_g) + H_2^+ (^2\Sigma_g^+) \rightarrow C\text{--}H\text{--}H^+$ 10.91 eV

$C_{\infty v}$	C_s	C_{2v}		$C_{\infty v}$	C_s	C_{2v}		$C_{\infty v}$	C_s	C_{2v}
$^3\Sigma_{(g)}^-$	$^3A''$	$^3B_1$		$^2\Sigma_g^+$	$^2A'$	$^2A_1$		$^{2,4}\Sigma^-$	$^{2,4}A''$	$^{2,4}B_1$
$^3\Pi_{(g)}$	$^3A'$	$^3B_2$						$^{2,4}\Pi$	$^{2,4}A'$	$^{2,4}B_2$
	$^3A''$	$^3A_2$							$^{2,4}A''$	$^{2,4}A_2$

$C^+ (^4P_g) + H_2 (^1\Sigma_g^+) \rightarrow C\text{--}H\text{--}H^+$ 12.05 eV

$C_{\infty v}$	C_s	C_{2v}		$C_{\infty v}$	C_s	C_{2v}		$C_{\infty v}$	C_s	C_{2v}
$^4\Sigma_{(g)}^-$	$^4A''$	$^4B_1$		$^1\Sigma_g^+$	$^1A'$	$^1A_1$		$^4\Sigma^-$	$^4A''$	$^4B_1$
$^4\Pi_{(g)}$	$^4A'$	$^4B_2$						$^4\Pi$	$^4A'$	$^4B_2$
	$^4A''$	$^4A_2$							$^4A''$	$^4A_2$

$C (^1D_g) + H_2^+ (^2\Sigma_g^+) \rightarrow C\text{--}H\text{--}H^+$ 12.17 eV

$C_{\infty v}$	C_s	C_{2v}		$C_{\infty v}$	C_s	C_{2v}		$C_{\infty v}$	C_s	C_{2v}
$^1\Sigma_{(g)}^+$	$^1A'$	$^1A'$		$^2\Sigma_g^+$	$^2A'$	$^2A_1$		$^2\Sigma^+$	$^2A'$	$^2A_1$
$^1\Pi_{(g)}$	$^1A'$	$^1B_2$						$^2\Pi$	$^2A'$	$^2B_2$
	$^1A''$	$^1A_2$							$^2A''$	$^2A_2$
$^1\Delta_{(g)}$	$^1A'$	$^1A_1$						$^2\Delta$	$^2A'$	$^2A_1$
	$^1A''$	$^1B_1$							$^2A''$	$^2B_1$

TABLE 4. $CH^+ + H$ or $CH + H^+$ (Products)

| $CH^+ (^1\Sigma^+)$ | + | $H (^2S_g)$ | $\leftarrow$ | $C -- H -- H^+$ | 7.17 eV |

| $C_{\infty v}$ | C_s | C_{2v} | | $C_{\infty v}$ | C_s | C_{2v} | | $C_{\infty v}$ | C_s | C_{2v} |

$^1\Sigma^+$ $^1A'$ $^1A_1$ $^2\Sigma^+_{(g)}$ $^2A'$ $^2A_1$ $^2\Sigma^+$ $^2A'$ $^2A_1$

| $CH^+ (^3\pi)$ | + | $H (^2S_g)$ | $\leftarrow$ | $C -- H -- H^+$ | 8.31 eV |

| $C_{\infty v}$ | C_s | C_{2v} | | $C_{\infty v}$ | C_s | C_{2v} | | $C_{\infty v}$ | C_s | C_{2v} |

$^3\pi$ $^3A'$ $\left.\begin{array}{l}^3B_2\\^3A''\end{array}\right\}$ or $\begin{array}{l}^3A_1\\^3A_2\end{array}\left.\right\}\begin{array}{l}^3B_1\end{array}$

$^2\Sigma^+_{(g)}$ $^2A'$ $^2A_1$

$2,4_\pi$ $2,4_{A'}$ $\left.\begin{array}{l}2,4_{B_2}\\2,4_{A''}\end{array}\right\}$ or $\begin{array}{l}2,4_{A_1}\\2,4_{A_2}\end{array}\left.\right\}2,4_{B_1}$

(π_g) (π_u) (π_g) (π_u)

| $CH (^2\pi)$ | + | $H^+ (^1S_g)$ | $\leftarrow$ | $C -- H -- H^+$ | 10.13 eV |

| $C_{\infty v}$ | C_s | C_{2v} | | $C_{\infty v}$ | C_s | C_{2v} | | $C_{\infty v}$ | C_s | C_{2v} |

$^2\pi$ $^2A'$ $\left.\begin{array}{l}^2B_2\\^2A''\end{array}\right\}$ or $\begin{array}{l}^2A_1\\^2A_2\end{array}\left.\right\}^2B_1$

$^1\Sigma^+_{(g)}$ $^1A'$ $^1A_1$

$^2\pi$ $^2A'$ $\left.\begin{array}{l}^2B_2\\^2A''\end{array}\right\}$ or $\begin{array}{l}^2A_1\\^2A_2\end{array}\left.\right\}^2B_1$

(π_g) (π_u) (π_g) (π_u)

{Note that for the linear intermediate even though it is $C_{\infty v}$ and does not retain the complete $D_{\infty h}$ symmetry we retain for information purposes the (g) and (u) subscript information. This is to enable us to ascertain the appropriate correlations going through C_s symmetry to the proper states in C_{2v} symmetry. This permits one to make the proper correlations such as the lowest $^2\Sigma^+$ state to the $^2B_2$ state [rather than to the $^2A_1$ state (44)]. Also there would otherwise be no a priori way to know whether the species $CH^+(^3\Pi)$ would lead to C_{2v} symmetry to $^3A_2 + {}^3B_2$ [giving a $^3\Pi_{(g)}$ state of linear (symmetrical) C-H-H$^+$] or to $^3A_1 + {}^3B_1$ [giving a $^3\Pi_{(u)}$ state of the linear (symmetrical) C-H-H$^+$].}

We performed large scale CI calculations for the linear approach and selected points for the off-angle approach. The results (34) indicated that the $^2\Pi$ state of C-H-H$^+$ was lower in energy than the $^2\Sigma^+$ state. It also correlated smoothly from the lowest state of C$^+$ + H$_2$ to CH$^+(^3\Pi)$ + H(^2S_g). [There is no switching of charge in diabatic correlations, thus the $^2\Pi$ state arising from C$^+(^2P_u)$+ H$_2(^1\Sigma_g^+)$ would not be expected to lead diabatically to CH($^2\Pi$) + H$^+$ as was indicated in reference 44. A C$^+$ reactant would lead diabatically to a CH$^+$ product and not to a neutral CH product.] In our CI calculations the $^{2,4}\Sigma^-$ and the $^4\Pi$ and second $^2\Pi$ state separated properly to precisely the same limiting energy value of the separated fragments C(^3P_g) + H$_2^+(^2\Sigma_g^+)$. This was gratifying since the $^{2,4}\Sigma^-$ and the $^4\Pi$ were the lowest roots of their symmetry and the $^2\Pi$ was the second lowest root of its symmetry and had been pulled by a modified Nesbet method (2). Until this calculation there had been expressed some doubts about the numerical accuracy of pulling higher roots by this method. However, this was both a numerical and physical proof of its accuracy. The vertical excitation energies of the excited states of C-H-H$^+$ at R_{HH} = 1.40 bohrs and $R_{CH(eq)}$ = 2.65 bohrs relative to the $\tilde{X}\,^2\Pi_i$ state (0.00 eV) were A $^2\Sigma^+$ (1.58 eV), a $^4\Sigma^-$ (3.26 eV), b $^4\Pi$ (3.57 eV), B $^2\Pi$ (6.69 eV), C $^2\Sigma^-$ (6.70 eV). The minima of the D $^2\Pi$ and C $^4\Pi$ states were shifted so far to the right that vertical excitations went to the steeply rising positions of the repulsive inner part of the potential curve.

C. Theoretical Explanation for the Non-Appearance of the Parent
    CF$_4^+$ Ion Upon Electron or Photon Impact on CF$_4$

The CF$_4$ molecule gives no parent ion under electron or photon impact (63). The onset of ionization of CF$_3^+$ from CF$_4$ has a long tail which was ascribed to the fact that the transition goes to a steeply rising part of a repulsive CF$_4^+$ state (63). However, in the photoelectron spectrum there were a number of peaks observed (4,29). The two peaks at lowest energy do not show any vibrational fine structure. This implies that these peaks are probably predissociated.

As a guide for the interpretation of this phenomenon we derived the symmetry and spin restrictions governing what intermediate states are permitted from the various states of the separated product fragments and calculated the molecular energies and symmetries of CF_4 and CF_4^+ as a function of r (CF_3-F') (32,36).

T_d CF_4 and CF_4^+ become molecules of C_{3v} symmetry when one F atom (F') is displaced axially along its bond direction. To determine the symmetry states of the resulting molecular states from the states of the separated parts, one must resolve the various states of CF_3 and F' into the species of point groups T_d or C_{3v}. This case is more involved than the examples presented in Herzberg (23k) so we shall indicate the analysis and present the results.

CF_3 is known experimentally to have a pyramidal structure (15). Its ground state is $^2A_1$ in C_{3v} symmetry. (Our calculations indicated that there are also fairly low lying excited CF_3 states of symmetries ^2E, $^2A_2$ and ^2E.) The photoionization experiments of Chupka and colleagues indicate that the geometry of the ground state of CF_3^+ is expected to be planar (63). The ground state of CF_3^+, if pyramidal and of C_{3v} symmetry, would be $^1A_1$; if planar and of D_{3h} symmetry, would be $^1A_2$".

CF_4 is T_d in its symmetrical equilibrium conformation and has a $^1A_1$ ground state. The highest occupied molecular orbital in our calculations turns out to be a triply degenerate f_2 orbital. [A more recent ab-initio Gaussian calculation (4) gave f_1 as the highest occupied molecular orbital with f_2 lower (but by only 0.2 eV)]. Thus CF_4^+ in its tetrahedral conformation would have either a $^2F_2$ ground state or a $^2F_1$ ground state. As the CF_3-F' distance is increased, the $^2F_2$ state splits into a $^2A_1$ and a ^2E state, and the $^2F_1$ state splits into a $^2A_2$ and a ^2E state.

The states of the F atom and its positive and negative ions were resolved into the species of point group T_d or C_{3v}. These were combined by group theory multiplication with the states of CF_3 and CF_3^+ to give the possible CF_4 and CF_4^+ states. (For details see reference 32).

As an illustrative example the following states of CF_4^+ are permitted from the combination of CF_3^+ + F and CF_3 + F^+.

Based on our calculated behavior of the CF_4^+ states, the relative thermochemical energies of the dissociated products, the spin and symmetry restrictions of the intermediate CF_4^+ states permissible from those products, the experimental appearance potential of CF_3^+ from CF_4 of 15.35 eV (63) and the complete absence of a CF_4^+ ion mass spectrometrically under any situation,

TABLE 5. CF_4^+ states permitted from the spin and symmetry combinations of the separated framents

(Energies relative to separated neutral fragments, eV)

CF_3^+ + F $\xrightarrow{\quad 9.17 \quad}$ CF_4^+

$^1A_1$ (pyramidal) ^2P_u T_d

T_d / \ C_{3v}

$^2F_2$ $^2A_1 + {}^2E$ $^2F_2$ $^2A_1 + {}^2E$

CF_3^+ + F $\xrightarrow{\quad 30.07 \quad}$ CF_4^+

$^1A_1$ (pyramidal) ^2S_g T_d / \ C_{3v}

T_d / \ C_{3v} $^2A_1$ $^2A_1$

$^2A_1$ $^2A_1$

CF_3 + F^+ $\xrightarrow{\quad 17.42 \quad}$ CF_4^+

$^2A_1$ (pyramidal) ^3P_g T_d C_{3v}

T_d / \ C_{3v}

$^3F_1$ $^3A_2 + {}^3E$ $^{2,4}F_1$ $^{2,4}A_2 + {}^{2,4}E$

or ^2E (pyramidal) $\xrightarrow{\quad 23.21 \quad}$ or $^{2,4}F_1 + {}^{2,4}F_2$ $^{2,4}A_2 + {}^{2,4}E$ $+ {}^{2,4}A_1 + {}^{2,4}E$

or $^2A_2$ (pyramidal) $\xrightarrow{\quad 24.34 \quad}$ or $^{2,4}F_2$ $^{2,4}A_1 + {}^{2,4}E$

we were able to sketch the following schematic curves for CF_4 and CF_4^+ states relative to their dissociation products. There is only one set of dissociated products $[CF_3^+ \ (\tilde{X} \ {}^1A_1) + F \ (\tilde{X} \ {}^2P_u)]$ that lies low enough in energy to predissociate both the $^2F_1$ and $^2F_2$ states of CF_4^+. Yet these separated fragments can correlate only to a CF_4^+ state of $^2F_2$ symmetry. (Figure 4) It is the fact that the lowest CF_4^+ state lies above the energy of these separated products that permits dissociation by the repulsive $^2A_1$ branch of the $^2F_2$ state. The doublet $^2F_2$ state dissociates to the repulsive $^2A_1$ curve and to a strongly attractive ^2E curve. This ^2E curve must be cut very quickly by the repulsive ^2E curve arising from the asymptotic splitting of the $^2F_2$ state into $^2A_1 + {}^2E$ states which leads to the

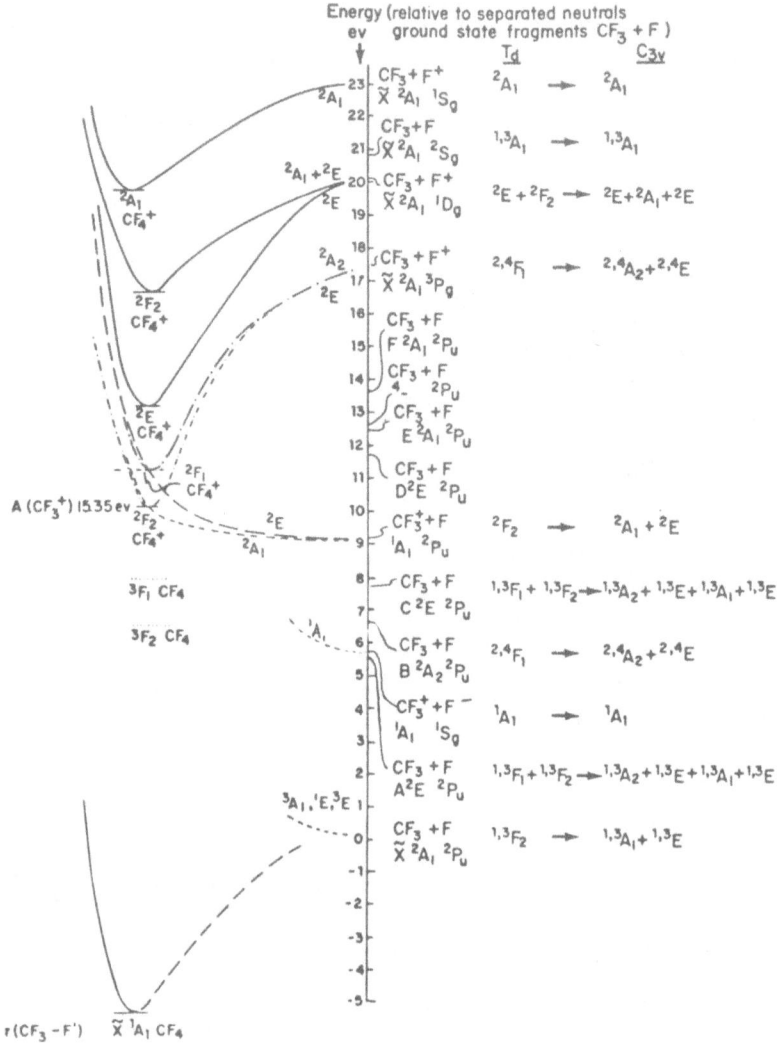

FIG. 4. CF$_4$ AND CF$_4^+$ POTENTIAL ENERGY CURVES (SCHEMATIC)

$^2F_1$ state and an avoided crossing should take place. This uneven splitting of the $^2F_2$ curve into the $^2A_1$ + ^2E curves explains how the one set of separated fragments which is only permitted to correlate to a T$_d$ CF$_4^+$ state of $^2F_2$ symmetry can predissociate both the $^2F_2$ and $^2F_1$ states of CF$_4^+$.

The theoretical symmetry and spin treatment outlined in this section is completely general and applicable to the entire field of molecular decompositions.

D. CI Calculations of the Excited States of H_3O^+

The excited states of H_3O^+ are pertinent to radiation chemistry, ion-molecule reactions, atmospheric processes, etc. Large scale ab-initio CI calculations are being carried out on the ground and excited states of H_3O^+ (53). For the final results several thousand configurations were included for each state. These were the best configurations selected by a perturbation procedure from up to twenty-five thousand configurations. Preliminary results utilized smaller configuration lists (still averaging several hundred to a thousand configurations).

The vertical energy separation of each excited state with relation to the $^1A_1$ ground state is given in the following table.

	Results VEE (eV)	Preliminary Results		Results VEE (eV)	Preliminary Results
$^1A_1$	0.00	0.00	$^3A_1$	11.36	11.86
	12.00	12.60		16.36	17.26
	17.12	18.06			
$^1A_2$	20.53	20.80	$^3A_2$		20.85
	23.60	24.54			24.34
^1E	14.64	15.09	^3E		14.57
	17.24	18.27			17.07

The final results for the calculated vertical excitation energies are a little lower than our preliminary results but not significantly so. The lowest excited state is the $^3A_1$, then the $^1A_1$ followed by the ^3E and ^1E states.

CONCLUSION

The most significant conclusion is the unambiguous confirmation of the overriding role that spin and symmetry restrictions play in any type of collisional process. Our original paper applying these restrictions to ion-molecule systems (37) states that these rules govern the course of all ion-molecule reactions as well as all reactive and non-reactive collisions. It does not matter if the reactants ever form even a transiently bound intermediate complex or not, the reactants and products must proceed along the symmetry and spin permitted paths, mitigated possibly by curve crossings, which are themselves also fixed by the potential energy surfaces. However, it should be stressed that it is not sufficient merely that the reactants have the proper spin and symmetry to permit them to combine to a particular electronic state of the intermediate, they must be the exact pair which does combine to this state. The same holds true for the products. There is an absolute uniqueness in the correspondence of an intermediate state to a pair of separated fragments. This uniqueness has permitted the derivation of a corresponding comprehensive general theory for molecular decompositions based on symmetry and spin restrictions.

ACKNOWLEDGEMENT

Special thanks are due to the Air Force Office of Scientific Research, Propulsion Division, and to its Chief, Dr. Joseph Masi, for their long time support of the author's research in theoretical and quantum chemistry which led to the understanding necessary to write this chapter.

The author should also like to thank Professor Walter S. Koski who collaborated in this research. Dr. Aaron Pipano collaborated in the CI computations of the $O^+ + N_2$ reaction. The configuration interaction calculations on the $O^+ + N_2$ potential energy surface were supported in part by BRL Contract No. DAAD05-70-0027 and in part by the Atomic Energy Commission, as was the CF_4^+ research. Dr. Richard C. Raffenetti and Dr. Harry J. T. Preston collaborated in the CI computations on H_3O^+, also supported by BRL.

REFERENCES

1a.   BATES, D. R., (1960) Proc. Roy. Soc. A257,22.
 b.   BATES, D. R., (1964) Proc. Roy. Soc. 84,517.
2.   BENDER, C. F., SHAVITT, I. AND PIPANO, A., unpublished.
3.   BORN, M. AND OPPENHEIMER, R., (1927) Ann. Physik. 84,457.
4.   BRUNDLE, C. R., ROBIN, M. B. AND BASCH, H., (1970) J. Chem. Phys. 53,2196.
5.   BRUNDLE, C AND TURNER, D. W., (1969) J. Mass Spectrometry and Ion Phys. 2,195.
6.   CALLOMON, J. H., (1959) Proc. Chem. Soc. p. 313.
7.   CATLOW, G. W., MCDOWELL, M. R. C., KAUFMAN, Joyce J., SACHS, L. M. AND CHANG, E. S., (1970) J. Phys. B: Atom. Molec. Phys. 3,833.
8.   CLEMENTI, E., (1968) Chem. Rev. 68,341.
9.   COULSON, C. A. AND ZALEWSKI, K., (1962) Proc. Roy. Soc. A268,437.
10a. CUSACHS, L. C., (1965) J. Chem. Phys. 43,S157.
  b. Ibid, (1966) 45,2717.
11.   DIBELER, V. H., WALKER, J. A. AND LISTON, S. K., (1967) J. Res. Natl. Bur. Stds. 71A,371.

12. DWORETSKY, S., NOVICK, R., SMITH, W. W. AND TOLK, N., (1967)
     Phys. Rev. Letters 18,939.
13. EBERHARDT, W. H., CRAWFORD, B., JR. AND LIPSCOMB, W. N., (1954)
     J. Chem. Phys. 22,989.
14. EYRING, H., WALTER, J. AND KIMBALL, G. E., (1944) Quantum
     Chemistry, John Wiley and Sons, Inc., New York.
15. FESSENDEN, R. W. AND SHULER, R. H., (1965) J. Chem. Phys.
     43,2704.
16. GERSHGORN, Z. AND SHAVITT, I., (1967) Int. J. Quantum Chem.
     1S,403.
17. GIESE, C. F., (1966) Advan. Chem. Ser. 58,20.
18. GILMORE, F., (1967) DASA Reaction Rate Handbook, Chapter 4.
19. GIOUMOUSIS, G. AND STEVENSON, D. P., (1958) J. Chem. Phys.
     29,294.
20. HARRIS, F. E., (1967) J. Chem. Phys. 46,2769.
21. HASTED, J. B. (1974) Lecture presented at Naval Research
     Laboratory, Washington, D.C., May.
22a. HERZBERG, G., (1950) Molecular Spectra and Molecular Structure.
     I. Spectra of Diatomic Molecules, Second Edition, D. Van
     Nostrand Co., Inc., New York, p. 315.
  b. Ibid, Table 30, p. 335.
  c. Ibid, Table 31, p. 336.
  d. Ibid, Table 32, p. 337.
23a. HERZBERG, G., (1966) Molecular Spectra and Molecular
     Structure. III. Electronic Spectra and Electronic
     Structure of Polyatomic Molecules, D. Van Nostrand Co., Inc.,
     New York. See Subject Index for tables for individual
     symmetry groups. (In the text of the present article this
     book is often referred to as "Herzberg Polyatomics".)
  b. Ibid, Table 21, p. 280.
  c. Ibid, Table 22, p. 283.
  d. Ibid, Table 24, p. 286.
  e. Ibid, Table 23, p. 285.
  f. Ibid, Table 25, p. 289.
  g. Ibid, Table 30, p. 331.
  h. Ibid, Table 31, p. 332.
  i. Ibid, p. 593.
  j. Ibid, p. 472.
  k. Ibid, p. 292.
24. HINZE, J. AND ROOTHAAN, C. C. J., (1967) Prog. Theor. Phys.
     Osaka, Suppl. 40,37.
25a. HOFFMAN, R., (1963) J. Chem. Phys. 39,1397.
  b. Ibid, (1964) 40, 2474 and 2480.
  c. HOFFMAN, R., (1966) Tetrahedron 22,521.
26a. HOFFMAN, R. AND LIPSCOMB, W. N., (1962) J. Chem. Phys. 36,2179.
  b. Ibid, (1962) 37,2872.
27a. IDEN, C. R., LIARDON, R. AND KOSKI, W. S., (1971) J. Chem.
     Phys. 54,2757.
  b. Ibid, (1972) 56,851.

28.   JAFFE, H. H., (1969) Accts. of Chem. Res. 2,136.
29.   JONAS, A. E., SCHWEITZER, G. K., GRIMM, F. A., AND CARLSON,
      T. A., (1972/73) J. Electron. Spectrosc. 1,29 and
      references cited therein.
30.   JUG, K., (1969) Theor. Chim. Acta 14,91.
31.   KAUFMAN, Joyce J., (1965) J. Chem. Phys. 43,S152.
32.   KAUFMAN, Joyce J., (1973)"The Role of Spin and Symmetry
      Restrictions in Reactive and Non-Reactive Collisions and
      Molecular Decomposition", chapter in the book WAVE MECHANICS
      the First Fifty Years, edited by W. C. Price, S. S. Chessick
      and T. Ravensdale, Buttersworth Co. Publ. Ltd., p. 208.
33.   KAUFMAN, Joyce J., "Potential Energy Surface Considerations
      for Reactions Involving Excited States," chapter in the
      book, Excited States in Chemical Physics, edited by J. W.
      McGowan, John Wiley Interscience Publishers, in press.
34.   KAUFMAN, Joyce J., (1972) Lecture presented at Batelle
      Memorial Institute, Columbus, Ohio, May.
35.   KAUFMAN, Joyce J., HARKINS, J. J. AND KOSKI, W. S., (1969)
      J. Chem. Phys. 50,771.
36a. KAUFMAN, Joyce J., KERMAN, Ellen, AND KOSKI, W. S., (1971)
      Int. J. Quantum Chem. 4S,391.
  b. KAUFMAN, Joyce J., KERMAN, Ellen AND KOSKI, W. S., (1970)
      "Implications of Photoelectron Spectroscopic Measurements
      for Compounds Which Produce No Parent Ion," Division of
      Physical Chemistry, 160th American Chemical Society National
      Meeting, September, Abstract #176.
37.   KAUFMAN, Joyce J. AND KOSKI, W. S., (1969) J. Chem. Phys.
      50,1942.
38.   KOOPMANS, T., (1933) Physica 1,104.
39a. LAIDLER, K. J., (1955) The Chemical Kinetics of Excited
      States, Clarendon Press, Oxford, England.
  b. Ibid, p. 79.
40.   LAIDLER, K. J. AND SHULER, K. E., (1951) Chem. Rev. 48,157.
41a. LANDAU, L., (1932) Soviet Phys. 1,89.
  b. LANDAU, L., (1932) Phys. Sowj. 2,46.
42.   LEWIS, J. K. AND HOUGEN, J. T., (1968) J. Chem. Phys. 48,5329.
43a. LICHTEN, W., (1963) Phys. Rev. 131,229.
  b. LICHTEN, W., (1965) Phys. Rev. 139A,27.
  c. FANO, U. AND LICHTEN, W., (1965) Phys. Rev. Letts. 4,627.
  d. LICHTEN, W., (1968) Advances in Chemical Physics, Vol. XIII,
      edited by I. Prigogine, Interscience Publishers, New York,
      p. 41.
  e. LICHTEN, W., (1967) Phys. Rev. 164,131.
44.   MAHAN, B. H. AND SLOANE, T. M., (1973) J. Chem. Phys. 59,5661.
45.   MCLEAN, A. D. AND YOSHIMINE, M., (1967) Supplement to IBM
      Journal Res. and Dev., November.
46.   NATALIS, P. AND COLLIN, J. E., (1969) J. Mass Spectrometry
      and Ion Physics 2,221.
47.   NICHOLSON, B. J., (1970) Adv. Chem. Phys. 18,249.

48. PIPANO, A. AND KAUFMAN, Joyce J., (1970) Chem. Phys. Letts. 7,99.

49a. PIPANO, A. AND KAUFMAN, Joyce J., (1971) Int. J. Quantum Chem. 5,233.

b. PIPANO, A. AND KAUFMAN, Joyce J., (1972) J. Chem. Phys. 56,5258.

50. PIPANO, A. AND SHAVITT, I., (1968) Int. J. Quantum Chem. II, 741.

51. POLANYI, J. C., (1971) Proceedings of the Conference on Potential Energy Surfaces in Chemistry, W. A. Lester, Jr., Ed., IBM Research Lab., San Jose, California, p. 10.

52. POPLE, J. A. AND BEVERIDGE, D. A., (1970) Approximate Molecular Orbital Theory, McGraw-Hill Book Co., New York.

53. RAFFENETTI, R. C., PRESTON, H. J. T. AND KAUFMAN, Joyce J., to be published.

54a. ROOTHAAN, C. C. J., (1951) Rev. Mod. Phys. 23,69.

b. Ibid, (1960) 32,179.

55. RUTHERFORD, J. A., AND VROOM, D. A., (1971) J. Chem. Phys. 55,5622.

56. SCHAEFER, H. F., III, (1972) The Electronic Structure of Atoms and Molecules. A Survey of Rigorous Quantum Mechanical Results, Addison-Wesley Publ. Co., Reading, Mass.

57. SCHMELTEKOPF, A. L., FERGUSON, E. E. AND FEHSENFELD, F. C., (1968) J. Chem. Phys. 48,2966.

58. SHULER, K. E., (1953) J. Chem. Phys. 21,624.

59. SLATER, J. C., (1963) Quantum Theory of Molecules and Solids, Vol. 1, McGraw-Hill Book Co.

60. Reference 59, page 108.

61. STUECKELBERG, E. G. C., (1932) Helv. Phys. Acta 5,369.

62. WALSH, A. D., (1953) J. Chem. Soc. 2260.

63. WALTER, T. A., LIFSHITZ, C., CHUPKA, W. A. AND BERKOWITZ, J., (1969) J. Chem. Phys. 51,3531 and earlier experimental references cited therein.

64. WIGNER, E. AND WITMER, E. E., (1928) Z. Physik. 51,859.

65. WOLFSBERG, M. AND HELMHOLZ, L., (1952) J. Chem. Phys. 29,827.

66. ZENER, C., (1932) Proc. Roy. Soc. A137,696.

REACTIONS OF ELECTRONICALLY EXCITED POSITIVE IONS

Walter S. Koski

The Johns Hopkins University
Department of Chemistry
Baltimore, Maryland 21218, U.S.A.

ABSTRACT

The electron bombardment of molecules generally gives rise to ions in more than one electronic state. The ion beam composition with respect to the electronic states can be determined by studying the attenuation of the beam as a function of the pressure of the attenuating gas. In this way the composition of B^+, C^+, F^+, H_2O^+ and H_3O^+ ion beams was determined. The difference between the reactive behavior of the ion beam in the ground state and first excited electronic state with simple target molecules and some of the factors that determine the amount of excited state produced were examined.

INTRODUCTION

Collisions between positive ions and neutral target gases have been studied extensively as a function of the projectile ion kinetic energy. However, until comparatively recently little attention was given to the role of internal energy in the projectile in such reactions. In these early studies it was frequently considered sufficient simply to use a mass analyzed beam of the correct chemical species and of known kinetic energy or a rationale was established indicating that negligible amounts of the excited ions were present or that they did not affect the final conclusion. As the state of the art advanced it was demonstrated in a number of experiments that the collision properties of ions of a given chemical type were dependent on their mode of preparation. In cases where the ion were prepared by electron bombardment frequently it was found that the collision properties of the ions were strongly dependent on the electron energy. An example of such an effect is the cross section variation of the reaction $O^+(N_2,O)N_2^+$ as a function of electron energy reported by Turner, Rutherford and Compton (1968). Since properties of ions in the excited states frequently differ dramatically from those of ions in the ground state it was desirable to be able to prepare ion beams in known electronic states especially if quantitative information was desired. One can determine the abundances of various ions produced by photo-ionization using the techniques of photoelectron spectroscopy. In principle such ions could be formed into a focussed beam, however, the terminal distribution of the electronic states frequently will differ from the initial distribution so a technique which can be used to determine the abundances of the electronic states from direct measurements on the beam is desirable. Such a technique has been developed by Turner, Rutherford and Compton (1968). Although it is not a precision technique its simplicity and ready adaptability to spectrometers used in collision studies make it a very useful and attractive technique. In this presentation we will outline the method and illustrate its applicability to the reactions of some monoatomic and polyatomic ions in electronically excited states. No attempt will be made to realize a complete literature coverage, rather illustrative examples will be presented.

PRINCIPLE OF THE METHOD

The method developed by Turner et al. (1968) for determining the beam compositions as far as electronic states is concerned is based on the attenuation of an ion beam on passage through a chamber containing a gas, the pressure of which can be varied.

When an ion beam containing ions in a single electronic state (i) passes through a reaction chamber of length ℓ containing a gas whose pressure can be varied, it will be attenuated in an exponential manner.

$$I_i = I_0 \exp (-n\sigma_i \ell) \tag{1}$$

where I_i is the ion current reaching the detector. I_0 is the ion current reaching the detector when the gas number density (n) is zero and σ_i is the sum of the cross sections for all processes by which ions of the ith type are lost. A semi logarithmic plot of I_i/I_0 vs. gas pressure gives a straight line. For another ion of the type j a similar equation holds and if $\sigma_j > \sigma_i$ a line of greater slope is obtained.

If a beam of ions containing both $\underline{i}$ and j type ions is passed through the same chamber the current I reaching the collector is

$$I = (1-f)I_0 \exp(-n\sigma_i \ell) + fI_0 \exp(-n\sigma_j \ell) \qquad (2)$$

where f is the fractional abundance of the jth type ion. A plot of I/I_0 would then give a curve. A curve of this type is indicative of two or more states in the ion beam and can in principle be resolved to give the concentration of these states. If a straight line is obtained at high pressure the second term in equation 2 is negligible in this region so I/I_0 can be extrapolated to zero pressure to give (1-f) directly.

APPARATUS AND PROCEDURE

The apparatus used by Turner et al. (1968) for ion attenuation studies is illustrated in Figure 1. The ions are produced by electron bombardment. The electron energy can be varied from zero to 200 eV. The ions are extracted with a small field, focussed and accelerated to 100 eV before momentum

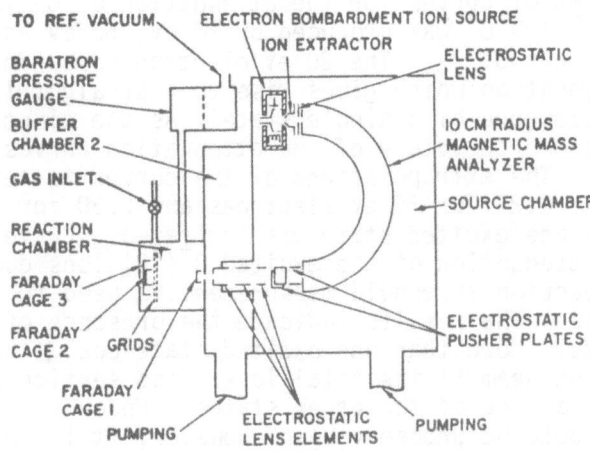

FIG. 1. SCHEMATIC DIAGRAM OF APPARATUS USED FOR ION BEAM
ATTENUATION STUDIES BY TURNER, RUTHERFORD AND COMPTON (1968).

analysis, and refocussed before leaving the source vacuum chamber through a 1/4 x 1/8 in. orifice. The ions then travel for about 1 cm through a separately pumped buffer chamber before entering the reaction chamber through a slightly smaller orifice. The base pressure in the source chamber is ~ 10^{-7} torr. The pressure in the buffer chamber is ~ 0.01 of that in the reaction chamber. The ion travel time is about 20 u sec. The reaction chamber is 10 cm long and has three Faraday cages. The one at the entrance to the chamber can be removed after I_0 is measured. The Faraday cages at the other end measure the ion current after passage through the gas. The latter Faraday cages subtend markedly different angles and can be used to assess the importance of the results of elastic scattering. Two grids are placed just before the Faraday cages one at + 9V and the other at - 9V. In this manner slow secondary ions cannot enter and the secondary electrons from the cages are suppressed.

     The experimental procedure is to obtain a stable beam of ions using a desired electron energy. The current is measured with the first Faraday cage which is then removed and the measurements are made with cages 2 and 3 as a function of pressure of the selected gas in the reaction chamber.

RESULTS

Electronic states of O^+, O_2^+ and NO^+ produced by electron bombardment

     A typical set of curves for the attenuation of O^+ in Argon is given in Fig. 2. The O^+ was produced by 20 eV, 50 eV and 100 eV electron bombardment of O_2. The 20 eV electron bombardment of O_2 gives an O^+ attenuation which gives rise to a straight line (A) indicating the presence of a single state. As the electron energy increases the curvature of the attenuation curves becomes more pronounced. The extrapolations of the curves to zero pressure give f = 0.27 for 50 eV electrons and 0.30 for 100 eV electrons. Only one excited state was indicated. Curve D represents the attenuation of the excited $O^+(^2D)$ ions due to all processes. A reaction threshold measurement is used to identify the excited state. The results indicate the presence of only one excited state. More than one excited state could, however, be present in the ion beam if its total loss cross section was nearly the same as that of one of the other states. Under such conditions, an excited ion would be unobservable. However, it is unlikely that the total loss cross sections for two states would be the same if different attenuating gases were used and it is advisable to make attenuation measurements in more than one gas in order to detect any such "hidden" states.

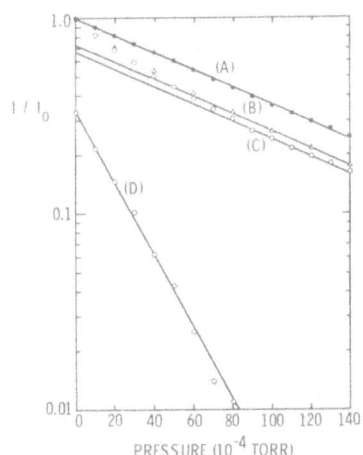

FIG. 2. ATTENUATION CURVE MEASURED FOR O^+ IN Ar. CURVES A,
B AND C ARE CURVES OBTAINED WHEN THE ELECTRON BOMBARDING
ENERGY WAS 20, 50 AND 100 eV RESPECTIVELY. CURVE D
REPRESENTS THE DECAY OF ONLY THE EXCITED IONS WHICH WERE
PRESENT IN THE ION BEAM FOR 100 eV ELECTRONS.

In actual practice there are other minor complications that
may have to be taken into account. An example is illustrated by
the work of Mathis, Turner and Rutherford, (1968), in which the
attenuation of NO^+ ions in O_2 was studied. In this study it was
found that there was a deviation from linearity in the attenuation
plot and as a result the slope at low pressures is significantly
smaller than the slope at high pressures. This has been
attributed to collisional de-excitation of $(A^+)*$ to A^+ in the
attenuating gas. It can be shown that when the excited ions are
collisionally de-excited into ground state ions, the transmitted
current, instead of being given by equation 2 is given by
a more complicated expression involving the collisional
de-excitation cross section.

It is clear that the above attenuation technique is not completely fool proof and frequently it is advisable to check the
attenuation measurements with some other type of experiment. The
work of Tiernan and Marcotte, (1970) on the role of kinetic and
internal energy in the collision dissociation of NO^+ and O_2^+
provides an interesting example since these two ions were also
studied by Mathis et al. Using an in-line tandem machine
Tiernan et al. obtained the data illustrated in Fig. 3a for the

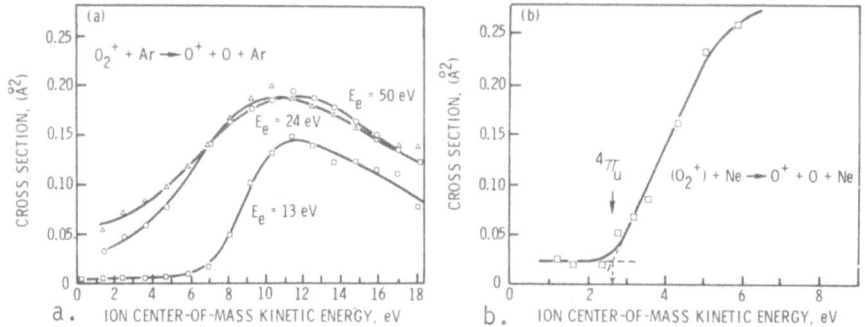

FIG. 3(a). CROSS SECTION FOR THE REACTION O_2^+(Ar,O,Ar)O^+ AS A
FUNCTION OF ION KINETIC ENERGY AT SEVERAL ELECTRON ENERGIES.
(b). KINETIC ENERGY DEPENDENCE OF THE CROSS SECTION FOR THE
REACTION O_2^{+*} (Ne,O,Ne)O^+.

reaction O_2^+(Ar,O,Ar)O^+. With 13 eV energy electron bombardment of
O_2 solely the reaction of the ground state ion is expected. This
is supported by the observed threshold value of 6.9 eV which is to
be compared with the adiabatic dissociation energy from spectroscopic data of 6.6 eV. The shift in threshold for dissociation of
O_2^+ at higher electron energies suggests the presence of an
electronically excited state in the ion beam. If the ground state
curve is subtracted from the curve obtained at the highest electron
energy (maximum amount of excited state) the resultant curve should
characterize dissociation reactions of the excited ion only. A
typical result reported by Tiernan et al., (1970) is shown in Fig.
3b. This curve was corrected to the population of the excited
reactant ion using the data of Turner et al. (1968). Results
obtained using Ar and Ne as target gases were reported to be
virtually identical. The maxima cross sections for dissociation
of excited ions were significantly higher than for the
corresponding ground states of O_2^+ and NO^+. In the case of
dissociation of excited O_2^+ by Ne the measured threshold of 2.7 eV
compared favorably with the adiabatic dissociation energy of the
metastable 4π state of O_2^+ (Gilmore (1965). The sharp onset
observed at threshold and the lack of structure in the plots
strongly indicate that only one excited state is present in
agreement with the conclusions of Turner et al., (1968). Similar
results were obtained in the collision induced dissociation of NO^+

Some reactions of C^+

Attenuation studies of C^+ ion beams produced by 100 eV electron bombardment of CO show that the C^+ ion beam composition was 32% ^4P and 68% ^2P (Lao, Rozett and Koski, (1968). Just as in the case of 0^+, 0_2^+ and NO^+ only two electronic states were observed. The composition of the ion beam was a function of not only the electron energy but also a function of the pressure in the primary ion source. If the source pressure was sufficiently high the C^+ ions produced were practically completely in their ground state (Sullivan, Rozett and Koski, (1970). By studying reactions of C^+ produced under conditions which gave rise to different abundances of the two states it was possible to disentangle the contributions of the ^2P and ^4P states of the ion to a reaction. For example, in the charge transfer reactions $C^+(N_2,C)N_2^+$ and $C^+(O_2,C)O_2^+$ when $C^+(^4P)$ was used as a projectile the cross sections were about 100 times greater than when the projectile ion was in the ^2P state. The reaction $C^+(O_2,O)CO^+$ gives a typical exothermic behavior, i.e., the cross section falls with kinetic energy, when $C^+(^4P)$ is the reactant. On the other hand when $C^+(^2P)$ is used the reaction cross section goes through a maximum indicating an activation energy although the reaction is exothermic. This apparently arises from the fact that spin and symmetry require the reaction to proceed through a high lying intermediate state imposing a significant activation energy on the process.

The dissociative charge transfer reactions $C^+(N_2,C,N)N^+$ and $C^+(O_2,C,O)O^+$ are endothermic for both states of C^+ and the cross section is highest for the reaction whose endothermicity is smaller.

Some reactions of B^+

The B^+ ion is an interesting projectile. Its ground state is a closed shell system (^1S_g) and its first excited state which lies 4.6 eV above the ground state has an open shell structure (^3P_u). Electron bombardment of BF_3 yields attenuation curves similar to those shown in Fig. 2. N_2O was used as the attenuating gas. It was clear that more than one electronic state was present and an examination of the intercept obtained by extrapolating the linear part of the curve showed that the abundance of the excited state decreased as the electron energy decreased, (Lin, Cotter and Koski, (1974). Consequently one can vary the amount of excited state of B^+ by lowering the electron energy and at energies close to the threshold the ground state predominates. If one is interested in obtaining as high an ion intensity as possible, as one may be in the use of the B^+ ion beam to study ion-molecule reactions, this approach to eliminating the excited state is not an attractive one because of the concomitant decrease in overall B^+ intensity.

Attempts to reduce the amount of excited state by collisional deactivation by increasing the pressure in the ion source have been successful in some other systems (Lindemann, Rozett and Koski, (1972) but this approach failed to eliminate the upper electronic state present in the B^+ case. The addition of paramagnetic impurities has also been used with some success to eliminate the excited electronic states in some ion beams (Cotter and Koski, (1973)) but this approach was not applicable here because of the high reactivity of the boron trihalides. However, it was found that the various boron trihalides gave different amounts of excited state on electron bombardment and BI_3 gave only ground state B^+ when 70 eV electrons were used. By studying reactions of B^+ produced using BF_3 and then using BI_3 as a target gas it was possible to determine the reaction cross sections for both the ground state and the first excited state of B^+. Examples of the difference in reactivity between ground and first excited state are given in Fig. 4 for the charge exchange reaction $B^+(N_2O, B)N_2O^+$. When B^+ is in its ground state, the energy defect for the reaction is considerably greater than for the case where B^+ is in the ^3P state in which case an almost resonant process might be expected. In the former case, one obtains a small cross section rising with energy and going through a broad maximum. In the case of B^+ produced from BF_3, the beam composition was 35.5% ^3P and 64.5% ^1S and using such an ion beam Fig. 4 shows that the cross section curve goes through a sharper and larger maximum displaced to the lower energy compared with the B^+ (^1S) curve. This is the expected behavior for a charge transfer process where the energy defect is changed in the manner indicated. In the case of B^+ from BF_3, if one subtracts the contribution of the ground state and then normalizes the curve, one obtains the third curve in Fig. 4 which represents the charge transfer reaction of B^+ (^3P) with N_2O.

In addition to charge exchange reactions with N_2O, the processes $B^+(N_2O,N_2)BO^+$, $B^+(N_2O,N,B)NO^+$ and $B^+(N_2O,N_2,B)O^+$ were observed and are graphically illustrated in Fig. 5a. All of these reactions are expected to be endothermic and the observations are in harmony with this expectation. The excited state reactions of B^+ behave as expected in comparison with the ground state reactions. For example, the maxima of the cross section curves and the apparent threshold values are shifted to lower energies when B^+ is in the excited state.

A more dramatic difference in the behavior of excited vs. ground state behavior is exhibited by the $B^+(D_2,D)BD^+$ and this arises from the fact that the thermodynamics of the reaction are such that the ground state reaction is endothermic with a threshold of 9.9 eV whereas the reaction involving the first excited state is exothermic.

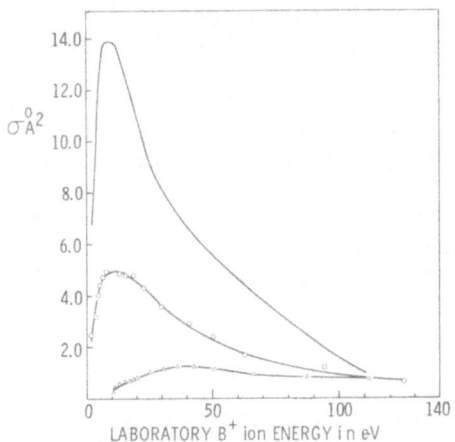

FIG. 4. CROSS SECTION VS. KINETIC ENERGY CURVES FOR THE CHARGE TRANSFER REACTION $B^+(N_2O,B)N_2O_3^+$ Δ - B^+ PRODUCED FROM BI_3,O - B^+ PRODUCED FROM BF_3 (35.5% ^3P and 64.5% ^1S). SOLID LINES GIVE THE CROSS SECTION VS. KINETIC ENERGY OF BOMBARDING ION FOR B^+ (^3P).

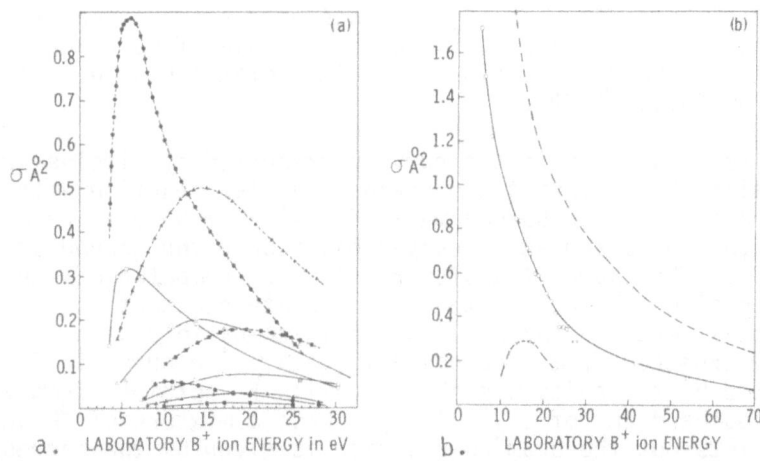

a. LABORATORY B^+ ion ENERGY in eV   b. LABORATORY B^+ ion ENERGY

FIG. 5a. CROSS SECTIONS OF VARIOUS MASS TRANSFER REACTIONS BE-TWEEN B^+ AND N_2O ● (UPPER) $B^+(^3P)(N_2O,N_2)BO^+$, ▲ (UPPER) $B^+(^3P)(N_2O,N,B)NO^+$ ⊡ UPPER $B^+(^3P)(N_2O,N_2,B)O^+$, O B^+ ($^3P + ^1S$) $(N_2O,N_2)BO^+$, Δ $B^+(^3P + ^1S)(N_2O,N,B)NO^+$, ⊡ $B^+(^3P + ^1S)$ $(N_2O,N_2,B)O^+$, ● LOWER $B^+(^1S)(N_2O,N_2)BO^+$, ▲ (LOWER) $B^+(^1S)$ $(N_2O,N,B)NO^+$, and ■ (LOWER) $B^+(^1S)(N_2O,N_2,B)O^+$. (b) CROSS SECTIONAL DEPENDENCE OF $B^+(D_2,D)BD^+$, $\Delta = ^1S$, O-($^3P+^1S$), -- ^3P STATE.

Typical cross sectional dependences of the $B^+(D_2,D)BD^+$ for ^1S and ^3P states of B^+ are given in Fig. 5b. The ground state reaction cross section gives a typical endothermic behavior starting at the expected threshold, going through a maximum and then declining. The B^+ ions in this case were prepared by 70 eV electron bombardment of BI_3. The reaction of ^3P state of B^+, on the other hand, gives a typical exothermic behavior. The ion beam was prepared by electron bombardment of BF_3 and the beam composition was determined by studying the attenuation of B^+ in N_2O gas as a function of gas pressure (Lin, Watkins, Cotter and Koski, (1974).

Some reactions of F^+

In the case of B^+ we had an opportunity to compare the reactivity of a closed shell ion and an open shell system. In F^+ the ground state is a ^3P state and the first excited is a ^1D state at 2.59 eV above the ground state. The next excited state is a ^1S state at 5.56 eV above ground. Here we have an opportunity to compare the reactivities of two open shell structures since 100 eV electron bombardment of CF_4 gives 55% $F^+(^3P)$ and 45% $F^+(^1D)$ in our ion source. The attenuation curves obtained were similar to the ones shown in Fig. 2. In order to produce ground state F^+ ions a 70-30 mixture of CF_4 and NO was irradiated with 100 eV electrons. The attenuation curves of F^+ so produced showed only ground state ions were present. Ne was used as an attenuating gas, (Lin, Cotter and Koski, (1974). The role of paramagnetic impurities in the deactivation of excited states has recently been reviewed by Chiu, (1972).

The results shown in the curve represented by circles in Fig. 6 were obtained by measuring the charge transfer reaction with F^+ obtained from electron bombardment of the CF_4 + NO mixture which judging from the attenuation curves gives only the ground state (^3P) of F^+. The charge transfer results are compatible with this conclusion since the use of the ground state reactant ion gives an experimental reaction threshold which is compatible with the value of 8.4 eV expected from thermodynamic data. The curve in Fig. 6a represented by triangles was obtained with F^+ obtained from 70 eV electron bombardment of CF_4. In this case a mixture of ^3P and ^1D states exists but the beam composition is given by the corresponding attenuation data. By subtracting the contribution of the ground state and normalizing one obtains the cross section behavior of the reaction $F^+(^1D)(Ne,F^+)Ne^+$. The appearance potential measurements

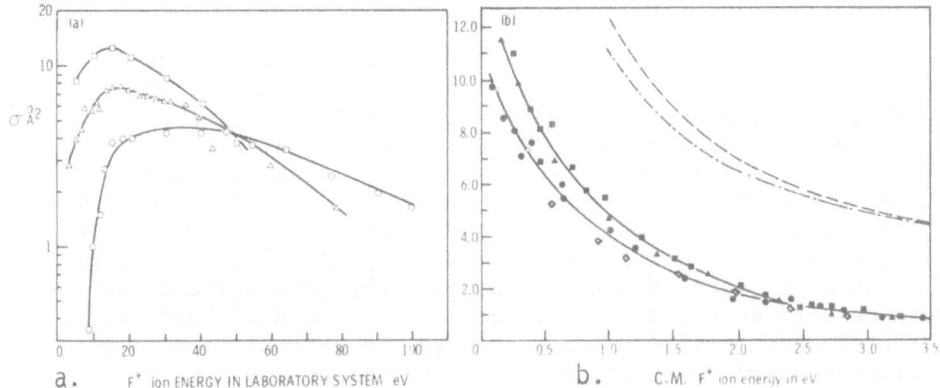

FIG. 6a.   CROSS SECTION VS. F^+ ION ENERGY FOR THE REACTION
$F^+(Ne,F)Ne^+$. (b) CROSS SECTION VS. F^+ ION ENERGY FOR THE
REACTIONS $F^+(^3P) + H_2, D_2$.   THE DASHED CURVES GIVE THE
CORRESPONDING RESULTS ASSUMING THE PHASE SPACE THEORY OF ION
MOLECULE REACTIONS. $F^+(^3P) + H_2$ (●) AND D_2 (□) AND $F^+(^1D) +$
H_2 (■) AND D_2 (▲).

showed that this state was the ^1D state of F^+ and the charge
exchange results are qualitatively consistent with this.   Both of
these reactions are expected to be endothermic.   The cross sections
as a function of energy both go through a broad maximum; however, the
excited state (^1D) reaction is the one which has a smaller energy
defect and its maximum occurs in a lower energy region as expected.
The measurement of the threshold for this reaction required a projectile kinetic energy lower than that accessible with our
instrument, consequently, we could not use such a measurement for
state confirmation in this case.

In Fig. 6b, the cross sectional variation with ion energy of
the reactions $F^+(^3P)$ and $F^+(^1D)$ are given for H_2 and D_2.
Experimental measurements indicate that the cross sections
corresponding to the two states are about the same at the higher
energies but diverge measurably at the lower energies.   The fact
that a divergence between the two states is to be expected is
borne out by the statistical phase space calculations which are
given by the dashed lines in the figure.

Electronic states of H_2O^+ and H_3O^+

Lindemann, et al. (1972) have studied the abundances of the electronic states of H_2O^+ produced by electron bombardment of H_2O vapor in the energy range 12-70 eV using the attenuation technique outlined earlier. It was found that at a pressure of 0.40 µ and 70 eV electrons the ground ($^2B_1$) state and the first excited state ($^2A_1$) were produced. When the H_2O pressure in the ion source was increased to 20 µ and 70 eV electrons were used only the $^2B_1$ state was produced. Lowering the electron energy to 14 eV at a pressure of 0.40 µ produced only ground state H_2O^+. These results show that the composition of the H_2O^+ beam, as far as the electronic states are concerned, is a function of the electron bombarding energy and the water vapor pressure in the primary ion source.

On irradiation of water vapor with electrons H_3O^+ as well as H_2O^+ is produced. The former arises from the ion-molecule reaction $H_2O^+(H_2O,OH)H_3O^+$. Cotter and Koski, (1973) have studied the reaction of this ion with D_2, i.e., $D_3O^+(D_2,D_2O)D_3^+$. The cross section behavior of this reaction as a function of beam energy is given in Fig. 7a. In the lower energy region the cross section falls with kinetic energy, a behavior indicative of an exothermic ion molecule reaction. This is the expected behavior if the D_3O^+ is in an excited electronic state. The peak in the curve arises from the opening of another reaction channel involving the ground electronic state of D_3O^+ which is expected to be an endothermic reaction.

It is not unreasonable for the D_3O^+ produced from the reaction $D_2O^+(D_2O,OD)D_3O^+$ to be in more than one state since there are two expected mechanisms by which the ion can be produced in this reaction. In one case, deuteron transfer can occur from the projectile to the target and in the other, atom transfer can take place from the target to the projectile. Furthermore, since the D_3O^+ ion has an even number of electrons, triplet as well as singlet states are possible. If the upper electronic state of the D_3O^+ ion produced in this study is indeed a triplet state, the experimental difficulty experienced in attempting to de-excite the ion to its ground states by techniques that were successful in the C^+ and H_2O^+ cases could be explained by the forbidden nature of the triplet-singlet transition. This reasoning suggested the incorporation of paramagnetic impurities in the water vapor being bombarded by electrons. This technique has been used successfully by Callear and Wood, (1971) in deactivating the triplet state of N_2 by NO. Introducing a combined D_2O - NO mixture at a total pressure of 20 µ in the ion source produced D_3O^+ ions whose reaction cross section with D_2 is shown in Fig. 7b and whose attenuation curve in NO was linear indicating the presence of only the ground state.

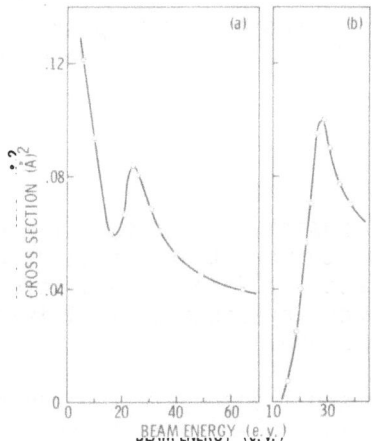

FIG. 7(a). CROSS SECTION VS. BEAM ENERGY FOR THE REACTION
$D_3O^+(D_2,D_2O)D_3^+$ WHEN D_3O^+ WAS PREPARED BY ELECTRON BOMBARD-
MENT OF PURE D_2O (b) CROSS SECTION VS. BEAM ENERGY WHEN
D_3O^+ WAS PREPARED BY A 70 eV ELECTRON BOMBARDMENT OF A 20 μ
MIXTURE OF D_2O - NO. THE MIXTURE WAS APPROXIMATELY 30% NO.

Dynamics of reactions involving electronically excited states

        Very few ion-molecule reactions have been studied in which
the dynamics of the reaction of the ground and excited states of
the cation have been examined. The reaction $O_2^+(D_2,D)O_2D^+$ is one
such reaction and it has been studied in two laboratories.
Gislasson, et al. (1969) studied this reaction with the projectile
ion in the ground state. In this state the reaction is endothermic
by 1.9 eV and at a relative energy of 3.86 eV the reaction appears
to proceed by a persistent complex mechanism as indicated by the
contour map for O_2D^+ as the projectile energy increased the
mechanism gradually changed to a direct one.

        Bosse, et al., (1971) on the other hand, used an O_2^+ beam
which was a mixture of the ground and excited states. The reaction
involving the excited state is exothermic by -2.0 eV. At a
relative energy of 1.0 eV which is below the threshold for the
ground state reaction only the excited state reacts and the contour maps indicated a direct mechanism, as the energy was increased both states participate in the reaction and the higher
energy contour maps were interpreted as arising from contributions
from both direct and complex mechanisms. Bosse, Ding and
Henglein (1971) also measured the velocity spectra and obtained
two peaks in the intensity vs. energy plot. A stripping peak due

to reaction with $H_2O_2^+$ ($^4\pi$) and a complex peak resulting from the reaction of $O_2^+(^2\pi_g^2)$ and μHD. The results were consistent with the contour maps.

CONCLUSION

The preparation of positive ion projectile by electron bombardment generally leads to more than one electronic state. The abundance of excited states is determined by electron energy, pressure in the ion source, presence of impurities as well as the nature of the target compound.

The presence of internal energy in the projectile ion can influence the course of ion-molecule reactions. The thermodynamics of the system frequently play an important role as in the $B^+(D_2,D)BD^+$ reaction. In most instances the higher the exothermicity the higher is the reaction cross section. This is the situation in reactions that do not exhibit an activation energy. In a few reactions certain restrictions make it necessary for the system to proceed through a high lying intermediate state. Such reactions exhibit a significant activation energy and spin, symmetry restrictions and details of the potential energy curves play an important role in the kinetics of the reaction. Examples of such reactions are $C^+(^2P) + O_2 \rightarrow CO^+ + O$ and $O^+ + N_2 \rightarrow NO^+ + N$.

Finally we have seen that in reactions such as $O^+ + D_2 \rightarrow O_2D^+ + D$ that the mechanism of the reaction can be influenced by electronic excitation.

ACKNOWLEDGEMENT

The work of the author as reported in this article was supported by the U.S. Atomic Energy Commission and the U.S. Army Research Office - Durham.

REFERENCES

BOSSE, G., DING, A. AND HENGLEIN, A. (1971). Ber. Bunsenges physik. Chem. 75, 413.
CALLEAR, A. B. AND WOOD, P. M. (1971). Trans. Faraday Soc. 67, 272.
CHIU, Y. N. (1972). J. Chem. Phys. 56, 4882.
COTTER, R. J. AND KOSKI, W. S. (1973). J. Chem. Phys. 59, 784.
GILMORE, F. (1965). J. Quant. Spectry. Radiative Transfer 5, 369.
GISLASSON, E. A., MAHAN, B.H., T'SAO, C. W. AND WERNER, A. S. (1969). J. Chem. Phys. 50, 5418.

LAO, R. C. C., ROZETT, R. W. AND KOSKI, W. S. (1968). J. Chem. Phys. 49, 4202.

LIN, K. C., COTTER, R. J. AND KOSKI, W. S. (1974). J. Chem. Phys. 60, 3412.

LIN, K. C., COTTER, R. J. AND KOSKI, W. S. (1974). In press.

LIN, K. S., WATKINS, H. P., COTTER, R. J. AND KOSKI, W. S. (1974). J. Chem. Phys. 60, 5134.

LINDEMANN, E., ROZETT, R. W. AND KOSKI, W. S. (1972). J. Chem. Phys. 56, 5490.

MATHIS, R. F., TURNER, B. R. AND RUTHERFORD, J. A. (1968). J. Chem. Phys. 48, 2051.

SULLIVAN, P. A., ROZETT, R. W. AND KOSKI, W. S. (1970). J. Chem. Phys. 52, 5321.

TIERNAN, T. O. AND MARCOTTE, R. E. (1970). J. Chem. Phys. 53, 2107.

TURNER, B. R., RUTHERFORD, J. A. AND COMPTON, D. M. J. (1968). J. Chem. Phys. 48, 1602.

TEMPERATURE EFFECTS IN ION-MOLECULE REACTIONS

Richard F. Porter

Dept. of Chemistry and the Materials Science Center

Cornell University, Ithaca, New York 14850

INTRODUCTION

Knowledge of the effect of temperature on reaction rates and equilibrium constants is fundamental to the studies of chemical kinetics and thermodynamics. From practical considerations temperature extrapolation of rate or thermodynamic data may be helpful in predicting chemical behavior under conditions not readily accessible in the laboratory. From theoretical considerations the mathematical form of the temperature dependence of rate data may be useful in interpreting and understanding reaction mechanisms. With the resurgence of interest in ion-molecule reactions it is appropriate to question whether present kinetic theories and models are applicable to reactions between a charged and neutral particle. In experiments on ion-molecule reactions it is possible to adjust the energy of a reactant ion by imposing electrostatic fields. This is an important feature unique to these processes. If an ion receives acceleration from an electrostatic field under conditions where kT<<ev, the reaction rate may depend on the potential and fail to exhibit a temperature dependence (Stevenson, 1963). It is important therefore that we limit our discussion to rate data obtained for thermal conditions in which the reactant ions and molecules have achieved a Maxwell-Boltzmann distribution of velocities. Henchman (Henchman, 1972) has discussed in some detail the difficult problem of measuring thermal rate constants. Another distinguishing feature of ion-molecule reactions is the experimental observations that a number of these processes proceed rapidly without apparent activation energy. In general, reaction cross sections for spontaneous exothermic ion-molecule reactions are one to two orders of magnitude larger than those corresponding to the fastest bimolecular reactions between two uncharged molecules. These factors must

be considered in any discussion of ion-molecule kinetics.

The following presentation is organized into two parts; the first will cover current developments in low temperature ion-molecule chemistry and the second will cover the kinetic implications of temperature effects on ion-molecule rate constants. Several reviews (Ferguson, 1972) and articles are relevant to our discussion. The two volumes (Franklin, 1972) on "Ion-Molecule Reactions" edited by J. L. Franklin have been especially helpful.

LOW TEMPERATURE ION-MOLECULE CHEMISTRY

One of the important recent technical advances in ion-molecule chemistry is the experimental capacity to study reactions in ion sources at low-temperatures. Pioneering efforts in this development are due to Field (Field, 1972) and coworkers and to Kebarle (Kebarle, 1972) and coworkers. At source temperatures between 80° and 300° K ion-molecule clustering reactions are observed. Examples of stabilized complexes include $C_2H_9^+$ and $BF_3 \cdot H_3^+$. The nature of the structure and reactivity of these species in terms of current bonding theories is of obvious interest. Presently there are no experimental data indicating geometrical structures of even the simplest complexes, H_3^+ and CH_5^+. Our current knowledge in this respect is limited to results of theoretical calculations (Comba, et.al, 1969).

For purposes of illustration of low temperature ion-molecule behavior we refer to ion-intensity data (Table I) obtained from a chemical ionization study of H_2-BF_3 and H_2-HBF_2 mixtures at low temperatures (Pierce, Thesis 1974). In these systems the following types of processes are observed.

(A) Proton Transfer

$$BF_3 + H_3^+ \rightarrow BF_3H^+ + H_2$$

$$HBF_2 + H_3^+ \rightarrow BF_2H_2^+ + H_2$$

(B) Clustering Reactions

$$BF_3 \cdot H^+ + 2H_2 \rightarrow BF_3 \cdot H_3^+ + H_2$$

$$BF_2^+ + 2H_2 \rightarrow BF_2 \cdot H_2^+ + H_2$$

$$BF_2 \cdot H_2^+ + 2H_2 \rightarrow BF_2 \cdot 2H_2 + H_2$$

These systems exemplify a number of features characteristic of low temperature ion-molecule behavior. Figure 1 illustrates how the rate of clustering reactions measured at constant pressure decreases

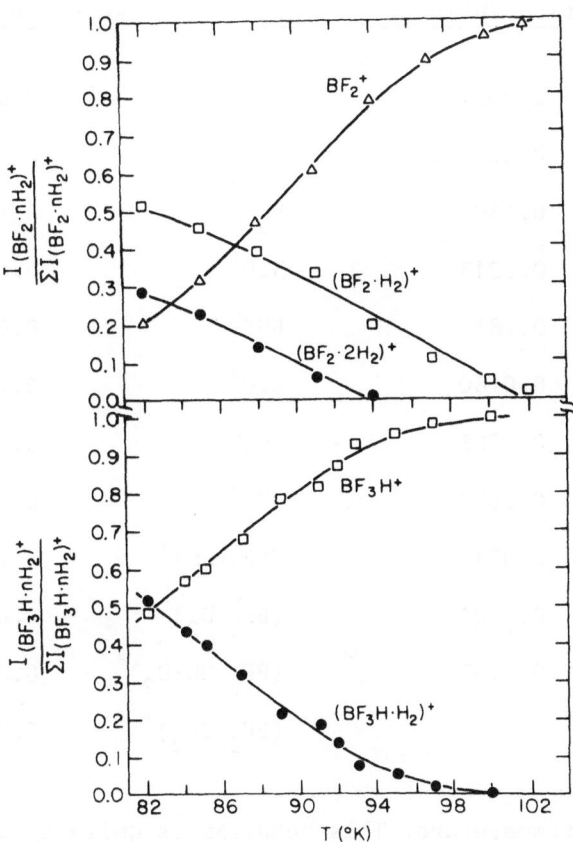

Fig. 1. Temperature dependence of positive ions formed by ion-molecule reactions in BF_3/H_2 and HBF_2/H_2 mixtures.

Table I

Ion-Molecule Mass Spectra of BF_3/D_2 and HBF_2/D_2 Mixtures

$BF_3:D_2$ = 1:1000 $HBF_2:D_2$ = 1:1000
(P_s = 0.20 mm; T = 84° K)   (P_s = 0.23 mm; T = 90° K)

Ion	Relative Intensity	Ion	Relative Intensity
D^+	0.0027	D^+	0.0152
D_2^+	0.0016	D_3^+	0.545
D_3^+	0.430	D_5^+	0.104
D_5^+	0.0243	N_2D^+	0.0596
N_2D^+	0.183	BHF^+	0.0355
BFD^+	0.0180	BDF^+	0.1260
BF_2^+	0.0295	$HClD^+$	0.00610
$(BF_2 \cdot D_2)^+$	0.0599	BF_2^+	0.00520
$(BF_2 \cdot 2D_2)^+$	0.0242	$(BF_2 \cdot HD)^+$	0.0758
BF_3D^+	0.1124	$(BF_2 \cdot D_2)^+$	0.00853
$(BF_3 \cdot D_3)^+$	0.1137	$(BF_2 \cdot HD \cdot D_2)^+$	0.0140
		$(BF_2 \cdot 2D_2)^+$	0.00144

with increasing temperature. This behavior is quite typical of three-body processes. The stoichiometry of these clustering reactions suggests that chemical bonding between hydrogen molecules and the reactant ion leads to stabilized structures with a maximum coordination of four groups (including the F atoms) about a boron site, assuming H_2 is counted as one unit. It is interesting to note that a species with composition $BF_2H_2^+$ is formed both by a clustering reaction at low temperatures (H_2-BF_3 system) and by proton transfer (H_2-HBF_2 system) at ambient source temperatures (~300° K). Based on the chemical evidence, it appears that two distinct isomeric structures for $BF_2H_2^+$ can be formed from different reaction pathways in these systems.

In principle thermodynamic data can be obtained from ion-molecule association reactions provided the system can attain a condition of equilibrium. The problem of demonstrating and measuring ion equilibrium are formidable. Experimental conditions must insure sufficient gas phase collisions to deactivate (quench) the product ion and sufficient reaction time for equalization of the reverse and forward reaction rates. For a reaction of the type

$$A^+ + B \rightleftarrows AB^+$$

the equilibrium constant may be written

$$K_{eq} = \left\{\frac{I_{AB^+}}{I_{A^+}}\right\} P_B \quad ,$$

where I_{AB^+}/I_{A^+} is the measured ratio of product/reactant ion currents when the reaction is at equilibrium and P is the pressure of component B. Instrument sensitivity places an upper and lower limit on the range of I_{AB^+}/I_{A^+} that can be determined experimentally. The standard free energy change for reaction can be determined directly from the relationship

$$\Delta G° = -RT \ln K_{eq} \ .$$

The standard enthalpy of reaction, $\Delta H°$, can be obtained from the temperature dependence of K_{eq} by the familiar Van't Hoff plot. The entropy change, $\Delta S°$, can be determined from the thermodynamic expression

$$\Delta S° = \frac{\Delta H° - \Delta G°}{T} \ .$$

Alternatively, $\Delta S°$ may be estimated with reasonable accuracy from procedures of statistical thermodynamics. The two independent methods of deriving $\Delta S°$ offer a check on the internal consistency of measurements. Recent literature data for a number of ion-molecule clustering reactions are summarized in Table II.

Table II

Ion-Molecule Clustering Reactions Studied at Low Temperature

System	T.Range(°K)	Reactions	$-\Delta H$ (kcal/mol)	Reference
Various positive and negative ion association reactions with atoms and molecules	82-280	$Ar^+ + Ar \rightarrow Ar_2^+$ $O^+ + N_2 \rightarrow ON_2^+$ $N^+ + N_2 \rightarrow N_3^+$		(Bohme, 1969)

Table II (continued)

System	T. Range ($^\circ$K)	Reactions	$-\Delta H$ (kcal/mol)	Ref.
		$O_2^+ + O_2 \rightarrow O_4^+$		
		$N_2^+ + N_2 \rightarrow N_4^+$		
		$O^- + CO_2 \rightarrow CO_3^-$		
		$Ar_2^+ + Ar \rightarrow Ar_3^+$		
O_2/H_2	300	$O_2^+ + H_2 \rightarrow H_2O_2^+$		(Adams, 1970)
O_2/N_2	200	$O_2^- + N_2 \rightarrow O_2N_2^-$		
		$O^- + N_2 \rightarrow N_2O^-$		
		$O^- + CO_2 \rightarrow CO_3^-$		
O_2	77-300	$O_2^+ + O_2 \rightarrow O_4^+$	10.8	(Conway, 1970)
		$O_4^+ + O_2 \rightarrow O_6^+$	6.8	
		$O_6^+ + O_2 \rightarrow O_8^+$	2.5	
		$O_8^+ + O_2 \rightarrow O_{10}^+$	2.46	
		$O_{10}^+ + O_2 \rightarrow O_{12}^+$	1.84	
CH_4	77-300	$CH_5^+ + CH_4 \rightarrow C_2H_9^+$	4.14	(Field, 1971)
		$C_2H_5^+ + CH_4 \rightarrow C_3H_9^+$	2.39	
		$C_2H_9^+ + CH_4 \rightarrow C_3H_{13}^+$	1.47	
CH_4/H_2O	270	$H_3O^+ + CH_4 \rightarrow H_3O \cdot CH_4^+$	8	(Conway, 1970)
		$C_2H_5^+ + H_2O \rightarrow C_2H_5 \cdot H_2O^+$		
		$H_3O \cdot CH_4^+ + CH_4 \rightarrow H_3O \cdot 2CH_4^+$	3.4	
		$C_3H_5^+ + H_2O \rightarrow C_3H_5 \cdot H_2O^+$		
NH_3/CH_4	113-193	$NH_4^+ + CH_4 \rightarrow NH_4 \cdot CH_4^+$	3.59	(Bennett, 1972)
		$C_2H_5^+ + NH_3 \rightarrow C_2H_5 \cdot NH_3^+$		
H_2S/CH_4	113-193	$H_3S^+ + CH_4 \rightarrow H_3S \cdot CH_4^+$	3.87	(Bennett, 1972)

Table II (continued)

System	T. Range (°K)	Reactions	$-\Delta H$ (kcal/mol)	Ref.
		$H_3S \cdot CH_4^+ + CH_4 \rightarrow H_3S \cdot 2CH_4^+$		
		$C_2H_5^+ + H_2S \rightarrow C_2H_5 \cdot H_2S^+$		
CF_4/CH_4	113-193	$CF_3^+ + CH_4 \rightarrow CF_3 \cdot CH_4^+$	4.55	(Bennett, 1972)
		$CF_3 \cdot CH_4^+ + CH_4 \rightarrow CF_3 \cdot 2CH_4^+$		
		$CH_5^+ + CH_4 \rightarrow C_2H_9^+$		
H_2	247-432	$H_3^+ + H_2 \rightarrow H_5^+$	9.7	(Bennett, 1972)
	102-139	$H_5^+ + H_2 \rightarrow H_7^+$	1.8	
C_2H_6	189-410	$C_2H_4^+ + C_2H_6 \rightarrow C_4H_{10}^+$		(Bennett, 1972)
		$C_2H_2^+ + C_2H_6 \rightarrow C_4H_8^+$		
		$C_2H_5^+ + C_2H_6 \rightarrow C_4H_{11}^+$		
O_2	80-300	$O_2^+ + O_2 \rightarrow O_4^+$		(Payzant, 1973)
		$O_4^+ + O_2 \rightarrow O_6^+$		
		$O_6^+ + O_2 \rightarrow O_8^+$		
BF_3/H_2	80-280	$BF_2^+ + H_2 \rightarrow BF_2 \cdot H_2^+$		(Pierce, 1974)
		$BF_2 \cdot H_2^+ + H_2 \rightarrow BF_2 \cdot 2H_2^+$		
		$BF_3H^+ + H_2 \rightarrow BF_3H \cdot H_2^+$		
HBF_2/H_2	80-280	$BF_2H_2^+ + H_2 \rightarrow BF_2H_2^+ \cdot H_2$		

KINETIC IMPLICATIONS IN THE TEMPERATURE DEPENDENCE OF ION-MOLECULE RATE CONSTANTS

We now examine the effects of temperature on reaction rate and will attempt to relate experimental rate constant data with existing kinetic theories and models. Specifically we will consider (A) Simple Collision Theory, (B) Absolute Rate Theory and (C) Other Approaches. The number of published articles reporting temperature dependent rate data for ion-molecule reactions is not large. In general, the bimolecular reactions that have been investigated are exothermic and have relatively high rate constants.

It is difficult experimentally to measure rate constants for slow
endothermic processes with high activation barriers because of
small product ion intensities and the problem of operating sources
at high temperatures. In our subsequent discussion we will consider
a steady state kinetic mechanism involving the reaction of an ion
A^+ with B, leading either to stabilized complex AB^+ or products,
C^+ and D.

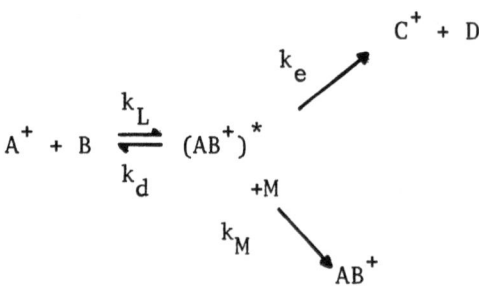

In this scheme $(AB^+)^*$ represents an activated intermediate whose
life time may vary from a single rotational period ($\sim 10^{-13}$ sec) to
the time required for collisional stabilization and k_L is assumed
to be of the magnitude of the collision rate constant. Using the
familiar steady state relationship

$$\frac{dI_{(AB^+)^*}}{dt} = k_L(I_A^+)(B) - \{k_d + k_e + k_m(M)\}(I_{(AB^+)^*}) = 0 \qquad (1)$$

we obtain for the intensity of the intermediate

$$I_{(AB^+)^*} = \frac{k_L(I_A^+)(B)}{k_d + k_e + k_m(M)} \qquad (2)$$

The rate constant, k_f(Expt.), for the formation of C^+ is obtained
from the measured rate,

$$\frac{dI_{C^+}}{dt} = k_f(\text{Expt.})(I_A^+)(B) \qquad (3)$$

usually under pseudo first order conditions when $(B) \gg (A^+)$. The
rate is also given by

$$\frac{dI_{C^+}}{dt} = k_e I_{(AB^+)^*} \qquad (4)$$

Replacing $I_{(AB^+)^*}$ with the right hand side of equation (2) we
have

$$\frac{dI_{C^+}}{dt} = \frac{\{k_L(I_A^+)(B)\}}{k_d+k_e+k_m(M)} \tag{5}$$

Comparing Equation 3 with 5 we obtain

$$k_f(\text{Expt.}) = \{\frac{k_L \cdot k_e}{k_d+k_e+k_m(M)}\} \tag{6}$$

When $k_e \gg k_m(M)$ and $k_e \gg k_d$, $k_f(\text{Expt.}) \sim k_L$ and we observe fast second order behavior. When $k_d \gg k_e$ and $k_d \gg k_m(M)$, $k(\text{Expt.})$ $\sim k_e k_L/k_d$ and we observe slow second order behavior.

The rate of formation of stabilized AB^+ follows the kinetic rate equation $dI_{AB^+}/dt = k_f(\text{Expt.})(I_A^+)(B)(M) = k_M I_{(AB^+)}*(M)$. Replacement of $I_{(AB^+)}*$ through equation 2 gives

$$k_f(\text{Expt.}) = \frac{k_m \cdot k_L}{k_d+k_e+k_m(M)} \tag{7}$$

At low densities of third body (M), with $k_d \gg \{k_e+k_m(M)\}$, $k_f(\text{Expt.})$ $\sim k_m k_L/k_d$ and the reaction follows third order behavior. At high densities of M with $k_m(M) \gg \{k_d+k_e\}$, $k_f(\text{Expt.}) \sim k_L$ and the reaction follows second order behavior. For the $O_2^+ + O_2$ and $N_2^+ + N_2$ association reactions the change in kinetic order from third to second with increasing gas density has been observed (Ferguson, 1972)

(A) Simple Collision Theory

From the basic considerations of collision theory, the dominant temperature effect in a rate constant is contained in the exponential term on the Arrhenius Equation, $k_f = A \exp(-E_a/kT)$ where A has the property of a collision frequency and E_a is the activation energy. The quantity $\exp(-E_a/kT)$ has the significance of a two-dimensional translational energy distribution function, $1/kT \exp(-E_a/kT)$ integrated over the energy range from the top of the activation barrier to infinity. The data available for ion-molecule reactions under consideration indicate that k_f changes either very slowly with temperature or decreases with increasing temperature (i.e. shows a negative temperature dependence). We will assume therefore that these effects are contained mainly in the pre-exponential term, A, and that activation barriers are zero or at least very small.

1) Hard sphere rate constant

We first formulate the problem by introducing a reaction distance within which a collision between A^+ and B inevitably leads to a reaction. The rate constant, which is averaged over the relative initial velocity of the particles, is just $k_f = \pi\sigma^2 \upsilon_{rel}$, which for a Maxwell-Boltzman distribution of velocities leads to the familiar expression $k_f = \pi\sigma^2 \{8kT/\pi\sigma\}^{\frac{1}{2}}$ where μ is the reduced mass of the colliding pair and $\pi\sigma^2$ is the reaction cross section. In this expression temperature does not enter into the cross section and the rate constant follows a slight positive, $(T)^{\frac{1}{2}}$, temperature dependence. An effective "interaction distance" defined by σ can then be obtained from a measured rate constant. For the classical reaction, $H_2^+ + H_2 \to H_3^+ + H$, $k_f = 2.0 \times 10^{-9}$ cm^3/molecules-sec. (Clow & Futrell, 1972). From this we calculate at 300° K, $\sigma \approx 5.0$ Å units, indicating that the interaction is occurring at distances large compared to molecular dimensions.

2) Effect of an ion-molecule interaction potential

Alternatively, we can use collision theory to obtain a bimolecular rate constant if the interaction potential between the particles is known. For the reaction of an ion with a symmetrical molecule, the ion-induced dipole potential, $V(r) = -(Ze)^2\alpha/2r^4$, leads to the Langevin rate constant, $k = 2\pi Ze(\alpha/\mu)^{\frac{1}{2}}$. In this equation α is the polarizibility of the molecule, Ze is the charge on the ion and μ is the reduced mass of the colliding pair. Su and Bowers (Su, 1973) have extended the theoretical treatment by including the effect of a permanent dipole moment in the reacting molecule. A derivation of the Langevin rate expression can be made using the approach of Johnston (Johnston, 1966), which applies to any interaction potential of the form $V(r) = -a/r^n$. For the special case of n = 4, the rate constant obtained is independent of initial particle velocity and temperature and the reaction cross section varies inversely with velocity. The Langevin equation predicts rate constant for single ion-molecule reactions without activation barriers to be of the order of 10^{-9} cm^3/molecule-seconds. Numerous examples of rate constants close to this value have been reported. Although it is clear that the Langevin model can not be applied generally, it has played a significant role historically. In table III temperature dependent data for some bimolecular processes are indicated by reference to an equation $k_f = C(1/T)^n$ where C is independent of temperature. In general reactions occurring with rate constants near the Langevin limit are expected to show a very slight temperature dependence. A few examples of reactions with n between 0 (Langevin approximation) and -0.5 (hard sphere approximation) are given in Table III. Recently temperature dependent rate data for two ion-molecule reactions of hydrocarbon species have been reported. These are an isothermal proton transfer reaction, $CH_4D^+ + CH_4 \to CH_5^+ + CH_3D$ and a hydride abstraction reaction, $t\text{-}C_4H_9^+ + i\text{-}C_5H_{12} \to t\text{-}C_5H_{11}^+ + i\text{-}C_4H_{10}$. These processes exhibit a pronounced negative temperature effect for which simple collision theory is not adequate.

(B) Absolute Rate Theory

We now consider the Absolute Rate Theory (or Transition State Theory) (Eyring, 1935) and its possible significance to ion-molecule reactions. In this approach the assumption is made that the complex intermediate, $(AB^+)^*$, is for a short duration in equilibrium with reactants, A^+ and B. The theoretical equation for the rate constant is

$$k_f = \kappa \left(\frac{kT}{h}\right) \exp(-\Delta G^*/RT) = \kappa \left(\frac{kT}{h}\right) \exp(\Delta S^*/R) \exp(-\Delta H^*/RT)$$

where ΔG^*, ΔH^* and ΔS^* are the free energy, enthalpy and entropy of activation, respectively, (kT/h) has the property of a frequency and κ is the dimensionless transmission coefficient. The equation may also be written:

$$k_f = \{\kappa (kT/h) Q^*_{(AB^+)} / (Q_{A^+})(Q_B)\} \exp(-\Delta E_o^*/RT)$$

where ΔE_o^* is equivalent to the activation energy, $Q^*_{(AB^+)}$ is the complete molecular partition function for the activated complex, excluding one degree of freedom along the reaction coordinate and Q_{A^+} and Q_B are the partition functions for the reactants. In general a complete partition function can be written

$$Q = Q_{trans.} Q_{rot.} Q_{vib.}$$

where Q_{vib} is further factored into $Q_{v_1} Q_{v_2} Q_{v_3}$....to include all vibrational degrees of freedom. Even when ΔE^* is zero, as we assume in this analysis, a temperature dependence in k_f may arise from the pre-exponential factors. (Gershinowitz, 1935) This is evident from the functional forms of the partition functions:

$Q_{trans.} = (2\pi mkT/h^2)^{3/2}$ for three degrees of freedom,

$Q_{rot} = (8\pi^2 IkT/h^2)$ for two degrees of freedom,

$Q_{vib} = 1/[1-\exp(-h\nu/kT]$ for one degree of freedom. Application of these equations requires a certain degree of subjectivity in chosing a model for the activated complex. For the simplest type of charge exchange process involving atomic species,

$$A^+ + B \rightarrow B^+ + A$$

the theory predicts a very slight temperature dependence in k_f since the Absolute Rate equation reduces to the hard sphere equation with the interaction distance arising in the moment of inertia, $I = \mu(r^*)^2$, chosen for the activated complex. For a process of the type

$$A^+ + BC \rightarrow AB^+ + C$$

where A is atomic and BC is diatomic, ART predicts a slight negative temperature dependence, $(1/T)^{0.5}$ in k_f if it is assumed that the vibration components to the partition functions for complex and reactant cancel. For the reaction $H_2 + H_2^+ \rightarrow H_3^+ + H$, Eyring, (Eyring, et.al., 1936) obtained a rate constant equal to the Langevin value. This result is not obvious by simple examination of the ART equation. It was assumed in this calculation that at large $H_2^+ \cdots H_2$ interaction distances, the particles were freely rotating and the rotational and vibrational partition functions for the separated species cancelled the corresponding internal functions in the complex. The only remaining partition function to evaluate, aside from the translational components, was that due to the rotation of the complex about its center of mass. A value for the interaction distance r^* was obtained by balancing the centrifugal forces with the attractive force between the ion and molecule. The result is $r^* = (8\pi^2\mu\alpha e^2/J(J+1)h^2)^{\frac{1}{2}}$ where J is the rotational quantum number and the energy of the complex is $E_J = E_{rot} + E_{pot} = J^2(J+1)^2h^4/128\pi^4\mu^2\alpha e^2$. Thus we note a distribution of r^* values as a function of J. Using a distribution function $(2J+1)\exp(-E_J/kT)$ we can obtain an averaged value of r^*, $\langle r^* \rangle$ over all J values. This quantity has the dimensionality of $(1/T)^{\frac{1}{4}}$ and the rate constant, which goes as $\langle r^* \rangle^2 \bar{v}$, is temperature independent. A complete derivation for k_f is also given in Mc-Daniel, (McDaniel, et.al, 1970).

The first application of Absolute Rate Theory to an ion-molecule reaction involving large molecules was discussed by Solomon, Meot-Ner and Field (Solomon, et.al., 1974) in their interpretation of the negative temperature dependence for the reaction

$$t\text{-}C_4H_9^+ + i\text{-}C_5H_{12} \rightarrow t\text{-}C_5H_{11}^+ + i\text{-}C_4H_{10}$$

These investigators observed the rate constant $k_f = C/T^3$ in the range 190°K to 570°K and that k_f was well below the Langevin limit. They accounted for the effect by the net loss in translational, rotational and internal rotational degrees of freedom in passing from reactants to activated complex. The strong temperature effect suggests that the intermediate complex is more rigid and structured than those in simple reactions ($H_2^+ + H_2$). Further studies of the temperature effect on hydrocarbon system will be helpful in testing the theory more generally.

Other Approaches

Recent experimental studies of ion-molecule clustering reactions have revealed pronounced negative temperature dependences for some third order rate constants (Table IV). For the reaction $O_4^+ + O_2 \rightarrow O_6^+$, k_f is proportional to $(1/T)^{5.1}$, an effect that is not readily explained by application of Absolute Rate Theory to reactions of small molecules. Three-body stabilization of an activated complex requires that the complex have an inherent life time of the order of a collision period. The rate of the overall reaction is coupled to the lifetime of the complex, τ, which is just $1/k_d$. Assuming the collisional processes in the mechanism have rate con-

Table III Temperature Dependence of Rate Constants for some Second Order Ion-Molecule Reactions; $k_f = C(1/T)^n$

Reaction	Class	T range($^\circ$K)	n	Ref.
$O^+ + N_2 \rightarrow NO^+ + N$	N atom transfer	80-600	0.5	Dunkin, et.al., 1968
$He^+ + N_2 \rightarrow He + N + N^+$	dissociative charge transfer	80-300	-0.5	Dunkin, et.al., 1968
$\rightarrow He + N_2^+$	charge transfer	300-600	0	
$CH_4D^+ + CH_4 \rightarrow CH_5^+ + CH_3D$	Proton transfer	80-300	6	Pierce & Porter, 1974
$t\text{-}C_4H_9^+ + i\text{-}C_5H_{12} \rightarrow$ $t\text{-}C_5H_{11}^+ + i\text{-}C_4H_{10}$	hydride transfer	190-570	3	Solomon, et.al., 1974

Table IV Temperature Dependence of Rate Constants for some Third Order Ion-Molecule Reactions; $k_f = C(1/T)^n$

Reaction	Third Body	T range($^\circ$K)	n	Ref.
$Ar^+ + Ar \rightarrow Ar_2^+$	He	80-300	1.9	Bohme, et.al., 1969
$O_2^+ + O_2 \rightarrow O_4^+$	He	80-200	2.7	Bohme, et.al., 1969 Adams, et.al., 1970
$O_2^+ + O_2 \rightarrow O_4^+$	O_2	80-300	3.2	Payzant, et.al., 1973
$O_4^+ + O_2 \rightarrow O_6^+$	O_2	80-300	5.1	Payzant, et.al., 1973

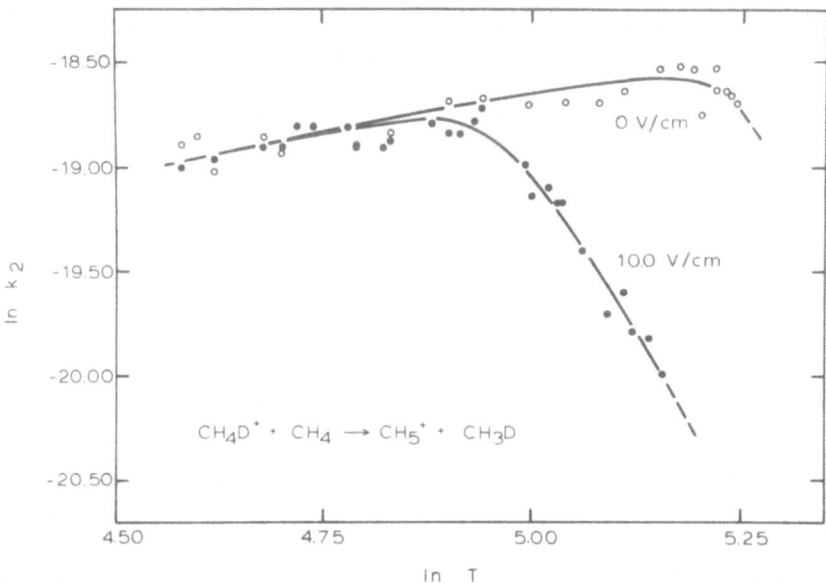

Fig. 2.   Temperature Dependence of Rate Constant for the reaction
$CH_4D^+ + CH_4 \rightarrow CH_5^+ + CH_3D$. Data are plotted in natural
logarithms.

stants k_L and k_M close to the Langevin limit, the rate constant
for cluster formation is then proportional to τ. The quantity
is contained in statistical theories of unimolecular reactions
(Robinson, 1972) which have been applied to ion fragmentation
mechanisms (Rosenstock, et.al., 1952). In the classical formulation
in the RRK approach

$$\tau \approx 10^{-13}(E/E-E_o)^{S-1}sec$$

where E_o is the critical energy for decomposition, and where E is
the total energy of the complex and S is the number of classical
oscillators (vibrational degrees of freedom) in the complex. When
an ion and molecule interact without an activation barrier, E will
be greater than E_o and the life-time of the complex will depend
on the redistribution of energy within its internal degrees of
freedom. When $E>>E_o$, τ is small and collisional stabilization is
not effective. This effect is noted by the decrease in clustering
when a reactant ion gains energy by a repeller potential. If, for
thermal conditions, we assume $E-E_o$ is b units of kT and $E_o >> b$ kT
then $\tau = (const) (E_o/T)^{S-1}$, the form for the temperature dependence
in k_f. In reality S is always greater than an experimental value
obtained by empirical use of the equation. This emphasises the

limitation in applying the theory with classical assumptions and the need to invoke quantum statistics and the newer RRKM approach.

A final example of an ion-molecule reaction with an unusual temperature dependence (Pierce, 1974) is the isothermal proton transfer process $CH_4D^+ + CH_4 \rightarrow CH_5^+ + CH_3D$. At low temperatures the clustering reaction

$$CH_4D^+ + CH_4 \xrightarrow{D_2} C_2H_8D^+$$

is also observed. Thus the system exemplifies a complexation mechanism involving bimolecular and trimolecular processes. At ambient source temperature (300° K). The rate constant for proton transfer is about $3 \times 10^{-11} cm^3$/molecule-sec, well below the Langevin limit (Abramson, 1966). As the temperature is lowered k_f increases to about 3×10^{-9} at 150° K and then decreases, slightly, below this temperature (Figure 2), with a small positive activation energy. Thus the reaction rate is apparently approaching a collisional limit as the temperature is decreased. It should be noted that, unlike most proton transfer reactions investigated, there is no highly exothermic channel for this process. The rate is markedly affected by repeller voltage, (See Figure 2), but at low temperatures this effect is diminished by quenching, presumably as $E-E_o$ approaches zero.

Concluding Remarks

In this lecture we have examined the effect of temperature on the chemistry and kinetics of ion-molecule reactions. We have considered existing theories of reaction kinetics and have searched for connections between theory and experiment. We have cited some interesting examples of the effect of temperature on rate behavior. Our analysis was necessarily limited by the relatively small volume of published literature on temperature dependent rate data especially for endothermic systems with activation barriers. We have not considered the possible quantum effects of barrier tunnelling which may be important in electron transfer and proton transfer reactions. It is hoped that, as more experiments are performed on the temperature effects of ion-molecule reactions, we will be able to identify the common features of the kinetic models in a more unified theory applicable to these interesting chemical processes.

Acknowledgement - I wish to thank the following Cornell graduate students for their assistance: Anthony DeStefano, Robert Pierce, Pui Po and Thomas Radus.

REFERENCES

Abramson, F.P. and Futrell, J.H. (1966), J. Chem. Phys. 45, 1925.
Adams, N.G., Bohme, D.K., Dunkin, D.B., Fehsenfeld, F.C. and
    Ferguson, E.E. (1970), J. Chem. Phys. 52, 3133.
Bennett, S.L. and Field, F.H. (1972), J. Am. Chem. Soc. 94, 5188.
Bennett, S.L. and Field, F.H. (1972), J. Am. Chem. Soc. 94, 6305.
Bennett, S.L. and Field, F.H. (1972), J. Am. Chem. Soc. 94, 8669.
Bennett, S.L., Lias, S.C. and Field, F.H. (1972), J. Phys. Chem.
    76, 3919.
Bohme, D.K., Dunkin, D.B., Fehsenfeld, F.C., and Ferguson, E.E.
    (1969), J. Chem. Phys. 51, 863.
Clow, R.P. and Futrell, J.H. (1972), Intern. J. Mass Spectrom.
    Ion Phys. 8, 119.
Conway, D.C., and Jamik G.S. (1970), J. Chem. Phys. 53, 1859.
Dunkin, D.B., Fehsenfeld, F.C., Schmeltekopf, A.L. and Ferguson,
    E.E. (1968), J. Chem. Phys. 49, 1365.
Eyring, H. (1935), J. Chem. Phys. 3, 107.
Eyring, H., Hirschfelder, J.O., and Taylor, H.S. (1936), J. Chem.
    Phys. 4, 479.
Ferguson, E.E. (1972). Flowing Afterglow Studies in Ion-Molecule
    Reactions (ed. J.L. Franklin) Plenum Press, New York.
Field, F.H. (1972). Chemical Ionization Mass Spectrometry in Ion-
    Molecule Reactions (ed. J.L. Franklin) Plenum Press, New York.
Field, F.H. and Beggs, D.P. (1971), J. Am. Chem. Soc. 93, 1585.
Franklin, J.L. (1972) ed. Ion-Molecule Reactions Plenum Press,
    New York.
Gamba, A., Morosi, G., and Simonetta, M. (1969), Chem. Phys. Lett.
    3, 20 (This paper gives a calculation for CH_5^+).
Gershinowitz, H. and Eyring, H. (1935), J. Am. Chem. Soc. 57, 985.
Henchman, M. (1972). Rate Constants and Cross Sections in Ion-mole-
    cule Reactions (ed. J.L. Franklin) Plenum Press, N.Y.
Johnston, H.S. (1966). Gas Phase Reaction Rate Theory, Ronald Press,
    New York.
Kebarle, P. (1972). Higher-Order Reactions-Ion Clusters and Ion
    Solvation in Ion-Molecule Reactions (ed. J.L. Franklin) Plenum
    Press, New York.
McDaniel, E.W., Cermak, V., Dalgarno, A., Ferguson, E.E. and
    Friedman, L. (1970). Ion-Molecule Reactions, Wiley-Interscience,
    New York
Payzant, J.D., Cunningham and Kebarle, P. (1973), J. Chem. Phys.
    59, 5615.
Pierce, R.C. (1974), Ph.D. Thesis, Cornell University, Ithaca, N.Y.
Pierce, R.C. and Porter, R.F. (1974), J. Phys. Chem. 78, 93.
Robinson, P.J. and Holbrook, K.A., (1972), Unimolecular Reactions,
    Wiley-Interscience, New York.
Rosenstock, H.M., Wallenstein, M.B., Wahrhaftig, A.L. and Eyring
    H. (1952), Proc. Nat. Acad. Sci. U.S.A. 38, 667.

Solomon, J.J., Neot-Ner, M. and Field, F. H. (1974), J. Am. Chem.
   Soc. 96, 3727.
Stevenson, D.P. (1963). Ion-Molecule Reactions in Mass Spectrometry
   (ed. C.A. McDowell), McGraw-Hill Book Company, Inc., N.Y.
Su, T. and Bowers, M.T. (1973). J. Am Chem. Soc. 95, 1370.

EFFECTS OF INTERNAL ENERGY ON ION-MOLECULE REACTIONS

William A. Chupka

Argonne National Laboratory

Argonne, Illinois 60439

EXPERIMENTAL METHODS

Electron impact

Electron impact has been the most commonly used method for production of reactant ions because of its simplicity, wide applicability and good intensity. However, the standard type of electron gun usually employed in electron impact experiments suffers from several disadvantages as a general method for controlling the internal energy of ions. The relatively wide energy spread (ca. 0.5 eV), the presence of space charge, the effects of charging of surfaces, sample pyrolysis and other experimental difficulties contribute to poor control of the ionization process. The threshold law for electron impact ionization is also unfavorable for control of closely spaced internal energy states. Nevertheless significant control of internal energy can be achieved in favorable cases. Thus preparation of ions in states of electronic excitation which are widely separated in energy can often be controlled. Indeed a number of ions (e.g. B^+, C^+, F^+) have been prepared in excited electronic states only by electron impact (Koski, 1974) since the necessary high energies are not readily available to other techniques such as photoionization. Metastable electronic states which differ in spin multiplicity from the neutral molecule by more than unity can often be prepared by electron impact but not by photoionization. Also it is often easy to insure preparation of ions in the ground electronic state only. The use of electron monochromators in the future may refine the degree of control of internal energy attainable by electron impact but only at the cost of experimental difficulty and of considerable loss of intensity.

Photoionization (Chupka, 1972)

As compared with electron impact, the photoionization technique offers a number of advantages one of which is much finer control over internal energy. The typical energy resolution attained by photoionization lies in the range 1-10 meV and the photoionization cross section threshold behavior is far more favorable than is the case for electron impact. Readily constructed continuum or quasi-continuum laboratory light sources and commercial vacuum ultra-violet monochromators provide enough photon flux for many kinds of ion-molecule reaction experiments up to photon energies of about 21 eV. At still higher energies synchrotron radiation is becoming available at a number of research centers.

Parent ions are usually very readily prepared in the ground vibronic state or with less than about 10 meV internal energy. Nearly pure preparations of single excited vibronic states are practical in only a few cases (e.g. H_2^+). More commonly one may prepare ions in a known distribution of energy states which can be varied with photon energy. The internal energy distribution may sometimes be inferred reliably from the experimental photoionization efficiency curve and/or the HeI photoelectron spectrum. Measurements of photoelectron spectra at the experimental wavelengths give unambiguous internal energy distributions and such data are now becoming available or can be measured by the experimenter without great difficulty. The still more powerful technique of photoelectron-photoion coincidence measurement has been applied by a number of workers to the study of unimolecular decomposition (Eland, 1972a, 1972b; Stockbauer, 1973) and its application to the study of ion-moleucle reactions appears feasible and imminent. Successful application of this technique would provide unambiguous knowledge and ready variability of the initial internal energy of reactant ions.

Charge transfer (Lindholm, 1972)

Charge transfer has been used extensively to investigate the fragmentation of ions as a function of internal energy. More recently the technique has also been applied to the preparation of reactant ions. In favorable cases reactant ions can be prepared with a narrow range of internal energy. However only a few ions useful for charge transfer have a single recombination energy. The possibility of several recombination energies in the more typical case and the possibility of non-resonant charge transfer impose serious limitations on the usefulness of this technique.

Photoexcitation

Production of excited reactant ions from electronic ground

state ions by photon absorption and the study of their subsequent
reactions is a promising new development in ion chemistry. The
technique has been used successfully in an ion cyclotron resonance
spectrometer with a xenon arc plus interference filters serving as
the light source to produce excited $C_2H_5^+$ ions which subsequently
were observed to react with ethylene (Kramer and Dunbar, 1972).
The photoexcitation technique has the advantage that the excitation
energy added to the ion is readily controlled. However, in order
to exploit this advantage fully, the internal energy distribution
of the absorbing ion should be very small. The example mentioned
above involved electronic excitation in the visible but pure
vibrational excitation by infra-red laser may be feasible in some
cases. In contrast to the other techniques, photoexcitation can
be used in general for fragment as well as parent ions. The
electronic states optically allowed for parent ions will not include
all those seen in the photoelectron spectrum but on the other hand
can include others which are too weak to be seen in the photoelectron
spectrum, e.g. states formed by formally two-electron processes. In
the case of electronic excitation the radiative lifetime of the excited state is of importance since for a strong absorption (not
followed by a radiationless transition) the lifetime will be at most
of the order of, and usually much shorter than, the mean collision
times for the gas densities typical of most ion-molecule reaction
studies. Radiative decay will often yield vibrationally excited
ions which may then react in a manned different from that of the
original unexcited ion. For larger molecular ions, the lifetime of
excited states may be increased by orders of magnitude by the occurrence of radiationless transitions (internal conversion and inter-
system crossing) to isoenergetic vibrational levels of lower lying
and especially the ground electronic states. Many larger hydrocarbon
ions with many high frequency C-H stretching modes have relatively
small energy gaps between electronic states and are quite likely to
undergo rapid radiationless transitions (Chupka, 1974). Obviously
this new field of ion photochemistry will exhibit the many complications of neutral molecular photochemistry but should add much to
understanding of the chemistry and spectroscopy of gaseous ions.

Excitation of Neutral Reactant

The excitation of neutral reactants, under conditions such that
the effect of a specific mode of excitation on an ion-molecule reaction can be studied, has been successfully accomplished only for
vibrationally excited N_2 and electronically excited O_2 (Ferguson,
1972). The excitation has been produced by passing the gases through
a microwave discharge.

GENERAL EFFECTS OF INTERNAL ENERGY

There are no generally valid, non-trivial rules governing the effects of internal energy on reaction rates. Usually endoergic reactions can be made to occur by sufficient internal excitation to overcome the endoergicity and very large effects of internal energy are often encountered in such reactions. The effect of internal energy on exothermic reactions is usually not as dramatic. In the ideal case of a collision complex of sufficiently long life and strong coupling among internal degrees of freedom all forms of energy become equivalent within the constraints imposed by conservation of linear and angular momentum and its decomposition becomes statistically controlled. For more direct reactions the observation of non-equivalence is to be expected and such observations can provide valuable information on the details of the mechanism of the reactions.

ROTATIONAL ENERGY

Very few experiments have successfully investigated the effects of rotational energy alone on an ion-molecule reaction. Sbar and Dubrin (1970) found that H_2 in the $J = 0$ state has a cross section 5% larger than that in the $J = 1$ state for the reaction with Ar^+ to produce ArH^+. These workers ascribe the effect to the reduction by rotation of the aligning ability of the long range force leading to reduction of the effective polarizability of H_2. Gislason (1972) has treated this reaction in some detail using a model which stresses the importance of potential surface crossings and avoided crossings. He attributes the Sbar-Dubrin result to a curve crossing for $Ar^+(^2P_{1/2})$ to a charge transfer product channel at their kinetic energy. Gislason has also predicted a fairly strong variation ($\sim 30\%$) with rotational state of the reaction cross section of $Ar^+(^2P_{1/2})$ with H_2 at thermal translational energy. This prediction is based on his model for which an avoided crossing of potential surfaces occurs at a distance which is rather sensitive to the rotational state of H_2. This sensitivity results from the fact that the difference in energy between the surfaces for $Ar^+(^2P_{1/2}) + H_2(v = 0)$ and $Ar + H_2^+(v = 2)$ varies fairly strongly with the rotational quantum numbers of H_2^+ because of the large difference in rotational constants for H_2^+ and H_2. While this prediction has not yet been unambiguously confirmed experimentally, this plausible model suggests the possibility of cases for which a reaction rate may be fairly strongly dependent on the rotational state of one or both reactants.

No detectable effect ($\lesssim \sim 10\%$) was found for small changes in the rotational energy of H_2^+ on its exothermic reaction with H_2 to form H_3^+ (Chupka, Russell and Refaey, 1968a). On the other hand, rotational as well kinetic energy of H_2^+ was found to contribute to overcoming the endoergicity of the reaction of H_2^+ with He to form HeH^+ (Chupka and Russell, 1968C).

Rotational energy of reactants is expected to play an important role in certain classes of reactions in which it constitutes a major fraction of the internal energy of a collision pair. Thus, in the association reaction of weakly bound diatomics such as $N_2 + 2N_2 \rightarrow N_4^+ N_2$, the rotational energy of the reactants has a strong effect on the lifetime of the N_4^{+**} collision complex and hence on the rate constant of the reaction.

VIBRATIONAL ENERGY

Exothermic reactions

For exothermic reactions with large (i.e. approximately Langevin) cross sections, the total reaction cross section has not been found to change very much with vibrational energy. In most cases the total cross section at thermal translational energies decreases with increasing vibrational energy of the reactant ion. Examples of such reactions are (Chupka et al, 1968a, 1968b): (1) $H_2^+(H_2, H) H_3^+$ and (2) $NH_3^+(NH_3, NH_2) NH_4^+$. In reaction (1) the cross section for ca. 1.0 eV vibrational energy drops to about 80% of the $v = 0$ value while for reaction (2) it drops to about 55% of the $v = 0$ value for ca. 1.0 eV vibrational excitation. Similar behavior has been observed by other workers for several other systems (LeBreton et al, 1974; Buttrill, 1974) involving $C_2H_4^+$ and $C_2H_2^+$ reactant ions. This behavior has been rationalized by a statistical phase space argument which has been discussed at length recently by Henchman (1972) and can be described briefly as follows. If the reactants have no internal energy and very little kinetic energy and the reaction is strongly exothermic, the ratio of the volume in phase space available in the product channel to that in the (backward) reactant channel is very large and the forward reaction is strongly favored. As the collision pair is formed with more internal energy, this phase space ratio decreases. Thus if the reaction is governed at least in part by phase space considerations, the overall reaction cross section may be expected to decrease with increasing internal energy although the decrease may not be large. Large effects may be expected in cases in which rearrangement is required for the forward reaction. In general large effects may be expected for cases in which there are two (or more) competing modes of decay of the collision pair (one of which may be the backward decay) one with both higher energy and entropy of activation than the other. Thus large effects on the branching ratio of certain competing forward reactions can be expected as for example in the reactions: $C_2H_2^+ + C_2H_2 \rightarrow C_4H_2^+ + H_2$ and $C_2H_2^+ + C_2H_2 \rightarrow C_4H_3^+ + H$ where the second reaction would be expected to increase relative to the first as the internal energy of the reactants is increased. This is the observed behavior (Buttrill, 1974).

The applicability of a statistical theory implies the indistinguishability of the various forms of energy which can be put into the collision pair. However, by varying the vibrational energy of a reactant rather than the relative translational (or rotational) energy, one can investigate rather simply the effect of the internal energy of the collision pair without having to consider the large effect of kinetic energy on the cross section for close collision and the effects due to the variation of the total angular momentum of the collision pair. While phase space theory (Light, 1967) properly conserves angular momentum, it cannot readily be applied to complicated systems and has only been used for three-atom and four-atom reactions. RRKM theory (Forst, 1973) on the other hand is readily applied to more complicated systems but does not rigorously conserve angular momentum, although some recent progress has been made to this end (Klots, 1972).

There are exceptions to the behavior of reaction cross section with vibrational energy described above. Thus the cross section for reaction (1) increases with increasing vibrational energy at higher kinetic energies, probably due to dominance of a direct proton transfer mechanism which would be favored by vibrational energy in H_2^+. Also the cross section for the reaction $NH_3^+(H_2O,OH)$ $NH4^+$ appears to be nearly independent of vibrational energy in NH_3^+ and may actually increase slightly with vibrational energy. Dominance of a direct H atom transfer mechanism which would be approximately independent of NH_3^+ vibrational energy is suggested.

Some very notable exceptions have been investigated recently. Albritton et al (1973) found that the apparently exothermic thermal energy charge transfer reaction $Ne^+ + N_2 \rightarrow N_2^+ + Ne$ increases drastically in rate with increasing vibrational temperature of N_2. The suggested explanation invokes a nearly resonant and Franck-Condon favorable charge transfer from $N_2(v = 2)$ to N_2^+ in a highly excited quartet state. The importance in charge transfer of the condition of near resonance and of Franck-Condon factors (possibly modified at close approach of reactants) implies a dependence of cross section on the vibrational states of the reactants. Schmeltekopf et al (1971) have found a very strong increase of the rate constant for the reaction $O^+(N_2,N)NO^+$ with increasing vibrational energy of N_2. While the interpretation of the many unusual characteristics of this reaction is still a matter of some controversy (Henchman, 1972), the negative temperature dependence of the thermal rate constant from

$80^{\circ}K$ to $600^{\circ}K$ and the positive effect of vibrational excitation suggest separate low and high energy mechanisms and in any case provide a rather stringent test of any proposed mechanism. These latter two examples show how the effects of vibrational energy give important clues regarding reaction mechanism.

Association reactions

This special class of exothermic reactions will be discussed separately since these are usually termolecular, requiring deactivation of a bimolecular collision complex. Since the third order rate constant is proportional to the lifetime of the complex which is in turn a rapidly decreasing function of internal energy, strong effects of internal energy are expected and a strong negative temperature coefficient is typical of such reactions. In a few cases, the effect of vibrational energy alone on an association reaction has been studied. One such case is the dimerization reaction of the benzene ion (or its perdeuterated analogue), $C_6H_6^+ + C_6H_6 \rightarrow C_{12}H_{12}^+$ which has been the subject of much controversy and apparently conflicting observations from which various investigators have claimed the order of the reaction to be second, third, fourth and even higher (Friedman and Reuben, 1971). A large part of the confusion results from the fact that the benzene ion as usually prepared by electron impact at $\sim$ 50 eV has a distribution in vibrational and electronic energy of about 4 eV. Photoionization experiments (Chupka and Russell, 1974) have shown conclusively that for benzene ions formed in the ground electronic state and for low conversion ($<$ 3%) the reaction is of third order and involves almost exclusively the ground vibrational state of $C_6H_6^+$ or $C_6D_6^+$. This is readily apparent from Fig. 1 which shows the photoionization efficiency curves for C_6D_6 and $C_{12}D_{12}^+$ in the region of the ionization potential. The curve for $C_{12}D_{12}^+$ is practically flat above the initial step corresponding to the production of $C_6D_6^+$ in its ground vibrational state while the curve for $C_6D_6^+$, as well as its photoelectron spectrum, shows the formation of the ion in a range of vibrationally excited states. The vibrationally excited monomer ions react in an approximately fourth order reaction, apparently requiring an initial deactivation, while there is evidence (Jones et al, 1974) that very highly excited benzene ions produced about 3.5 eV above the ground state dimerize in a bimolecular reaction to form a dimer which is presumably much more strongly bound and thus of different structure from that formed by the ground state reactant.

Endoergic reactions

For endoergic reactions the effects of vibrational energy can be spectacularly large. Of particular interest is the determination of the relative importance of vibrational as compared with transla-

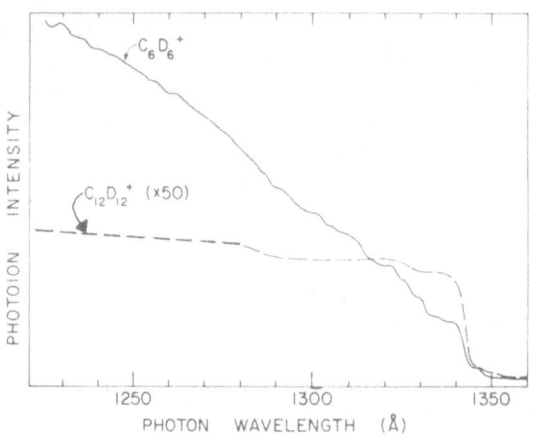

Fig. 1.   Photoionization efficiency curves for $C_6D_6^+$ and for the
dimerization reaction product $C_{12}D_{12}^+$.

tional energy in surmounting the potential barrier. The reactions
of H_2^+ (ν = 0—5) with He have been extensively investigated (Chupka
1972) with this goal in mind. The cross section for the reaction
$H_2^+(He,H)$ HeH$^+$ has been measured as a function of kinetic energy
for H_2^+ with ν = 0 to ν = 5. The results show that vibrational
energy is much more effective than translational energy in causing
reaction. The magnitude of the effect may be seen in Table 1

Table 1.   Comparison of theoretical and experimental cross sections
for the reaction $H_2^+(He,H)$ HeH$^+$ as a function of vibrational
quantum number of H_2^+ for a total internal energy of 2.0 eV in the
collision complex.

ν	σ (Å^2)	
	calc.	exp.
0	0.64	0.10
1	0.76	0.34
2	0.95	0.93
3	1.17	1.70
4	1.49	2.35
4	1.94	2.49

(Truhlar 1972) which compares reaction cross sections calculated by
phase space theory (which does not distinguish between translational

and vibrational energy except as required by conservation of linear
and angular momentum) with experimental values for a total energy
of 2.0 eV in the collision complex. While the calculation yields
an enhancement of a factor of about three from v = 0 to v = 5 the
experimental data show enhancement by a factor of twenty-five. The
greater effectiveness of vibrational energy is expected from simple
theoretical considerations (Levine and Bernstein, 1974) which
indicate that for reactions with a "late" barrier (i.e. in the exit
part of the potential surface for the reaction) as is this reaction,
vibrational energy should be more effective than translational while
an early barrier should have the opposite effect. However, the
detailed shape of the potential surface in the reaction region can
influence the characteristics in a complicated fashion as can be
seen even for very primitive idealized potential surfaces (Mahan,
1974). The potential surface for the H_2^+ -He reaction has been
calculated with good accuracy (Brown and Hayes, 1971) and has been
used in various quantal but primitive theoretical calculations of
reaction rates with some modest success in reproducing experiment
(Cooper et al, 1974).

Another case of a very strong effect of vibrational energy on
an endoergic reaction is illustrated by the data of Fig. 2 for the

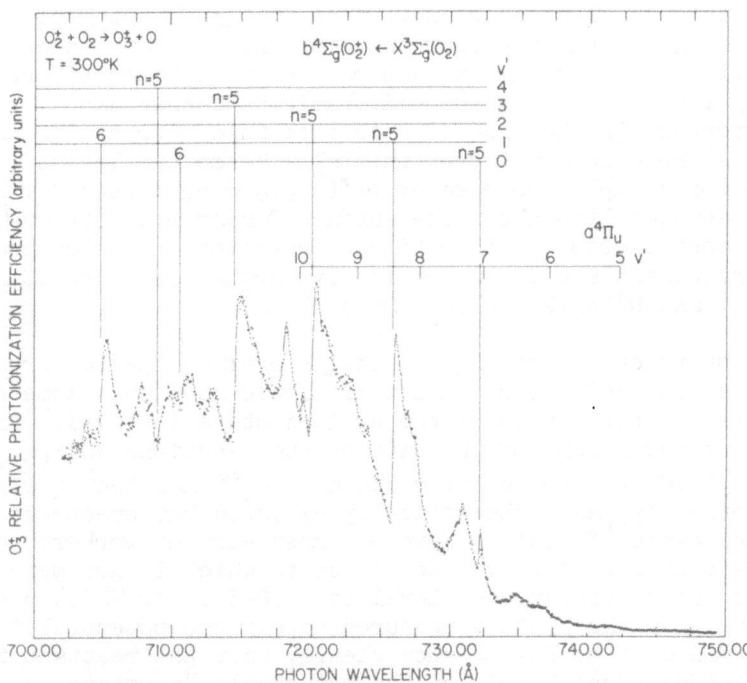

Fig. 2. Photoionization efficiency curve for O_3^+ produced by the
ion-molecule reaction $O_2^+(O_2,O)$ O_3^+.

reaction $O_2^+(^4\pi_\mu) + O_2 \rightarrow O_3^+ + O$ (Dehmer and Chupka, 1974). The reactant O_2^+ ion is produced by photoionization in the long-lived metastable excited state but the reaction only becomes energetically allowed for the $v = 5$ level of that state. The product ion intensity rises very strongly from the $v = 5$ threshold through the $v = 9$ level. The Franck-Condon factors for the formation of these vibrational states are nearly constant and the strong rise is a measure of the strong increase in reaction cross section with vibrational energy. The presence of pronounced autoionization structure allows only a rough quantitative measurement of the effect but it is clear that the reaction cross section rises by between one and two orders of magnitude in the range. The explanation for the magnitude of the effect is not yet clear in this case. It may be in accord with statistical phase space theory for the endoergic reaction. However, the situation is complicated by the fact that the reactants are initially on an excited potential surface of the O_4^+ system which may not correlate adiabatically with the ground state products.

Electronic energy

While the effects of translational, rotational and vibrational energy can often be discussed in terms of a single potential surface for the reactions, electronic excitation in a reactant implies a different potential surface from that for ground state reactants and non-adiabatic ("surface jumping") processes must be considered and are usually important. Such non-adiabatic processes can of course also be important in the case of reactants in electronic ground states since these need not even initially be on the lowest potential surface as for example the case of $Ar^+(^2P_{3/2}$ ground state$) + H_2$ which at large distances lies above the surface for $Ar + H_2^+$ ($v = 0$). Indeed two of the most thoroughly investigated reactions in which non-adiabatic processes are important involve ground state reactants: $H^+(H_2,H)$ H_2^+ and $O^+(N_2,N)$ NO^+(Henchman, 1972).

Data on reactions of electronically excited species are relatively sparse and often consist of little more than sometimes questionable identification of the excited state involved. Franklin (1972) has reviewed some of the data on the reactions $N_2^{+*}(N_2,N)$ N_3^+ and $O_2^{+*}(O_2,O)$ O_3^+. The reacting state of N_2^{+*} has been assigned, probably correctly, as a theoretically expected but spectroscopically unknown metastable $^4\Sigma$ state. However, most earlier workers considered the O_2^{+*} reactant ion to be in the $^2\pi_\mu$ state which is not metastable but has its ground vibrational level at 17.045 eV or 727.4 Å which seemed to be near the crudely measured appearance potential for the reaction. The data of Fig. 2 show clearly that the reactant ion is in excited vibrational levels of the metastable $^4\pi_\mu$ state as correctly inferred by Leventhal and Friedman (1967).

The complexity of reactions involving excited states can be illustrated by contrasting two reactions which at first sight might be expected to exhibit similar behavior. They are $Ar^+(H_2,H)$ ArH^+ and $Kr^+(H_2,H)$ KrH^+. Both reactions occur exothermically with ground state $^2P_{3/2}$ ions as well as with the ions in the long-lived metastable $^2P_{1/2}$ state. In the case of Ar^+, the excited state reacts with about 1.3 times larger cross section than that of the ground state (Chupka and Russell, 1968C), whereas in the case of Kr^+ the excited state reacts with only about 0.4 times the cross section of the ground state (Chupka, 1974). The explanation for this behavior is not yet established. Gislason (1972) has proposed an explanation for the Ar^+ behavior based on the theoretical work of Kuntz and Roach (1972) who showed that the Ar^+-H_2 and $Ar-H_2^+$ surfaces cross in the entrance valley. Gislason attributes the enhanced reaction cross section of the $^2P_{1/2}$ state to the fact that it is in near resonance with the state $A_r + H_2^+(\nu = 2)$ and that a favorable crossing to the latter more attractive surface occurs at larger distances than a similar crossing for the $^2P_{3/2}$ state. This suggestion could be tested by carrying out the reaction with D_2 and HD for which the relative positions of energy levels would be different. Gislason's model is not applicable to the case of Kr^+ since the ionization potential of H_2 is about 0.8 eV above that of even the $^2P_{1/2}$ excited state of Kr^+ and such crossings cannot occur at large distances. The potential surface for the Kr^+ reaction given by Kuntz and Roach (1972) is not very reliable since it is just scaled from the Ar^+ surface and predicts an energy barrier to the reaction which is in fact absent since the reaction occurs readily with ground state reactants at 78°K.

The reaction of O^+ with H_2(and D_2) has been studied extensively by Chiang et al (1971) and by Bosse et al (1971). Endoergic reactions forming HO_2^+, OH^+ and H_2O^+ from the ground electronic state of O_2^+ proceed via a long-lived complex at low kinetic energies. However, O_2^+ in the metastable $^4\pi_u$ state forms HO_2^+ with high cross section in a direct reaction. Mahan (1971) has argued on the basis of molecular orbital correlations that the O_2^+ ground state ion should react partly via a ground electronic state long-lived $H_2O_2^+$ complex to give the same products as observed in the ionization and fragmentation of the hydrogen peroxide molecule and partly by a direct mechanism to give HO_2^+. The $^4\pi_u$ state on the other hand would be expected to react predominantly by a direct process to produce HO_2^+ possibly in an excited electronic configuration. Photoionization studies show that the $^4\pi_u$ state reacts exclusively to form HO_2^+ at low kinetic energies and can form small amounts of the other products only above a kinetic energy threshold. Fig. 3 (Chupka and Russell, 1974) shows some of the photoionization data which suggests that the formation of OH^+ by $O_2^+(^4\pi_u)$ may involve formation of excited products or result from decomposition of an initial HO_2^+ product. The latter process should have a threshold at about 40 eV (Lab) for the

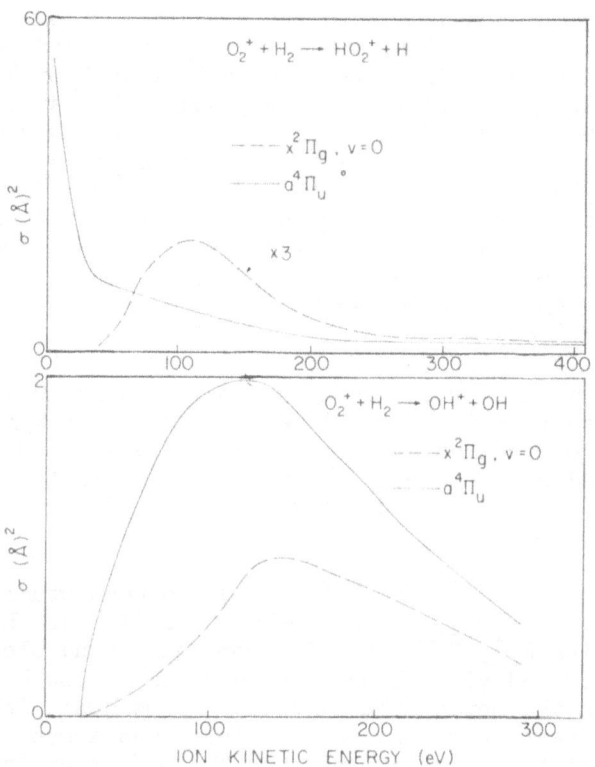

Fig. 3. Cross section as a function of laboratory kinetic energy
for the reaction of O_2^+ in the ground and $^4\pi_\mu$ metastable excited
state with H_2 to produce HO_2^+ and OH^+.

vibrational ground state and at about 20 eV for the large vibrational distribution actually used in the photoionization data of Fig.3. The data are in reasonable agreement with the calculated threshold.

The few examples of the effects of electronic excitation given above serve mainly to emphasize the great need for a knowledge of relevant potential surfaces for further understanding. In many cases some guidance may be obtained from a consideration of molecular orbital correlations.

REFERENCES

Albritton, D. L., Bush, Y. A., Fehsenfeld, F. C., Ferguson, E. E., Govers, T. R., Mc Farland, M. and Schmeltekopf, A.L. (1973). J. Chem. Phys. 58, 4036

Bosse, G., Ding, A. and Henglein, A. (1971). Ber. Bunsenges. Physik. Chem. 75, 413

Brown, P. J. and Hayes, E. F. (1971). J. Chem. Phys. 55, 922

Chiang, M. H., Gislason, E. A., Mahan, B. H., Tsao, C. W. and Werner, A. S. (1971). J. Chem. Phys. 75, 1426

Chupka, W. A. Russell, M. E. and Refaey, K. (1968a). J. Chem. Phys. 48, 1518.

Chupka, W. A. and Russell, M. E. (1968b). J. Chem. Phys. 48, 1527

Chupka, W. A. and Russell, M. E. (1968c). J. Chem. Phys. 49, 5426

Chupka, W. A. (1972). Ion-Molecule Reactions by Photoionization Techniques in "Ion-Molecule Reactions" Vol 1 (Ed. J. L. Franklin) Plenum Press, New York.

Chupka, W. A. (1974a). Photoionization and Fragmentation of Polyatomic Molecules in "Chemical Spectroscopy and Photochemistry in the Vacuum-Ultraviolet" (Ed. C. Sandorfy, P. J. Ausloos and M. B. Robin) D. Reidel Publishing Co., Boston.

Chupka, W. A. (1974b) Unpublished data

Chupka, W. A. and Russell, M. E. (1974). To be published.

Cooper, W. G., Evers, N. S., Kouri, D. J. (1974). Molec. Phys. 27, 707

Dehmer, P. M. and Chupka, W. A. (1974). To be published.

Eland, J. H. D. (1972a). Int. J. Mass Spectrom. Ion Phys. 8, 143

Eland, J. H. D. (1972b). Int. J. Mass Spectrom. Ion Phys. 9, 397

Ferguson, E. E. (1972). Flowing Afterglow Studies in "Ion-Molecule Reactions" Vol. 2 (Ed. J. L. Franklin). Plenum Press, New York.

Franklin, J. L. (1972). Positive-Ion-Molecule Reaction Studies in a Single Electron-Impact Source in "Ion Molecule Reactions" Vol 1 (Ed. J. L. Franklin) Plenum Press, New York.

Friedman, L. and Reuben, B. G. (1971). Adv. Chem. Phys. 19, 33

Gislason, E. A. (1972). J. Chem. Phys. 57, 3396

Henchman, M. (1972). Rate Constants and Cross Sections in "Ion-Molecule Reactions", Vol. 1 (Ed. J. L. Franklin). Plenum Press, New York.

Jones, E. G., Bhattacharya, A. K. and Tiernan, T. O. (1974).
    Submitted to Int. J. Mass Spectrum. Ion Phys.
Klots, C. E. (1972). Z. Naturforsch. 27a, 553
Koski, W. (1974). Reactions of Electronically Excited Ions. This
    volume.
Kramer, J. M. and Dunbar, R. C. (1972). J. Am. Chem. Soc. 94, 4346
Kuntz, P. J. and Roach, A. C. (1972). Trans. Faraday Soc. II 68,
    259
Le Breton, P. R. Williamson, A. D., Beauchamp, J. L. and Huntress,
    W. T. (1974). J. Chem. Phys. (in press).
Leventhal, J. J. and Friedman, L. (1967). J. Chem. Phys. 46, 997
Levine, R. D. and Bernstein, R. B. (1974). "Molecular Reaction
    Dynamics," Oxford University Press, New York. Chapter 4.
Light, J. C. (1967). Disc. Faraday Soc. 44, 14
Lindholm, E. (1972). Mass Spectra and Appearance Potentials Studied
    by Use of Charge Exchange in a Tandem Mass Spectrometer in "Ion-
    Molecule Reactions" Vol. 2, (Ed. J. L. Franklin). Plenum Press,
    New York.
Mahan, B. H. (1971). J. Chem. Phys. 55, 1436
Mahan, B. H. (1974). J. Chem. Educ. 51, 377
Sbar, N. and Dubrin, J. (1970). J. Chem. Phys. 53, 842
Schmeltekopf, A. L., Fehsenfeld, F. C., Gilman, G. I. and Ferguson,
    E. E. (1971). Planet. Space Sci. 15, 401
Stockbauer, R. (1972). J. Chem. Phys. 58, 3800
Truhlar, D. G. (1972). J. Chem. Phys. 56, 1481

COLLISIONAL DEACTIVATION OF EXCITED IONS

A. G. Harrison

University of Toronto

Toronto, Canada

Energy transfer on collision between vibrationally excited molecules and ground state molecules is a field of considerable theoretical and experimental activity, in part because such energy transfer processes are of fundamental importance in chemical kinetics. When a vibrationally excited molecule collides with an atomic species, such as a rare gas, the vibrational energy may be transferred to relative translational energy (V-T) or to molecular rotational energy (V-R). If the collision partners are both molecular, vibrational-vibrational (V-V) energy transfer also is possible. Although it is now possible, in many instances, to obtain detailed information on each specific type of energy trans- fer process, many of the earlier techniques and, thus, many of the earlier studies provided only vibrational relaxation times or vibrational deactivation efficiencies without determination of the energy transfer mechanism. In the present review we will be concerned primarily with vibrational deactivation processes which result in stabilization of highly excited species which otherwise would undergo a distinctive chemical change. Most of the studies of such processes to date have involved neutral systems and only relatively recently has work involving gaseous ions been reported. In the present paper we will summarize briefly the significant conclusions to be drawn from the neutral work and review the results for collisional stabilization of gaseous ions in the light of these conclusions.

ENERGY TRANSFER IN NEUTRAL SYSTEMS

Vibrational energy transfer from molecules in low-lying states (primarily v=1) has been investigated most extensively, both experimentally and theoretically. Qualitatively, one might expect

that the most effective interaction between the intra-molecular
vibration and the force exerted between the colliding partners
would occur when the collision duration, or period of action of
the force, is comparable to the vibration period. This is confirmed by a semi-classical treatment of the collision between a
diatomic molecule and a monatomic collision gas (see Cottrell and
McCoubrey (1961)). The collision duration depends on the steep-
ness of the repulsive potential curve (the "hardness" of the
collision) and on the relative velocity of the colliding pair. At
ordinary temperatures (for simple gases) average relative velocities
are $\sim 5 \times 10^4$ cm sec^{-1} and, assuming an interaction distance of
1×10^{-8} cm, the collision duration is $\sim 2 \times 10^{-13}$ sec. A typical
molecular vibration ($\nu = 1000$ cm^{-1}) has a period of 3×10^{-14} sec
with the result that only encounters with higher than average
relative velocities are likely to be effective in collisional deactivation. Both semi-classical and quantum-mechanical theoretical
treatments (Cottrell and McCoubrey (1961), Herzfeld and Litovitz
(1959)) predict that the probability of energy transfer per collision should range from 10^{-3} to 10^{-9} and increase with increasing
temperature. This is in agreement with much of the experimental
data which has been derived primarily from ultrasonic dispersion
studies. For polyatomic molecules the energy transfer probabilities are higher because of the lower frequency vibrations. In
addition, the probability increases with decreasing mass of the
collision gas, because of the higher relative velocity.

For species with considerably higher vibrational excitation
somewhat different results have been obtained in that it frequently is observed (Borrell (1967)) that the efficiency for
energy transfer increases with increasing collision duration. The
most detailed results have been obtained for vibrational energy
transfer from I_2($B^3\pi$) excited to v=15 and v=25 (Brown and Klemperer
(1964), Steinfeld and Kemperer (1965), Steinfeld (1966)). They
found that the collision efficiencies were high, even for rare
gases, and initially increased with $\mu^{\frac{1}{2}}$ (μ = reduced mass of colliding pair), and, thus with collision duration given by
where d is the interaction distance. The collision $d/\left(\dfrac{8kT}{\pi\mu}\right)^{\frac{1}{2}}$
efficiencies reached a maximum when the collision
duration equalled the vibration period and then decreased for
heavier collision partners. This apparent difference compared to
the low energy systems results from the much longer vibration
periods ($\nu \sim 80$ cm^{-1}). Note that in this case the lighter collision gases are less effective since the average collision duration
is less than the period of vibration. The total variation in
collision efficiencies was approximately a factor of five, much
less than the variation observed in the low energy regime.

The most detailed studies of the collisional deactivation of
highly vibrationally excited polyatomic molecules, such as those
arising during chemical reactions, have been carried out by

Rabinovitch and co-workers. In one aspect of this work the effect of added collision gases on the limiting low pressure rate coefficient of unimolecular reactions was investigated. For the unimolecular reaction

[1] A → B

the limiting low pressure first-order rate coefficient in the presence of a second collision gas is given in the Lindemann formulation by

[2] $k_o = k_A^a C_A + k_M^a C_M$

where k_A^a and k_M^a are the average rate constants for activation by A and M respectively to states above the critical threshold for reaction. From plots of k_O vs C_M average rate coefficients for activation by M can be obtained which by detailed balancing lead to average rate coefficients for collisional deactivation of A* excited to states above the critical threshold.

In this manner Chan et al (1970) have determined rate coefficients for deactivation of CH_3NC* for approximately 100 collision gases. These rate constants were converted to relative deactivation efficiencies per collision by taking account of the differences in the collision rates of CH_3NC* with the various M. In agreement with the more limited results for $I_2(B^3\pi)$ (see above), the entire range of efficiencies was approximately a factor of six. However, in contrast to the I_2 results the collision efficiencies did not correlate with $\mu^{\frac{1}{2}}$ or the collision duration since, for example, the rare gases He to Xe gave practically identical efficiencies. Correlations with a number of physical properties of the collision gas (such as boiling point) were found, all of which could be interpreted as an increase in deactivation efficiency with increasing molecular complexity.

Lin and Rabinovitch (1970) have rationalized these results by a model of collisional deactivation, based on RRKM theory, which assumes that the activated molecule (the substrate) and the chemically inert bath gas form a complex in which energy redistribution occurs among the internal modes of the substrate and the transitional modes of the complex; the transitional modes are vibrational modes of the complex correlating with relative translational and rotational motion of the colliding partners. Three classes of bath gases can be distinguished, monatomic, diatomic and small linear, and polyatomic molecules according to the number of transitional modes (i.e., the number of translational and rotational degrees of freedom lost in forming the complex). The experimental results are in generally good agreement with the model; in particular, energy redistribution involving vibrational

modes of the bath gas appears to occur only for very efficient
deactivators and even in these cases contributes only to a small
extent to the overall energy transfer process.

A further aspect of the work of Rabinovitch and co-workers
has involved the study of multi-step deactivation processes to
determine probability distributions for ΔE the energy lost from
the substrate per collision. The most accurate data have come
from "chemical activation" studies, in which by chemical reaction,
a species is produced with more internal energy than that necessary
for decomposition but which may be stabilized by collisional deactivation to below the critical energy. From variation of the
ratio of stabilization products to decomposition products with
pressure of the collision gas, they have calculated, using RRKM
theory, the average amount of energy removed per collision
(Kohlmaier and Rabinovitch (1963), Georgakakos and Rabinovitch
(1972)). For inefficient collision gases, such as the rare gases,
H_2, or D_2, an exponential type distribution is found for ΔE, with
small values being most probable. The results clearly show that
there are not a large number of elastic collisions with a few inelastic collisions removing a large amount of energy, but rather
that most collisions remove only a small amount of energy. By
contrast, efficient collision gases (most polyatomic molecules)
remove, on average, considerably larger amounts of energy with
the distribution of ΔE being such that very small deactivation
steps are less probable than larger ones. The distribution has
been approximated by a "step-ladder" model, the equivalent of a
Gaussian distribution having a small standard deviation.

ENERGY TRANSFER INVOLVING GASEOUS IONS

In contrast to the extensive studies in neutral systems there
have been few studies of energy transfer processes resulting from
collision of vibrationally excited ions with neutral molecules.
Those studies which have been made provide information primarily
on the relative collisional deactivation efficiencies of various
bath gases and there are practically no data pertaining to the
average amount of energy transferred per collision.

Limited information concerning collisional deactivation efficiencies can be obtained from the magnitudes of the rate coefficients for three-body ion-neutral association reactions, reviewed
recently by Ferguson (1972) and by Kebarle (1972). Following
Ferguson we may write the following general scheme for the association of A^+ with B.

[3] $\quad A^+ + B \underset{k_{-3}}{\overset{k_3}{\rightleftharpoons}} (AB^+)^*$

$$[4] \quad (AB^+)^* + M \xrightarrow{k_4} AB^+ + M$$

$$[5] \quad (AB^+)^* + M \xrightarrow{k_5} A^+ + B + M$$

Assuming a steady-state in $(AB^+)^*$ we may write

$$[6] \quad \frac{\partial [A^+]}{\partial t} = \frac{k_3 k_4 [A^+][B][M]}{k_{-3} + (k_4 + k_5)[M]}$$

from which we derive the low-pressure third-order rate coefficient

$$[7] \quad k^{(3)} = \frac{k_3 k_4}{k_{-3}} = k_3 \rho k_4' \tau$$

where k_4' is the rate coefficient for collision of $(AB^+)^*$ with M, ρ is the fraction of these collisions resulting in stabilization of $(AB^+)^*$ and $\tau(= \frac{1}{k_{-3}})$ is the lifetime of $(AB^+)^*$.

Table I lists third order rate coefficients at 300°K for a number of association reactions where more than one M has been employed; these data are taken from the review by Kebarle (1972). To convert these rate coefficients, for each reaction system, to relative deactivation efficiencies we have divided each $k^{(3)}$ by a calculated k_4', the collision rate coefficient of $(AB^+)^*$ with M, assumed to be given by the Langevin ion-induced dipole model ((Gioumousis and Stevenson(1958)) and for each reaction system normalized the resultant numbers to $H_e = 1$. The relative stabilization efficiencies so obtained are given in the final column.

The collision gases Ne, Ar, and N_2 all show collisional stabilization efficiencies approximately three to four times larger than He, although O_2 appears to have an efficiency equal to that of He. There is no clear dependence on the reduced mass of the colliding pair (collision duration model), nor do the results clearly agree with the collision complex model of Lin and Rabinovitch (1970), which predicts that all monatomic species should have the same stabilization efficiencies. The rate coefficients listed in Table I are derived from several different original sources, and some of the variations observed may reflect differing experimental errors in measuring the third order rate coefficients. Obviously a systematic study of one reaction by the same technique using a variety of collision gases would provide more reliable data.

The first such study was reported by Anicich and Bowers (1974). Using ICR techniques they measured the third-order rate coefficients for the association reactions

TABLE I

Three-Body Ion Association Reactions

Reaction	M	$k^{(3)}$ (300°K)*	Rel. Collision Eff.
$Ar^+ + Ar \rightarrow Ar_2^+$	He	1.3 (-31)	1.0
	Ne	3 (-31)	3.4
$O_2^+ + H_2O \rightarrow O_2^+ \cdot H_2O$	He	8.5 (-29)	1.0
	Ar	3.5 (-28)	3.6
	N_2	4.5 (-28)	4.0
	O_2	1 (-28)	1.0
$N_2^+ + N_2 \rightarrow N_4^+$	He	1.9 (-29)	1.0
	N_2	8 (-29)	3.2
$NO^+ + H_2O \rightarrow NO^+ \cdot H_2O$	He	0.6 (-28)	1.0
	Ar	1.9 (-28)	2.7
	N_2	2.2 (-28)	2.8

*cm^6 molecule^{-2} sec^{-1}

[8] $CH_2CF_2^+ + CH_2CF_2 \xrightarrow{M} C_4H_4F_4^+$

[9] $C_6H_6^+ + C_6H_6 \xrightarrow{M} C_{12}H_{12}^+$

using the parent molecules, the rare gases, and for reaction [8] N_2 and CO as collision gases. Interpreting their results by the scheme, reactions [3] to [5], from the measured values of $k^{(3)}$ they derived (equation [7]) values of $k_3\rho\tau$ by assuming that k_4 equalled the calculated rate coefficient for collision of the complex with M. As shown in Fig. la for the CH_2CF_2 system they found a linear correlation of the resulting $k_3\rho\tau$ values, and hence the collision efficiency, with the square root of the reduced mass of the colliding pair, indicating that the probability of collisional deactivation increases with increasing collision duration, estimated as the ion fly-by time. Anicich and Bowers calculated that, for the $C_4H_4F_4^+$-Xe collision pair, the "collision duration" was between 0.6×10^{-13} and 4.3×10^{-13} sec. Since the deactivation efficiency was either still increasing or had maximized for Xe they concluded that the vibration period of the relevant mode in $(C_4H_4F_4^+)^*$ must be $< 0.6 \times 10^{-13}$ sec ($\nu < 550$ cm^{-1}).

For the benzene system (Fig. lb) the total variation in collisional deactivation efficiencies is small for the rare gases.

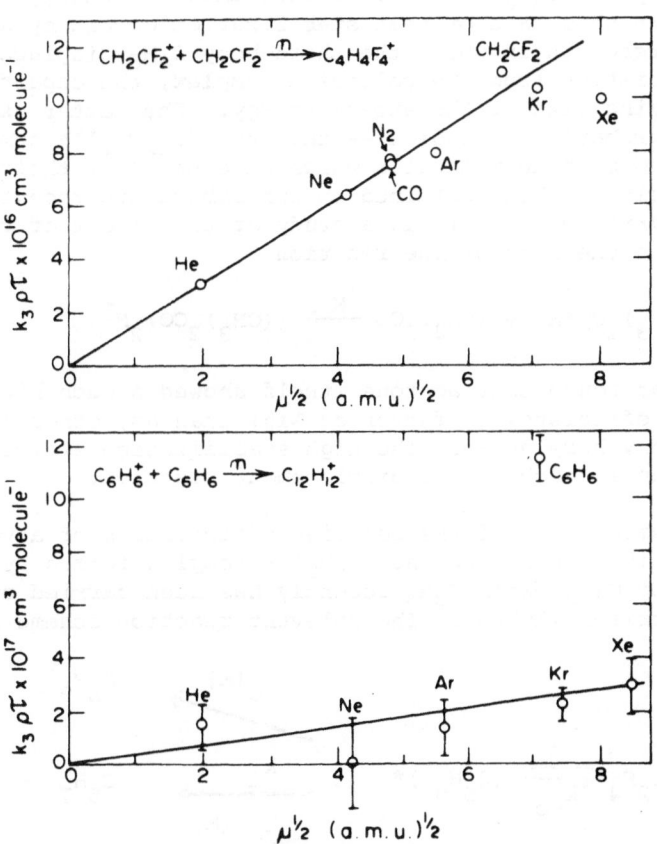

Fig. 1. Collision efficiencies as a function of reduced mass.

Although a correlation with $\mu^{\frac{1}{2}}$ is shown it is equally valid to conclude that, within experimental error, the efficiencies are identical. The latter result would be predicted by the collision complex model of Lin and Rabinovitch (1970) discussed above. The most notable result for the benzene system is that C_6H_6 itself shows a collision efficiency approximately 5 times larger than the rare gases. By contrast, in the difluorethylene system $C_2H_2F_2$ does not show an abnormally high stabilization efficiency. The high efficiency for C_6H_6 must mean either that V-V energy transfer is important in this case or that stabilization occurs by a displacement reaction in which the third-body C_6H_6 displaces one of the C_6H_6 moieties from the collision complex, the departing molecule carrying with it the excess energy. The latter explanation is more probable and indicates that the $(C_{12}H_{12}^+)^*$ complex consists of two benzene molecules which have not lost their identity. A similar result has been obtained in our laboratory recently (Miasek and Harrison (1974a)) in a study of the effect of various third bodies on the rate of the reaction

[10] $(CH_3)_2CO \cdot H^+ + (CH_3)_2CO \xrightarrow{M} ((CH_3)_2CO)_2H^+$

in which it was found that acetone itself showed a much higher stabilization efficiency (a factor of ~15) than any other collision gas. Presumably here as well the high stabilization efficiency results from an interchange or displacement reaction.

An extensive study of the relative efficiencies of a variety of bath gases in stabilizing the $(C_5H_9^+)^*$ complex formed by condensation of $C_3H_5^+$ with C_2H_4 recently has been carried out by Miasek and Harrison (1974b). The relevant reaction scheme is

$$C_3H_5^+ + C_2H_4 \underset{k_{-c}}{\overset{k_c}{\rightleftharpoons}} (C_5H_9^+)^* \quad \begin{array}{c} \xrightarrow{k_s[M]} C_5H_9^+ \\[2mm] \xrightarrow{k_d} C_5H_7^+ + H_2 \\[2mm] \xrightarrow{k_{cd}[M]} C_5H_7^+ + H_2 \end{array}$$

Both the unimolecular (k_d) and collision-induced ($k_{cd}[M]$) route to formation of $C_5H_7^+$ are necessary to rationalize the pressure dependence of the $[C_5H_9^+]/[C_5H_7^+]$ product ratio. From the above reaction scheme one may derive for R the observed $[C_5H_9^+]/[C_5H_7^+]$ ratio

[11] $R = \dfrac{[C_5H_9^+]}{C_5H_7^+} = \dfrac{k_s[C_2H_4]}{k_d + k_{cd}[C_2H_4] + k'_{cd}[M]} + \dfrac{k'_s[M]}{k_d + k_{cd}[C_2H_4] + k'_{cd}[M]}$

From equation [11] one observes that the initial slopes of plots of R vs [M] at constant low pressures of ethylene $(k_{cd}[C_2H_4] + k'_{cd} [M] \ll k_d)$ yield k'_s /k_d and thus relative values of k'_s the stabilization rate coefficient.

TABLE II

M	$(C_5H_9^+)*$ Rel. Stabilization		$(CH_3NC)*$
	Rate Coeff(k_s)	Effic(k_s/k_{LANG})	Rel. Stabiliz. Effic.
He	(1.00)	(1.00)	(1.00)
Ne	0.75	1.11	1.13
Ar	1.31	1.22	1.15
Kr	1.06	0.98	1.01
H_2	2.37	0.86	0.98
D_2	2.27	1.15	1.07
N_2	2.11	1.69	1.56
CO_2	3.76	2.86	2.26
CH_4	5.57	2.97	2.51
CD_4	5.26	3.03	2.35
CF_4	4.54	3.97	> 2.47
SF_6	5.62	4.32	2.67
C_2H_4	10.3	5.21	2.47
C_2H_6	10.5	5.32	3.13
C_2D_6	9.28	5.11	2.86

The relative stabilization rate coefficients (k_s) obtained for non-polar collision gases are shown in Table II, column 2, relative to the value for He which has been assigned a value of unity. In column 3 these rate coefficients have been converted to relative stabilization efficiencies by dividing the k_s values by the rate coefficient calculated for collision of $(C_5H_9^+)*$ with M by the Langevin ion-induced dipole model. The results have again been normalized to the value for He.

A number of points are immediately evident from the results of Table II. First, the stabilization efficiencies for the rare gases He, Ne, Kr, and Ar, are practically identical despite a variation in $\mu^{\frac{1}{2}}$ (and, thus, by inference, in the collision duration, as defined by the ion fly-by time) of a factor of 3. This is in contrast to the results for the difluoroethylene system (Fig. 1a) and indicates that, for the ethylene system, the probability of collisional stabilization is not determined by the "collision duration" so defined. Further evidence for this is obtained by comparison of CH_4 with Ne, CO_2 with Ar, and C_2H_6 with N_2 where each pair has essentially the same reduced mass, but where for each pair the more complex collision gas has a considerably higher collisional stabilization efficiency.

The results obtained are, however, in general agreement with the predictions of the collision complex model of Lin and Rabinovitch (1970) in which the collision efficiency for energy transfer depends largely on the number of transitional modes available in the complex formed between $(C_5H_9^+)*$ and M. As shown in Table II there is a clear distinction between monatomic, diatomic, and polyatomic collision gases, as predicted by the theory. For comparison we show in the final column of Table II relative efficiencies for stabilization of $(CH_3NC)*$ measured by Chan et al (1970). The same general separation according to molecular complexity is observed. The total range of relative efficiencies is greater for the ionic system than for the neutral system. For the neutral $(CH_3NC)*$ system the average energy above the critical threshold is $\sim$2 kcal mole^{-1} while for $(C_5H_9^+)*$ the average energy above the critical threshold is probably much larger (see below). Current and Rabinovitch (1964) have shown that inefficient collision gases such as the rare gases, H_2, and D_2 show a much reduced efficiency relative to efficient collision gases when the energy above the critical threshold is large. This is probably the explanation of the larger range in efficiencies observed for the ionic system.

In their study of the ethylene system Miasek and Harrison (1974b) also included a number of polar bath gases. The results obtained are instructive in examining the effect of ion-dipole interactions on the rate of ion-neutral collisions. The relative stabilization rate coefficients obtained are given in Table III. One might anticipate that N_2O and SO_2 would have stabilization efficiencies similar to CO_2, CH_4, or CF_4, while CH_3F and CH_2F_2 would have stabilization efficiencies between those observed for CH_4 and CF_4. However division of the k_s values by the $C_5H_9^+$-M collision rate coefficient calculated from ion-induced dipole interactions only leads to relative stabilization efficiencies (column 3) which are much larger than those observed for the comparable non-polar molecules. Division of the k_s values by a collision rate coefficient estimated by the average dipole orientation (ADO) theory (Su and Bowers (1973)) at thermal energies leads to the relative stabilization efficiencies in column 4. Again these are considerably higher than for the comparable non-polar molecules. Recently Blair and Harrison (1973) have found that rate coefficients measured for strongly exothermic proton transfer reactions to polar molecules, where the reaction might be expected to occur with unit collision efficiency, are in agreement with rate coefficients calculated using the locked-dipole model at a

TABLE III

Stabilization of $(C_5H_9^+)$* by Polar Molecules

M	k_s*	k_s/k_{LANG}	k_s/k_{ADO}	k_s/k_{LD} 0.4eV	k_s/k_{LD} 0.3eV	"Best" Value
CO	2.63	2.00	2.00	1.79	1.74	1.74
N_2O	4.54	3.26	3.26	2.84	2.79	2.79
SO_2	9.90	7.00	4.16	3.42	3.16	3.26
CH_3F	11.96	8.47	4.21	3.11	2.84	3.00
CH_2F_2	12.99	10.32	4.84	3.95	3.58	3.73
CH_2Cl_2	6.19	3.89	3.68	3.16	3.05	3.11

*relative to k_s(He) = 1.00

relative kinetic energy of 0.4 eV (lab scale). Thus, columns 5 and 6 show relative stabilization efficiencies calculated from the relative k_s values using the locked-dipole expression to estimate the $C_5H_9^+$-M collision rate coefficient. In this case the stabilization efficiencies for the polar molecules are in good agreement with those obtained for the comparable non-polar molecules. Thus it appears that the ADO theory underestimates the collision rate for ion-polar molecule systems although the collision rates can be estimated from the locked-dipole expression if an appropriate (arbitrary) ion energy is chosen. The comparison of collisional stabilization efficiencies for similar polar and non-polar molecules may prove to be a sensitive probe of the effect of ion-dipole interactions. However one cannot preclude the possibility that the stronger ion-dipole interactions increase the lifetime of the complex sufficiently to lead to enhanced energy redistribution involving the vibrational modes of the collision gas.

The results summarized above show that, in general, the various third bodies show less than a factor of five difference in collisional stabilization efficiencies unless specific processes such as interchange are operative. Unanswered, however, is the question whether for efficient collision gases the absolute stabilization efficiency is close to unity or is much smaller than unity. From the measured third-order rate coefficients, $k^{(3)}$, one can, from equation [7], obtain lower limits for $\rho\tau$, the product of the lifetime of the activated ion and the fraction of collisions resulting in deactivation by assuming that k_3 and k_4 equal the collision rate coefficients calculated from appropriate theoretical models. The relevant data for a number of reactions with He third body are summarized in Table IV where it is seen that $\rho\tau$ varies from 3.6×10^{-13} sec for the Ar^+ - Ar system to 4.5×10^{-7} sec for the $C_3H_5^+ + C_2H_4$ system. The major part of this variation undoubtedly results from variation in τ. Ferguson (1972) has discussed in

TABLE IV

$\rho\tau$ for Association Reactions (He third body)

Reaction	$k^{(3)}$	k_4	k'_4	$\rho\tau$
$Ar^+ + Ar \rightarrow Ar_2^+$	1.3 (-31)	6.7(-10)	5.5(-10)	3.6(-13)
$N_2^+ + N_2 \rightarrow N_4^+$	1.9(-29)	8.3(-10)	5.5(-10)	4.2(-11)
$O_2^+ + O_2 \rightarrow O_4^+$	2.4(-30)	7.4(-10)	5.8(-10)	5.9(-12)
$C_2H_2F_2^+ + C_2H_2F_2 \rightarrow C_4H_4F_4^+$	1.7(-25)	$\sim$ 2(-9)	5.4(-10)	1.6(-7)
$C_3H_5^+ + C_2H_4 \rightarrow C_5H_9^+$	2.9(-25)	1.2(-9)	5.5(-10)	4.5(-7)

some detail both the experimental data and the calculations supporting the lifetimes $\tau(O_4^+) \approx 7 \times 10^{-9}$ sec at 80°K and $\tau(N_4^+) \approx 2 \times 10^{-9}$ sec at 200°K. Assuming a T^{-3} dependence for both we estimate $\tau(O_4^+) \approx 1.3 \times 10^{-10}$ sec and $\tau(N_4^+) \approx 6 \times 10^{-10}$ sec at 300°K. These lifetimes, along with the data in Table IV, indicate that ρ (He) for the nitrogen and oxygen systems is approximately 0.05 to 0.07 indicating that in these cases roughly 1 in 20 collisions with He results in stabilization of the excited ion. This value for ρ (He) does not appear inappropriate for the other systems. Since the most efficient collision gases are only $\sim$5 times more efficient than He this would imply that roughly 1 in 4 collisions of efficient third bodies results in stabilization of the complex.

While the magnitudes of the third order rate coefficients can be interpreted as indicating that from 1 in 4 to 1 in 20 collisions are effective in stabilizing the excited ions they provide no significant information on ΔE the average amount of energy removed per collision. For the ion-neutral association reactions of Table I and Fig. 1 where the only decomposition channel is reversion to reactants the removal of a relatively small amount of energy by collision would lead to lifetimes for the gaseous ions which are long compared to the typical observation times (rarely longer than milliseconds).

On the other hand, the $(C_5H_9^+)*$ complex formed in the ethylene system has an exit channel to $C_5H_7^+ + H_2$ as well as reversion to reactants. Recent measurements by Lossing (1974) indicate that the most stable structure for $C_5H_7^+$ is the cyclopentenyl cation with an enthalpy of formation of 200 kcal mole^{-1}. Thus reaction of $C_3H_5^+$ (ΔH_f = 226 kcal mole^{-1}) with ethylene (ΔH_f = 12.5 kcal mole^{-1}) produces $(C_5H_9^+)*$ with approximately 39 kcal mole^{-1} excess energy referred to the most stable $C_5H_7^+ + H_2$ decomposition products.

This suggests that in the stabilization of $(C_5H_9^+)*$ relatively large amounts of energy must be removed on each collision. However we cannot discount the possibility that either there is a significant activation barrier for decomposition of $C_5H_9^+$ to $C_5H_7^+ + H_2$ or a less stable isomer of $C_5H_7^+$ is formed. In the absence of definitive information concerning the ion structures more detailed considerations are not warranted.

Clearly there is a need for careful studies of the collisional stabilization of excited ions of known structures and energy contents where it should be possible to derive information concerning the average amount of energy removed per collision.

The only relevant study to date has been that of Gill, Inel, and Meisels (1971) who have studied in detail the collisional deactivation of the $C_4H_8^{+*}$ excited species produced in the vacuum ultraviolet photolysis of ethylene. The $C_2H_4^+$ ions formed initially reacted with ethylene to produce $C_4H_8^{+*}$ with an internal energy $\sim$58 kcal mole^{-1} above the most stable 2-butene or methylpropene structure. Neutralization of the $C_4H_8^{+*}$ by suitable charge acceptors yielded neutral 2-butene and methylpropene. It was observed that the ratio methylpropene: 2-butene decreased from 0.58 at the lowest pressure of an added collision gas to a constant minimum of 0.08 at higher pressures. The authors assumed that the relative yields of the two neutral products reflected the relative concentrations of the ionic precursors which were assumed to be in equilibrium with each other, the equilibrium distribution being dependent on the excess energy above the barrier for isomerization ($\sim$ 30 kcal mole^{-1}). Within these assumptions they calculated from the variation of the product ratio with pressure of the collision gas that the average energy lost per collision with neon as a third body was not greater than 0.7 kcal mole^{-1} and for ethylene as a third body was not greater than 2 kcal mole^{-1}. This latter number is less than that observed in many neutral systems for efficient collision gases however Gill et al pointed out that the apparent energy lost per collision is on average weighted heavily by the small energy loss collisions at large impact parameters. It should be noted, however, that more recent work by Lias and Ausloos (1971) suggests that the variation of the ratio of product butenes results, at least in part, from competition between interception of the ions by the charge acceptor and reaction of i-$C_4H_8^+$ (but not 2-$C_4H_8^+$) with ethylene itself. This casts serious doubts on the detailed model proposed by Gill et al and, thus, on the results obtained for the average energy transferred per collision.

SUMMARY

A beginning has been made on the detailed study of energy transfer between excited ions and ground state neutral collision

gases. These studies should have the distinct advantage that it
is much easier to calculate reasonably accurate rate coefficients
for ion-neutral collisions than it is for neutral-neutral collisions. The two most detailed studies to date differ in that the
relative collision efficiencies reported by Anicich and Bowers
(1974) can be interpreted by a model where the probability of
energy transfer is determined by the collision duration, defined
as the ion fly-by time. The results of Miasek and Harrison (1974b)
cannot be interpreted by this model but rather require formation
of a complex in which energy redistribution occurs involving the
transitional modes of the complex. It is not clear whether this
difference reflects a fundamental difference in the systems investigated or whether it is related to the different experimental
techniques employed in the two studies. There is a clear need for
further studies, particularly those involving stabilization of
ions of known structure and energy content where it should be
possible to obtain further information concerning the amount of
energy transferred per collision.

REFERENCES

Anicich, V.G. and Bowers, M.T. (1974). J. Am. Chem. Soc., 95,
1279.

Blair, A.S. and Harrison, A.G. (1973). Can. J. Chem., 51, 1645.

Borrell, P. (1967). Advan. Molec. Relax. Processes, 1 69.

Brown, R.L. and Klemperer, W. (1964). J. Chem. Phys., 41, 3072.

Chan, S.C., Rabinovitch, B.L., Bryant, J.T., Spicer, L.D. Fujimoto,
T. Lin, Y.N., and Pavlou, S.P. (1970). J. Phys. Chem., 74, 3160.

Cottrell, T.L. and McCoubrey, J.C. (1961). Molecular Energy
Transfer in Gases., Butterworths Scientific Publications, London.

Current, J.H. and Rabinovitch, B.L. (1964), J. Chem. Phys., 40,
2742.

Ferguson, E.E. (1972). Flowing Afterglow Studies. Ion-Molecule
Reactions (Ed. J.L. Franklin). Plenum Press, New York.

Georgakakos, J.H. and Rabinovitch, B.L. (1972). J. Chem. Phys.,
56, 5921, and references contained therein.

Gill, P.S., Inel, Y., and Meisels, G.G. (1971). J. Chem. Phys.,
54, 2811.

Gioumousis, G. and Stevenson, D.P. (1958). J. Chem. Phys., 29,
294.

Herzfeld, K.F. and Litovitz, T.A. (1959). Absorption and Dispersion
of Ultrasonic Waves, Academic Press, New York.

Kebarle, P. (1972). Higher-Order Reactions - Ion Clusters and Ion
Solvation. Ion-Molecule Reactions (Ed. J.L. Franklin). Plenum
Press, New York.

Kohlmaier, G.H. and Rabinovitch, B.L. (1963). J. Chem. Phys., 38,
1692.

Lias, S.G. and Ausloos, P. (1971). J. Res. Nat. Bur. Stds., 75A,
591.

Lin, Y.N. and Rabinovitch, G.L. (1970). J. Phys. Chem., 74, 3151.
Lossing, F.P. (1974). Paper presented at Twenty-Second Annual
Conference on Mass Spectrometry, Philadelphia.
Miasek, P.G. and Harrison, A.G. (1974a). Unpublished results.
Miasek, P.G. and Harrison, A.G. (1974b). J. Am. Chem. Soc., in
press.
Steinfeld, J.L. and Klemperer, W. (1965). J. Chem. Phys., 42, 3475.
Steinfeld, J.L. (1966). J. Chem. Phys., 44, 2740.
Su, T. and Bowers, M.T. (1973). Int. J. Mass Spectrom. Ion Phys.,
12, 347.

MEASUREMENTS OF ION-MOLECULE REACTION RATES

USING ION SWARM METHODS

J. B. HASTED

Department of Physics, Birkbeck College

Malet Street, London WC1E 7HX England

Introduction

This contribution will be concerned primarily with measurements of ion-molecule reaction rates using ion swarm methods; however we shall view this technique in its perspective against the background of competing types of experiment, keeping constantly under view the relative importance of different parameters of the collision process which we need to measure. For the purpose of experiment no distinction need be made between the charge transfer and the interchange processes. We shall also discuss certain features of the data reported, particularly the shallow minima displayed by the cross-section functions.

There are two contrasting methods of measuring ion-molecule reaction rates in a drift tube; each of them involves the establishment of a known configuration of ion swarm, drifting under the action of a uniform electric field in a buffer gas of a pressure of the order of 1 torr. In one method (Bloomfield and Hasted 1966) a pulsed grid measurement of drift velocity is carried out, so as to produce a measurement of a sharp pulse or 'peak' of ions at the detector which terminates the drift tube; since product ions are formed continuously along the length of the drift tube, their detected peak will not be sharp, but will have a log-linear onset whose slope is proportional to their rate of formation. In the other method (Kaneko et al 1966) the pulsed

grids do not operate during the rate determination,
although their presence is a necessity for a subsequent
determination of drift velocity, from which the mean
impact energy can be calculated (Wannier 1951). The
projectile ion current i_1 and product ion current i_2
are related to the rate constant k and to the reactant
gas density n in the manner:

$$k = \frac{v_d}{n\ell} \quad \ell n \quad (\frac{i_2}{i_1} + 1) \qquad (1)$$

for projectile ion drift velocity v_d and drift tube
length ℓ. This applies only to the case of a single
reaction converting species 1 into species 2.

In both of these methods the radial diffusion
(and to a smaller extent the axial diffusion) of the
ions must be taken into account. The difference
between the diffusion coefficients of the projectile
and product ions gives rise to a difference in the
efficiency of axial monitoring of ion currents (Kaneko
et al 1966). A correction is applied to equation 1,
based on a knowledge of the mobilities and a knowledge
of the radial ionic profile at the entrance to the drift
tube. A variable length of drift tube enables the
understanding of this correction to be checked experimentally.

Establishment of the drifting ions

A variety of methods have been used to establish
a swarm of ions drifting in a buffer gas. In general,
electron impact ionization of the appropriate molecules
is necessary. This can take place in the drift tube
itself, or in a region adjacent to it; but the system
must be designed in such a way as to minimize the
proportions of unwanted ions arising from the reactant
gas or the buffer gas. It is possible to allow the
ions to spend sufficient time in the pre-drift region
for long lifetime excited states to decay. However,
most techniques of establishing swarms of ions have
the disadvantage that proportions of unwanted ions
enter the reaction region. The alternative method used
in the present work is to inject a mass-separated ion
beam through a small orifice on the axis of the drift
tube. The variable-length drift tube with injection
facility is shown schematically in Fig. 1. The
injection technique has the great advantage that

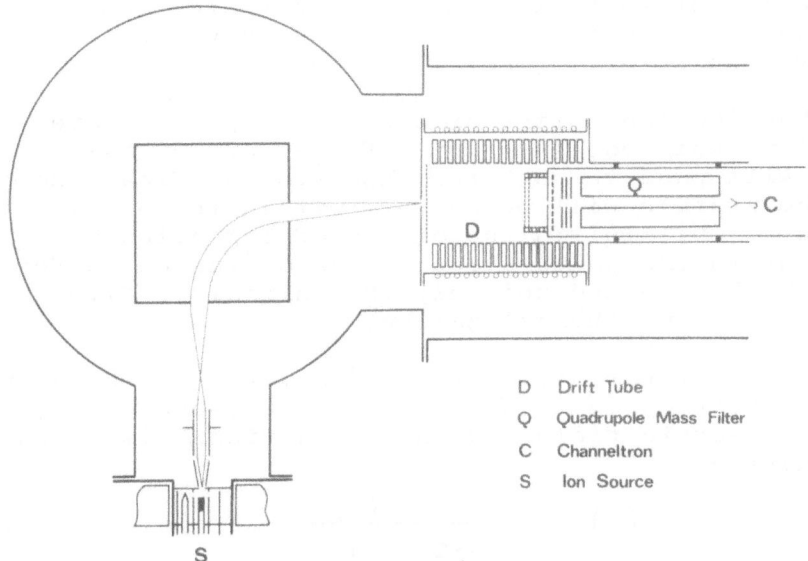

D   Drift Tube
Q   Quadrupole Mass Filter
C   Channeltron
S   Ion Source

Fig. 1

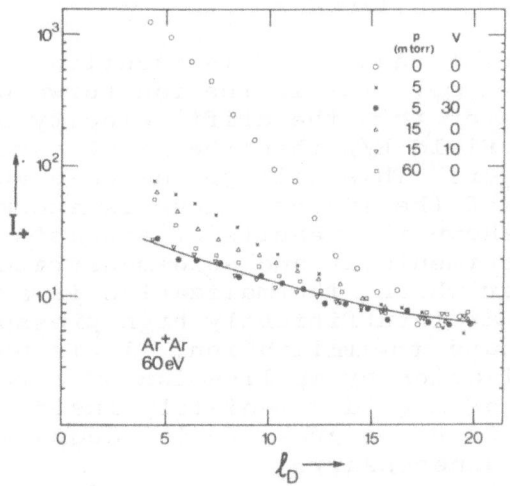

Fig. 2

projectile ion populations of high purity can be
achieved;   but long-lifetime excited states must
be avoided in the ion source,   since they are not
destroyed on entering the drift tube.

When the ion optics of the mass separator are
correctly designed, only ions of the correct mass
number will pass through the injection orifice, and
the beam profile will be uniform within the orifice
diameter. This initial condition of injection is
important to the calculation of the diffusion correction,
in which are assumed not only the initial profile
but also instant thermalization.

The axial variation of ion density $N(z)$ for a swarm
of ions drifting in a buffer gas at pressure p in
uniform electric field E with no reactions, is given by
the equation

$$N(z) = \frac{N(0)}{\pi \delta^2 + (\frac{4\pi D}{v_d})z} \qquad (2)$$

where δ is the diameter of injection orifice, z the
axial dimension, D the diffusion coefficient and v_d
the drift velocity.   Since in the Nernst-Einstein
relation

$$D = \frac{T v_d}{11600 E} \qquad (3)$$

the denominator of equation 2 is identical for all ions
of identical charge. But if the ion forward velocities
were to be greater than the drift velocity appropriate
to the reduced field E/p then the axial variation $N(z)$
would be stronger. This will be the case when the
thermalization of the injected ions is incomplete.
In Fig. 2 are shown the results of attempts to verify
equation 2 experimentally and so demonstrate the
conditions under which thermalization is adequate. It
will be seen that insufficiently high pressure gives
rise to inadequate thermalization. It is possible to
assist thermalization by application of a retarding
field by means of a grid immediately inside the orifice,
but when the buffer gas pressure is adequate this
retardation is unnecessary.

Long-lived excited ions

It is of great importance that the ions injected
into the drift tube should not contain an appreciable
proportion of long-lived excited states. Such states
will not necessarily be easily deactivated by
collisions with buffer gas, and will often possess
different charge transfer collision cross-sections
from those of the ground state ions which it is desired
to measure.

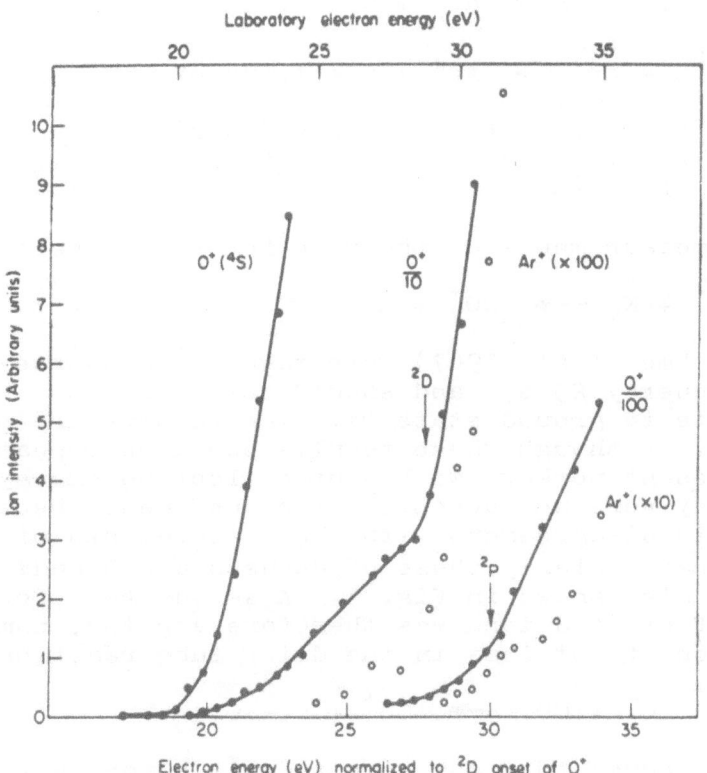

Fig. 3

In previous work (Bohme et al. 1967) it was found
that the proportions of excited states of O⁺ could be
varied by variation of the electron impact energy in the
ion source. As this energy is increased, the ionization
cross-section rises approximately linearly with energy;
at the onset energies for the production of excited ions,
the cross-section energy dependence rises faster, after
the manner of Fig. 3. At electron energies higher than
these onsets the ion beam will contain both ground and
excited states, but at lower energies only ground state
ions should be present.

Ions injected into the drift tube and allowed to
drift in the lowest possible field will be unable to
undergo endothermic collision processes. A reaction
which is exothermic to excited state ions and
endothermic to ground state ions will only be observed
above the onset; this should be the case for

$$O^+ + Ar \longrightarrow Ar^+ + O - 2.15 \quad eV \tag{4}$$

$$O^+ \, ^2D + Ar \longrightarrow Ar^+ + O + 1.17 \quad eV \tag{5}$$

and the effect is in fact clearly observed in Fig.3
where the Ar⁺ ions do not appear strongly until above
the onset of $O^+ \, ^2D$.

The measurements of the rate for the process

$$O^+ + N_2 \longrightarrow NO^+ + N \tag{6}$$

made by Bohme et al (1967) were taken with source
electron energy 25 eV, and should therefore be
appropriate to ground state O⁺. The results are shown
in Fig. 4. Although these results could be repeated
by the present workers with source electron energy
25 eV, they were not accepted with confidence because
they are in disagreement with the measurements of other
workers, particularly those of Johnson and Biondi (1972)
which are also shown in Fig. 4. A second test for the
absence of excited ions was therefore applied, namely
the absence of CO⁺ ions in the drift tube reaction

$$O^+ + CO \longrightarrow CO^+ + O - 0.4 \; eV \tag{7}$$

This test showed that some excited $O^+ \, ^2D$ ions were
present in the beam even when the ion source electron
acceleration energy was such that no $O^+ \, ^2D$ ions were
shown up on the argon test. And under such conditions

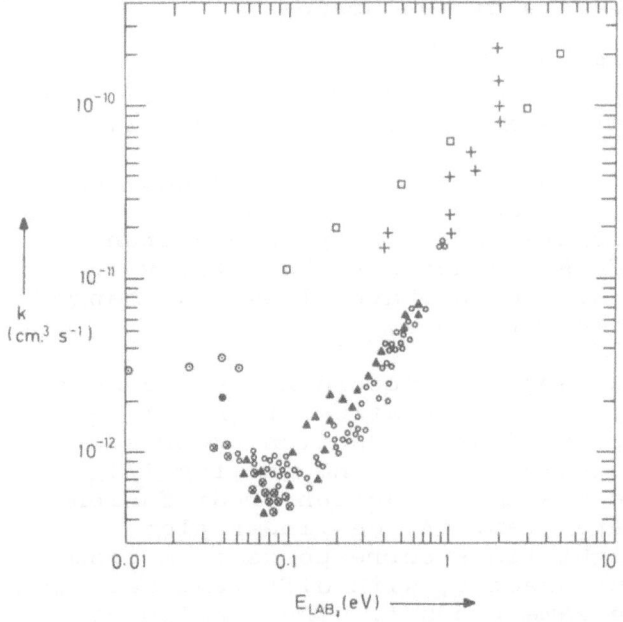

+ Giese (1966)
⊕ Dunkin et al (1968)
○ Johnsen & Biondi (1973)
□ Bohme et al (1967)
◎ Nakshbandi & Hasted (1967
● Copsey et al (1966)

Present work ▲

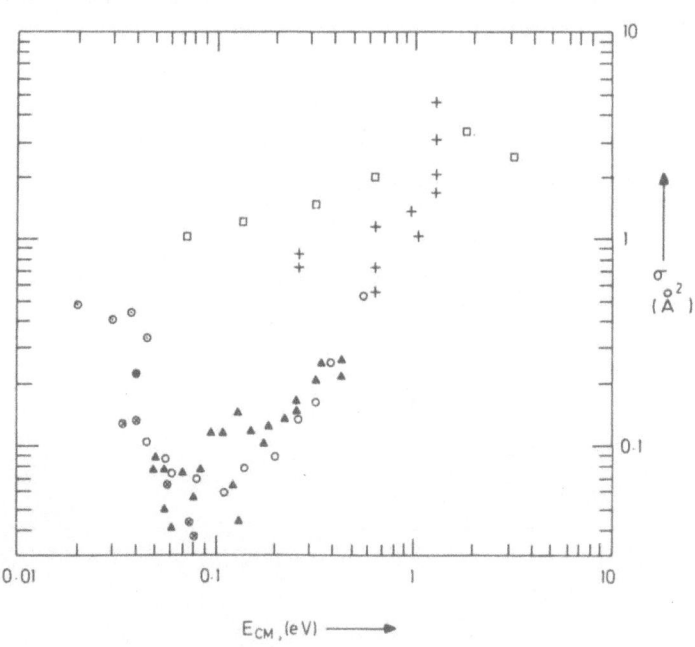

O⁺ + N₂ ⟶ NO⁺ + N

Fig. 4

the measurements of the rate of the process

$$O^+ + N_2 \longrightarrow NO^+ + N \qquad\qquad (8)$$

still yield the same values as those of Bohme et al.

   The clue to this anomaly is that the presence of excited states is extremely sensitive to ion source gas pressure, in addition to its sensitivity to electron accelerating energy. This may not have been fully understood by Bohme et al who may have allowed a change of pressure during the measurements.

   A further test was applied, according to the method of Turner et al (1968). The attenuation of the O^+ beam in pure nitrogen was measured as a function of pressure; a pure exponential ('log-linear') dependence $\log I/I_0$ versus p corresponds to a single reaction or diffusion rate, therefore to a pure beam. A log-linear plot consisting of two straight lines corresponds to a beam containing two different species, with different reaction rates which lead to the same product. The results of this test for O^+ ions in pure N_2 are shown in **Fig. 5**, confirming the existence of O^+ ^2D in the ion beam

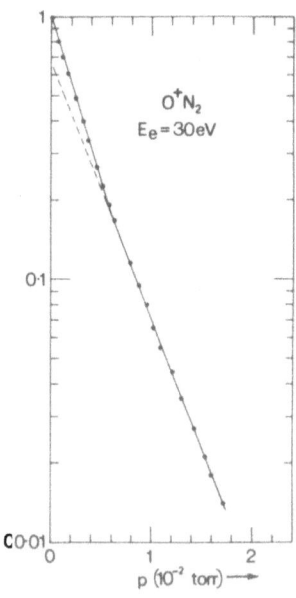

Fig. 5

under conditions where it was not previously suspected. Thus we are confident of being able to produce pure ground state $O^+ X \, ^4S$ beams, with which the present measurements have been made.

Mass Discrimination at the sampling orifice

Since the mass-spectrometric sampling of a high pressure drift tube is not fully understood, some experiments were carried out to investigate variables which might be expected to introduce or eliminate mass-discrimination: the sampling orifice diameter r_o and the ion accelerating potential V_A applied behind it. The results of these experiments are shown in Fig. 6,

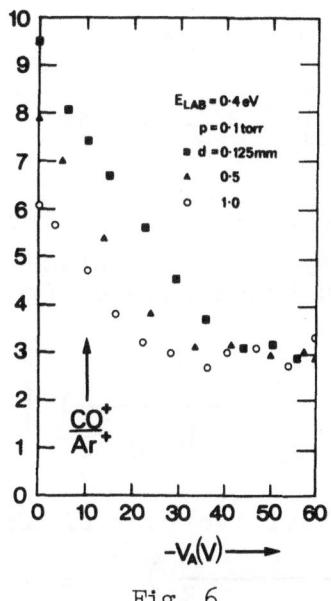

Fig. 6

displaying ratios of CO^+ counts to Ar^+ counts as functions of V_A, for different values of r_o, under identical drift conditions of Ar^+ in Ar with added CO. Only for sufficiently large r_o and V_A is mass-discrimination eliminated, so that the ratio becomes independent of the sampling conditions.

One possible contributory factor to the mass-discrimination is convective flow, which has been considered by Kebarle (1966) and Parkes (1971). The collisions suffered by the ions in the region inside the orifice ensure that the sampling is governed by competition between drift and convective flow. An ion will only be sampled from a certain point if the time τ_d taken to drift from that point to the collector plate exceeds the time τ_f taken for it to pass to the hole by convective flow.

The convective flow rate F through the hole develops around its centre in hemispherical symmetry, with radial velocity

$$v = F / 2 \pi r^2 \tag{9}$$

For helium

$$F \simeq 0.2 \, A \, s^{-1} \tag{10}$$

with orifice area A mm^2. Now

$$\tau_f = \int_{r_o}^{r} \frac{2 \pi r^2}{F} \, dr \tag{11}$$

and $\tau_d = z/v_d$; with $r^2 = x^2 + z^2$. By setting $\tau_f = \tau_d$ to obtain the boundary of the sampled volume, its maximum radius x_m can be calculated from

$$x_m^2 = r_o^2 \left\{ R^2 - \left(\frac{F}{2 \pi R v_d r_o} \right)^2 \right\} \tag{12}$$

where R is the real positive root of the equation

$$R^4 - R - 3 \left(\frac{F}{2 \pi v_d r_o^2} \right)^2 = 0 \tag{13}$$

Only when both terms within brackets ( ) are sufficiently small will R become unity and the apparent sampling area $\pi \, x_m^2$ equal the orifice area $\pi \, r_o^2$. Otherwise the sampling efficiency will be dependent upon drift velocity, and since this is different for primary and product ions, the mass-discrimination will introduce an error into the measurements.

The drift velocity of Ar^+ in Ar is considerably smaller than that of CO^+ in Ar, on account of the high probability of symmetrical resonance charge transfer. It follows that according to convective flow theory it is the CO^+ and not the Ar^+ which should be lost by drift to the collector plate; yet the experiments show a contrary discrimination, so that the convective effect must be small compared with a contrary effect, for which we propose the following.

There is evidence that ion-optical effects play an important role in ion sampling; the orifice has been discussed (Hasted 1974) in terms of a single aperture or Davisson-Calbick lens (1931) for which the focal length f is given by

$$\frac{1}{f} = \frac{-(E_2 - E_1)}{4\ V} \tag{14}$$

with E_2 the electric field outside the sampling orifice, E_1 the field inside it and V the effective potential through which the ions reaching the lens have been accelerated. For drifting ions of mass m_+ it would seem appropriate to take $V = \frac{1}{2}\, m_+\, v_d^2$, where v_d is their drift velocity. E_1, the drift tube field, is of course known, but E_2 is comparatively small because of the peculiar configuration of the quadrupole mass-spectrometer used in these early experiments. The quadrupole rods, separated from the orifice axis by $r_q = 0.8$ cm, are spaced only a = 1 mm from the orifice plate; but the field on the axis is very much smaller than V_a/a; E_2 is usually much smaller than E_1, so that the lens is strongly divergent, and even more so for ions of smaller drift velocity; thus more Ar^+ ions than CO^+ ions are lost. But when a sufficiently large potential V_a is applied the divergence of the orifice lens is weakened, and only a small proportion of ions are lost, so that the mass discrimination disappears. A smaller potential V_a is required to neutralise the lens when the orifice is larger, since a larger potential is found at the orifice circumference when its diameter is increased.

Nevertheless a smaller diameter orifice, which
enables higher gas pressures to be attained in the
drift tube, can be employed with sufficiently large V_A.
The dependence of the ratios CO^+/Ar^+ on V_A under
different conditions has been shown in Fig.6.

Energy distributions of ions in the drift tube

We have measured by retardation analysis the
energy distribution of the ion species injected into the
drift tube and emerging from the sampling orifice. For
this measurement two planar grids are mounted
respectively 1 mm and 2mm behind the sampling orifice
plate;   the quadrupole mass spectrometer is a further
7 mm behind the second grid. The first grid is
maintained at a retarding potential + V_R with respect
to the sampling orifice. The second grid is connected
to the quadrupole mass spectrometer, which is held at
– 50 V, so as to collect the great majority of the ions
which are able to pass through the first grid. The
collected ion count functions $I_+(V_R)$, are differentiated
graphically to obtain energy distributions d I^+ (E)
such as those shown (the closed circles, upper curves)
in Fig 7.

It is obvious that there is ion-optical
discrimination in this experiment;   the lowest energy
ions are preferentially collected, but for the faster
ions there appears to be less discrimination. The
The sampling orifice, as discussed previously, acts as
a Davisson-Calbick (1931) thin lens of focal length f
given by equation 14. V and E_1 are as before, but E_2
is now given by – $V_R/0.1$ cm. Now the magnification
of a thin lens is approximately inversely proportional
to f when the image distance is taken to be constant and
much larger than the object distance. We must apply to
the distributions a correction inversely proportional
to the magnification;   when we multiply the currents by
f, then the low energy anomaly of Fig. 7 disappears,
without the high energy ends of the energy distributions
being greatly distorted;   the modified distributions
are also shown in Fig. 7. The efficiency of the sampling
can be considered to be approximately independent of V_R
when this correction is applied.

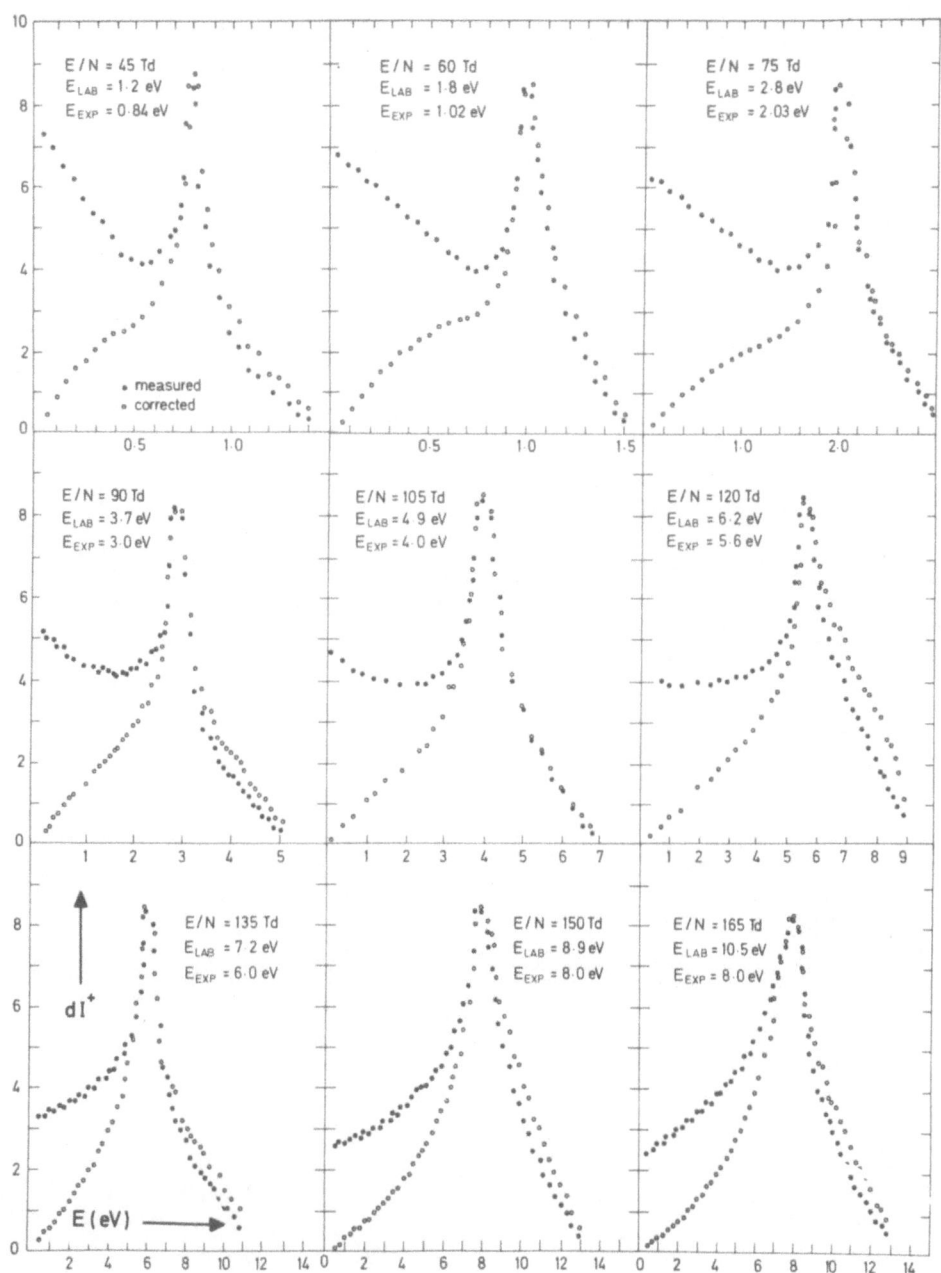

Fig. 7.   O⁺ in He; Energy Distribution

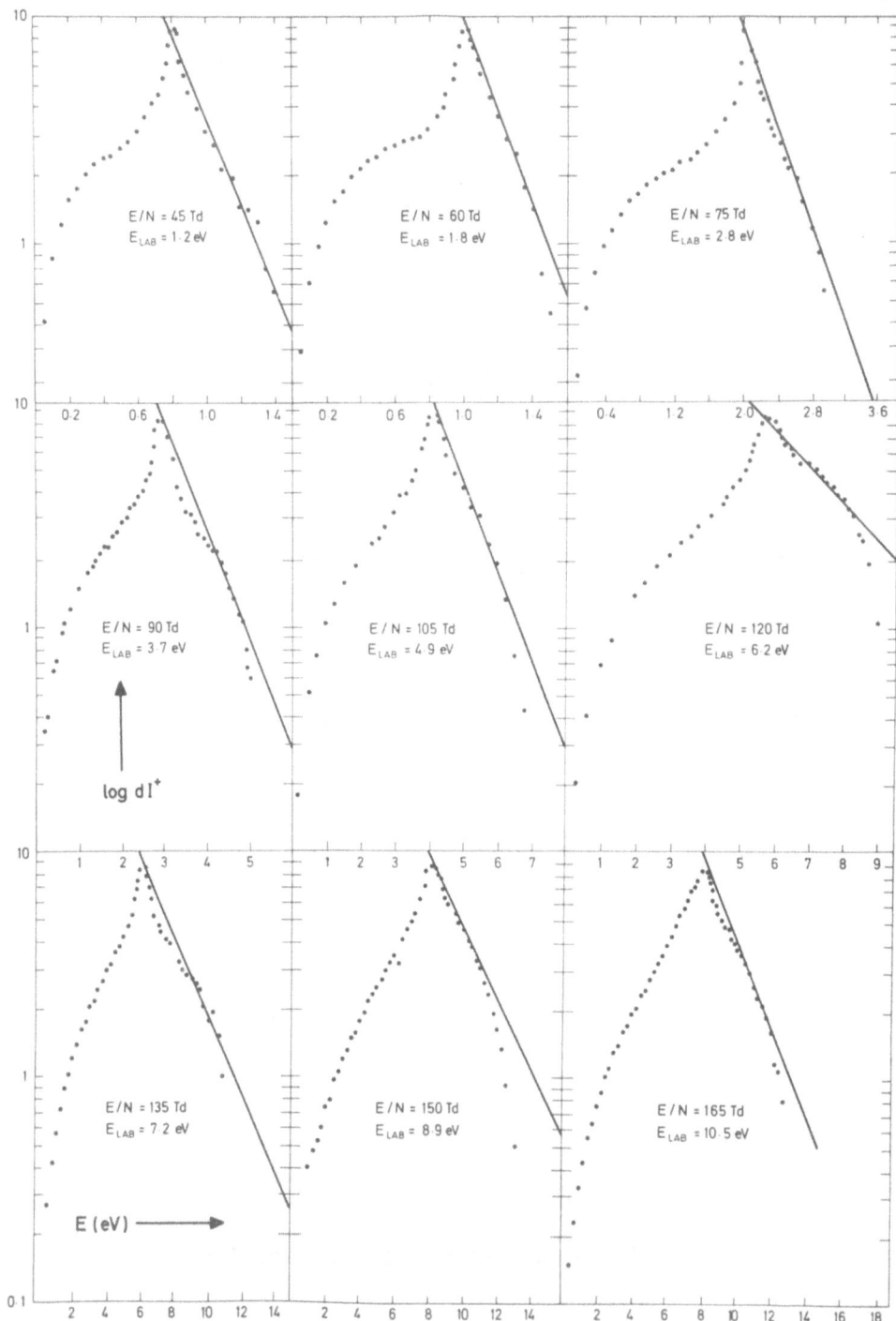

Fig. 8.   O$^+$ in He; Log of Energy – Distribution

In Fig. 8 the energy distributions for O^+ ions in
He so obtained are plotted on a semi-logarithmic scale,
in order to show the magnitude of the departure from
Maxwellian high energy tail. They are in semi-quantitative agreement with the Wannier formula (Wannier 1951)
for the mean energy $\bar{E}$ of ions of mass m_+ and drift
velocity v_d, drifting in gas of mass m_g;

$$\bar{E} = \tfrac{1}{2} m_+ v_d^2 + \tfrac{1}{2} m_g v_d^2 + \tfrac{3}{2} kT \qquad (15)$$

In Fig. 9, a comparison is made between E_{exp}, the
energies at which the measured distributions
pass through maxima, and v_d^2.

In Fig. 10 is shown $\Delta E_{\frac{1}{2}}$ the full width at half
height of the energy distribution, as a function of
E_{exp}.

A rather wide distribution appears to persist down
to very low drift fields and velocities. The collision
process

$$O^+ \,{}^4S + CO \; X \,{}^1\Sigma^+ \longrightarrow \; CO^+ \; X \,{}^2\Sigma^+ + O \,{}^3P \qquad (16)$$

is endothermic by 0.4 eV (0.6 eV in the laboratory
scale). Yet the onset of the rate function is as low
as 0.1 eV (Fig.13), nevertheless the cross-section is
one half of its maximum value for $E_{CM} = 0.2$ eV, which
is consistent with the FWHM curve of Fig. 10, as
indicated by the arrow.

New measurements of O^+ and N^+ rates

Kosmider and Hasted (1974) are currently reporting
drift tube measurements of collision processes of O+
and N^+ ions with reactant gases O_2, N_2, NO, CO; the
buffer gas was in all cases helium.

The measured rate constants k, are shown as
functions of mean energy E_L (laboratory frame) and the
cross-sections σ calculated therefrom as functions of
centre of mass energy in Figs. 4, 11-17. The values
of E_L are calculated from the measured drift velocities
using equation 15. The measured drift velocities
O+ in He and N^+ in He are shown as functions of E/n
in Fig. 18. They are somewhat higher than the values

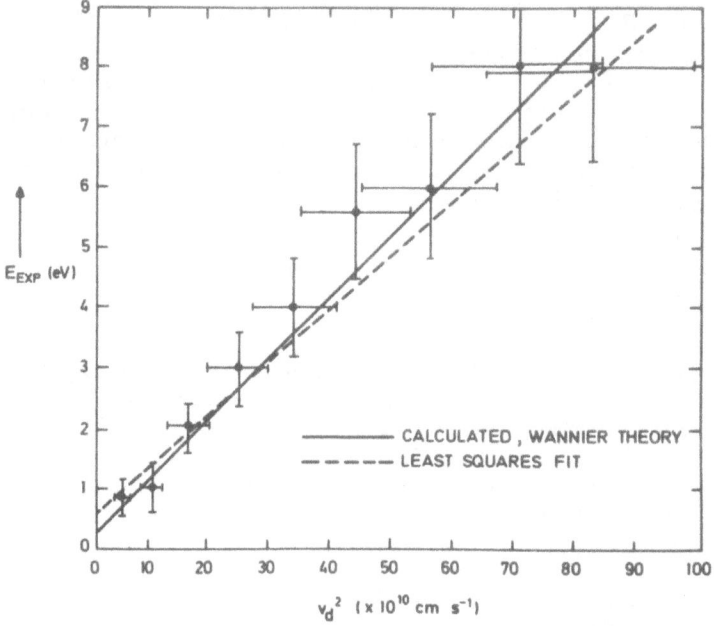

Fig. 9.   Experimental Energy versus Drift Velocity O⁺ in He

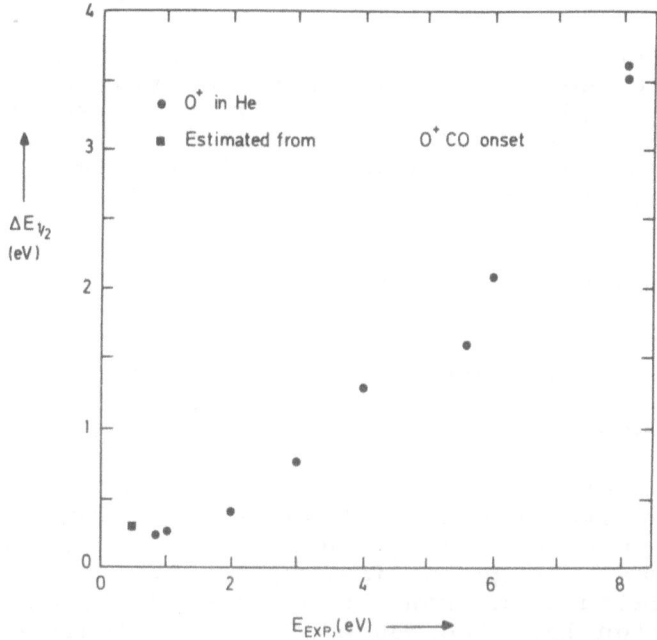

Fig. 10.   FWHM of Distirbution as Function of Experimental Energy
           O⁺ in He

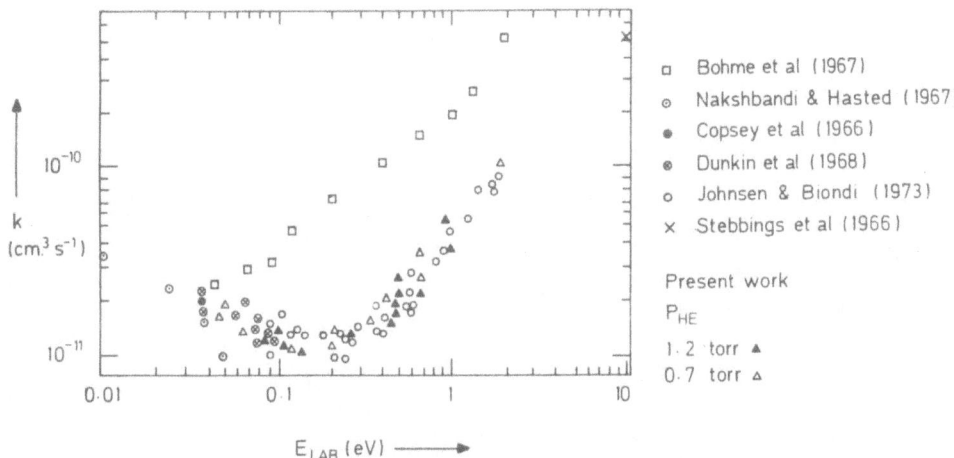

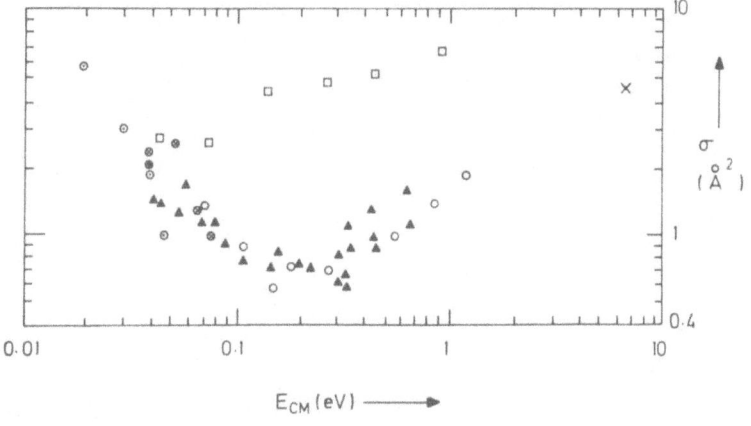

$$O^+ + O_2 \longrightarrow O_2^+ + O$$

Fig. 11

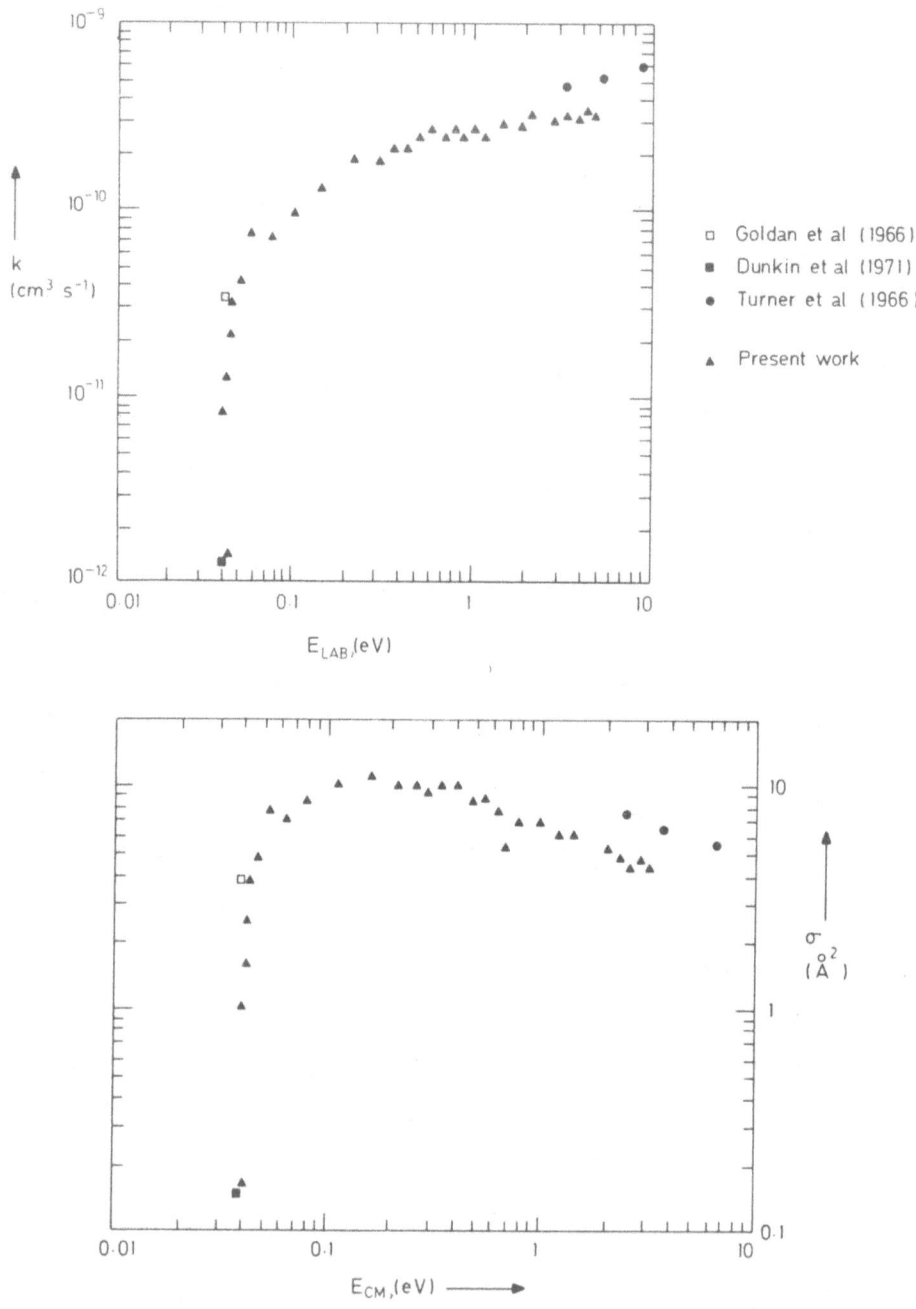

$O^+ + NO \longrightarrow NO^+ + O$

Fig. 12

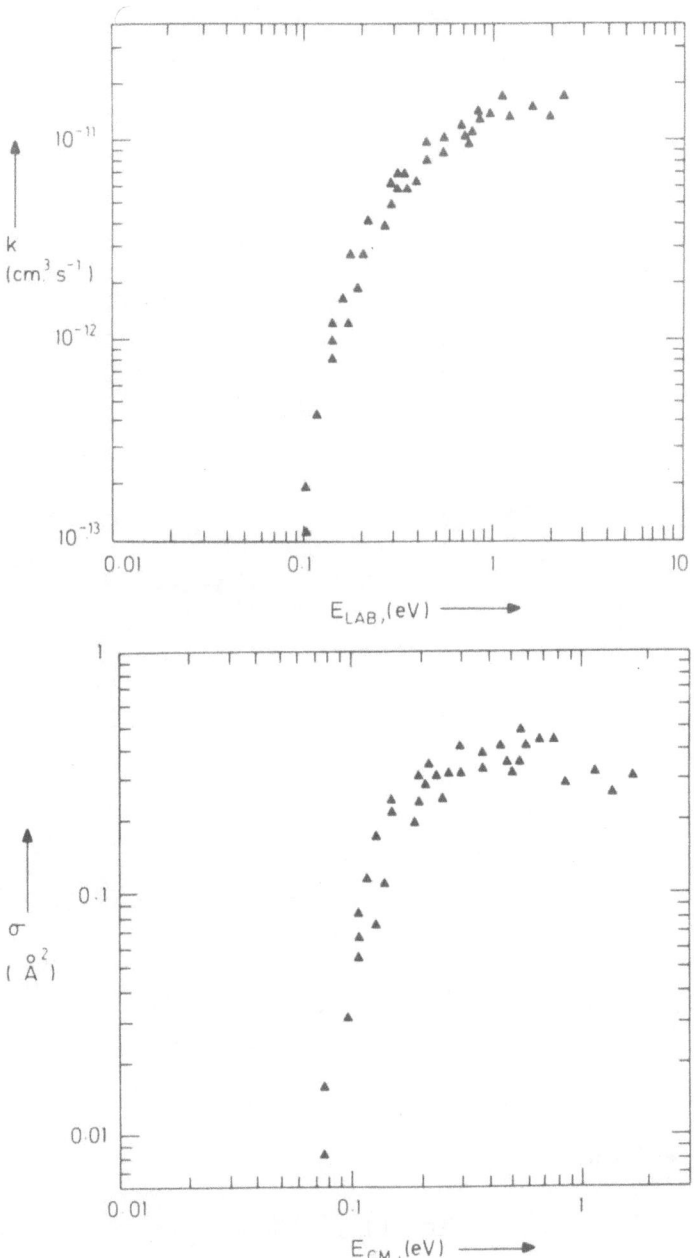

$$O^+ + CO \longrightarrow CO^+ + O$$

Fig. 13

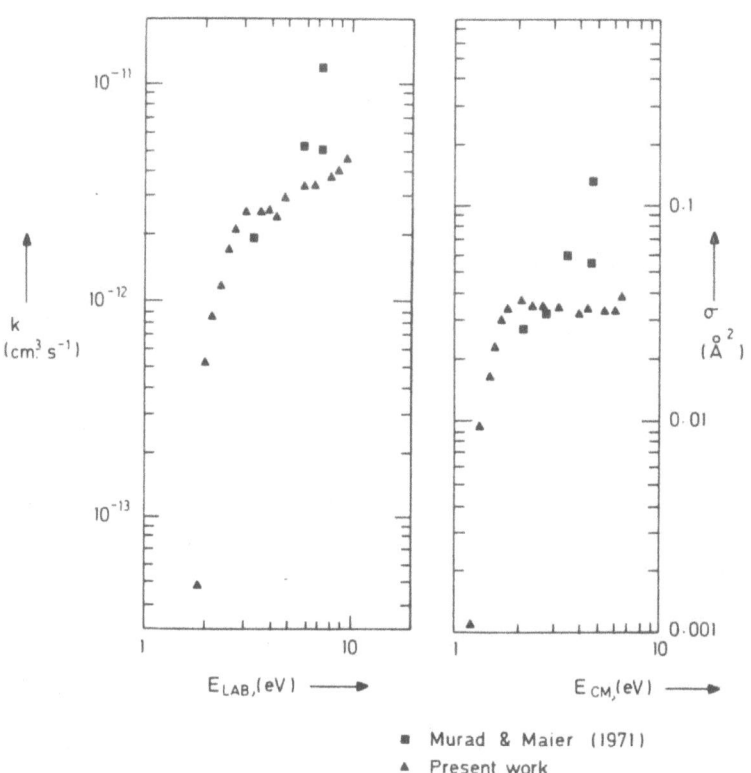

■ Murad & Maier (1971)
▲ Present work

$N^+ + N_2 \longrightarrow N_2^+ + N$

Fig. 14

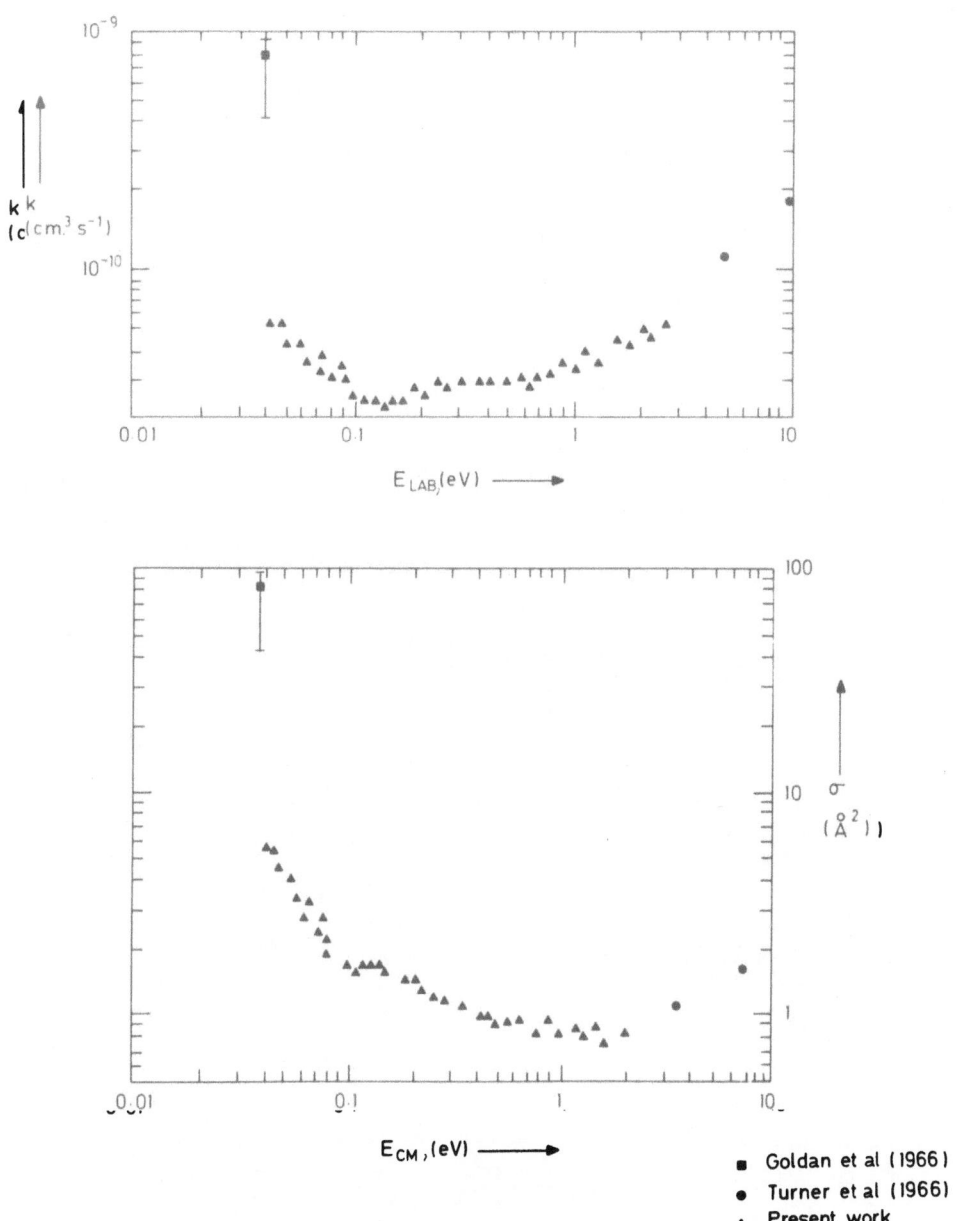

k
k
(c(cm.³ s⁻¹)

E$_{LAB}$,(eV) ⟶

σ
(Å²))

E$_{CM}$,(eV) ⟶

■ Goldan et al (1966)
● Turner et al (1966)
▲ Present work

N⁺ + NO ⟶ NO⁺ + N

Fig. 15

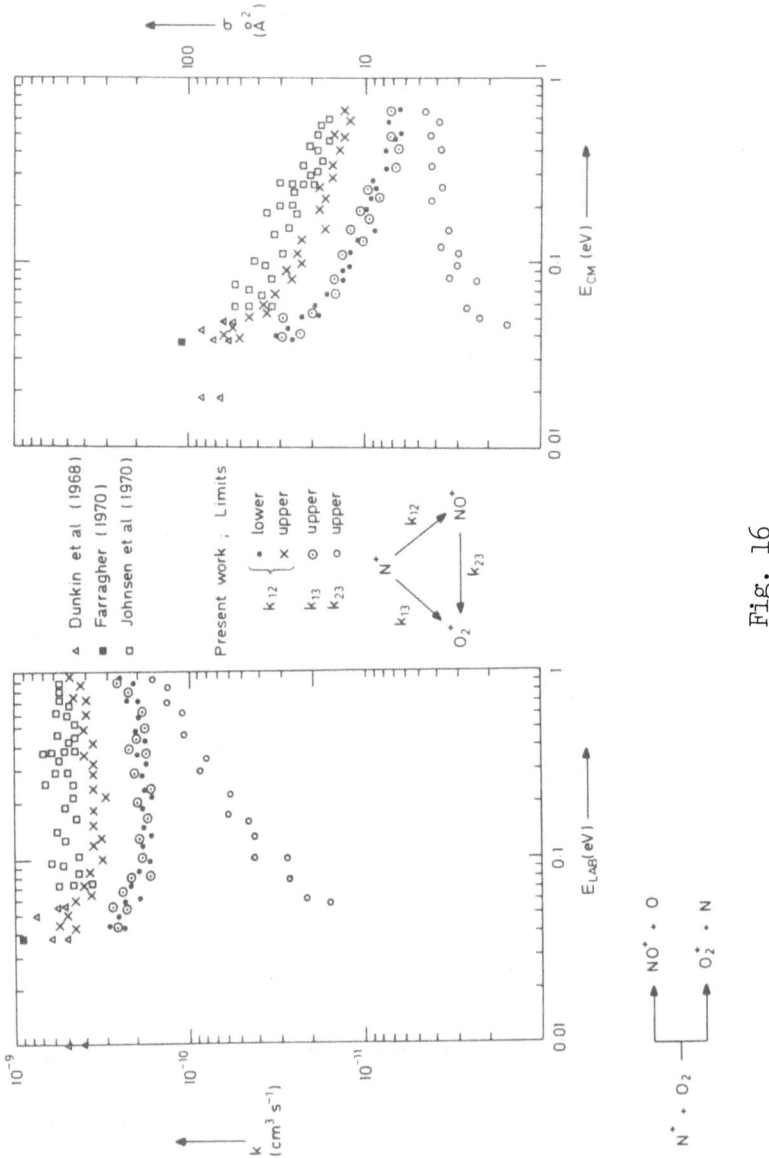

Fig. 16

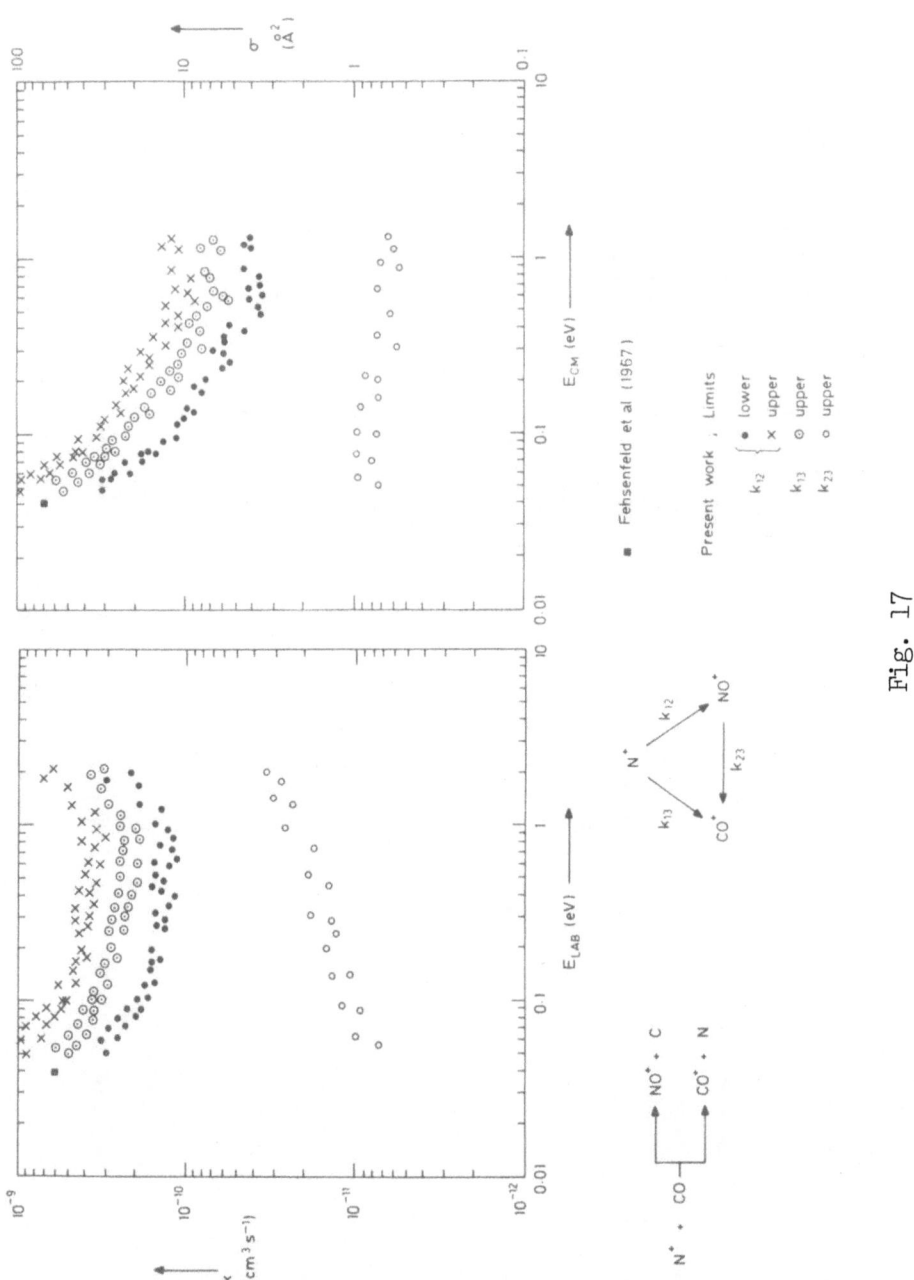

Fig. 17

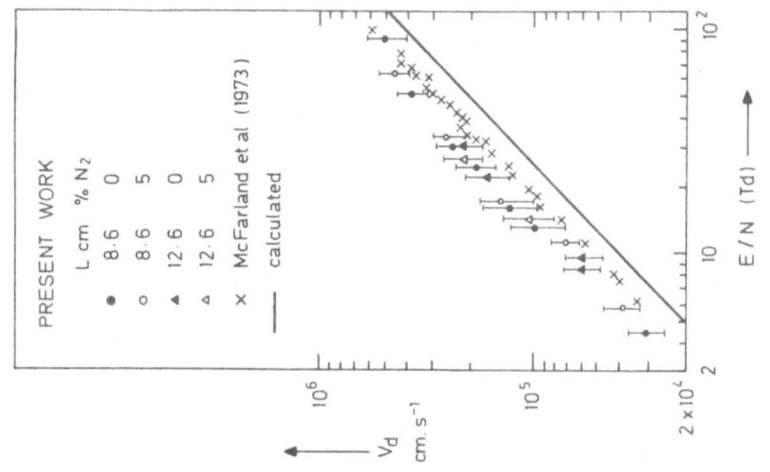

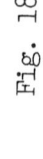

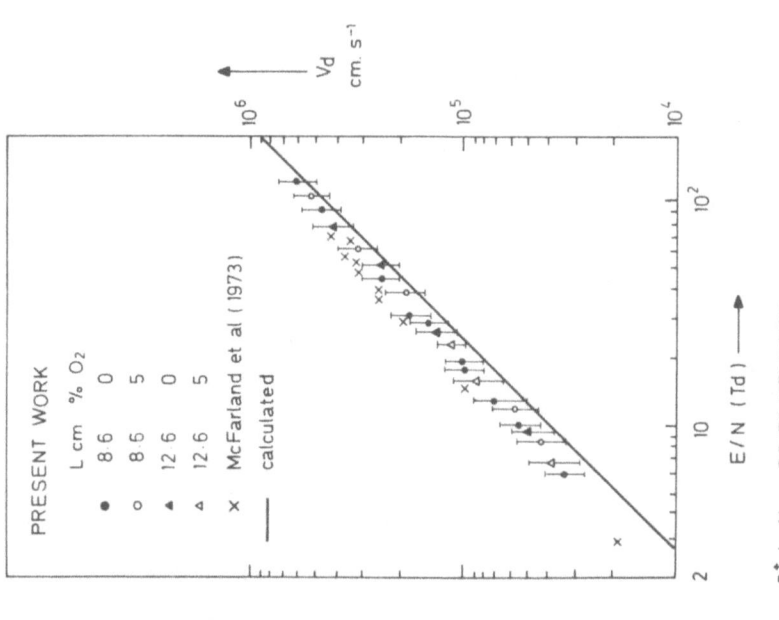

Fig. 18

calculated from the classical formula (Dalgarno et al 1958)

$$K = \frac{v_d}{E} = \frac{35.9}{\sqrt{\alpha\mu}} \tag{16a}$$

where α is the gas polarizability and μ the reduced mass of the ion-gas combination.

Now that additional investigations of the causes of the discrepancies between previous drift tube measurements of the O^+ N_2, and also of O^+ O_2 collisions have eliminated this discrepancy, there remain very few important disagreements between the rates reported for these eight collision processes. Of course an accurate comparison between drift tube rates and rates calculated from beam measurements of cross-section is not possible, because of the wide and only approximately known ion energy distribution obtaining in the drift tube. Nevertheless the drift tube and beam rates for several processes compare reasonably well.

The present rates for the processes

$$O^+ + N_2 \longrightarrow NO^+ + N \tag{17}$$

$$O^+ + O_2 \longrightarrow O_2^+ + O \tag{18}$$

are seen from Figs. 4, 11 to agree reasonably well not only with the flowing afterglow measurements (Dunkin et al 1968) but also over the entire energy range with the drift tube measurements of Johnsen and Biondi (1973). As has already been mentioned it is possible to reproduce the data of Bohme et al (1967) by running the ion source under conditions such that a proportion of long-lived excited ions O^+ ^2D are present in the injection beam. In the experiment of Johnsen and Biondi the long period spent by the ions under swarm conditions before the measurement region is entered precludes the presence of O^+ ^2D ions.

We may consider the remainder of the collision processes under two headings; those in which only one product ion species is produced, and those for which more than one product ion species is observed. The former include O^+O_2, O^+NO, O^+CO, N^+N_2 and N^+NO. The second group includes N^+O_2 and N^+CO.

The O^+O_2 collision process leads to the production
of O^+_2 X $^2\Pi$

$$O^+ {}^4S + O_2 \text{ X } ^3\Sigma \longrightarrow O_2^+ \text{ X } ^2\Pi + O \text{ }^3P + 1.55 \text{ eV (19)}$$

with products in the ground states. The present data
are in agreement with the drift tube data of Johnsen
and Biondi (1973), and lie markedly below the drift
tube data of Bohme et al (1967); the difference, as
has been discussed above, is attributed to the presence
of excited O^+ in the earlier work.

This collision process, together with that between
O^+ and N_2, possesses only a small rate at low impact
energies, but a larger rate at higher energies. It
might be thought that the O^+ O_2 collision in its ground
states is in violation of the Wigner spin rule, since
the spin quantum numbers do not add up $(3/2 + 1 \neq \frac{1}{2} + 1)$.
However it is the moduli of the spin quantum numbers
which are significant in the pseudo-molecule of collision.
For this process the possible total spin quantum numbers
are 5/2, 3/2, $\frac{1}{2}$, and it is perfectly possible for the
second and third of these to separate into the right-
hand side products, without any violation.

The small thermal energy rates for the O^+N_2 process
have been thought to arise from repulsive configurations
of the potential energy surfaces, which make it so that
transitions between these surfaces can only take place
at impact energies sufficiently high for the system
to reach the appropriately tight geometrical configuration. Fig. 19 displays the potential energy curves
calculated by Kaufman (1972); at sufficiently high
impact energies the interchange is able to take place
entirely on the $^4\Sigma$ surface.

There remains the interesting question of why the
cross-section function rises with decreasing energy at
the lowest impact velocities of all. Presumably there
is a second route through which transitions, albeit less
likely, might take place, without involving repulsive
surfaces. Such a route could be $^4\Sigma \longrightarrow ^2\Pi (\longrightarrow ^4\Pi) \longrightarrow ^4\Sigma$.
At impact energies less than about 0.5 eV the attractive
interaction energy dominates most of the collision time;
dynamical calculations (Dugan and Magee 1971) show that
the species pursue complicated trajectories, somewhat
after the pattern of rosettes. The higher the impact
energy, the smaller the proportion of "rosetting", and

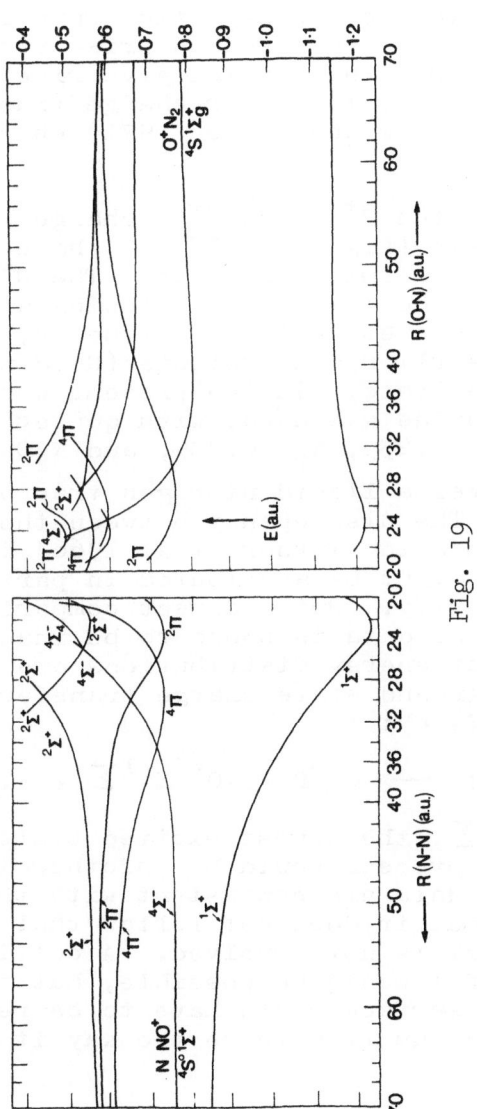

Fig. 19

the smaller the number of transitions through this
second channel, so that the cross-section function falls
until there is competition from the repulsive potential
energy surface channel.

The minimum in the cross-section function for the
O^+N_2 collision is thus supposed to arise from quite
different causes from those which give rise to minima
in near-thermal energy molecular charge transfer
processes (Birkinshaw and Hasted 1971) which will be
discussed below.

The data for the O^+NO and O^+CO charge transfer
collision processes (Figs. 12, 13)   have the
appearance of endothermic functions. The O^+NO data are
in agreement with the most recent flowing afterglow
measurements (Dunkin et al 1968), although, owing to gas
impurities, the earlier measurements (Goldan et al 1966)
gave a much higher rate. In the present work 99%
Matheson nitric oxide was used, with quoted impurities
as follows: NO_2 0.25%, N_2 0.25%, and N_2O 0.5%. The
gas was passed over a liquid nitrogen trap on its way to
the drift tube. The discrepancy between the present
data and the beam data (Turner et al 1966) is not
significant, and might be attributed in part to
inadequate gas purity. But a direct comparison between
beam and drift tube data is bound to be inaccurate,
because the impact energy distributions are very
different. The ground state charge transfer process is
exothermic by 4.24 eV:

$$O^+ \; {}^4S + NO \; X \; {}^2\Pi \longrightarrow O \; {}^3P + NO^+ \; X \; {}^1\Sigma + 4.24 \text{ eV}$$

However if $NO^+ \; {}^3\Sigma$, the lowest excited state of the ion
were formed, the process would be endothermic by 0.5 eV.
The experimental data are consistent with the dominance
of this channel but it does not follow that the
$NO^+ \; X \; {}^1\Sigma^+$ channel is not involved. A collisional spin
quantum number of 1 would be possible, but the appropriate
potential energy surface would have to be repulsive for
the measured rate function to be the way it is.

The process

$$O^+ \; {}^4S + CO \; X \; {}^1\Sigma^+ \longrightarrow CO^+ \; X \; {}^2\Sigma^+ + O \; {}^3P$$

with all constituents in their ground states, is
endothermic by 0.4 eV (0.6 eV in the laboratory

coordinate system). This is consistent with the data of Fig. 13 , bearing in mind the drifting ion energy distribution discussion above. Unfortunately no other experimental data for this process is available for comparison.

The charge transfer process

$$N^+ \; {}^3P + N_2 \; X \; {}^1\Sigma \longrightarrow N_2^+ \; X \; {}^2\Sigma + N_2 \; X \; {}^1\Sigma - 1.05 \; eV$$

is endothermic by 1.05 eV (1.58 eV in the laboratory coordinate system), which is consistent with the data of Fig. 14. A further 1 eV would be necessary for the N_2^+ to be produced in its first excited state A ${}^2\Pi$. The present data are compared with beam measurements of Murad and Maier (1971). It is not clear whether any structure is present in the cross-section function.

The charge transfer process

$$N^+ \; {}^3P + NO \; X \; {}^2\Pi \longrightarrow NO^+ \; X \; {}^1\Sigma^+ + N \; {}^4S + 5.26 \; eV$$

is exothermic by 5.25 eV when proceeding into ground states. Formation of N in excited states is possible, and if the NO^+ is formed in its first excited state a ${}^3\Sigma$, the process is endothermic by only 0.26 eV. The minimum in the function (Fig. 15) is consistent with the idea that more than one pair of channels is involved. There is no discrepancy between the present data and beam measurements of Turner et al (1968). However the thermal energy rate reported by Goldan et al (1966) is much higher than the present data, and it is likely that gas impurities may have been responsible for the high rate in that early work. Turner et al found that the cross-section was insensitive to ion source electron energy - i.e. long lifetime excited N^+. In our experiments the ion source electron energy was held sufficiently low for the threshold of the N^+ N_2 process to be clearly observed. But the form of the N^+ NO function did not appear to be strongly dependent on the ion source electron energy.

There are two N^+ collision processes in which two product ions are observed. In both cases it is possible for one product ion to be converted into the other by collision with the reactant gas. It follows that the three rate constants k_{12}, k_{23}, k_{13} cannot be deduced from two measured product ion currents. If one is assumed zero, then limits can be placed on the other two.

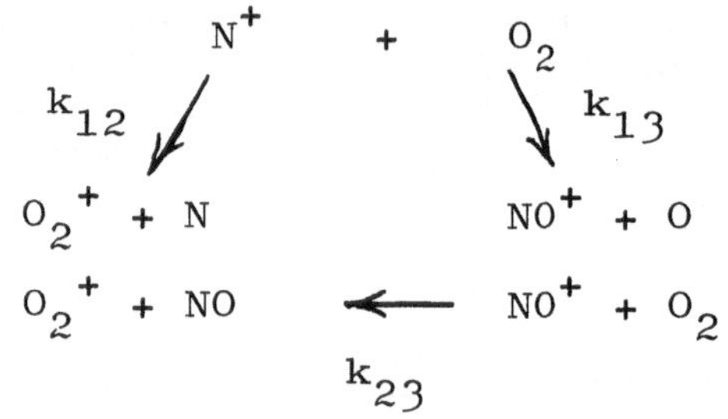

$$
\begin{array}{ccc}
& N^+ \quad + \quad O_2 & \\
k_{12} \swarrow & & \searrow k_{13} \\
O_2^+ + N & & NO^+ + O \\
O_2^+ + NO & \longleftarrow & NO^+ + O_2 \\
& k_{23} &
\end{array}
\qquad (20)
$$

and in the other case

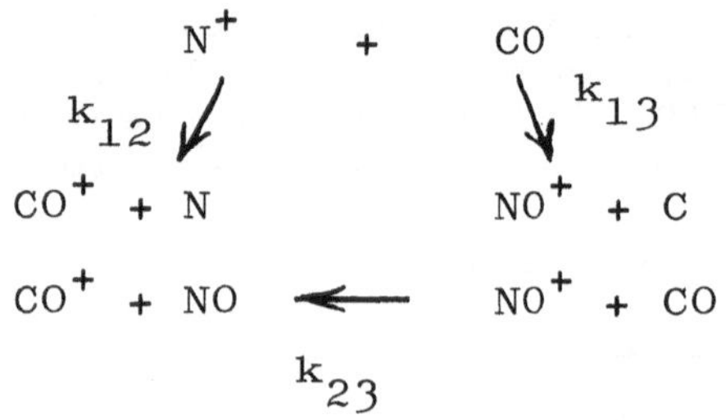

$$
\begin{array}{ccc}
& N^+ \quad + \quad CO & \\
k_{12} \swarrow & & \searrow k_{13} \\
CO^+ + N & & NO^+ + C \\
CO^+ + NO & \longleftarrow & NO^+ + CO \\
& k_{23} &
\end{array}
\qquad (21)
$$

If it is assumed that process 23 has zero rate, then the upper limit to k_{12} is calculated from

$$
k_{12} = \frac{v_{d_1}}{n\,\ell}\ \ell n\ \left(\frac{i_2 + i_3}{i_1} + 1\right) \qquad (22)
$$

and the upper limit to k_{13} from

$$
k_{13} = \frac{v_{d_1}}{n\,\ell}\ \ell n\ \left(\frac{i_3}{i_1} + 1\right) \qquad (23)
$$

If it is assumed that process 13 has zero rate then the lower limit to k_{12} is calculated from

$$
k_{12} = \frac{v_{d_1}}{n\,\ell}\ \ell n\ \left(\frac{i_2}{i_1} + 1\right) \qquad (24)
$$

and the upper limit to k_{23} from

$$
\frac{i_2 + i_3}{i_3} = \exp\left(\frac{n\,\ell\,k_{23}}{v_{d_2}}\right) - \exp\left(\frac{n\,\ell\,k_{12}}{v_{d_1}}\right) \qquad (25)
$$

The calculation of the small diffusion correction to k_{23} is complicated and has been omitted.

The limiting rate constants calculated from equations 22-25 are displayed in Figs. 16,17.

For ground state products the 13 process

$$N^+ \; ^3P + O_2 \; X \; ^3\Sigma_g^- \longrightarrow O_2^+ \; X \; ^2\Pi + N \; ^4S + 2.47 \text{ eV} \quad (26)$$

is exothermic by 2.47 eV. The process involving the formation of O_2^+ a $^4\Pi$, the lowest excited state, would be endothermic; but the formation of the lowest excited state N ^2D would be exothermic by 0.99 eV. The 12 process is exothermic by 6.6 eV in the ground states

$$N^+ \; ^3P + O_2 \; X \; ^3\Sigma_g^- \longrightarrow NO^+ \; X \; ^1\Sigma^+ + O \; ^3P + 6.6 \text{ eV} \quad (27)$$

and if O ^1D were formed it would still be exothermic by 4.63 eV, and for O ^1S, 2.41 eV. For excited NO^+,

$$N^+ \; ^3P + O_2 \; X \; ^3\Sigma_g^- \longrightarrow NO^+ \; a \; ^3\Sigma^+ + O \; ^3P \quad (28)$$

the process is exothermic by 0.29 eV.

The state in which NO^+ is formed is of importance so far as process 23 is concerned. For ground states throughout, the process would be endothermic by 2.8 eV, but for NO^+ a $^3\Sigma^+$ it would be exothermic by about 2.5 eV. It is quite probable that in process 12 some of the NO^+ is formed in this state. The rate for the process 23 should not be taken as that for ground state NO^+.

One further exothermic process is possible:

$$N^+ \; ^3P + O_2 \; X \; ^3\Sigma_g^- \longrightarrow O^+ \; ^4S + NO \; X \; ^2\Pi + 2.32 \text{ eV} \quad (29)$$

but in the present experiments no O^+ count could be detected; nor would the fast removal of O^+ by collisions with O_2 be likely, at any rate below 1 eV.

The total rate for the loss of N^+ ions, forming both NO^+ and O_2^+ is in agreement with the flowing

afterglow and drift tube data (Dunkin et al 1968,
Farragher 1970, Johnsen et al 1970). It is in cases
where competing reactions can occur that the ion
injection technique offers the advantage of avoiding
competing projectile ions.

The collision processes between N^+ and CO are
similar in principle. If all reactants and products
are considered to be in their ground states, then the
charge exchange (process a) will be exothermic by
0.53 eV, whilst the ion-atom transfer (process b) will
be endothermic by 0.13 eV. It would seem likely that
in the drift tube only ground state processes are
involved, and the limits to the data are shown in
Fig. 17. The experiments have not been conducted at
energies above 2 eV, in order that the production of
C^+, the onset of which is around 3 eV, be avoided.
The total rate for the loss of N^+ ions, forming both
CO^+ and NO^+, is in agreement with flowing afterglow
measurements (Fehsenfeld et al 1967); this is rather
surprising, since N_2^+ ions must have been present
in the afterglow, to contribute to the CO^+ count by
charge transfer. However, the rate for the process
is $\sim 10^{-11}$ cm3/sec, which is ten times smaller than
the equivalent $N_2^+ O_2$ rate, which was proposed to
account for the $N^+ O_2$ discrepancy.

An important and interesting feature of several
of the charge transfer cross-section functions
$(N^+O_2, O^+O_2, N^+NO, N^+CO)$ as well as previously reported
functions $(Ar^+CO, Ar^+O_2, Ar^+NO, Ar^+N_2, N_2^+CO, CO^+O_2,$
$CO_2^+O_2, Kr^+O_2, Kr^+NO, N_2^+O_2)$ and certain ion-atom
interchange functions $(N^+ + O_2 \rightarrow NO^+, N^+ + CO \rightarrow NO^+,$
$H^+ + D_2 \rightarrow D^+)$ is the shallow minima they exhibit.

We believe this feature to be characteristic of
the procedure of the collision by transitions at the
crossing of potential energy surfaces. The quantum
probability of transition is markedly dependent on
the rate of change dr/dt of separation r of the ion and
the molecule in a certain critical range of r. This is
true of the Landau-Zener approximation, and can also
be seen very clearly in the formulation due to
Demkov (1964), in which the probability of transition
P is given by the equation

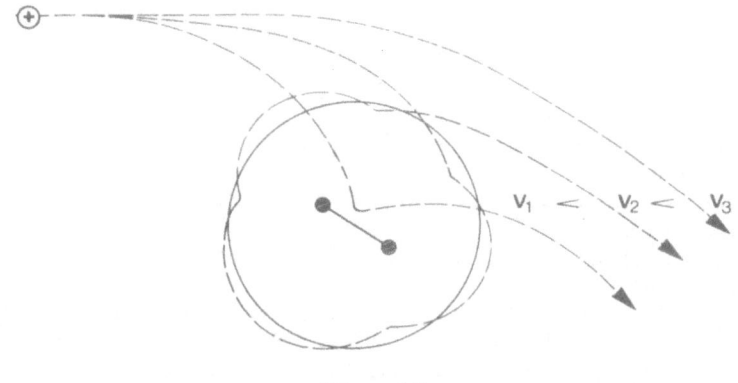

Fig. 20

$$P = \left| \frac{1}{2} \ \text{sech} \ \frac{\pi \Delta E}{2 \ (2 \ m \ E_i)^{\frac{1}{2}}} \ \left(\frac{dr}{dt} \right)_{r_c}^{-1} \right|^2$$

where ΔE is the energy defect, E_i the mean ionization
potential, m the electron mass, r_c the critical
separation at which the exchange energy U_{12} is equal
to the difference between the initial and final
potential energy curves $(U_{12} = U_{11} - U_{22})$.

Although the trajectories of ion-molecule
collisions are highly complicated (Dugan and Magee 1971)
it is likely that both at high and at low impact
energies dr/dt never reaches very low values for long
periods of time. However at intermediate impact
energies these low values are often reached during the
trajectory, which can have a 'rosetting' form. Indeed
in the ion-atom collision there is a Langevin orbiting
velocity at which, for a particular impact parameter,
dr/dt becomes zero. The high impact energy trajectory
(small angle scattering), the low impact energy trajectory
(close approach and scattering) and the intermediate
energy trajectory (complicated path remaining at
intermediate r for a relatively long period of time)
are shown schematically as broken lines for identical
impact parameter in Fig. 20.

REFERENCES

Birkinshaw,K. and Hasted, J.B. (1971) J.Phys.B. $\underline{4}$, 1711
Bloomfield C.H. and Hasted,J.B.(1966) Brit. J. Appl. Phys.
        $\underline{17}$, 449
Bohme, D.K., Ong,P.P. and Hasted, J.B. (1967) Planet.
        Space Sci. $\underline{15}$, 1777
Dalgarno, A.,McDowell,M.R.C. and Williams, A.(1958) Phil.
        Trans. $\underline{A250}$, 411
Davisson,C.J. and Calbick,C.J. (1931) Phys.Rev. $\underline{38}$, 585
Demkov, Yu.N. (1964) Atomic Coll. Proc. Proc. 3rd Int.
        Conf.Phys.Electrn.Atom.Colls. London 1963,
        Amsterdam:North-Holland, 831
Dugan, J.V. Jr. and Magee,J.L. (1971) Chemical Dynamics
        (Ed.Hirschfelder),John Wiley and Sons Inc. p 207
Dunkin, D.B.,Fehsenfeld,F.C.,Schmeltekopf,A.L. and
        Ferguson,E.E. (1968) J.Chem.Phys. $\underline{49}$, 1365
Farragher,A.L. (1970) Trans.Faraday Soc. $\underline{66}$, 1411

Fehsenfeld,F.C.,Schmeltkopf,A.L. and Ferguson,E.E.
        (1967) J.Chem.Phys. $\underline{46}$, 2019
Goldan,P.D.,Schmeltkopf,A.L., Fehsenfeld,F.C.,Schiff,H.I.
        and Ferguson,E.E. (1966) J.Chem.Phys. $\underline{44}$, 4095
Hasted, J.B. Adv.Mass Spectrometry, (1974) $\underline{6}$, 901.
        Elseviers
Johnsen, R., Brown,H.L. and Biondi,M.A. (1970)
        J. Chem. Phys. $\underline{52}$, 5080
Johnsen, R. and Biondi,M.A. (1972) J.Chem.Phys. $\underline{57}$ 1975
Johnsen, R and Biondi, M.A. (1973) SRCC Report No. 188,
        University of Pittsburgh, June 1973
Kaneko, Y., Megill, L.R. and Hasted, J.B. (1966)
        J. Chem. Phys. $\underline{45}$, 3741
Kaufman, J. (1972) J. Chem. Phys. $\underline{56}$, 5256
Kebarle, P. (1966) Adv. Chem. $\underline{58}$, 210
Kosmider, R.G. and Hasted, J.B. 1974. In course of
        Publication
Murad, E. and Maier W.B. (1971) VII Int.Conf.Phys.
        Electrn.Atom Colls. Amsterdam, North-Holland
        p 369
Parkes, D. (1971) Trans. Faraday Soc. $\underline{64}$, 711
Turner, B.R., Mathis, R.F. and Rutherford, J.A. (1968)
        J. Chem. Phys. $\underline{45}$, 2051
Wannier, G.H. (1951) Phys. Rev. $\underline{38}$, 281

IONOSPHERIC ION-MOLECULE REACTIONS

Eldon E. Ferguson

Aeronomy Laboratory, Environmental Research Laboratories
National Oceanic and Atmospheric Administration
Boulder, Colorado 80302 U.S.A.

INTRODUCTION

Aeronomy is one of the interesting areas of application of ion-molecule chemistry. Ideally, ionospheric ion chemistry would be pursued in the following way. All of the relevant atmospheric parameters would be measured and all of the necessary laboratory data would be acquired and models would be developed to test whether a consistent and comprehensive picture of the physics and chemistry involved emerged. If it did we would assume that we understood the atmospheric processes and proceed to a new endeavor. If it did not we would then search for errors in our measurements of atmospheric parameters or for errors in our laboratory data or we would seek additional processes that we had failed to consider in our initial model. In practice we almost never have the complete set of either atmospheric observations or laboratory data and our understanding of atmospheric processes proceeds by a converging series of improvements in in situ observation, laboratory data and theoretical models.

What we would like to have specifically for planetary ion chemistry would be the ion composition, the neutral composition and the ionizing sources (primarily the solar ultraviolet flux) as functions of altitude and time. Then using laboratory (or theoretically derived) cross sections for ionization, photon absorption cross-sections, ion-molecule reaction rate constants, electron attachment and recombination rate constants, photodetachment cross sections, etc., we would see if we could produce theoretically the observed ion composition profiles and thereby test the accuracy of our measurements and the state of our understanding of the relevant processes.

313

We seldom have this ideal situation, although it is approached rather well in the earth's upper ionosphere and the situation steadily improves. Even when ion composition measurements are not available, other observations involve ion chemistry in such a way as to apply useful constraints on ion chemistry and thus supply motivation for laboratory measurement. In particular an electron density profile can provide a valuable clue to the ion chemistry by determining the ion density. Airglow observations can also provide information on the ion composition.

Ion compositions of planetary atmospheres have been determined directly by experiments utilizing rocket and satellite borne mass spectrometers in the earth's ionosphere and indirectly by optical spectrometers and radio transmitters on space craft to Mars and Venus and recently Jupiter. Some day presumably mass spectrometers will also be carried to these and other planets as well. Along with these heroic efforts of space scientists in obtaining planetary ion compositions and the necessary related data to study the ion chemistry, it is also quite essential to have laboratory reaction rate measurements in order to understand planetary ion chemistry. Acquiring the necessary laboratory data and coming to some understanding of ionospheric chemistry has been an interesting challenge to many people over the past ten years or so in which a substantial amount of ionospheric data has been available.

Many of the reactions requiring study for ionospheric purposes, perhaps even most of these reactions, required different techniques than had been developed by the mass spectroscopists, radiation chemists, and chemical kineticists who were in the ion-molecule reaction field at the time when developments in space physics opened up ionospheric chemistry. Thus it happened that new techniques were developed specifically to study ionospheric reactions. For the most part these techniques were developed by physicists drawn from the discipline called gaseous electronics. This has led to somewhat of a separation between this specialized part of ion chemistry and the conventional chemical kinetics activity in this field. Fortunately this communication barrier has been disappearing in recent years and this NATO conference is certainly a great help in that direction.

Some of the pioneers in laboratory aeronomical ion chemistry were Sayers and Hasted in Great Britain, and Fite in the U.S., all of whom employed stationary discharge afterglows for reaction rate measurements. Fehsenfeld and Schmeltekopf then developed the flowing afterglow just ten years ago and this broke a bottleneck, leading to literally hundreds of ionospheric ion-molecule reaction rate constants within a few years. Mass spectrometry has subsequently made an important contribution, particularly the high pressure mass spectrometry work of Kebarle and his students. Theoretical ionospheric chemistry in the pioneering stages was almost entirely a

British effort, in particular the work of Massey, Bates, and Dalgarno.

Recently Biondi and Johnsen have made the drift tube a very powerful aeronomical tool and McFarland, Albritton, Fehsenfeld and Schmeltekopf have combined a drift tube with a flowing afterglow to extend the chemical versatility of this approach. The drift tube and drift tube - flowing afterglow combination allows the determination of reaction rate constants as a function of ion kinetic energy from thermal to several electron volts.

The powerful capability of the ion cyclotron resonance technique is generally appreciated. For the most part this technique has been applied to basic chemical problems of a non-aeronomical nature, with a notable exception being the work of Huntress at the Jet Propulsion Laboratory on the ionospheric chemistry of Jupiter.

Beam experiments and experiments with tandem mass spectro meters have made an important contribution, particularly the work of Rutherford, Stebbings, Turner, and Vroom and the work of Tiernan and his colleagues.

E- AND F-REGION ION CHEMISTRY

The charged component of planetary atmospheres at high altitudes (and hence low pressures) is composed of simple positive ions and electrons. Concentration measurements are simpler in this low pressure regime and the required reaction rate constants are easier to obtain for the most part than is the case for higher pressure ionospheric regimes. (This is to some extent offset because we ask more detailed questions about E- and F-region reactions, like their energy dependences, product states, etc.) For the reason of simplicity the ion chemistry of the earth's high ionosphere was studied before that of the lower ionosphere. In the case of other planets, only low pressure ionospheric data is available at present. For the earth this low pressure range consists of the so-called E-region from about 90 - 140 km and the F-region above 140 km. This ion chemistry involves binary positive ion reactions almost exclusively.

$O^+ + H \rightarrow H^+ + O$

The accidentally resonant charge-transfer reaction

$$O^+ + H \rightleftarrows H^+ + O \tag{1}$$

is of interest as the major source of H^+ in the high ionosphere. It was earlier proposed that because of the near resonance both forward (1) and reverse (-1) reactions would be very fast and an

equilibrium distribution would prevail at certain altitudes, allowing the atomic hydrogen concentration to be deduced from the more
readily measurable O^+, H^+ and O concentrations.

The question of whether accidental (or non-symmetric) resonance
charge-transfer rates stay large at thermal energy has been a con-
tentious one in atomic theory, some theorists invoking the adiabatic
hypothesis to predict small rate constants. The results of measurements are shown in Fig. 1 which gives recent beam data of Rutherford
and Vroom (1974) on (-1) down to 2 eV ion energy and a thermal
Flowing Afterglow measurement at $300^{\circ}K$. The symmetric resonance
formula $\sigma^{\frac{1}{2}} = a - b \log E$ fits the data over a very wide energy
range. This particular result is of considerable importance both
for aeronomy and charge-transfer theory.

$He^+ + N_2$

The problem of the loss of terrestrial helium ions by the reaction

$$He^+ + N_2 \rightarrow N^+ + N + He \qquad (2a)$$
$$\rightarrow N_2^+ + He \qquad (2b)$$

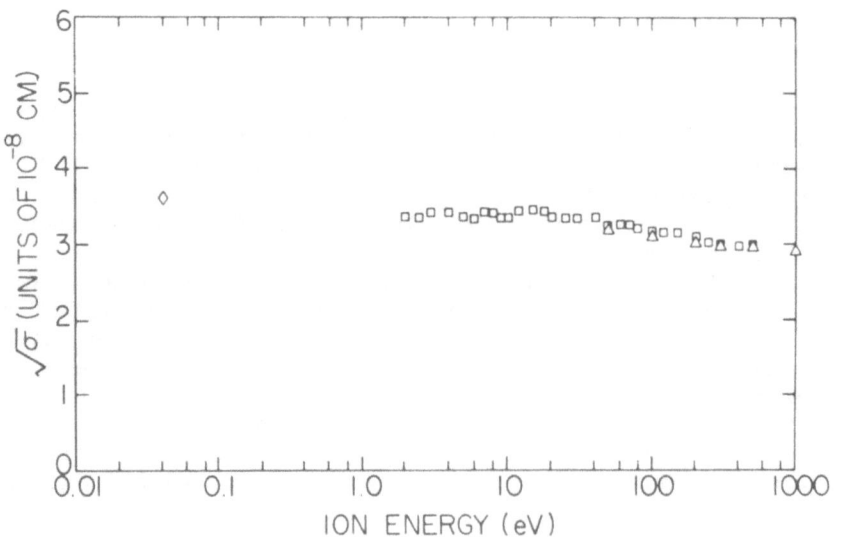

Figure 1. Square root of the cross section for the reaction $H^+ + O \rightarrow$
$O^+ + H$ as a function on ion kinetic energy. $\Diamond$ represents the
thermal 300° measurement of Fehsenfeld and Ferguson, J. Chem. Phys.
56, 3066 (1972), $\square$ is the beam data of Rutherford and Vroom, J
Chem. Phys. 59, xxxx (1974), and $\triangle$ is the beam data of Stebbings,
Smith and Ehrhardt, J. Geophys. Res. 49, 2349 (1964).

has had a very interesting history. This reaction has been widely
studied in the last decade and numerous measurements (Ferguson,
1973) all agree on a very large rate constant approximately equal
to the Langevin or orbiting rate constant, $k_2 \approx k_L = 2\pi e \sqrt{\alpha/\mu} = 1.2$
x 10^{-9} cm^3/sec. The measurements followed theoretical predictions
(Bates and Patterson, 1962) that k_2 would be small. Many aeronom-
ers (Taylor et al., 1963; Hanson, 1962; Bauer, 1966a, 1966b) believed that ionospheric observations of large He^+ concentrations
were inconsistent with a large value for k_2. In addition to thermal
energy rate constant measurements, it has been found that k_2 does
not depend on temperature (Dunkin et al., 1968; Lindinger et al.,
1974), ion kinetic energy (Heimerl et al., 1969), or nitrogen vibrational state (Schmeltekopf, 1968) although the branching ratio $k_{2a}/$
k_{2b} does. From combined optical and mass spectroscopic studies it
was believed (Albritton et al., 1968) that this reaction was understood in detail, being a resonant charge-transfer into the C state
of N_2^+, followed either by predissociation to give N^+ (2a) or radiation to the N_2^+ ground state to give N_2^+ (2b). However, recently
Kemper and Bowers (1973) made the surprising observation in some low
pressure ICR studies that k_{2a}/k_{2b} depends on pressure and from this
they deduced the involvement of longer lived N_2^+ excited quartet
states as primary products. Bowers has subsequently reported that
this pressure dependence is much less than originally reported.
Govers et al., (1974) find from studies involving the $^{14}N_2$ and $^{15}N_2$
isotopes that a strong isotope effect exists. The conclusions of
Govers et al., and Kemper and Bowers both support the role of additional N_2^+ product states in this reaction.

O^+ Reactions

The reactions of O^+ with N_2 and O_2 are of major importance in
the F-region,

$$O^+(^4S) + N_2(^1\Sigma) \rightarrow NO^+(^1\Sigma) + N(^4S) \qquad (3)$$

and

$$O^+ + O_2 \rightarrow O_2^+ + O. \qquad (4)$$

Because they convert atomic ions with very small electron recombination coefficients to molecular ions with very large ones, these
reactions determine the maximum electron density of the earth's
ionosphere. The dissociative recombination products of the molecule
ions are excited state N and O atoms which then have important roles
in ionospheric chemistry and airglow and this provides further in-
centive to acquire precise values of k_3 and k_4. The recent data on
the energy dependences of k_3 and k_4 are shown in Figs. 2 and 3.

The analogous reactions to (3) and (4) in the CO_2 planet (Mars
and Venus) ionosphere is

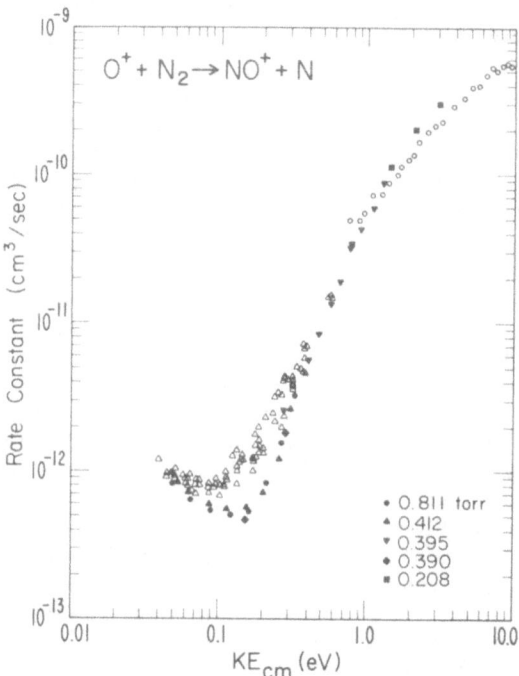

Figure 2. Rate constant for $O^+ + N_2 \rightarrow NO^+ + N$ as a function of
center-of-mass kinetic energy. Solid symbols from McFarland et
al., J. Chem. Phys. 61, xxxx (1974); △ from Johnsen and Biondi,
J. Chem. Phys. 59, 3504 (1973); O from Rutherford and Vroom,
J. Chem. Phys. 55, 5622 (1971).

$$O^+(^4S) + CO_2(^1\Sigma) \rightarrow O_2^+(^2\Pi) + CO(^1\Sigma) \tag{5}$$

Interestingly enough reactions (3) and (4) are relatively slow,
k_3, $k_4 \ll k_L$ and (5) is fast $k_5 \approx k_L$ which has the immediate implication that the Mars and Venus ionospheres will be relatively
less dense than that of the earth. It is interesting chemically
to note that the fast reaction (5) is spin forbidden while the slow
reactions (3) and (4) are spin allowed. The relevant spin consideration in thermal energy ion molecule reactions is not overall spin
conservation but spin conservation in the formation of the intermediate complex.

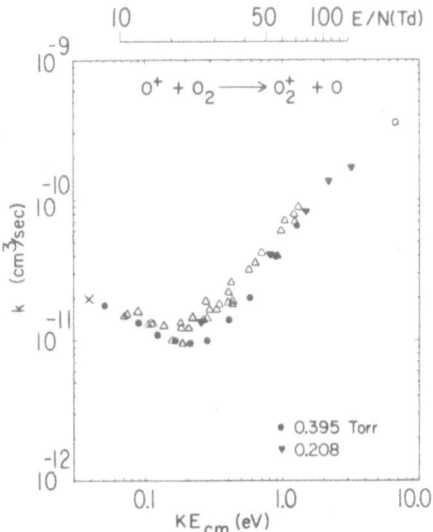

Figure 3. Rate constant for $O^+ + O_2 \rightarrow O_2^+ + O$ as a function of relative kinetic energy. Solid symbols, McFarland et al., J. Chem. Phys. 59, 6620, (1973); △Johnsen and Biondi, J. Chem. Phys. 59, 3504, (1973); O Stebbings, Turner and Rutherford, J. Geophys. Res. 71, 771 (1966); ✕ Dunkin et al., J. Chem. Phys. 49, 1365 (1968).

A reaction which is not ionospherically important because it is very slow is the charge-transfer

$$O^+ + NO \rightarrow NO^+ + O + 4.35 \text{ eV} \qquad (6)$$

which has the anomalous behavior shown in Fig. 4, appearing to have an endothermic threshold. We have rationalized this with the correlation diagram shown in Fig. 5 in which it is seen to be endothermic for $O^+ + NO$ to adiabatically produce NO_2^+. Presumably (6) does not occur via non-adiabatic electron jump because the Franck-Condon factor is so low.

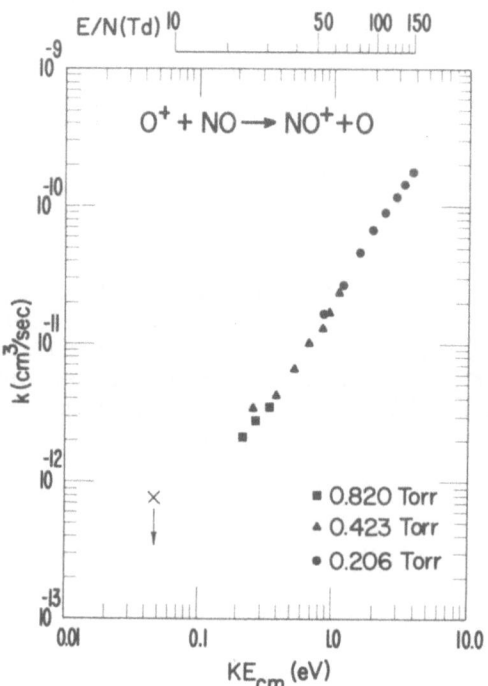

Figure 4. Rate constant for $O^+ + NO \rightarrow NO^+ + O$ as a function of
relative kinetic energy, McFarland et al., J. Geophys. Res. <u>79</u>,
2005 (1974). ✗ represents an upper limit thermal energy value.

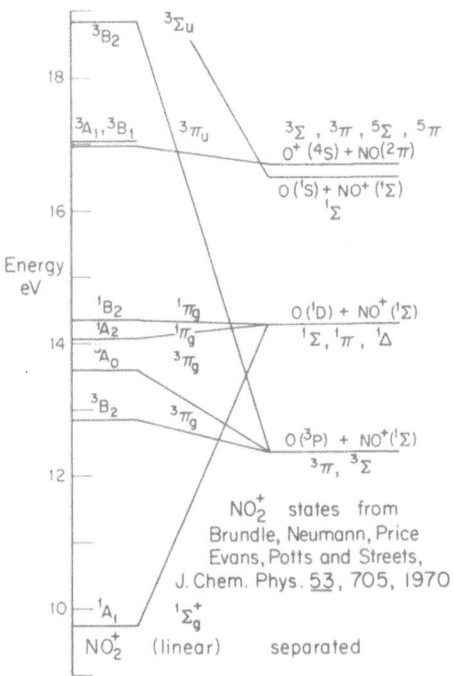

Figure 5. Correlation diagram for NO_2^+ and separated O and NO states.

$\underline{N_2^+ + O}$

The reaction

$$N_2^+ + O \rightarrow O^+ + N_2 \qquad (7a)$$

$$\rightarrow NO^+ + N \qquad (7b)$$

is the major N_2^+ loss process in the high ionosphere. Only recently has the energy dependence as shown in Fig. 6 and the branching ratio as shown in Fig. 7 been determined in the thermal energy range. The ability to react ions with chemically unstable neutrals such as O atoms in the thermal range is a unique capability of flowing after-glow systems so far. Rutherford and Vroom now obtain data at such low energies in their beam experiments that this advantage is not as striking as it would have been a few years ago. The ion-atom interchange channel of (7) dominates the charge-transfer channel, at thermal energy.

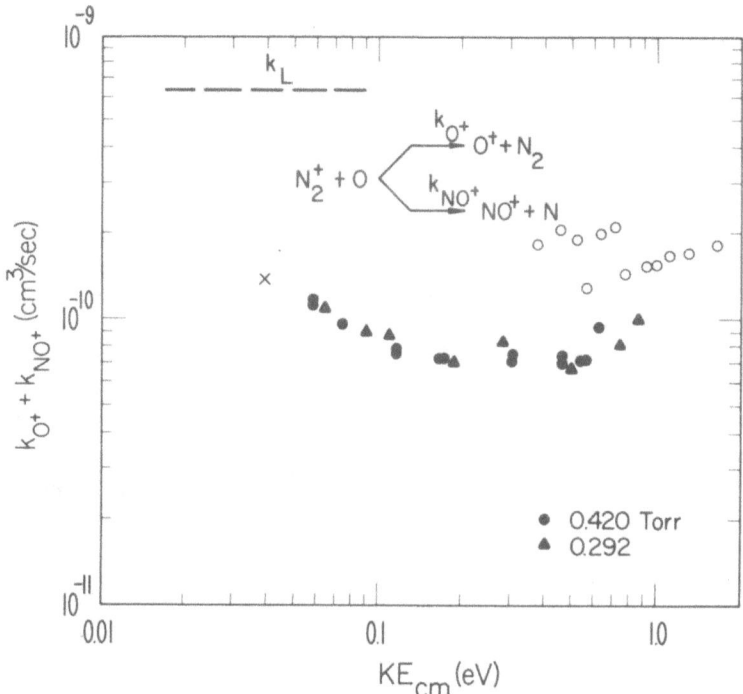

Figure 6. Rate constant for $N_2^+ + O$ as a function of relative
kinetic energy. Solid symbols McFarland et al., J. Geophys. Res.
<u>79</u>, 2925 (1974); O Rutherford and Vroom, J. Chem. Phys.,
X Fehsenfeld et al., Planet. Space Sci. <u>18</u>, 1267 (1970). k_L
is Langevin or orbiting collision rate constant.

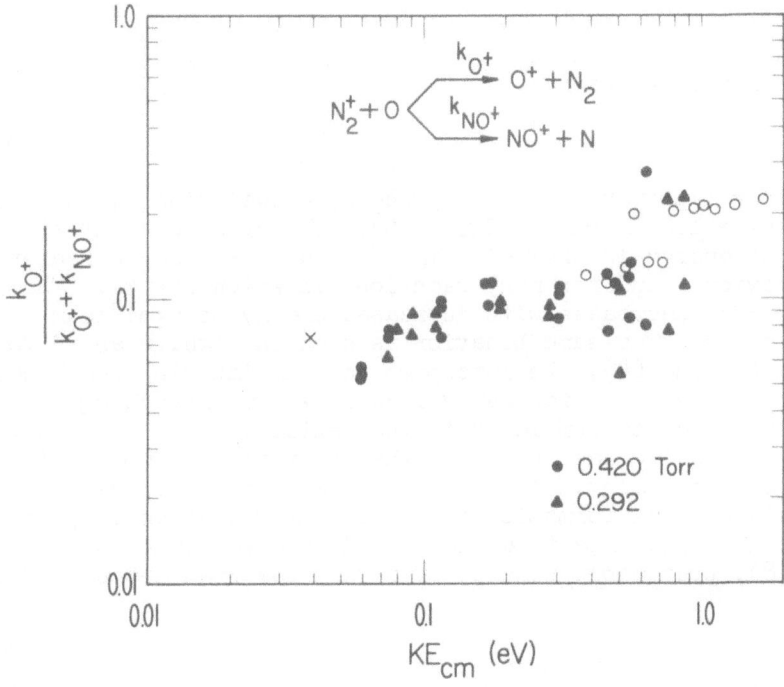

Figure 7. Branching ratio for N_2^+ + O reaction as a function of relative kinetic energy. X is a conventional afterglow measurement, other symbols same as for Fig. 6.

Reaction (7b) is an example of a reaction in which it would be of great value to know the reaction products. Energetically the nitrogen atom could be either ground state $N(^4S)$ or the long-lived metastable $N(^2D)$. Because of the Norton and Barth (1970) theory of ionospheric NO production which utilizes the reaction $N(^2D) + O_2 \rightarrow$ NO + O, a knowledge of the ionospheric sources of $N(^2D)$ is of great importance. Product state determinations in thermal energy reactions is a very difficult task, largely beyond the present state of the art.

$N_2^+ + O_2$

The charge-transfer

$$N_2^+ + O_2 \rightarrow O_2^+ + N_2 \tag{8}$$

is the other important N_2^+ loss process, outweighing (7) at altitudes below ~ 160 km where $\lceil O_2 \rceil > [O]$. The rate constant as a function of energy is shown in Fig. 8. This has the characteristic behavior found for many reactions in which $k < k_L$. The rate constant first decreases with increased energy to a minimum and then increases. The same behavior is seen in Figs. 2 and 3 for reactions (3) and (4). We interpret this as implying that these reactions, charge-transfer as well as ion-atom interchange, proceed through intermediate complexes in the region where k decreases with energy and this decrease reflects the decrease in complex lifetime with increasing energy. In order for charge-transfer reactions to proceed via complex formation it is necessary that an electron jump not occur at large impact parameter. This seems to be the situation in (8), presumably because of a very low Franck-Condon factor.

$N^+ + O_2$

The N^+ produced in the ionosphere by dissociative photoionization and reaction (2a) reacts rapidly with O_2,

$$N^+ + O_2 \rightarrow NO^+ + O \tag{9a}$$
$$\rightarrow O_2^+ + N. \tag{9b}$$

The reaction channels are comparable, however the branching ratio has not been as well measured as the overall reaction rate constant and different experiments do not agree well on the branching ratio. This is a rather common situation in aeronomy and arises from several problems. A problem for some experiments is the question of sampling discrimination. To measure k_9 in an afterglow experiment for example only relative N^+ concentrations versus O_2 added is required. In order to determine the branching ratio however the relative sampling efficiencies for the two product ions must be

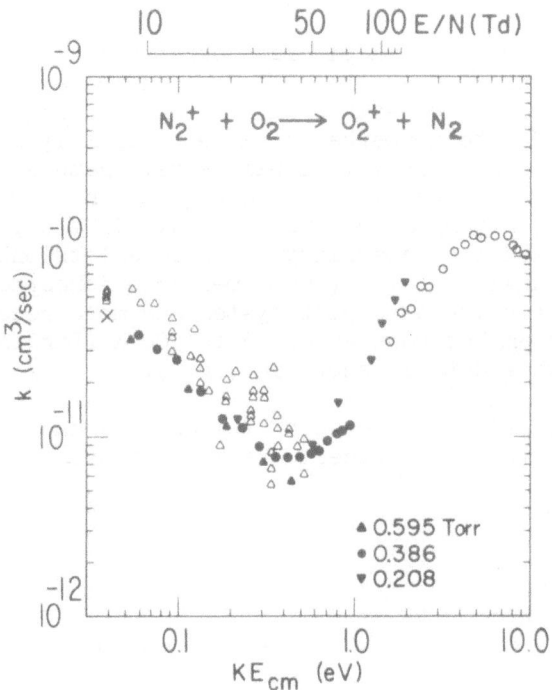

Figure 8. Rate constant for $N_2^+ + O_2 \to O_2^+ + N_2$ as a function of
relative kinetic energy. Solid symbols McFarland et al., J. Chem.
Phys. <u>59</u>, 6620 (1973); $\triangle$ Johnsen, Brown and Biondi, J. Chem.
Phys. <u>52</u>, 5080 (1970); $\bigcirc$ Neynaber, Rutherford and Vroom, Gulf
Radiation Technology Report A12209, July 1972; $\times$ Dunkin et al.,
J. Chem. Phys. <u>49</u>, 1365 (1968).

known. Mass discrimination can be serious, is difficult to measure
and has certainly led to published errors. The ICR technique offers
some promise in this regard. Reaction (2) is another case in which
the branching ratio is in some doubt. Due to the greater disparity
in product masses, the discrimination can be worse for (2) and (7)
than for (9).

Another problem which can be very serious, particularly for
the more complex D-region reactions involving hydrated positive and
negative ions, is due to consecutive reactions in the experiment.
If water vapor is added to the system to produce hydrated reactant
ions, for example, it will also hydrate product ions to some extent
making it difficult to deduce which the initial product ions are.
The product ions can undergo normal reactions with the neutral reactant under some conditions leading to confusion about the products.

Energy Dependence

Figures 2-8 show ionospheric rate constants as a function of relative kinetic energy. As these figures indicate, the capability of obtaining such data now exists and the energy range covered above 300°K exceeds the ionospheric range. Normally in the ionosphere, of course, the energy variable is temperature, i.e. vibrational, rotational and translational energy and not translational energy alone. It is very much more difficult in the laboratory to vary temperature and the capability to do so exists only over a limited range. Dunkin et al., (1968) measured ionospheric reactions up to 600°K in a flowing afterglow system and more recently the range of this system has been extended to 900°K (Lindinger et al., 1974). Some of this data is shown in Fig. 9.

We have compared the effects of temperature and ion kinetic energy alone on several ionospheric rate constants. For reactions

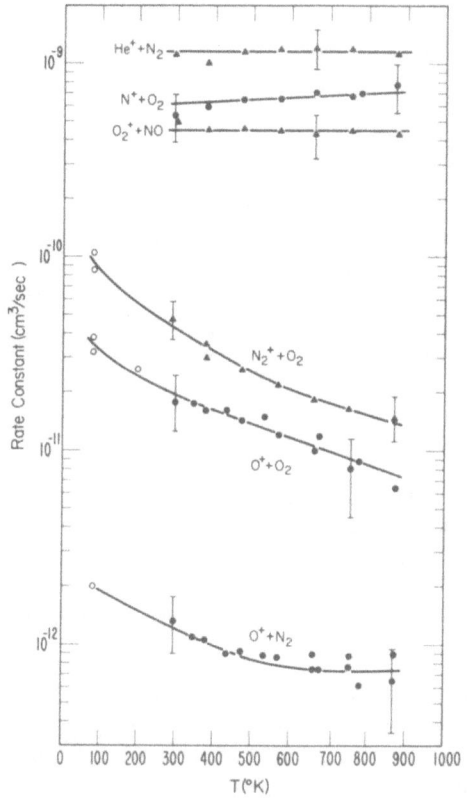

Figure 9. Rate constants for several ionospheric reactions from 300°K (in some cases from 80°K) to 900°K from Lindinger, Fehsenfeld, Schmeltekopf, and Ferguson, J. Geophys. Res. (1974).

(2), (3), (4) and (5) the effect of ion kinetic "temperature" defined as $KE = 3/2 \ kT_{KE}$ is the same as that of true temperature within the accuracy of the measurements. In the case of reaction (8) there seems to be a stronger dependence of k on T than on T_{KE}. This indicates the role of internal energy (vibrational and rotational) on the lifetime of the complex, according to our simple theory, and the difference is in the correct direction.

Doubly Charged Ions

An important bit of laboratory data still lacking is the loss rate of doubly charged ions with atmospheric neutrals. Both O^{++} and N^{++} have been observed in the F-region and their charge-transfer rate constants with He, O, O_2 and N_2 are of interest and would allow more detailed analyses of the ion profiles.

Excited State Ion Reactions

An important area in which very little work has been done concerns the reactions of electronically excited ions. Substantial fractions of the O^+ and O_2^+ produced in the upper ionosphere are produced in metastable excited states and very little thermal energy data on their loss rates is available. Of particular interest are the reactions

$$O^+(^2D) + N_2 \rightarrow N_2^+ + O, \tag{10}$$

$$O^+(^2D) + O_2 \rightarrow O_2^+ + O, \tag{11}$$

$$O_2^+(a\,^4\Pi_u) + N_2 \rightarrow N_2^+ + O_2, \tag{12}$$

as well as the quenching of these metastables by O, O_2 and N_2. The quenching of course competes with reaction and so must be known. Ryan (1969) has reported measurements on reaction (12).

METAL ION CHEMISTRY

Metal ion chemistry is a rather specialized aspect of the earth's ion chemistry which is largely uncoupled from the ion chemistry of the non-metallic atmospheric gases and it is usually considered quite separately. Metal ions (and a few metal oxide ions) are routinely observed by mass spectrometers in D- and E-layers of the ionosphere. The neutral metal atoms from which these ions arise are deposited in this region of the atmosphere ($\sim$ 100 km) by the ablation of meteors incident upon the earth's atmosphere. Neither the production rates nor loss rates of the metal ions are known and no detailed quantitative analyses of the metal ion chemistry have been carried out as yet. A great deal of the relevant ion chemistry has been measured in the laboratory although a great deal remains to be done. There has been relatively little interest or activity in this field to date.

The peculiarities of atmospheric dynamics in the E-region, related to changes in neutral wind directions and the resulting $\bar{v} \times \bar{B}$ forces have led to a very striking ionospheric phenomenon known as Sporadic-E layers, occurring around 105-120 km on certain occasions. The Sporadic-E layer is a thin dense layer of ionization, sometimes exhibiting an order of magnitude increase in electron density above ambient in a few kilometer thick layers. These layers of ionization have been known because of their anomalous radio wave reflection. Only with the advent of rocket borne mass spectrometers was it learned that the ions are predominately atomic metal ions.

The atomic metal ions are concentrated by dynamic forces and can be concentrated in high densities, along with the electrons, because of the low recombination coefficient of atomic ions with electrons. The fast dissociative recombination with electrons prevents the normally dominant NO^+ and O_2^+ ions in this altitude range from being concentrated to the same extent by these dynamic forces.

The production of metallic ions from metallic neutrals is quite straightforward; because of their low ionization potentials the normal ionospheric ions can charge-transfer with the metals and they appear to readily do so. A large amount of data has been published by Rutherford and Vroom in recent years from beam studies down to about 1 eV. The have measured the charge-transfer of NO^+ and O_2^+ with Na, Mg, Ca and Fe for example and they are all quite fast. The extrapolation to thermal energy for a fast charge-transfer seems to be reasonably safe from our experience. However, the N_2^+ and O_2^+ charge-transfers with Na measured by Farragher et al. (1969) in a flowing afterglow do not agree very well with these extrapolations, differing by factors of 3 and 2 respectively. At the recent Philadelphia Mass Spectrometry meeting, Fite indicated that new measurements were being carried out at Pittsburgh using an ion trapping method.

The manner of loss of the metallic ions is not so straightforward. Because of their low energy content, metallic ions cannot react with the major atmospheric species O_2 and N_2. One possibility that was suggested by several people rather early was reaction with ozone, e.g.

$$Mg^+ + O_3 \rightarrow MgO^+ + O_2. \tag{13}$$

This converts an atomic ion to a molecular ion and thus might have been the rate controlling step for deionization of Sporadic-E. The reactions of Mg^+, Ca^+, Fe^+, Na^+ and K^+ with ozone were all measured in the flowing afterglow (Ferguson and Fehsenfeld, 1968) and all but Na^+ and K^+ found to be fast, e.g. $k_{13} = 2.3 \times 10^{-10}$ cm^3/sec. However, the effect of (13) was vitiated in the E-region because we also found that

$$MgO^+ + O \rightarrow Mg^+ + O_2 \qquad\qquad (14)$$

was fast, $k_{14} = \sim 1 \times 10^{-10}$ cm^3/sec and the ratio of atomic oxygen to ozone is very large. Reactions of FeO$^+$ and CaO$^+$ with O have not been measured but presumably are also fast. The reaction

$$SiO^+ + O \rightarrow Si^+ + O_2 \qquad\qquad (15)$$

is fast, $k_{15} \sim 2 \times 10^{-10}$ cm^3/sec (Fehsenfeld, 1969) which is of interest since Narcisi has apparently observed SiO$^+$ in the E-region and the fast loss implied by (15) puts an extreme burden on the unknown SiO$^+$ source. The observation that (15) is fast and hence exothermic led to a revision in the dissociation energy of SiO$^+$ by about a volt, incidentally, which was traced back to an error in the reported SiO ionization potential.

The failure of Na$^+$ and K$^+$ to react with ozone may imply that these reactions are endothermic, i.e. the bond energies of NaO$^+$ and KO$^+$ are less than 1 eV. Thermodynamic data is not definitive on this point.

The metal ions will associate with neutrals in three-body reactions as will any ion due to the electrostatic forces which are not chemically specific. There is data on the association of Mg$^+$, Ca$^+$, Fe$^+$, Na$^+$ and K$^+$ with O$_2$ (Ferguson and Fehsenfeld, 1968), Na$^+$ with O$_2$ and CO$_2$ and K$^+$ with CO$_2$ (Keller and Beyer, 1971) and Na$^+$ and K$^+$ with H$_2$O (Johnsen et al., 1971). Presumably then the common meteoric ions, Mg$^+$, Ca$^+$, Fe$^+$, Si$^+$, Na$^+$ etc. are lost by diffusion down into the D-region where the pressure is large enough for three-body reactions to play a role but the details of this loss have not been worked out.

After three-body association, it is expected that various exothermic switching reactions would occur with essentially Langevin rate constants, e.g.

$$Na^+ \cdot O_2 + H_2O \rightarrow Na^+ \cdot H_2O + O_2. \qquad\qquad (16)$$

One type of reaction which has not been studied at all is

$$MO_2^+ + O \rightarrow MO^+ + O_2 \qquad\qquad (17)$$

for various metals M. A review of current laboratory data on atmospheric metal ion reactions is available (Ferguson, 1972).

D-REGION ION CHEMISTRY

Below ~ 80 km in the earth's ionosphere, the so-called D-region, the nature of the ion chemistry changes drastically from its behavior at higher altitude. This is a consequence of the higher pressure and significant concentrations of minor neutral species, H_2O, NO, O_3, etc. The relatively high pressure introduces three-body reactions in a very significant role and introduces negative ion reactions (as a consequence of three-body electron attachment to O_2).

D-region ion chemistry is in a much less satisfactory situation than E- and F-region ion chemistry. The laboratory reaction rate measurements are a great deal more difficult and so are the in situ ion composition measurements. The presently known D-region positive ion chemistry is inadequate to quantitatively explain the observed ion composition and the observations to date are insufficient to critically test the laboratory derived negative ion schemes.

The laboratory measurements required involve reactions of ions with atomic oxygen as in the E- and F-region ion chemistry and also with the neutrals, H_2O, O_3, NO_2, H, OH, and $O_2(^1\Delta_g)$ and few methods have this chemical versatility.

Positive-Ion Chemistry

The most significant aspect of D-region positive ion chemistry is that the major ion reported (Narcisi and Bailey, 1965; Krankowsky et al., 1972; Goldberg and Aiken, 1971) is $H_5O_2^+$, along with other hydrated oxonium ions whereas the major ion produced is NO^+ with a lesser production of O_2^+. The problem posed then is to convert NO^+ to $H_5O_2^+$ in the appropriate time scale. The quantitative D-region positive ion composition remains somewhat uncertain due to sampling problems, particularly a severe break-up problem of the weakly bound ambient cluster ions. Recent measurements of Arnold and Krankowsky (1974) found $H^+(H_2O)_4$ to be the most abundant ion below 85 km.

The method of conversion of NO^+ to hydrated oxonium ions has been established and studied quantitatively by four different groups using three different techniques, with good agreement. The D-region positive ion chemistry is shown schematically in Fig. 10. The rate constants are given in Table I. The conversion step is the reaction

$$NO^+(H_2O)_3 + H_2O \rightarrow H_3O^+(H_2O)_2 + HNO_2 \qquad (18)$$

where $k_{18} = 7 \times 10^{-11}$ cm^3/sec. The difficulty in explaining ionospheric observations is that it takes too long to hydrate NO^+ with

TABLE I

D-Region NO^+ Reaction Sequence, $296°K$

Reaction	Rate Constant	Reference
$NO^+ + H_2O + N_2 \rightarrow NO^+(H_2O) + N_2$	1.6×10^{-28} cm^6/sec	a,b,c[*],h[**]
$NO^+(H_2O) + H_2O + N_2 \rightarrow NO^+(H_2O)_2 + N_2$	1.1×10^{-27} cm^6/sec	a,b,c[*],h[**]
$NO^+(H_2O)_2 + H_2O + N_2 \rightarrow NO^+(H_2O)_3 + N_2$	1.9×10^{-27} cm^6/sec	a,b,c[*],h[**]
$NO^+(H_2O)_3 + H_2O \rightarrow H^+(H_2O)_3 + HNO_2$	7×10^{-11} cm^3/sec	a,b,c
$NO^+(H_2O) + H \rightarrow H_3O^+ + NO$	$< 7 \times 10^{-12}$ cm^3/sec	d,i
$NO^+ + CO_2 + N_2 \rightarrow NO^+(CO_2) + N_2$	$2.5 \pm 1.5 \times 10^{-29}$ cm^6/sec (200°K)	e
$NO^+ + CO_2 + NO \rightarrow NO^+(CO_2) + NO$	$2.4 \pm 0.7 \times 10^{-29}$ cm^6/sec	f
$NO^+(CO_2) + NO \rightarrow NO^+ + CO_2 + NO$	$< 1 \times 10^{-10}$ cm^3/sec	f
$NO^+ + N_2 + NO \rightarrow NO^+(N_2) + NO$	$2.0 \pm 0.6 \times 10^{-31}$ cm^6/sec	f
$NO^+(N_2) + NO \rightarrow NO^+ + N_2 + NO$	$< 3 \times 10^{-11}$ cm^3/sec	f
$NO^+ + O_2 + NO \rightarrow NO^+(O_2) + NO$	$9 \pm 3 \times 10^{-32}$ cm^6/sec	f
$NO^+(O_2) + NO \rightarrow NO^+ + O_2 + NO$	$< 2 \times 10^{-11}$ cm^3/sec	f
$NO^+(H_2O) + OH \rightarrow H_3O^+ + NO_2$	$< 1 \times 10^{-10}$ cm^3/sec	i
$NO^+(H_2O) + HO_2 \rightarrow H_3O^+ + NO_3$	$< 10^{-9}$ cm^3/sec	***
$NO^+(CO_2) + H_2O \rightarrow NO^+(H_2O) + CO_2$	$\sim 10^{-9}$ cm^3/sec	e
$NO^+(N_2) + CO_2 \rightarrow NO^+(CO_2) + N_2$	$\sim 10^{-9}$ cm^3/sec	f
$NO^+ + O_3 \rightarrow NO_2^+ + O_2$	$< 1 \times 10^{-14}$ cm^3/sec	g

a.   Howard, Rundle and Kaufman, J. Chem. Phys. 55, 4772 (1971).
b.   Fehsenfeld, Mosesman and Ferguson, J. Chem. Phys. 55, 2120 (1971).
c.   Puckett and Teague, J. Chem. Phys. 54, 2564 (1971).
d.   Fehsenfeld and Ferguson, Radio Sci. 7, 113 (1972).
e.   Dunkin, Fehsenfeld, Schmeltekopf and Ferguson, J. Chem. Phys. 54, 3817 (1971).
f.   Heimerl and Vanderhoff, J. Chem. Phys., 60, 4362 (1974)
g.   Fehsenfeld, Ferguson and Howard, J. Geophys. Res. 78, 327 (1973).
h.   French, Hills and Kebarle, Can. J. Chem. 51, 456 (1973).
i.   Fehsenfeld, Howard, Harrop and Ferguson, J. Geophys. Res. to be published.
*   Reference c used NO third body, results are same as with N_2 in refs. a and b.
**   Reference h includes equilibrium constant data.
*** Reaction is unmeasured, this is a theoretical upper limit.

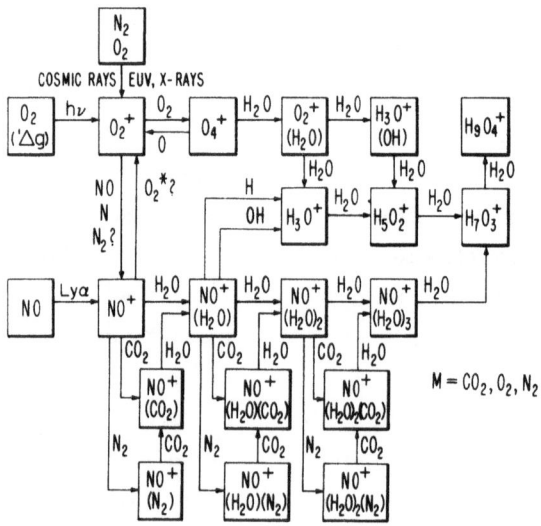

Figure 10. Schematic diagram of positive ion reactions in the
    D-region.

three water molecules and also the product of (18) is $H_7O_3^+$ rather than $H_5O_2^+$. Several proposals have been made to alleviate this problem, none of which have been found to be sufficient.

Burke (1970) proposed the possible reaction

$$NO^+(H_2O) + H \rightarrow H_3O^+ + NO + 2 \text{ eV} \tag{19}$$

but this turns out to be much too slow, $k_{19} < 7 \times 10^{-12}$ cm^3/sec (Fehsenfeld et al., 1974).

Heimerl et al., (1974) proposed two other possible reactions,

$$NO^+(H_2O) + OH \rightarrow H_3O^+ + NO_2 + 0.68 \text{ eV} \tag{20}$$

and

$$NO^+(H_2O) + HO_2 \rightarrow H_3O^+ + NO_3 + 0.02 \text{ eV} \tag{21}$$

These reactions cannot solve the problem however, (20) is too slow, $k_{20} < 1 \times 10^{-10}$ cm^3/sec and there is too little HO_2 in the D-region for (21) to be effective even if $k_{21} \sim 10^{-9}$ cm^3/sec (Fehsenfeld et al., 1974).

Dunkin et al., (1971) showed that the first hydrate of NO^+ is formed faster by

$$NO^+ + CO_2 + M \rightarrow NO^+ \cdot CO_2 + M \tag{22}$$

$$NO^+ \cdot CO_2 + H_2O \rightarrow NO^+ \cdot H_2O + CO_2 \tag{23}$$

than by direct hydration

$$NO^+ + H_2O + M \rightarrow NO^+ \cdot H_2O + M \tag{24}$$

due of course to the large $[CO_2]/[H_2O]$ ratio ≈ 100.

Heimerl et al., (1974) similarly proposed that

$$NO^+ + N_2 + M \rightarrow NO^+ \cdot N_2 + M \tag{25}$$

$$NO^+ \cdot N_2 + CO_2 \rightarrow NO^+ \cdot CO_2 + N_2 \tag{26}$$

is a more effective way to produce $NO^+ \cdot CO_2$ than (22), due to the large $[N_2]/[CO_2]$ ratio $= 2.6 \times 10^3$. There is some uncertainty about this, k_{25} is so small that it is extremely difficult to measure and somewhat uncertain. The $NO^+ \cdot N_2$ bond energy is so low, ~ 0.2 eV, that collisional breakup, the reverse of (25), dominates the $NO^+ \cdot N_2$ loss and therefore must be known. Since the breakup rate constant is a strong function of temperature, fitting an Arhennius expression, it must be known at the appropriate D-region temperature. It should be remarked that since the dominance of $H_5O_2^+$ ions in the D-region is observed under a variety of conditions (altitude, latitude, season) encompassing a wide range of temperatures, no explanation invoking a peculiarly high or low temperature for $H_5O_2^+$ production could solve the general problem.

The ratio of $NO^+ \cdot CO_2$ production rates by (25) + (26) to (22) in the D-region is given by $P_{25+26}/P_{22} \sim k_{26}K_{25}/k_{22}$. At $\sim 200^\circ K$ this ratio is $\leq 10^{-9} \times 10^{-19}/3 \times 10^{-29}$ or ≤ 3. This is a very marginal and still somewhat uncertain enhancement in rate and not sufficient to relieve the overall D-region positive ion problem. At higher temperatures the ratio will of course be less and at lower temperatures more. The role of clusters $NO^+ \cdot H_2O \cdot N_2$ and $NO^+(H_2O)_2 \cdot N_2$ can safely be neglected. The role of $NO^+ \cdot H_2O \cdot CO_2$ in accelerating the formation of $NO^+(H_2O)_2$ remains unmeasured but is likely significant.

There are also O_2^+ ions produced in the D-region, by cosmic ray ionization of O_2 and N_2 (N_2^+ charge transfers rapidly with O_2) and by solar UV photoionization of metastable $O_2(^1\Delta_g)$ molecules which are subject to ionization by radiation that is not absorbed by ground state O_2. The reaction scheme which converts O_2^+ to hydrated hydronium ions is shown in Table II. Under disturbed ionospheric conditions, where there is intense particle bombardment (such as produces auroras) the O_2^+ production can exceed the NO^+ production but normally this is not the case.

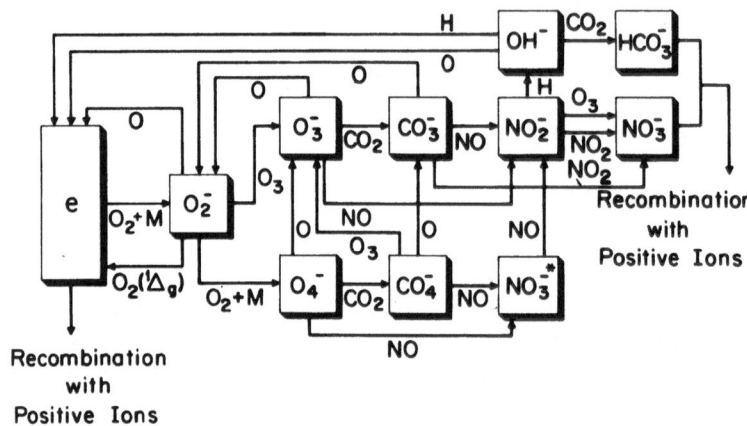

Figure 11. Schematic diagram of negative ion reactions in the D-region, neglecting hydration.

TABLE II

D-Region O_2^+ Reaction Sequence, $296°K$

Reaction	Rate Constant	Reference
$O_2^+ + 2O_2 \rightarrow O_4^+ + O_2$	$2.6 \times 10^{-30}(\frac{T}{300})^{3.2}$ cm^6/sec	a
$O_4^+ + O \rightarrow O_2^+ + O_3$	$3 \pm 2 \times 10^{-10}$ cm^3/sec	b
$O_4^+ + H_2O \rightarrow O_2^+(H_2O)$	1.5×10^{-9} cm^3/sec	c,d,e
$O_2^+(H_2O) + H_2O \rightarrow H_3O^+(OH) + O_2$	1.0×10^{-9} cm^3/sec	c,d,e
$\rightarrow H_3O^+ + OH + O_2$	2×10^{-10} cm^3/sec	c,d,e
$H_3O^+(OH) + H_2O \rightarrow H^+(H_2O)_2 + OH$	1.4×10^{-9} cm^3/sec	c,d,e
$H^+(H_2O)_2 + H_2O + N_2 \rightarrow H^+(H_2O)_3 + N_2$	2.3×10^{-27} cm^6/sec	d*
$H^+(H_2O)_3 + H_2O + N_2 \rightarrow H^+(H_2O)_4 + N_2$	2.4×10^{-27} cm^6/sec	d*
$H^+(H_2O)_4 + H_2O + O_2 \rightarrow H^+(H_2O)_5 + O_2$	9×10^{-28} cm^6/sec	d

a.   Payzant, Cunningham and Kebarle, J. Chem. Phys. 59, 5615 (1973).

b.   Fehsenfeld and Ferguson, Radio Sci. 7, 113 (1972).

c.   Howard, Bierbaum, Rundle and Kaufman, J. Chem. Phys. 57, 3491 (1972).

d.   Good, Durden and Kebarle, J. Chem. Phys. 52, 212, 222 (1970).

e.   Fehsenfeld, Mosesman and Ferguson, J. Chem. Phys. 55, 2115 (1971).

*   Rate constants approximately same for O_2 third body, ref. d. Equilibrium constants for $H^+(H_2O)_n + H_2O \rightleftharpoons H^+(H_2O)_{n+1}$ given in: Cunningham, Payzant and Kebarle, J. Am. Chem. Soc. 94, 7627 (1972); Kebarle, Searles, Zolla, Scarborough and Arshadi, J. Am. Chem. Soc. 89, 6393 (1967).

The various laboratory studies agree very well on the details of the reaction scheme in Table II. A rather critical ionospheric reaction is

$$O_4^+ + O \rightarrow O_2^+ + O_3, \qquad k_{27} = 3 \times 10^{-10} \ cm^3/sec \quad (27)$$

which has been measured in a flowing afterglow system and whose role in the ionosphere was not initially recognized. The association of O_2 with O_2^+ to form O_4^+, followed by (27) represents an interesting ion catalyzed recombination of O with O_2 to produce ozone.

Negative Ion Reactions

The rather complex negative ion reaction scheme of the D-region is illustrated schematically in Fig. 11. The negative ion reaction scheme is initiated by the three-body attachment of electrons to O_2 to produce O_2^-. It is terminated most efficiently by the associative-detachment reactions

$$O_2^- + O \rightarrow O_3 + e, \qquad k_{28} = 3.3 \times 10^{-10} \ cm^3/sec \quad (28)$$

Reaction (28) and many of the other reactions in the scheme of Fig. 11 have only been measured in the NOAA flowing afterglow system. It would be very desirable to have checks made by other methods and other laboratories. The direct D-region negative ion composition measurements have been relatively few, largely qualitative, and in some disagreement so that there has been no substantial test of the reaction scheme to date.

The O_2^- ions which survive (28) can either charge-transfer with ozone to form O_3^- or associate with O_2 to form O_4^-. Both of these ions react rapidly with the relatively abundant CO_2 in the atmosphere making CO_3^- and CO_4^- very important ions (which have been observed). NO in the D-region (or NO_2 at lower altitudes) converts the CO_3^- and CO_4^-, eventually leading to NO_3^- as the stable (or terminal) D-region negative ion, in agreement with the observations of Narcisi and his colleagues (1971). The stable NO_3^- ion will readily hydrate. No reactions have yet been found which convert NO_3^- to another species however. The electron affinity of NO_3 is 3.9 eV so that NO_3^- is a very stable negative ion.

It is interesting to note in Fig. 11 that there are two distinct forms of NO_3^-, with very different energies and reactivities. They have been speculated to be the stable nitrate ion NO_3^- in which the three oxygen atoms are each bonded to the nitrogen and are equivalent and the metastable NO_3^{-*} peroxy ion O-N-O-O$^-$ with a relatively weak O_2^- - NO bond (Ferguson et al., 1972).

The HCO_3^- is a fairly stable ion although the concentration of atomic hydrogen is too low for this to be a dominant one. CO_2

displaces H_2O clustered to OH^- in a fast reaction and CO_2 associates with OH^- more readily than H_2O in three-body reactions.

A question of some concern in D-region negative ion chemistry has been the possible effect which hydration might have on the over-all reaction scheme. Recent measurements (Fehsenfeld and Ferguson, 1974) suggest that this is probably of little importance. The charge-transfer rate constant of O_2^- with O_3 to produce O_3^- is little effected by one or two waters of hydration. The associative-detach-ment of $O_2^- \cdot H_2O$ by O atoms or the collisional-detachment by $O_2(^1\Delta_g)$ on the other hand become endothermic and will not occur. However, the lifetime of O_2^- is too short in the D-region for it to become appreciably hydrated.

The reaction of O_3^- with CO_2,

$$O_3^- + CO_2 \rightarrow CO_3^- + O_2, \qquad k_{29} = 5.5 \times 10^{-10} \text{ cm}^3/\text{sec} \qquad (29)$$

is only slightly reduced by O_3^- single hydration. Two water mole-cules however substantially quench this reaction $k < 10^{-10}$ cm3/sec, but the large abundance of CO_2 precludes O_3^- hydration.

The most vulnerable points in the scheme for hydration effects are CO_3^- and CO_4^- because of their relatively long lifetimes due to low NO concentrations and small reaction rate constants. The reac-tion of $CO_3^- \cdot H_2O$ with NO to produce NO_2^- is about as fast as for the unhydrated CO_3^- however. The reaction of $CO_4^- \cdot H_2O$ with NO will not arise since the reaction

$$CO_4^- \cdot H_2O + H_2O \rightarrow O_2^-(H_2O)_2 + CO_2 \qquad (30)$$

is exothermic and fast and there is a much larger concentration of H_2O than NO.

REFERENCES

Albritton, D. L., A. L. Schmeltekopf and E. E. Ferguson, Spectro-
    scopic investigation of the ion-molecule reaction of He^+ and
    N_2, Bull. Am. Phys. Soc. 14, 212 (1968).
Arnold, F., and D. Krankowsky, New aspects of D-region positive
    cluster ion composition inferred from an improved mass spec-
    trometer measurement, J. Geophys. Res. 1974, in press.
Bates, D. R., and T. N. L. Patterson, Helium ions in the upper
    atmosphere, Planet. Space Sci. 9, 599 (1962).
Bauer, S. J., Chemical processes involving helium ions and the be-
    havior of atomic nitrogen ions in the upper atmosphere, J.
    Geophys. Res. 71, 1508 (1966a).
Bauer, S. J., Hydrogen and helium ions, Annales de Geophysique 22,
    247 (1966b).
Burke, R. R., Hydrogen atom participation in D-region ion chemistry,
    J. Geophys. Res. 75, 1345 (1970).
Dunkin, D. B., F. C. Fehsenfeld, A. L. Schmeltekopf and E. E. Fer-
    guson, Ion-molecule reaction studies from 300° to 600° in a
    temperature controlled flowing afterglow system, J. Chem. Phys.
    49, 1365 (1968).
Dunkin, D. B., F. C. Fehsenfeld, A. L. Schmeltekopf and E. E. Fer-
    guson, Three-body association reactions of NO^+ with O_2, N_2 and
    CO_2, J. Chem. Phys. 54, 3817 (1971).
Farragher, A. L., J. A. Peden and W. L. Fite, Charge-transfer of
    N_2^+, O_2^+ and NO^+ to sodium atoms at thermal energies, J. Chem.
    Phys. 50, 287 (1969).
Fehsenfeld, F. C., Si^+ and SiO^+ reactions of atmospheric importance,
    Can. J. Chem. 47, 1808 (1969).
Fehsenfeld, F. C., and E. E. Ferguson, Laboratory studies of nega-
    tive ion reactions with atmospheric trace constituents, J.
    Chem. Phys. 61, Oct. 1 (1974).
Ferguson, E. E., and F. C. Fehsenfeld, Some aspects of the metal
    ion chemistry of the earth's atmosphere, J. Geophys. Res.
    73, 6215 (1968).
Ferguson, E. E., Atmospheric metal ion chemistry, Radio Sci. 7,
    397 (1972).
Ferguson, E. E., D. B. Dunkin and F. C. Fehsenfeld, Reactions of
    NO_2^- and NO_3^- with HCl and HBr, J. Chem. Phys. 57, 1459 (1972).
Ferguson, E. E., Rate constants of thermal energy binary ion-mole-
    cule reactions of aeronomic interest, Atomic Data and Nuclear
    Data Tables 12, 159 (1973).
Goldberg, R. A. and A. C. Aiken, Studies of positive-ion composi-
    tion in the equatorial D-region ionosphere, J. Geophys. Res.
    76, 8352 (1971).
Govers, T. R., F. C. Fehsenfeld, D. L. Albritton, P. G. Fournier
    and J. Fournier, Molecular isotope effects in the thermal-
    energy charge exchange between He^+ and N_2, Chem. Phys. Letters
    26, 134 (1974).

Hanson, W. B., Upper-atmosphere helium ions, J. Geophys. Res. $\underline{67}$, 183 (1962).

Heimerl, J., R. Johnsen and M. A. Biondi, Ion-molecule reactions, $He^+ + O_2$ and $He^+ + N_2$, at thermal energies and above, J. Chem. Phys. $\underline{51}$, 5041 (1969).

Heimerl, J. M., and J. A. Vanderhoff, Rate constants for the clustering of CO_2, N_2, and O_2 to NO^+, J. Chem. Phys. $\underline{60}$, 4362 (1974).

Johnsen, R., H. L. Brown and M. A. Biondi, Reactions of Na^+, K^+ and Ba^+ ions with O_2, NO and H_2O molecules, J. Chem. Phys. $\underline{55}$, 186 (1971).

Keller, G. E. and R. A. Beyer, CO_2 and O_2 clustering of sodium ions, J. Geophys. Res. $\underline{76}$, 289 (1971).

Kemper, P. R., and M. T. Bowers, Mechanism of thermal energy gas phase charge exchange reaction: $He^+ + N_2$, J. Chem. Phys. $\underline{59}$, 4915 (1973).

Krankowsky, D., F. Arnold, H. Wieder, J. Kissel and J. Zähringer, Positive-ion composition in the lower ionosphere, Radio Sci. $\underline{7}$, 93 (1972).

Lindinger, W., F. C. Fehsenfeld, A. L. Schmeltekopf and E. E. Ferguson, Temperature dependence of some ionospheric ion-neutral reactions from 300°-$900^\circ K$, J. Geophys. Res. in press (1974).

Narcisi, R. S., and A. D. Bailey, Mass spectrometric measurements of positive ions at altitudes from 64 to 112 kilometers, J. Geophys. Res. $\underline{70}$, 3687 (1965).

Narcisi, R. S., A. D. Bailey, L. Della Lucca, C. Sherman and D. M. Thomas, Mass Spectrometric measurements of negative ions in the D- and lower E-regions, J. Atmos. Terr. Phys. $\underline{33}$, 1147 (1971).

Norton, R. B., and C. A. Barth, Theory of nitric oxide in the earth's atmosphere, J. Geophys. Res. $\underline{75}$, 3903 (1970).

Ryan, K. R., Ionic collision processes in gaseous mixtures of oxygen and nitrogen, J. Chem. Phys. $\underline{51}$, 4136 (1969).

Schmeltekopf, A. L., E. E. Ferguson and F. C. Fehsenfeld, Afterglow studies of the reactions He^+, $He(2^3S)$ and O^+ with vibrationally excited N_2, J. Chem. Phys. $\underline{48}$, 2966 (1968).

Taylor, H. A., L. H. Brace, H. C. Brinton and C. R. Smith, Direct measurement of helium and hydrogen ion concentration and total ion density to an altitude of 940 kilometers, J. Geophys. Res. $\underline{68}$, 5339 (1963).

THE FORMATION OF INTERSTELLAR MOLECULES BY ION-MOLECULE REACTIONS

A. Dalgarno

Center for Astrophysics

Cambridge, Massachusetts 02138

A description is given of various chemical schemes involving
ion-molecule reactions that are significant in the formation of
interstellar molecules and attention is drawn to the basic chemical
data that are required before a quantitative assessment of the
chemistry can be made.

1. Introduction

In the interstellar medium, the radiation from the stars
ionizes the regions around them and all the stellar photons that
are capable of ionizing atomic hydrogen are contained within these
H II regions. Outside the H II regions the medium contains neutral
hydrogen and the only ionization produced by the stellar radiation
is from elements such as carbon, silicon, iron, sulphur and mag-
nesium which have ionization potentials less than that of atomic
hydrogen. However we know from pulsar dispersion data that the
fractional ionization of the interstellar gas is about 0.2, much
larger than the value of about 5×10^{-4} provided by the ionization
of the minor constituents. Some of the additional ionization may
exist in H II regions surrounding stars that have not been observed
in the visible region of the spectrum and some arises from ionization by cosmic rays and X-rays.

In any event there is a large amount of neutral hydrogen in
the H I regions that lie beyond the H II regions. The H I regions
consist apparently of a warm intercloud medium with a particle
density approaching 1 cm^{-3} and a temperature that may be nearly
10^4 $^{\circ}$K containing condensations or clouds with densities greater
than 10 cm^{-3} and temperatures less than 100 $^{\circ}$K. There may exist

a third phase of the interstellar medium with a density of the order of 10^{-2} cm^{-3} and a temperature of the order of 5 x 10^5 °K.

We are interested here in the clouds which can be character-ized as diffuse or dense or dark. The diffuse clouds are observable in the 21 cm radiation of atomic hydrogen and by their absorption of visible and ultraviolet radiation from hot stars. They have densities in the range from 10 cm^{-3} to 10^3 cm^{-3}, and temperatures typically near 80 °K. Dense clouds are associated with H II regions. They have temperatures also around 80 °K but the densities are greater than 10^3 cm^{-3}. Dark clouds have similar densities but are much colder with temperatures less than 20 °K. In the diffuse clouds, only diatomic molecular species have been detected. The dense and dark clouds are the locations of the very complex molecules.

The interstellar medium also contains dust grains which shield the molecules from the destructive stellar ultraviolet radiation field and which many believe provide the surfaces on which all the interstellar molecules are formed (cf. Watson and Salpeter 1972). The recent detection of N_2H^+ (Turner 1974) tends to support the grain hypothesis. However gas-phase mechanisms involving ion-molecule reactions do play an important role. Their importance follows from the recognition that molecular hydrogen exists in even diffuse interstellar clouds (where it is formed presumably by association of hydrogen atoms on the surface of grains).

2.   Interstellar Molecules

The conditions for the formation of interstellar molecules are not favorable. The densities are so low that three-body collisions are negligible and all particles are in their lowest energy states. The environment is a hostile one and the molecules are subjected to interstellar ultraviolet photons, to cosmic rays and to X-rays. Yet a large number of molecules, many quite complex, have been detected (cf. Rank and Townes 1972, Solomon 1973). Table 1 is a list of the observed molecules. Various isotopic species have also been discovered in interstellar clouds. All the molecules discovered by radio astronomy have permanent dipole moments; molecules such as carbon dioxide are presumably present in the dense and dark clouds, but remain undetected.

3.   Ion-Molecule Reactions

The measurements by the ultraviolet spectrometer on the Copernicus satellite of absorption by molecular hydrogen in several rotational states of the lowest vibrational level

(Spitzer et al. 1973; Spitzer, Cochran and Hershfeld 1974) strongly
suggest that the relative population of the j = 1 and j = 0 levels
is characterized by the kinetic temperature of the cloud so that
the levels must be coupled by collisions. Because odd j levels are
ortho-hydrogen and even j levels are para-hydrogen the collisions
involve an exchange process. Neutral hydrogen atoms are inefficient at interstellar cloud temperatures and a natural suggestion
is the ion-molecule reaction (Dalgarno, Black and Weisheit 1973)

$$H^+ + o\text{-}H_2 \leftrightarrows H^+ + p\text{-}H_2. \tag{1}$$

The deuterated molecule HD is also observed by the Copernicus instrument. After correction for the self-shielding of H_2 had been
made (Spitzer et al. 1973), it was clear that there is a source of
HD in interstellar clouds other than that which forms H_2. Dalgarno
et al. (1973) and Watson (1973a) suggested the ion-molecule reaction
analogous to (1),

$$D^+ + H_2 \rightarrow H^+ + HD, \tag{2}$$

the D^+ ions being produced by the charge transfer reaction

$$H^+ + D \rightarrow H + D^+. \tag{3}$$

The interstellar abundance ratio of deuterium to hydrogen $[D]/[H]$
has been measured to be 1.38×10^{-5} (Rogerson et al. 1973) in the
direction of β Cen. If the value is typical of the galaxy, it may
be used in conjunction with the observed ratios of HD to H_2 to
derive the H^+ densities, the rate coefficient of (2) having been
measured (Fehsenfeld et al. 1973). Given the H^+ densities, the
ionizing flux that produces the photons can be determined once the
H^+ equilibrium chemistry has been established (Black and Dalgarno
1973a).

    The tentative chemical scheme is shown schematically in
figure 1 (Dalgarno, Oppenheimer and Berry 1973; Herbst and Klemperer
1973; Black and Dalgarno 1973; Watson 1973b). The chemical scheme
for the removal of H^+ leads to the production of H_2O and in particular to the production of OH. A quantitative study of OH in interstellar clouds has been given by Black and Dalgarno (1974). The
OH is removed preferentially by reactions with C^+:

$$C^+ + OH \rightarrow CO + H^+ \tag{4}$$

$$\rightarrow CO^+ + H. \tag{5}$$

If (4) is the main reaction path, an H^+ is produced which can
participate in the cycle. Reaction (5) is often followed by

$$CO^+ + H \rightarrow CO + H^+ \tag{6}$$

Table   1

Interstellar molecules

CH^+	CH	CN	CO	H_2	
OH	CS	SO	SiO	SN	
H_2O	HCN	OCS	H_2S	N_2H^+	C_2H
NH_3	H_2CO	H_2CS	HNCO		
$NCOOH^*$	HC_3N				
CH_3OH	CH_3CN	NH_2HCO			
CH_3C_2H	CH_3HCO	CH_3NH_2			
$(CH_3)_2O$					

*The detection of formic acid is not certain.   There
are also unidentified emitters at 89.189GHz and 90.665GHz.
That at 89.189GHz has been attributed to HCO^+ (Klemperer
1970) but the isotopic molecule $HC^{13}O^+$ has not been found
(Buhl and Snyder 1973); that at 90.665GHz may be HNC.

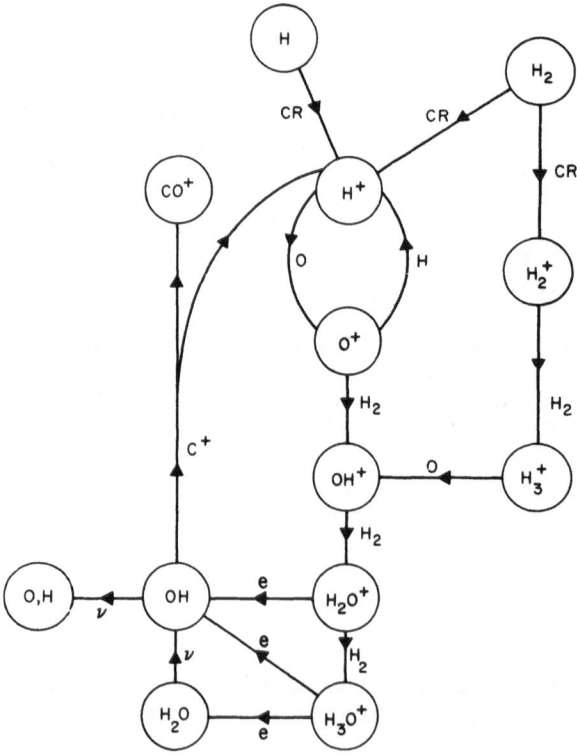

Fig. 1   The H^+ chemistry and the formation of OH and CO.

and the destruction of OH by C^+ is probably the largest source of CO in interstellar clouds (Oppenheimer and Dalgarno 1974a).

The largest uncertainty in the quantitative description of the chemical scheme leading to the removal of H^+ and the formation of OH and CO is probably the rate coefficient of the reaction

$$O(^3P_2) + H^+ \rightarrow O^+(^4S) + H. \tag{7}$$

There are interesting differences between the gas phase chemistries of the similar systems oxygen and sulphur which may lead to an elucidation of the comparative roles of gas phase and grain surface reactions in the formation of interstellar molecules (Oppenheimer and Dalgarno 1974b). Thus sulphur is ionized in diffuse clouds but unlike O^+ S^+ does not react with H_2 and nor do SH^+ and H_2S^+.

Jura (1974) has pointed out that the ionization potential of Cl is less than that of H and that the reaction of Cl^+ with H_2,

$$Cl^+ + H_2 \rightarrow HCl^+ + H, \tag{8}$$

is exothermic. The rate coefficient has been measured by Fehsenfeld and Ferguson (1974) who also found that

$$HCl^+ + H_2 \rightarrow H_2Cl^+ + H \tag{9}$$

is rapid. The resulting chemistry in interstellar clouds (Jura 1974, Dalgarno et al. 1974) is represented in figure 2. It appears that in dense clouds most of the chlorine will be in the form HCl.

The sequences that form OH and HCl consist of hydrogen abstractions, the importance of which has been stressed particularly by Herbst and Klemperer (1973) and by Watson (1973b). A sequence of particular interest is that leading to the formation of ammonia (Herbst and Klemperer 1973). It is summarised in figure 3.

The scheme also leads to the formation of HCN and CN. The branching ratio for the dissociative recombination of H_2CN^+ with an electron is unknown, as is usually the case. The branching ratios play a critical role in ion-molecule schemes for forming interstellar molecules and they cause major uncertainties in quantitative studies.

Several other reaction paths lead to HCN and CN, one of which arises from a study of the formation of CH^+ and CH. In a classic paper on the formation of interstellar molecules, Bates and Spitzer (1951) suggested that CH^+ and CH might be produced by radiative association

$$C^+ + H \rightarrow CH^+ + h\nu \tag{10}$$

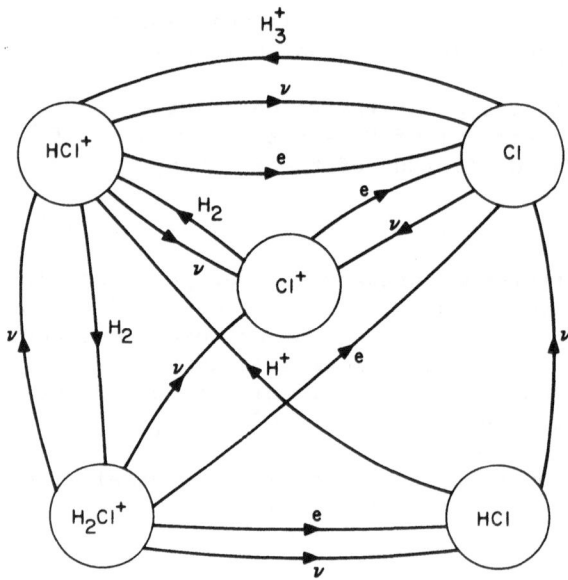

Fig. 2   The Cl chemistry

Fig. 3 The NH₃ chemistry and
the formation of HCN and CN.

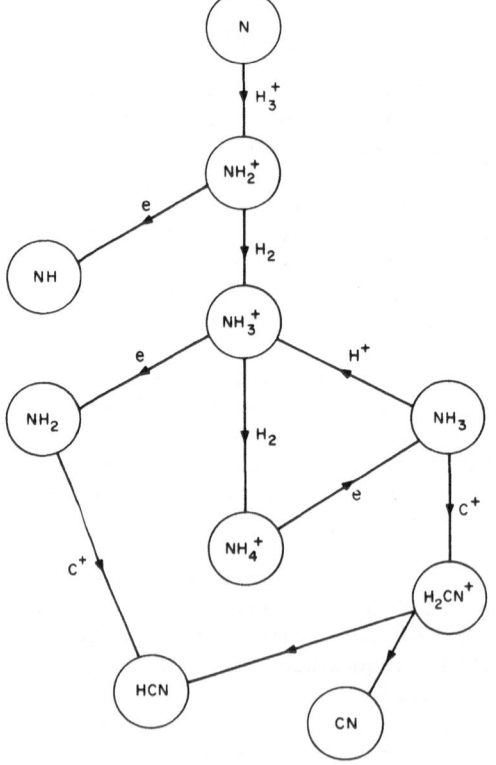

$$C + H \rightarrow CH + h\nu \tag{11}$$

and by radiative recombination

$$CH^+ + e \rightarrow CH + h\nu \; . \tag{12}$$

Solomon and Klemperer (1972) have argued more recently that the observed abundances of CH^+ can be explained only if the dissociative recombination process

$$CH^+ + e \rightarrow C + H \tag{13}$$

is slow, contrary to theoretical expectation (Bardsley and Junker 1973, Krauss and Julienne 1973). The difficulty is apparently compounded by the recognition (Black and Dalgarno 1973b, Watson 1974a) that

$$CH^+ + H_2 \rightarrow CH_2^+ + H \tag{14}$$

is rapid (Kim, Theard and Huntress 1974).

More severe difficulties attend the chemical scheme suggested for CH (cf. Smith et al. 1973) which may be resolvable by the substitution of H_2 for H in (10). An extract from the chemical scheme is illustrated in figure 4.

According to Black and Dalgarno (1973b), a rate coefficient between 10^{-16} and 10^{-15} cm^3 s^{-1} for the radiative association

$$C^+ + H_2 \rightarrow CH_2^+ + h\nu \tag{15}$$

is necessary to reproduce the measured CH abundances. An experimental test of this assumption would be invaluable.

The chemical scheme outlined by Black and Dalgarno (1973b) is efficient for the production of CH but it does not yield the measured CH^+ abundance. It appears however that the inclusion of the photodissociation process

$$CH_3^+ + h\nu \rightarrow CH^+ + H_2 \tag{16}$$

may be sufficient. Calculations are in progress (Dalgarno et al. 1974) but measurements of the photodissociation cross sections of CH_x^+ ions are needed before the scheme can be made quantitative.

The production of substantial concentrations of CH_3^+ ions, which occurs also in dense clouds, opens reaction paths for the formation of other molecules. Thus

$$CH_3^+ + N \rightarrow H_2CN^+ + H \tag{17}$$

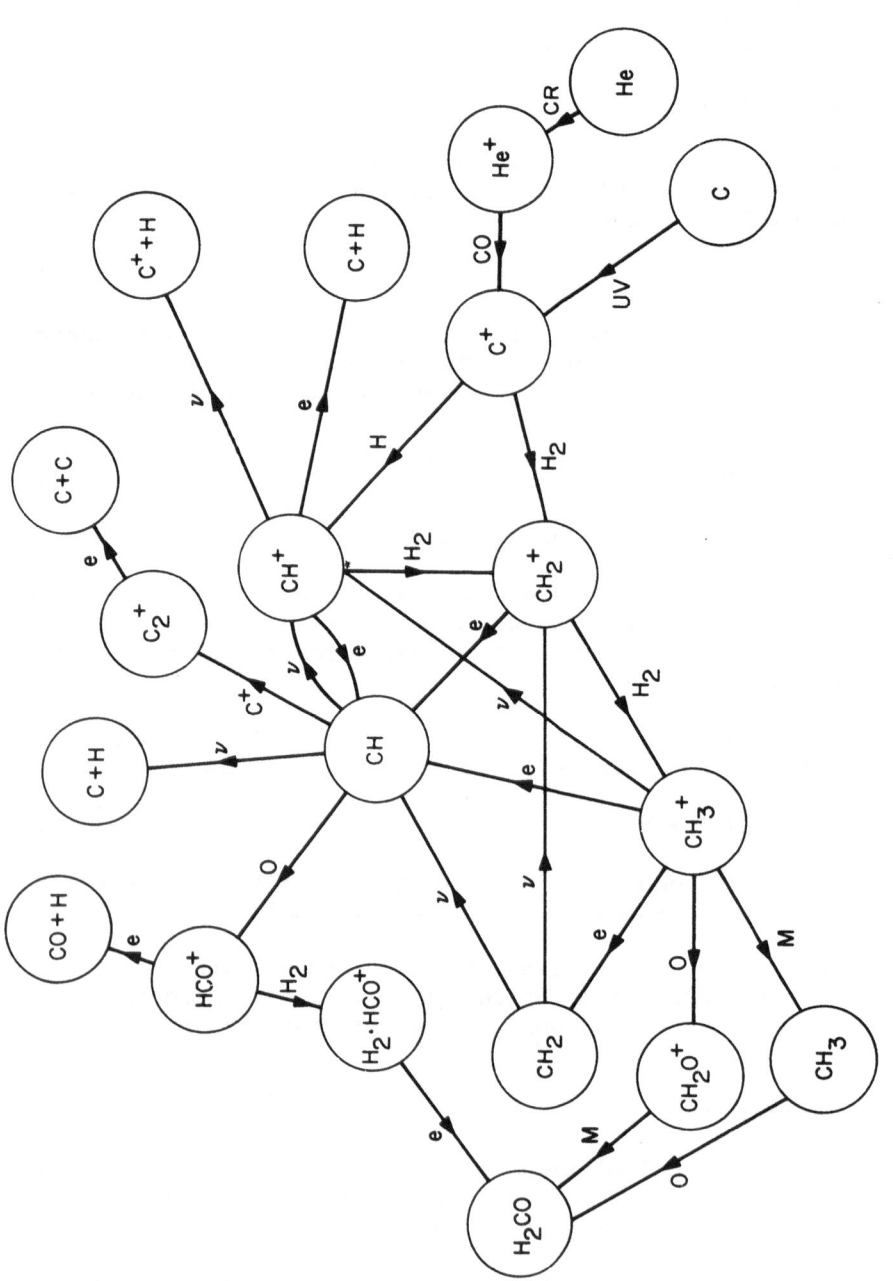

Fig. 4 The chemistry of CH and CH⁺.

leads to HCN and CN. Presumably the reaction

$$CH_3^+ + O \rightarrow H_2CO^+ + H \tag{18}$$

also occurs. The molecule HCO has not been detected. Possibly H_2CO^+ recombines dissociatively according to

$$H_2CO^+ + e \rightarrow CO + H + H \tag{19}$$

and not to HCO + H.

Formaldehyde can also be produced from CH_3^+ (Dalgarno, Oppenheimer and Black 1973), though the mechanisms are not very efficient. Thus

$$CH_3^+ + M \rightarrow CH_3 + M^+ \tag{20}$$

$$CH_3^+ + e \rightarrow CH_3 + h\nu , \tag{21}$$

where M is some neutral minor constituent of the cloud, produce CH_3 and may be followed by

$$CH_3 + O \rightarrow H_2CO + H \quad . \tag{22}$$

Alternatively H_2CO can be formed by

$$CH_3^+ + O \rightarrow H_2CO^+ + H \tag{23}$$

followed by charge transfer of H_2CO^+ with M or radiative recombination with electrons.

Other mechanisms are possible. Herbst and Klemperer (1973) have suggested the radiative association

$$HCO^+ + H_2 \rightarrow H_2 \cdot HCO^+ + h\nu \tag{24}$$

followed by

$$H_2 \cdot HCO^+ + e \rightarrow H_2CO + H \quad . \tag{25}$$

The photon in (24) is emitted in a pure vibrational transition (unlike (15) which is an electronic transition). Herbst and Klemperer (1973) argue on theoretical grounds that a rate coefficient of $10^{-17} cm^3 sec^{-1}$ is plausible for (24) which is apparently rapid enough to form the formaldehyde observed in dense clouds. It is probably too slow in diffuse clouds where photodissociation is an important destruction process. Another reaction that leads to H_2CO is

$$H_3O^+ + CH_2 \rightarrow H_3CO^+ + H_2 \tag{26}$$

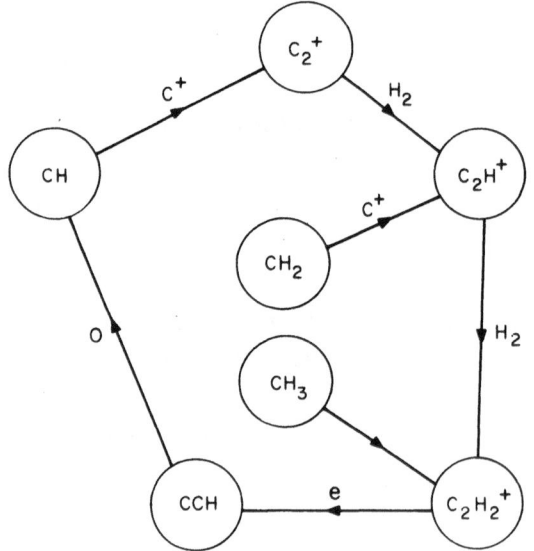

Fig. 5 The chemistry of CCH.

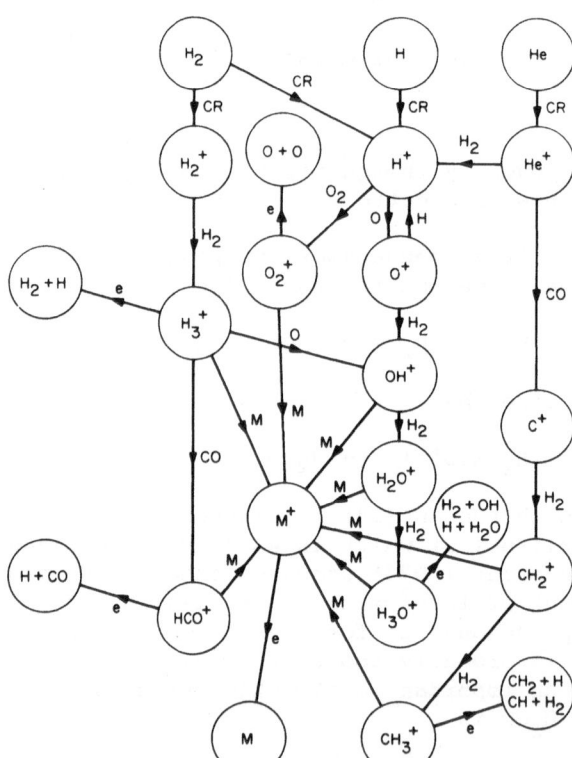

Fig. 6 The chemical
scheme that controls the
ionization in a dense
cloud.

followed by dissociative recombination and other possibilities
exist. Many basic molecular data are needed in order to carry
through a quantitative evaluation of the various schemes.

The ethynyl radical C_2H has been detected recently (Tucker
et al. 1974). A possible scheme for its formation (Watson 1974b)
is illustrated in figure 5.

The fractional ionization in a cloud is critical to the
collapse of the cloud. If the ionization is high, the magnetic
field is trapped and its increasing intensity will retard and
ultimately prevent the collapse. The degree of ionization produced in a cloud by cosmic rays is finally a question of chemistry.
Figure 6 shows the general scheme (Oppenheimer and Dalgarno 1974c)
that occurs. Most of the complex molecular ions are removed
rapidly by dissociative recombination but some fraction charge
transfer to heavier species such as sodium which react slowly and
recombine slowly.

The study of the chemistry of interstellar molecules is now
entering a quantitative phase where it should be possible to
construct detailed schemes of formation and destruction that will
establish correlations between different molecules and that will
predict molecules as yet undetected. A large array of basic
chemical data on complex neutral and basic species are needed.

References

J. N. Bardsley and B. R. Junker 1973 Ap. J. Letters <u>183</u>, 135.
D. R. Bates and L. Spitzer 1951 Ap. J. <u>113</u>, 441.
J. H. Black and A. Dalgarno 1973a Ap. J. Letters <u>184</u>, 101.
J. H. Black and A. Dalgarno 1973b Astrophys. Letters <u>15</u>, 79.
J. H. Black and A. Dalgarno 1974 Ap. J. in press.
D. Buhl and L. E. Snyder 1973 Ap. J. <u>180</u>, 791.
A. Dalgarno, J. H. Black and M. Oppenheimer 1974 in preparation.
A. Dalgarno, J. H. Black and J. C. Weisheit 1973 Astrophys.
     Letters <u>14</u>, 77.
A. Dalgarno, T. de Jong, M. Oppenheimer and J. H. Black 1974
     Ap. J. Letters in press.
A. Dalgarno, M. Oppenheimer and R. S. Berry 1973 Ap. J. Letters
     <u>183</u>, 21.
A. Dalgarno, M. Oppenheimer and J. H. Black 1973 Nature <u>245</u>, 100.
F. C. Fehsenfeld, D. B. Dunkin, E. E. Ferguson, and
     D. L. Albritton 1973 Ap. J. <u>183</u>, 25.
F. C. Fehsenfeld and E. E. Ferguson 1974 J. Chem. Phys. in press.
E. Herbst and W. B. Klemperer 1973 Ap. J. <u>185</u>, 505.
M. Jura 1974 Ap. J. Letters <u>190</u>, 33.
J. K. Kim, L. P. Theard and W. T. Huntress 1974 in press.
M. Krauss and P. S. Julienne 1973 Ap. J. Letters <u>183</u>, 139.

M. Oppenheimer and A. Dalgarno 1974a in preparation.

M. Oppenheimer and A. Dalgarno 1974b Ap. J. $\underline{187}$, 231.

M. Oppenheimer and A. Dalgarno 1974c Ap. J. in press.

D. M. Rank and C. H. Townes 1972 Science $\underline{174}$, 1083.

J. B. Rogerson, L. Spitzer, J. F. Drake, K. Dressler, E. B. Jenkins,
    D. C. Morton and D. G. York 1973 Ap. J. Letters $\underline{181}$, 97.

W. H. Smith, H. S. Liszt and B. L. Lutz 1973 Ap. J. $\underline{183}$, 69.

P. Solomon 1973 Physics Today $\underline{26}$, 32.

L. Spitzer, W. D. Cochran and A. Hirshfeld 1974 Ap. J. Suppl.
    in press.

L. Spitzer, J. F. Drake, E. B. Jenkins, D. C. Morton,
    J. B. Rogerson and D. G. York 1973 Ap. J. Letters $\underline{181}$, 116.

K. D. Tucker, M. L. Kutner and P. Thaddeus 1974 Ap. J. Letters
    in press.

B. E. Turner 1974 Ap. J. Letters in press.

W. D. Watson 1973a Ap. J. Letters $\underline{182}$, 73.

W. D. Watson 1973b Ap. J. Letters $\underline{183}$, 17.

W. D. Watson 1974a Ap. J. 189, 221.

W. D. Watson 1974b Ap. J. $\underline{188}$, 35.

W. D. Watson and E. E. Salpeter 1972 Ap. J. $\underline{174}$, 321.

REACTIONS OF NEGATIVE IONS

T. O. Tiernan

Aerospace Research Laboratories

Wright-Patterson AFB, Ohio 45433, U.S.A.

INTRODUCTION

As a review of the literature demonstrates, markedly fewer investigations of the reactions of negative ions have been reported than for positive ions. This is likely due in part to the fact that positive ions are somewhat easier to produce and study by the techniques conventionally employed. Even in the generous time allotted for this lecture however, it would not be possible to present an exhaustive survey of negative-ion reaction studies, and no attempt will be made to do so. Rather, in keeping with the expressed purpose of the conference, a sampling of recent work from both our own and other laboratories will be presented, which will illustrate some of the many interesting features of negative-ion processes. These processes have important consequences in certain regions of the upper atmosphere, as already described in presentations by Ferguson, Arnold, and Narcissi at this conference. Negative-ion reactions also have broad significance for lasers, discharges, and a variety of other energetic media. The results to be discussed here will be restricted to the low energy regime (less than 100eV), and will be mainly concerned with bimolecular reactions. For more comprehensive discussions of negative-ion reactions, the reader is referred to several recent review articles (Calcote, 1972; Craggs, 1971; Ferguson, 1969, 1970, 1972; Futrell and Tiernan, 1972; Kebarle, 1972; Paulson, 1972; Thynne, 1972).

EXPERIMENTAL TECHNIQUES

Most of the experimental methods described at this conference

have been applied for negative-ion reaction studies. As was the
case for positive ion reactions, many of the earliest investigations were accomplished in mass spectrometer ion sources. Subsequently, beam devices, drift tubes, high pressure mass spectrometers, flowing afterglows and ion cyclotron resonance spectrometers have been utilized for such experiments. Applications of
flowing afterglow techniques to negative ion processes have been
described at this conference by Bohme, Fehsenfeld and Ferguson,
and Kebarle has discussed solvation and clustering reactions of
negative ions which are observed with a high pressure mass
spectrometer.

Much of the experimental data on negative ion reactions to be
presented here was obtained with the ARL tandem mass spectrometer,
which has been described previously (Tiernan and Marcotte, 1970;
Futrell and Tiernan, 1972). This instrument is shown schematically
in Figure 1. Briefly, the first stage or ion gun of the apparatus
is a double-focusing mass spectrometer which produces a well-
collimated mass-selected ion beam having a kinetic energy of ∿170eV
in the laboratory scale, the half-width of the energy distribution
at half-maximum being 0.3eV. This beam passes through a thick-
element field lens, which retards the translational energy to the
desired interaction energy without introducing any additional
spread, and then into a shielded field-free collision chamber.
Product ions which emerge from the collision chamber are reaccel-
erated by a Soa-type lens, mass analyzed in a second sector mass
spectrometer and detected using an electron multiplier operated
in pulse-counting mode. The application of the instrument for
measuring energy thresholds of endothermic ion-neutral processes
and the methods employed to quantitatively define the energy scale
have been discussed in previous papers from our laboratory,
(Tiernan and Marcotte, 1970). The procedures applied for measuring reaction cross sections, rate constants and excitation
functions with this apparatus, and the limitations on these
measurements were described at the panel discussion on Ion-Molecule
Reaction Rates and appear in another section of this volume.

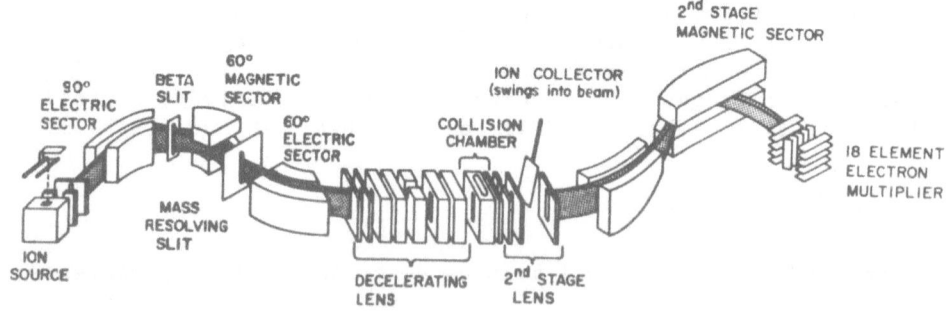

Fig. 1. Schematic diagram of the ARL tandem mass spectrometer.

PRODUCTION OF NEGATIVE IONS

    Negative ion reactants are usually produced by electron
impact, and there are several well-known mechanisms, which are
discussed briefly below.

Resonance Electron Attachment

$$AB + e \rightarrow AB^- \tag{1}$$

    This process is important only for a very narrow range of
electron energies, typically below 10eV. The incident electron
excites the molecule to an upper vibrational level from which
capture to form a vibrationally excited molecular ion may then
occur. If the molecular ion is not stabilized by some means, it
may undergo autodetachment.

Dissociative Attachment

$$AB + e \rightarrow A^- + B \tag{2}$$

    This mechanism is significant at lower pressures and for the
electron energy range up to about 15eV. The process involves a
vertical transition from the ground state of the neutral molecule
to the molecular negative ion state. The latter than decomposes
by autodetachment or dissociates into an ion and a neutral fragment. The ions formed upon dissociation will have a kinetic
energy distribution which peaks at,

$$E_o = \left(1 - \frac{m_A^-}{m_{AB}} \right) [E_e - D(AB) + E.A.(A) - E(B)] \tag{3}$$

where m_A^- and m_{AB} are the masses of the negative ion and neutral
molecule respectively, E_e is the electron energy, $D(AB)$ is the
dissociation energy of AB, E.A.(A) is the electron affinity of A
and E(B) is the excitation energy of B (Chantry and Schulz, 1967).
Equation (3) is actually an approximation since the appearance
potentials for associative detachment processes have been shown
to be temperature dependent, and corrections must be applied to
correct the data to an effective temperature of 0^oK (Spence and
Schulz, 1969). It has also been demonstrated in some cases, that
the most probable kinetic energy with which the ion fragment is
formed is independent of the electron energy over a range of a few
volts above threshold. Chantry (1969) has established, for example,
that for O^- ions formed by dissociative attachment to N_2O, E_o has
a value of 0.38eV which is invariant for electron energies in the
range 1.5 to 2.5eV above threshold. This requires that the excess
energy appear as vibrational and rotational excitation of the

N_2 fragment. From this discussion, it is apparent that under
source conditions where collisional deactivation does not occur,
negative ions formed by dissociative attachment will generally
have excess kinetic energy (that is, above thermal energy).
Obviously, this is important in experiments where negative ion
formation and reaction occur in the same chamber, at pressures
which are insufficient for collisional deactivation to take place.
In beam experiments, where the reactant negative ion undergoes
energy analysis prior to collision with the neutral target, the
energy distribution imparted to the ion by the dissociative attachment process presents no complication.

Three-body Attachment

$$AB + e + M \rightarrow AB^- + M \tag{4}$$

This mechanism is the dominant one for negative ion formation
under conditions where pressures are relatively high and the
electron energies are quite low. Such conditions are characteristic of discharges and low-density plasmas and three-body
attachment is insignificant for conventional electron-impact
sources. The rates of these processes appear to be strongly
dependent upon the nature of the third-body, however. (Phelps,
1969).

Pair Production

$$AB + e \rightarrow A^- + B^+ + e \tag{5}$$

In this case, the electron released can carry off excess
kinetic energy. The appearance potential of ions formed by this
process is given by the relation,

$$E(A^-) = E(B^+) = D(AB) + I(B) - E.A.(A) + \Delta E \tag{6}$$

where $I(B)$ is the ionization potential of B, ΔE is the excess
energy released and the other terms are as defined above. This
process is nonresonant and occurs over a wide energy range above
threshold (typically 10eV and higher). Cross sections for pair
production have been measured in relatively few cases.

Ion-Molecule Reactions

Many negative-ion reactants of interest cannot be prepared by
electron impact in sufficient intensities for practical use.
In some instances, such reactants are produced in good abundance
from ion-molecule reactions, however. An example is the formation

of alkoxide negative ions from alcohols. When mixtures of N_2O
and alcohols are introduced into the primary ion source of the
tandem mass spectrometer, it can be shown that the reaction
sequence involves dissociative attachment to N_2O,

$$N_2O + e \rightarrow O^- + N_2 \tag{7}$$

followed by reaction of O^- with the alcohol,

$$O^- + ROH \rightarrow RO^- + OH \tag{8}$$

where R represents an alkyl group. In a similar fashion a
variety of organic and inorganic negative ions can be obtained for
use as reactants. Such processes will be discussed further in a
later section.

RESULTS AND DISCUSSION

Endothermic Negative-Ion Charge-Transfer Reactions

   Negative ion reactions of the type,

$$X^- + AB \rightarrow AB^- + X \tag{9}$$

where X^- is an atomic negative ion and AB is a molecule, are of
interest because of the information which they can provide about
molecular electron affinities. If the electron affinity of X is
greater than that of AB, then reaction (9) is endothermic for
thermal energy reactants and is not observed under these conditions.
Experiments in our laboratory (Lifshitz, Hughes and Tiernan, 1970;
Tiernan, Hughes and Lifshitz, 1971) have shown however, that by
increasing the collision energy (that is, the translational
energy in the center of mass), such charge transfer processes can
be driven and energy thresholds can be measured. If the electron
affinity of the atomic species X is accurately known, then it
can be used in conjunction with the energy threshold to calculate
the electron affinity of AB from the relation,

$$E.A.(AB) = E.A.(X) - E_{threshold} \tag{10}$$

   The electron affinities of many atoms are well-known from
photodetachment studies and can even be calculated with reasonable
accuracy (Moisewitsch, 1965; Hotop, Patterson and Lineberger,
1974). Molecular electron affinities are much less reliably
known and, in general, a wide range of values is reported in the
literature for any given species. For a molecule, the minimum of
the stable molecular negative ion potential curve is usually
displaced with respect to the minimum of the neutral ground state

curve, and so the vertical detachment energy (as obtained by photodetachment) does not necessarily correspond to the adiabatic electron affinity.

Data obtained thus far, using the charge-transfer methods described, suggest that these procedures are generally applicable for obtaining accurate electron affinities of a large number of molecules. The accuracy is limited in part by the assumptions which are made with respect to these experiments. It must be assumed that the reactant ions are in the ground state, that the product ions are formed at threshold without excess internal or translational energy, and that there is no activation energy barrier. Some indication that the reactant ions are indeed in the ground state is provided by our observations that for a given reaction, no differences in the cross section or threshold behavior can be detected when the reactant ion is generated from several different source molecules. We are unable to determine the energy states of the product ions formed in the present experiments, however, which means that from a rigorous standpoint, the electron affinities derived are only lower limits. However, the fact that a consistent value of the electron affinity for a given molecule can be obtained from several different reactions lends support to the assumption that this is actually the true electron affinity. The possibility that there is an activation energy for the negative ion-molecule processes with which we are concerned cannot be entirely ruled out, but this seems unlikely in view of the strong attractive ion-induced dipole forces which characterize such reactions, and which should dominate any small repulsive forces.

It should also be noted that the threshold behavior observed in our experiments could be affected by changes in the collection efficiency of the apparatus with increasing reactant ion energy, a problem which we have discussed previously (Tiernan, Hughes and Lifshitz, 1971; Tiernan and Marcotte, 1970). In an earlier paper, we reported measurements of the cross section for the O^-/O_2 endothermic charge transfer reaction, over the laboratory energy range from ~ 0.3 to 170eV, and showed that our results were in very good agreement with other beam observations on this reaction by Rutherford and Turner (1967), and by Bailey and Mahadevan, (1970). Similar comparisons between our measured cross-sections as a function of energy and those reported from other laboratories have also been made for positive ion processes, and consistency was also demonstrated in these cases. It is reasonable, therefore, to assume that discrimination effects are relatively minor in so far as these affect the actual thresholds determined. Such effects may influence the shape of the cross sections at energies above that at which the maximum cross section is observed, and may account for the skewing on the high energy side. However, we are unable to assess these effects in detail with our existing instrumentation.

A sample of the type of negative-ion charge-transfer data obtained in these experiments is shown in Figure 2, where the relative cross section for the reaction of I^- with O_2 yielding O_2^- is plotted as a function of energy. It is seen that at lower energies only the background level of O_2^- is detected, while at threshold onset occurs and the O_2^- signal rises rather sharply to a maximum. Some tailing is evident in the threshold region, and this introduces uncertainty in defining the true onset energy. In fact, if the extrapolation indicated by the straight lines drawn through the data points on the figure is taken to represent the threshold energy, then the calculated electron affinity of O_2 is obviously much too high. The problem here arises from the fact that the energy scale specified in Figure 2 has been determined by considering that the target molecules in this experiment are effectively stationary. Actually, since the target molecules

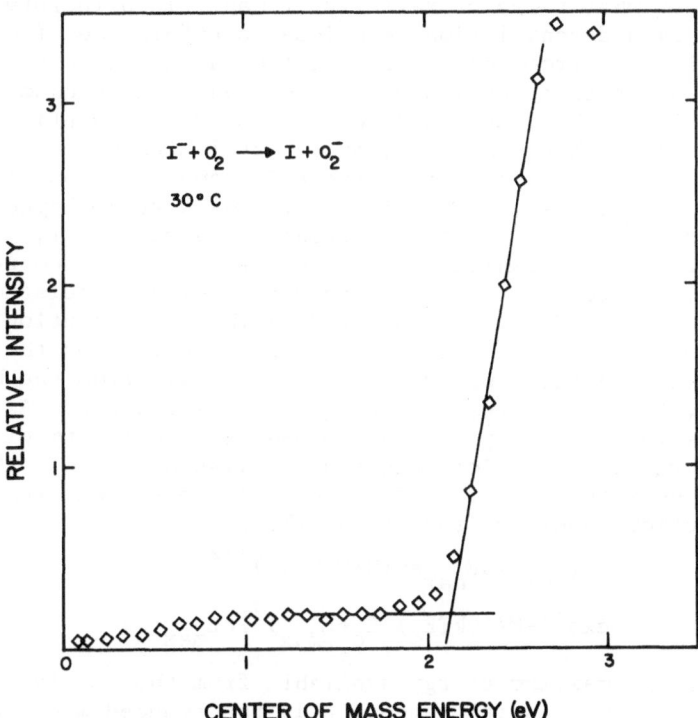

Fig. 2. Relative cross section as a function of nominal center-of-mass energy for the reaction $I^- + O_2 \rightarrow I + O_2^-$ at collision chamber temperature of $30^\circ C$, showing details of onset region.

are contained in a collision chamber, they have the normal
distribution of thermal velocities appropriate to the temperature
of the chamber. This results in a broadening of the effective
collision energy - so called Doppler broadening - with a resultant
smearing of the excitation function for the reaction. The sever-
ity of this broadening and the magnitude of the error inherent
in the uncorrected or linearly extrapolated threshold can be
shown to depend on both the width of the effective energy distribution, (which can be calculated assuming a Boltzmann distribution
of neutral particle velocities), and on the shape of the true
threshold function. It is currently not possible to specify "a
priori" the nature of the threshold law for endothermic ion-
neutral collisions. The statistical phase-space theory of reactions predicts that the cross section for such a process should
rise sharply from threshold with no delayed onset and that it
should exhibit an exponential dependence on the excess translational energy available for the reaction (Pechukas and Light,
1965; Truhlar, 1969). Little experimental data has been obtained,
however, to test the predictions of the statistical theory. In
addition, it is not clear, even if the threshold law can be
predicted, just what range of validity it has. In principle,
one could apply a deconvolution technique to unfold the effective
energy distribution from the experimental data and thereby arrive
at the residual or true threshold function (which is presumably
that applicable to the reaction at $0^{\circ}K$). Successful application
of deconvolution methods usually requires preliminary smoothing
of the data however, and even then there is some question as to
the uniqueness of the solution. Therefore, we have employed a
somewhat simpler method, already described in some detail,
(Tiernan, Hughes and Lifshitz, 1971), to effect Doppler corrections
where these are necessary. This procedure entails calculation of
the effective energy distribution, folding this distribution
into an assumed or trial threshold function, and fitting the curve
thus synthesized to the experimental data points. From the best
fit achieved, the true threshold energy is thus deduced. As
described previously, an approximate one-dimensional form of the
effective energy distribution function, derived by Bethe, (1937),
can be employed with sufficient accuracy for the present results.
This distribution function is given by the expression,

$$\omega(E_{max}) dE_{max} = \tfrac{1}{2}(M/\pi\mu kTE_0)^{1/2}$$
$$\exp[(-M/4\mu kTE_0)(E_0-E_{max})^2] dE_{max} \tag{11}$$

where E_{max} is the maximum energy available from the collision,
M is the mass of the neutral target, μ is the reduced mass, and
E_0 is the center-of-mass energy for stationary neutral targets,
that is,

$$E_0 = (M/m + M) E_{lab}. \tag{12}$$

The convolution integral which gives the calculated cross section
may then be expressed as

$$f(E_o)dE_o = \int_{-\infty}^{+\infty} f(E_{max})\omega(E_{max})dE_{max}, \qquad (13)$$

where $f(E_{max})$ is the assumed threshold function.

Typical results of such computations are shown in Figure 3,
which illustrates the calculated relative energy distribution for
the I^-/O_2 reaction at 13.0eV(lab) and a temperature of 30^oC, and
the calculated excitation functions obtained by convoluting this
function with three types of trial threshold functions, a step
function, a function increasing linearly with energy above the
threshold, and a delta function. The latter function was included
because all the observed experimental cross-section dependences
are of quasi-Gaussian shape, and it is evident that only with such
a threshold function can the higher energy behavior be approximated
by the calculations. The possible significance of the decay in

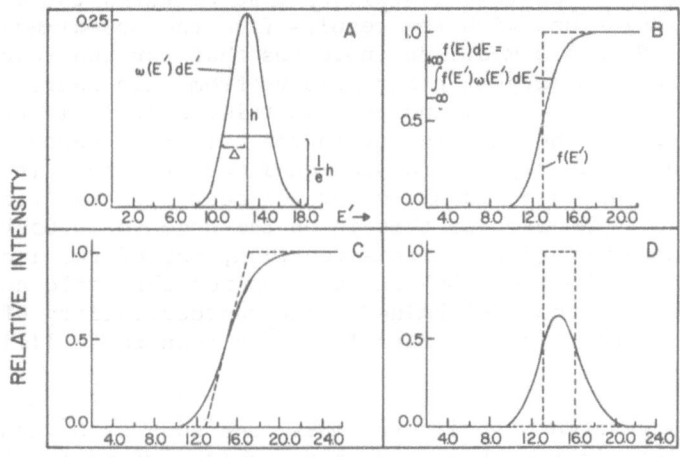

I^- LABORATORY ION ENERGY (eV)

Fig. 3. (a) Calculated relative energy distribution, $\omega(E)'$, for
the reaction pair I^-/O_2 at 13.0eV I^- laboratory energy; (b), (c),
and (d) assumed threshold laws, $f(E')$, are represented by dashed
lines; computed curves, obtained by convoluting $f(E')$ with $\omega(E')$
of Part (a), are represented by solid curves. (b) $f(E')$ is a step
function; $f(E') = 0.0$ below the threshold energy, E_T, and $f(E') =$
1.0 for $E' > E_T$. (c) $f(E')$ is a linear function: $f(E') = 0.0$ for
$E' < E_T$; $f(E') = 1.0$ for $E' > (E_T+4.0eV)$, and $f(E') - 1/4(E'-E_T)$ for
$E_T < E' < (E_T + 4.0eV)$. (d) $f(E')$ is a delta function; $f(E')-0.0$ for
$E' > E_T$ and for $E' > (E_T + 3eV)$ and $f(E') = 1.0$ for $E_T < E' < (E_T +$
3eV).

the experimental cross sections at higher energies will not be
discussed here.

An actual fit of excitation functions calculated in this manner to experimental data points is illustrated in Figure 4 for the
reaction of Br^- with O_2. The best fit is achieved for the calculated curve which was obtained by assuming a threshold energy
which would correspond to an electron affinity of 0.40eV for O_2.
This is in good agreement with the recent laser photodetachment
result of 0.43 $\pm$ 0.03eV (Celotta et al., 1971).

For those reactions which involve the interaction of heavy
ions such as I^- and Br^- with relatively light target molecules,
Doppler broadening is the principle source of error in the
measured energy thresholds. This error is substantially reduced
by treating the data as described above. However, there is still
a small residual error even in these cases, as a result of our
application of an approximate form of the energy distribution
function. An exact treatment of the thermal effect has recently
been presented by Chantry, (1971) who also compared the more
accurate calculations with the results from the one-dimensional
derivation. Such a comparison indicates that for the reactions
reported here, the largest error arising from this source would
be for the I^-/NO reaction, and even in this case, it is very
small ($\sim$0.06eV). The uncertainty in the effective reaction
energy which is due to the translational energy spread in the
reactant ion beam is obviously minor for heavy ion reactions;
thus, for the I^-/NO process this is on the order of 0.05eV. The
cumulative uncertainties for this reaction, therefore, indicate
error limits of about $\pm$ 0.1eV on the measured threshold and on the
electron affinity thus determined. The reproducibility of the
measured thresholds is considerably better than these limits
indicate.

For the reactions of lighter mass ions such as O^- and S^-,
the uncertainty in the threshold energy which originates from the
translational energy distribution of the incident ion beam becomes more important than that due to Doppler smearing of the
threshold. This is the case, of course, because the width of the
effective energy distribution which results from the thermal
motion of the target molecules is much smaller for reactions in
which the ion and neutral are of more nearly equal mass. Also,
the threshold functions which best describe many of the lighter
ion reactions are linear functions, and as described in our
earlier paper, under these circumstances, the uncorrected linearly extrapolated threshold energy is essentially the true threshold
energy. For these reactions therefore, the error limits on the
thresholds (and on the electron affinities), are specified on
the basis of the uncertainty in the effective energy arising from

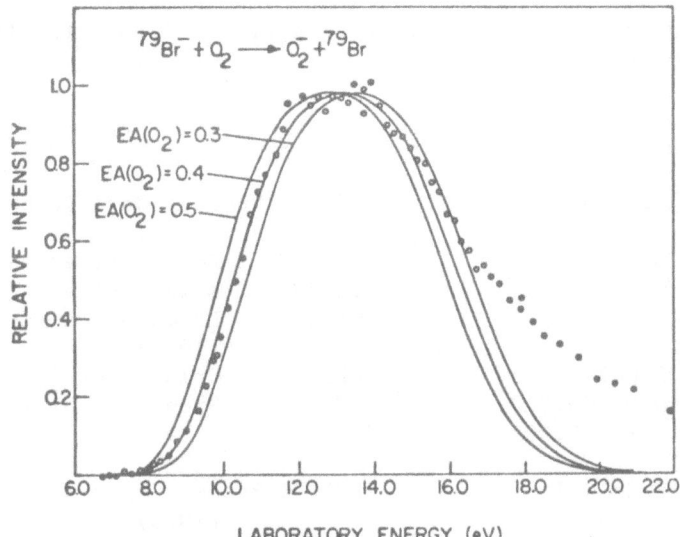

Fig. 4. Calculated cross section as a function of laboratory energy (eV) for the reaction $Br^- + O_2 \rightarrow Br + O_2^-$ at 300°K, for three different assumed values of the threshold energy, (E_T), corresponding to E.A. $(O_2) = 0.3$, 0.4, and 0.5eV (from right to left), respectively. A delta function threshold law was used in the calculations. Points are experimental data.

the reactant ion translational energy spread. We have not attempted to make corrections for the latter factor.

A summary of electron affinity data for various small molecules which has been derived using the negative-ion charge-transfer methods outlined here, (Hughes, Lifshitz and Tiernan, 1973), is presented in Table I. Recent electron affinity values reported in the literature, as determined by various techniques are also given for comparison.

The mechanism by which the negative-ion charge-transfer reactions under consideration proceed has a direct bearing on the question of whether or not the threshold measured for these reactions correspond to their actual endothermicities. It is possible for example, that a significant rotational barrier exists in the outgoing charge transfer channel and that the measured thresholds are therefore larger than the minimum reaction energy. As discussed at the panel on Charge-Transfer Processes (and reported elsewhere in this volume), we have observed isotopic scrambling in the products of the Br^-/Br_2 and Cl^-/Cl_2 reactions which suggests that these reactions proceed via an X_3^- intermediate (X=Cl,Br) in which the two X-X bonds are equivalent. This means that at

TABLE I.   Comparison of ARL Electron Affinity Data with Other Reported Results.

Molecule	Electron affinity(eV)	Method[a]	Reference
NO	0.015 ± 0.1	ECT,BG	Present data–Hughes et al (1973)
	0.09	ECT,BG	Berkowitz et al, (1971)
	0.1 ± 0.1	CI,BG	Nalley et al, (1973)
	$\sim$0	CI,BG	Lacmann & Herschbach (1970)
	0.047 ± 0.005	PES	Celotta et al (1970)
	$0.024+0.010$ -0.005	PES	Siegel et al (1972)
NO_2	2.28 ± 0.1	ECT,BG	Present data–Hughes et al (1973)
	2.30 ± 0.15	ECT,BG	Lifshitz et al (1970)
	2.04	ECT,BG	Berkowitz et al (1971)
	2.70 ± 0.2	CI,BG	Leffert et al (1973)
	2.50 ± 0.1	CI,BG	Nalley et al (1973)
	2.45	CI,BB	Lacmann & Herschbach (1970)
	2.50 ± 0.1	CI,BB	Baede (1972)
	2.54 ± 0.05	CI,BB	Warneck (1969)
	3.9	M	Page and Goode (1969)
	<3.9	P	Milligan et al (1970)
	3.1	P	Lacmann & Herschbach (1971)
	2.38 ± 0.06	TBCT	Dunkin et al (1972)
SO_2	0.99 ± 0	ECT,BG	Present data–Hughes et al (1973)
	1.0 ± 0.05	P	Feldmann (1970)
	1.0 <E.A. <1.12	TBCT	Kraus (1961)
CS_2	0.50 ± 0.2	ECT,BG	Present data–Hughes et al (1973)
	1.0 <E.A. <1.12	TBCT	Kraus et al (1961)
Cl_2	2.32 ± 0.1	ECT,BG	Present data–Hughes et al (1973)
	2.38 ± 0.1	ECT,BG	Chupka et al (1971)
	2.5 ± 0.5	CI,BG	Lacmann & Herschbach (1970)
	2.3; 2.45; 2.5	CT,BG	Baede (1972)
	2.52 ± 0.17	DEC	DeCorpo and Franklin (1971)
	2.35 ± 0.15	TBCT	Dunkin et al (1972)

TABLE I.   (Continued)

Molecule	Electron affinity(eV)	Method[a]	Reference
Br_2	2.62 ± 0.2	ECT,BG	Present data–Hughes et al (1973)
	2.51 ± 0.1	ECT,BG	Chupka et al (1971)
	2.6; 2.55	CI,BG	Baede (1972)
	2.3 ± 0.1	CI,BB	Helbing & Rothe (1969)
	2.87 ± 0.14	DEC	DeCorpo & Franklin (1971)
I_2	2.42 ± 0.2	ECT,BG	Present data–Hughes et al (1973)
	2.58 ± 0.1	ECT,BG	Chupka et al (1971)
	2.5; 2.55; 2.3	CI,BG	Baede (1972)
	2.54	CI,BB	Lacmann & Herschbach (1971)
C_2H	$2.2.\pm0.4$	ECT,BG	Present data–Hughes et al (1973)
	3.73 ± 0.05	P	Feldmann (1970)
	2.2	TBCT	Bohme et al (1972)
	2.7	M	Page and Goode (1969)

[a]ECT–Endothermic negative-ion charge transfer; CI–Chemioniza-tion; PES–Photodetachment electron spectroscopy; P–Photodetach-ment; DEC–Dissociative electron capture; M–Magnetron, TBCT–Thermal bracketing negative ion charge transfer or ion–molecule reaction; CT–Charge transfer; BG–Beam–gas; BB–Beam–beam.

threshold, the only collisions contributing to reaction are those having a very low impact parameter, or even head-on collisions (b=0). It follows therefore, that the orbital angular momentum for the reactants, $L=\mu\nu b$, is rather low. Conservation of angular momentum in the reaction demands that

$$L + I = L' + I',\tag{14}$$

where I is the rotational angular momentum and the primed quantities are for the products. Thus, since both L and I are small in this case, L' and I' must also be small and so a very low rotational barrier is expected for product formation. Only at higher translational energies, where long-range charge transfer becomes more prominent, do large impact parameter collisions contribute to reaction. In this instance, however, the available energy is already greater than the sum of the endothermicity and the rotational barrier. It seems unlikely that very long-range (high impact parameter) charge transfer collisions contribute at all to the reactions in question. None of the molecules discussed capture electrons directly, since the resulting negative ions which would be formed are unstable with respect to autodetachment. Thus, during charge transfer, the collision must be close enough to permit stabilization of the molecular negative ion by vibrational energy transfer. That such vibrational energy transfer does occur for collision at energies near threshold is also indicated by the halogen isotope scrambling results, since for these energies the two X-X bonds in the X_3^- intermediate are apparently equivalent (that is, the energy is equally distributed between them). Good "mixing" of the internal energy therefore occurs in these cases. A similar situation is expected for the other negative-ion charge-transfer reactions studied here, and indeed, in separate experiments, the corresponding intermediates have been observed to be stable (NO_2X^- and SO_2X^-, where X=Cl,F). From these considerations, we believe that the translational energy thresholds measured in the present experiments correspond to the true endothermicities of the reactions. Detailed scattering data now being obtained in our laboratory for these processes should confirm this.

In addition to the negative-ion reactions involving small molecules which were described above, we have also investigated reactions of several atomic negative ions with larger molecules, including SF_6 and a series of perfluorocarbons (Lifshitz; Tiernan and Hughes, 1973), which are known to have positive electron affinities and which form long-lived negative ions (Compton et al, 1966; Naff et al, 1968). It was observed that the thresholds measured for charge transfer from various atomic negative ions to a given polyatomic molecule did not always yield consistent values of the electron affinity. These differences can presumably be attributed to the existence of several stable molecular negative ion states for these molecules, which is not at all surprising. Little is known about the structures or energy states for such gaseous negative ions at this point, however, and so it is not possible to definitively interpret these results. Reasonably consistent data were obtained from several reactions with SF_6. The major processes observed were,

$$X^- + SF_6 \rightarrow SF_6^- + X \tag{15}$$

$$X^- + SF_6 \rightarrow SF_5^- + F + X \tag{16}$$

At still higher collision energies, an F^- fragment was detected. The threshold behavior typically exhibited for these reactions is similar to that shown earlier for other such reactions (Figure 4).

Applying straightforward thermodynamic considerations, it can readily be shown that the electron affinities of SF_6 and SF_5 can be calculated from the thresholds for such reactions, along with the known atomic electron affinities, and the bond dissociation energy of SF_6, by using the relations,

$$E.A.(SF_6) = E.A.(X) - \Delta H_{R15} \tag{17}$$

$$E.A.(SF_5) = E.A.(X) + D_o(SF_5-F) - \Delta H_{R16} \tag{18}$$

Again, such calculations presume that the measured threshold energies represent the true endothermicities of reactions (15) and (16), and that the products are formed without excess kinetic or internal energy. Similarly, the bond dissociation energy of the SF_6^- ion can be calculated from the relation,

$$D_o(SF_5^- - F) = \Delta H_{R16} + E.A.(SF_6) - E.A.(X) \tag{19}$$

And finally, the endothermicity of the dissociative electron attachment reaction with SF_6,

$$e + SF_6 \rightarrow SF_5^- + F \tag{20}$$

can be obtained from the expression,

$$\Delta H_{R20} = \Delta H_{R16} - E.A.(X) \tag{21}$$

Data obtained for SF_6 by utilizing these relations is summarized in Table II for several different reactions investigated. The SF_6 dissociation energy used here, (3.3eV) was taken from Cottrell (1958). It is seen that a quite consistent set of values is obtained from the several reactions, with the exception of those noted in the table and as discussed in more detail in our earlier paper (Lifshitz, Tiernan and Hughes, 1973), the discrepancies for the I⁻ reaction can be explained by the formation of other products which complicate the threshold behavior. For the Br^-/SF_6 reaction, the threshold for SF_6^- production appears to indicate a higher-lying SF_6^- state.

TABLE II. Thermodynamic Data Derived From Energy Thresholds
Measured for Electron Transfer and Dissociative Electron
Transfer Reactions with SF_6.

Reactant ion	E.A.(SF_6)(eV)	E.A.(SF_5)(eV)	$D_0(SF_5^- - F)$(eV)	ΔH[Reaction (21)](eV)
S⁻	0.6	2.8	1.1	0.5
O⁻	0.67	2.9	1.0	0.4
F⁻	---	2.9	1.1	0.5
Br⁻	0.1[a]	2.7	1.2	0.6
I⁻	---	3.1[b]	0.8[b]	0.2[b]

[a]. May indicate formation of an excited SF_6^- state.

[b]. Data is questionable owing to the formation of other products which interfere with threshold measurements for the charge-transfer products.

The thermodynamic data obtained for SF_6 in these experiments can be conveniently displayed in a qualitative quasi-diatomic potential diagram, as shown in Figure 5, which shows the overall consistency of the results. No excited states of SF_6^-, such as suggested by Fehsenfeld (1970), have been included on this diagram, although such states may certainly exist. Indeed, if the excited state of SF_6^- which is apparently formed in the Br^-/SF_6 reaction is a stable electronic state, then this state has a minimum about 0.5eV above that of the SF_6^- ground state, which is presumably that formed in most of these reactions.

An apparent exception to the threshold behavior described above for endothermic negative-ion charge-transfer reactions which has been reported in the literature is the reaction,

$$O^- + H_2O \rightarrow OH^- + OH \tag{22}$$

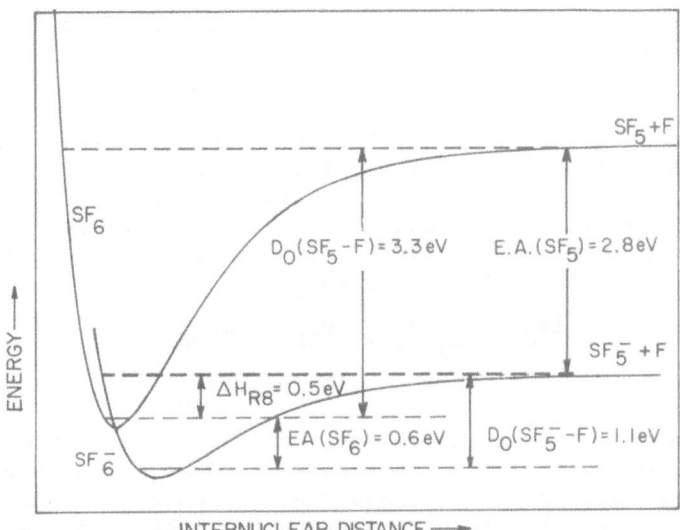

Fig. 5. Qualitative representation of the potential energy curves for SF_6 and SF_6^- , showing thermodynamic data derived in the present investigation.

Vogt (1969, 1971) observed reaction (22) as a function of energy, and found that the cross section increases monotonically with decreasing energy. Although the lowest energy attained in his experiments was indicated to be about 2eV, Vogt concluded from the data that reaction (22) is exothermic. As Paulson (1972) has noted however, on the basis of currently available thermodynamic data, reaction (22) should be endothermic by 0.35eV, and from our experience it would therefore be expected to exhibit a threshold. Vogt (1971) also conducted differential scattering studies of reaction (22) and concluded that the reaction proceeds via a direct H-atom stripping mechanism over the entire energy range accessible to his experiments. Isotopic studies by Vogt, appeared to confirm this conclusion, since he found that the cross section for reaction (23a) was more than 100 times larger than that for reaction (23b).

$$^{18}O^- + H_2{}^{16}O \rightarrow {}^{18}OH^- + {}^{16}OH \tag{23a}$$

$$^{18}O^- + H_2{}^{16}O \rightarrow {}^{16}OH^- + {}^{18}OH \tag{23b}$$

Durup (1971) pointed out that this is unexpected, since one would anticipate that, at sufficiently low ion energies, the duration of the collision would be sufficient for the two nearly identical particles to lose their identity, and the cross sections for H^+ stripping and H-atom stripping should then become equal.

In an attempt to clarify these questions, the reactions of
O^- with H_2O were studied in the ARL tandem mass spectrometer,
which has a considerably lower energy capability than Vogt's
apparatus. The excitation function which we observed for reaction
(22) is shown in Figure 6 and does indeed exhibit a threshold
which appears to be very near the calculated endothermicity of the
process. Also of interest is the structure in the excitation
function at about 1.5eV(CM), which suggests a second onset and
could be interpreted as resulting from a change in the reaction
mechanism. Cross sections for the reactions,

$$^{16}O^- + H_2^{18}O \rightarrow {}^{16}OH^- + {}^{18}OH \qquad (24a)$$

$$\rightarrow {}^{18}OH^- + {}^{16}OH \qquad (24b)$$

were also measured in our laboratory, and the ratio of these as a
function of energy is plotted in Figure 7. Also shown in Figure
7, is an O^- product from the O^-/H_2O reaction which can only be
detected in our experiments by observing the isotopically-labelled
species,

$$^{16}O^- + H_2^{18}O \rightarrow {}^{18}O^- + H_2^{16}O \qquad (25)$$

This product was not reported by Vogt. It is clear from the data
shown in Figure 7 that the ratio of H^+ pickup to H-atom pickup is
very near unity at energies up to about 1.5eV, where a sharp
transition occurs and H-atom pickup begins to dominate. It should

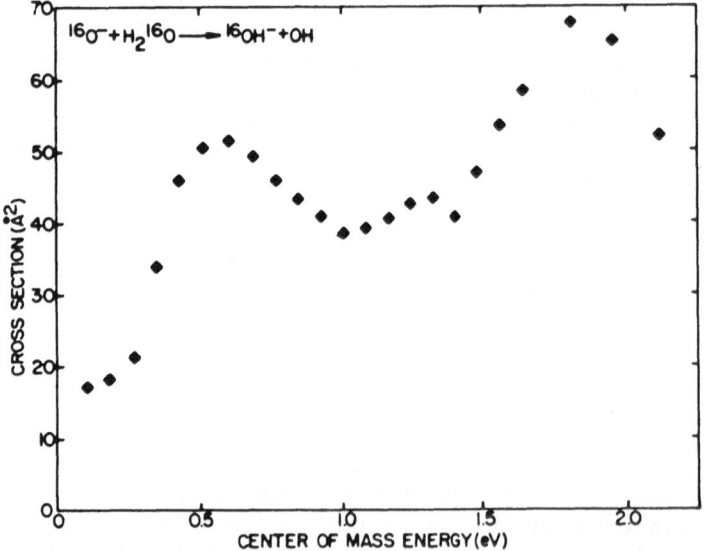

Fig. 6. Translational energy dependence of the reaction $O^- + H_2O$
$\rightarrow OH^- + OH$ as observed in the ARL tandem mass spectrometer.

be noted that this transition energy corresponds precisely to the energy at which the second onset is apparent in Figure 6. It can also be seen that the $^{18}O^-$ product from reaction (25) is observed only in the low energy regime where H-atom and H^+ pickup are equally important. On the whole this data would appear to provide presumptive evidence for the formation of an $[H_2O_2]^-$ intermediate in this reaction, although obviously detailed differential scattering data at low energies is required to confirm this.

Negative Ion Reactions with N_2O and CO_2

Nitrous oxide and carbon dioxide are both linear 16-valence-electron systems. The Walsh rules (Herzberg, 1966) predict however, that the 17 valence-electron molecular negative ions of these should be bent in their ground states. This was confirmed for CO_2^- by electron spin resonance (Bennett, Mile and Thomas, 1965) and infrared studies (Hartman and Hisatsune, 1966) of the negative ion in solids, which established that the bond angle is $134°$ and that the C-O internuclear distance in CO_2^- is about 10% greater than in the neutral molecule. "Ab initio" calculations for CO_2^- by Krauss and Neumann indicate that attachment of a $2\Pi_\mu$

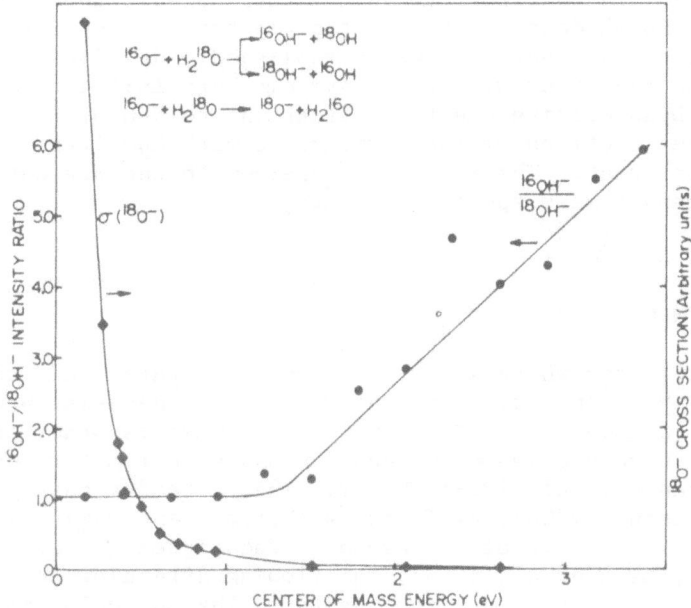

Fig. 7. Ratio of the relative cross sections for H-atom and H^+ stripping in the reaction of $^{16}O^-$ with $H_2^{18}O$ as a function of energy.

electron to CO_2 yields a $^2\Pi_u$ configuration which in turn yields
two states when the molecule is bent symmetrically, a lower energy
$^2A_1$ state and a $^2B_1$ state. An SCF calculation of the $^2A_1$ state
shows that it is bound relative to the neutral $^1A_1$ curve at the
ion equilibrium angle. From these results, it is possible to
deduce that the CO_2^- negative ion would be metastable in the gas
phase. In recent experimental studies, Cooper and Compton (1972)
have indeed observed long-lived metastable negative ions of CO_2 in
the gas phase, as direct products of electron and cesium atom
collisions with organic molecules which contain "bent" CO_2. With
respect to N_2O^-, theoretical considerations by Bardsley (1969)
indicate that the $^2A'$ ground state of N_2O^- should be bound relative to the neutral curve. Experiment also supports the contention
that N_2O^- is a stable negative ion, since it has been observed as
a product from the reaction of NO^- with N_2O by both Chantry (1969)
and Paulson (1970a).

From the discussion just presented, it might be anticipated
that the formation of N_2O^- and CO_2^- by negative-ion charge-transfer
processes such as described earlier would be difficult. At the
very least, a large activation energy barrier would be expected
for such reactions, and this would prevent reliable estimation
of the electron affinities of these species from energy threshold
measurements. In an effort to verify these predictions and to
obtain additional data bearing on the electron affinities of N_2O
and CO_2 (which have not yet been established), detailed studies of
negative ion reactions with these systems were initiated at ARL.

N_2O. Dissociative electron attachment to N_2O produces O^- and
NO^-, and the reactions of these fragments with N_2O itself are of
principal interest. The reactions observed in the present study
using the tandem mass spectrometer are,

$$O^- + N_2O \rightarrow NO^- + NO \tag{26}$$

$$NO^- + N_2O \rightarrow N_2O^- + NO \tag{27}$$

Reaction (26), for which we have determined a rate coefficient,
$k = 1.1 \times 10^{-10}$ cm^3/molecule sec. at 0.3eV(lab), has also been
proposed to account for NO^- production by numerous other investigators, who have applied single-source mass spectrometers (Paulson,
1966; Stockdale et al, 1969; Chantry, 1969), tandem mass spectrometers (Paulson, 1970a), drift tubes (Moruzzi and Dakin, 1968;
Parkes, 1972), ion cyclotron resonance techniques (Schaefer and
Henis, 1965; Marx et al, 1973), and flowing afterglow devices
(Marx et al, 1973) to study this system. The molecular negative
ion, N_2O^-, has only been observed in single-source (Paulson, 1966;
Chantry, 1969) and tandem mass spectrometric experiments (Paulson,
1970a), and reaction (27) was also proposed from these earlier
studies. Production of N_2O^- ions from cesium atom collision with

N_2O has also been reported recently by Nalley et al (1973). The
failure to observe N_2O^- in experiments such as drift tubes, afterglows and ICR spectrometers, where the residence times are relatively long, has generally been explained on the basis of a fast
collisional detachment reaction of NO^-, which was presumed to
compete effectively with reaction (27). Recent measurements by
Parkes (1972), however, indicated that the rate coefficient for
collisional detachment of NO^- on N_2O is on the order of 5×10^{-12}
cm^3/molecule sec. Since the rate coefficient determined for reaction (27) in the present study is $1.9 \times 10^{-11} cm^3$/molecule sec.
(at 0.3eV), the former process should not occur to the exclusion
of reaction (27). The present study gave no indication of a
reaction of O^- with N_2O yielding O_2^-, as was reported by Paulson
(1970a).

    The cross section for reaction (26) was observed to decrease
sharply with increasing translational energy in the present study,
as expected for an exothermic process. This is in general
agreement with the findings of Paulson (1970a), although considerable scatter is exhibited in the latter data. Rate coefficients
calculated from the measured cross sections are plotted as a
function of energy in Figure 8, and data from other investigations
is also shown for comparison. There is good agreement between our
measured rate coefficient and that of Marx et al (1973) in the
neighborhood of 0.5eV, and our results would appear to extrapolate
reasonably to the thermal energy values of Marx et al (1973)
and Parkes (1972). Our data shows a somewhat more rapid decrease
in the rate coefficient at higher energies (above 0.5eV) than does
the ICR data (Marx et al, 1973). The line indicated for Paulson's
data (1970) is a rough average through the scattered points and is
perhaps not a fair representation.

    The excitation function which was observed in the present
experiments for reaction (27) is markedly different from that found
by Paulson (1970a), in that the latter results suggest an energy
threshold, characteristic of an endothermic reaction, while our
data indicates a monotonically decreasing cross section with increasing energy, as usually observed for an exothermic process
(see Figure 9). These results are important because of their implications for the electron affinity of N_2O. Data presented earlier in this lecture indicated that the electron affinity of NO
is about 0.02eV. Our results would therefore suggest that
E.A.$(N_2O) > 0.02$eV and this is further supported by the collisional
dissociation data to be discussed below. Also shown in Figure 9
is the cross section for a reaction of NO^- with N_2O yielding O^- as
a product, which appears at higher energies. Experiments with
isotopically labelled reactants established that this product is
produced by both collision induced dissociation of NO^- and dissociative charge transfer to N_2O.

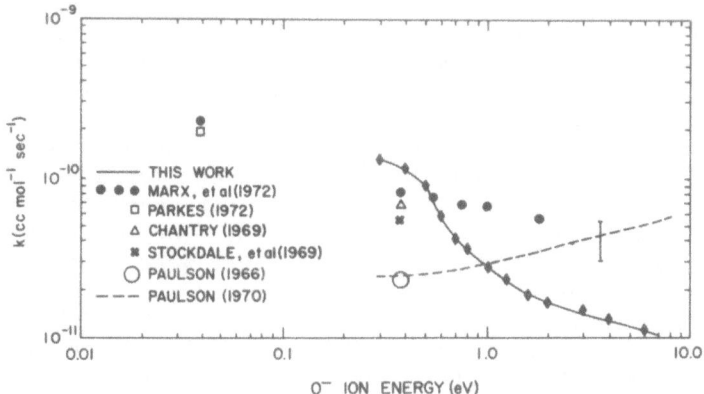

Fig. 8. Rate coefficients as a function of translational energy for the reaction $O^- + N_2O \rightarrow NO^- + NO$. Comparison of ARL data with that from other sources.

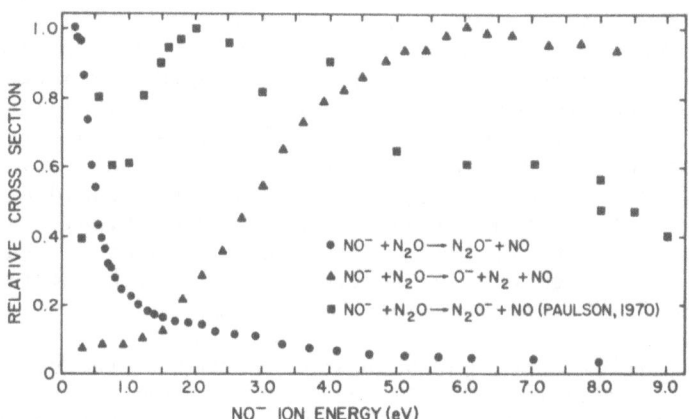

Fig. 9. Relative cross sections as a function of energy for reactions of NO^- with N_2O. Comparison of ARL data with that reported by Paulson (1970).

Other negative-ion reactions with N_2O which were observed by using isotopically-labelled reactants are shown below.

$$^{16}O^- + N_2{}^{18}O \rightarrow N^{16}O^- + N^{18}O \tag{28a}$$

$$\rightarrow N^{18}O^- + N^{16}O \tag{28b}$$

$$^{16}O^- + N_2{}^{18}O \rightarrow {}^{18}O^- + N_2{}^{16}O \tag{29}$$

$$N^{16}O^- + N_2{}^{18}O \rightarrow N_2{}^{18}O^- + N^{16}O \tag{30}$$

The relative cross sections for reactions (28a) and 28b) were found to be equal at energies up to 0.7eV(CM). The cross section for reaction (29) is large at the lowest energy attainable in these experiments, being on the order of 5.4Å^2 at 0.2eV(CM). Taken together, these observations suggest the involvement of a symmetrical intermediate of the form, [O–N=N–O]$^-$, and a product corresponding to this intermediate $N_2O_2^-$, has been observed in drift tube experiments, under conditions where collisional stabilization is possible. Reaction (30) shows that the N_2O^- product from the NO^-/N_2O reaction is not formed by N atom transfer, but only by charge transfer. This is in agreement with the earlier conclusions of Paulson (1970a).

Attempts were made to produce N_2O^- by charge transfer reactions of several other negative ion species with N_2O, including S^-, SH^-, Cl^-, Br^-, I^- and O_2^-. All such reactions yielded no N_2O^- product and gave predominantly an O^- product from dissociative charge transfer. The thresholds measured for these processes were consistently higher than expected from the calculated endothermicities of the reactions, which is likely related to the manner in which the potential curves for the N_2O neutral and the N_2O^- ion intersect. No detailed data on these curves is presently available, however.

Collision-induced dissociation reactions of N_2O^- were also explored in the present study in the hope of obtaining data which could provide an indication of the N_2O electron affinity. These reactions are of the type,

$$N_2O^- + X \rightarrow O^- + N_2 + X \tag{31}$$

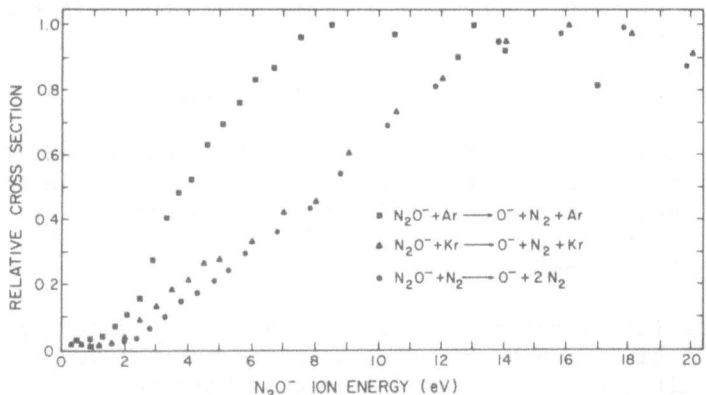

Fig. 10. Excitation functions for collision-induced dissociation of N_2O^- on Ar, Kr and N_2 target molecules.

TABLE III. Thresholds for Collision-Induced Dissociation Reactions
of N_2O^- as Measured with the ARL Tandem Mass Spectrometer

Reaction	Measured Energy Threshold(eV,CM)
$N_2O^- + Ar \rightarrow O^- + N_2 + Ar$	0.8
$N_2O^- + Kr \rightarrow O^- + N_2 + Kr$	1.0
$N_2O^- + N_2 \rightarrow O^- + 2N_2$	0.9

where X is the neutral target, and in the present experiments
thresholds were measured for dissociation with three different
target species, Ar, Kr, and N_2. Typical excitation functions
observed for these processes are shown in Figure 10. Since the
N_2O^- reactant ions used in these experiments should be in the
ground electronic state, the measured dissociation energy should
yield an electron affinity which approximates the adiabatic value.
Table III summarizes the threshold energies measured for the indicated reactions, and as can be seen, these are in excellent agreement. Averaging these values, leads to a dissociation energy of
0.9eV for N_2O^-.

Applying conventional thermodynamic considerations and
assuming that the dissociation products are formed at threshold
without excess energy, the dissociation energy of N_2O^- may be
expressed as,

$$\Delta H_R = 0.9eV = \Delta H_f(O) - E.A.(O) - \Delta H_f(N_2O) + E.A.(N_2O) \quad (32)$$

where ΔH_f denotes the respective heats of formation and E.A. is
the electron affinity. The electron affinity of N_2O is thus given
by the relation,

$$E.A.(N_2O) = 0.9 - \Delta H_f(O) + E.A.(O) + \Delta H_f(N_2O) \quad (33)$$

Inserting the appropriate data for heats of formation and the known
electron affinity of O, the value calculated for the electron
affinity of N_2O from the measured dissociation energy is,

$$E.A.(N_2O) = 0.6eV \quad (34)$$

Nalley et al (1973) recently reported a value of −0.1eV for the
electron affinity of N_2O which was obtained from cesium atom collisional ionization experiments. These authors noted however,
that the N_2O^- ion is probably stable and that the negative electron
affinity may arise from the activation energy for the process. The
latter would presumably correspond to the energy required to bend

N_2O from the linear neutral configuration to the bent ion configuration.

CO_2. Dissociative electron attachment to CO_2 yields O^- as the only negative ion fragment. This ion does not undergo any ion-molecule reaction at low energies, but as the collision energy is increased, two products are observed,

$$O^- + CO_2 \rightarrow CO_2^- + O \tag{35}$$

$$O^- + CO_2 \rightarrow O_2^- + CO \tag{36}$$

The excitation functions for these reactions are shown in Figure 11, which also displays the data reported by Paulson (1970b). The general features of Paulson's curves are in agreement with our results, but much less detail can be discerned. The excitation functions obtained with the ARL tandem mass spectrometer exhibit distinct structure which was shown to be quite reproducible. The fine structure observed in the excitation function for reaction (35) suggests a second threshold at 5.2eV(CM) The plot for reaction (36) also gives evidence of a shoulder in this region, while at much higher energies, a third threshold is apparent for the CO_2^- product from reaction (35). The initial threshold for reaction (35) was determined to be 2.0eV(CM), while that for reaction (36) is at 3.2eV(CM).

Experiments with isotopically-labelled CO_2 provide additional insight into the source of the structure discussed above. The reactions observed are,

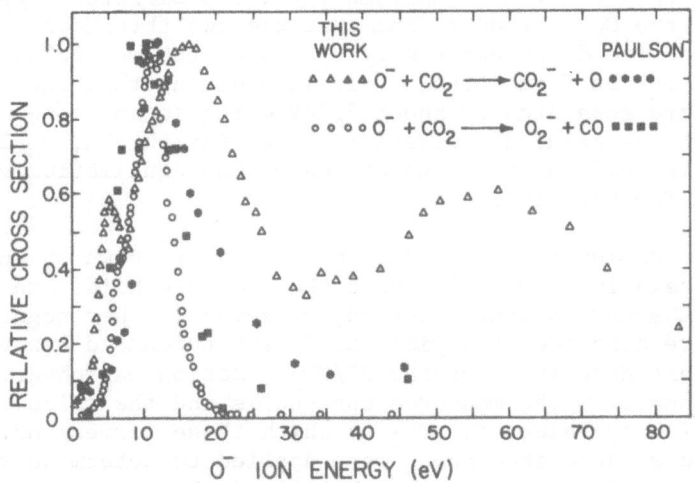

Fig. 11.   Excitation functions for the reactions of O^- with CO_2. Comparison of ARL data with that of Paulson (1970b).

$$^{16}O^- + C^{18}O_2 \rightarrow C^{16}O^{18}O^- + {}^{18}O \qquad (37a)$$

$$\rightarrow C^{18}O_2{}^- + {}^{16}O \qquad (37b)$$

$$\rightarrow {}^{18}O^- + C^{16}O^{18}O \qquad (37c)$$

$$^{16}O^- + C^{18}O_2 \rightarrow {}^{16}O^{18}O^- + C^{18}O \qquad (38)$$

It is apparent from these results, that in contrast to the situation described above for N_2O, the $CO_2{}^-$ charge transfer product is formed by two mechanisms. One mechanism is a pure charge transfer process in which the reactant ions are not incorporated in the product, while the isotopically mixed charge transfer product (reaction 37a), and the O^- product (reaction 37c) apparently result from a $[CO_3]^-$ intermediate which survives long enough for new bonds to be established. The ratio of the cross section for $C^{16}O^{18}O^-$ production to that for $C^{18}O_2{}^-$ production from reactions (37a) and (37b) is plotted as a function energy in Figure 12.

In the energy region from threshold, 2.0eV(CM) to about 5eV(CM), the two charge transfer mechanisms are in competition. At ∿5eV, both charge transfer products become equally probable, and at still higher energies, the ratio of $C^{16}O^{18}O^-/C^{18}O_2{}^-$ declines sharply. The plateau value is due to a small $C^{16}O^{18}O$ impurity in the $C^{18}O_2$. It is particularly noteworthy that the energy at which the ratio maximizes, 5eV, is exactly the energy at which the structure in the excitation functions of Figure 11 is apparent. It appears therefore that this structure is due to a transition in the mechanism of the charge transfer process. It is conceivable that the two mechanisms yield two different configurations of the $CO_2{}^-$ product, and that the two thresholds actually correspond to two different energy states of the molecular anion. Although it may be fortuitous, it is interesting that the two thresholds are separated by about 3.2eV which is in close agreement with the calculated excitation energy for the ${}^2A_1 \rightarrow {}^2B_1$ transition in $CO_2{}^-$ in the neighborhood of the equilibrium geometry (Krauss and Neumann, 1972).

The excitation function for reaction 35 is shown on an expanded energy scale in Figure 13. Here the two thresholds are readily apparent. Charge transfer reactions of several other negative ions with CO_2 were also found to yield $CO_2{}^-$, and threshold behavior analogous to that described for the O^-/CO_2 reaction was observed. Table IV summarizes the measured thresholds and the calculated electron affinity values for CO_2 to which these correspond. The same procedures described above were applied to determine the electron affinities from the thresholds and the known electron affinities of the reactant species. The electron affinities derived from the several experiments show reasonable consistency. The lower value, −0.5eV, is in good agreement with the value

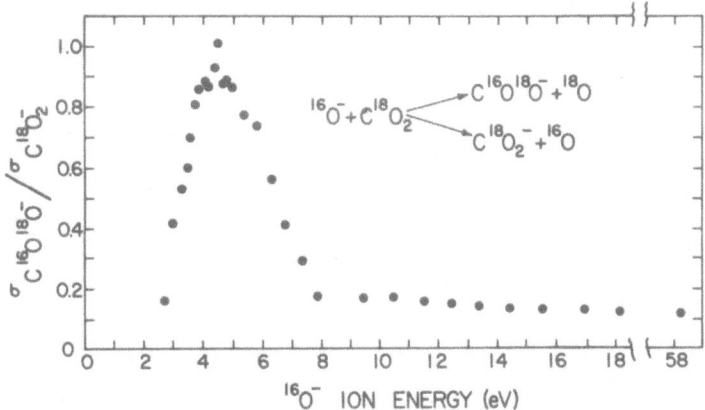

Fig. 12.   Ratio of the cross section for $C^{16}O^{18}O^-$ production to
that for $C^{18}O_2^-$ production from the reaction of $^{16}O^-$
with $C^{18}O_2$ as a function of translational energy.

recently obtained for the CO_2 electron affinity by Compton (1974)
from cesium atom chemiionization experiments.

Associative Detachment Reactions

    Associative detachment reactions of negative ions are of the
form,

$$A^- + B \rightarrow AB + e \tag{39}$$

and cannot usually be studied in beam devices because no charged
product results from the reaction. Fehsenfeld has described at

TABLE IV.   Energy Thresholds for Endothermic Negative-Ion Charge-
Transfer Reactions with CO_2 and Calculated Electron
Affinities for CO_2.

Reaction	Threshold Energy(eV,CM)	Electron Affinity(eV)
$O^- + CO_2 \rightarrow CO_2^- + O$	2.0	−0.5
	5.2	−3.7
$O_2^- + CO_2 \rightarrow CO_2^- + O_2$	1.2	−0.8
	4.3	−3.9
$S^- + CO_2 \rightarrow CO_2^- + S$	3.3	−1.2
	6.0	−3.9
$NO^- + CO_2 \rightarrow CO_2^- + NO$	3.6	−3.6

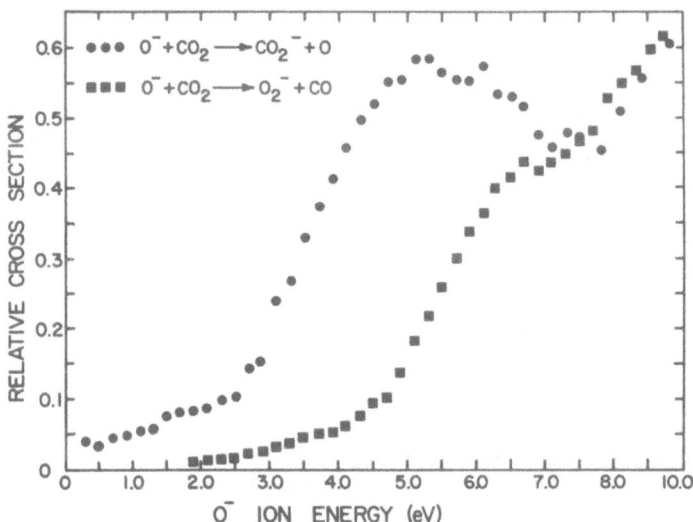

Fig. 13.   Excitation functions for the reactions of O^- with CO_2
           plotted on an expanded energy scale.

some length studies of these reactions using flowing afterglow and
combined afterglow-drift tube techniques, and this discussion
appears in another section of this volume. The results presented
here will be limited to a brief description of a method which we
have developed to investigate associative detachment in a tandem
mass spectrometer. This method is based on the use of SF_6 as a
detector of the electrons released in reaction (39), and the resulting SF_6^- ions are monitored as a measure of the associative
detachment process. Capture of the detached electrons by SF_6 is
surprisingly efficient, which implies that the electrons released
are for the most part low energy electrons. Studies by Mauer
and Schulz (1973) have shown that the energy distributions of the
electrons produced by several of the associative detachment processes to be discussed here peak at quite low energies (typically
0.1-0.2eV for reactions with low energy ions). Also, the widths
of the energy distributions are quite narrow. Since the resonance
electron attachment cross section for SF_6 is very sharp and peaks
quite close to 0eV, this means that substantial fraction of the
electrons produced by associative detachment can be captured.

     The experimental procedure applied in the present studies in-
volves the utilization of mixtures of SF_6 and the target gas
(typically in the ratio, SF_6/target gas = 0.5), in the collision
chamber of the tandem mass spectrometer. The method is somewhat
complicated by the fact that there is a small direct reaction of
the incident ion with SF_6 which also produces SF_6^-, and smaller

intensities of SF_5^- and F^-. For low energy collisions however, this direct reaction constitutes less than 5% of the total product ions monitored. Pressure dependence studies indicate that the only ionic product formed by the capture of the electrons released in the associative detachment reaction is SF_6^-. By maintaining constant conditions for a series of experiments with different target gases and monitoring the SF_6^- product, it is possible to obtain relative cross sections for a number of associative detachment processes. If one such reaction is then selected as a reference, for which an absolute rate is known independently from other experiments, then rate coefficients can be derived for the entire series of reactions in the manner discussed earlier. Data obtained in the present study for various associative detachment processes is shown in Tables V and VI and data from other sources where available, is presented for comparison. Direct correlation of our results with those obtained in flowing afterglow and drift tube experiments is not expected, of course, since our data refer to reactant ion energies of 0.3eV, while the other data applies to thermal energy ions.

The energy dependences of two associative detachment reactions (O^-/H_2 and O^-/CO), were also measured in the present study using the techniques described above. Such energy dependence studies

TABLE V. Rate Coefficients for Associative Detachment Reactions

$$k(X10^{10}cc/molecule\ sec)$$

Reaction	Present Results (at 0.3eV)	Flowing Afterglow	Drift Tube
$O^- + CO \rightarrow CO_2 + e$	$7.0^{a\cdot}$	$4.4^{b\cdot}$	$6.5^{d\cdot};7.3^{e\cdot}$
$O^- + H_2 \rightarrow H_2O + e$	3.6	$6.0^{b\cdot}$	$7.5^{d\cdot};7.0^{e\cdot}$
$O^- + D_2 \rightarrow D_2O + e$	4.3	--	$4.8^{e\cdot}$
$O^- + SO_2 \rightarrow SO_3 + e$	9.3	$7.0^{b\cdot}$	--
$O^- + O_2 \rightarrow O_3 + e$	<0.01	$<0.01^{c\cdot}$	
$O^- + NO \rightarrow NO_2 + e$	1.4	$1.6^{b\cdot}$	$2.2^{d\cdot}$
$S^- + O_2 \rightarrow SO_2 + e$	0.07	$0.3^{b\cdot}$	--

a. Reference reaction
b. Ferguson (1970)
c. Fehsenfeld, Ferguson and Schmeltekopf (1966)
d. Moruzzi, Ekin, Jr., Phelps (1968)
e. Parkes (1973)

TABLE VI.   Rate Coefficients for Associative Detachment Reactions
$k(10^{10}$ cc/molecule sec)

Reaction	Present Results (at 0.3eV)	Flowing Afterglow	Drift Tube
$O^- + CO \rightarrow CO_2 + e$	$7.0^{a.}$	$4.4^{b.}$	$6.5^{c.};7.3^{d.}$
$O^- + CH_4 \rightarrow CH_4O + e$	<0.01	--	not obsd.
$O^- + C_2H_6 \rightarrow C_2H_6O + e$	0.2($\pm$0.04)	not obsd.	--
$O^- + C_2H_4 \rightarrow C_2H_4O + e$	4.7($\pm$0.2)	$7.7^{b.}$	$4.0^{d.}$
$O^- + C_3H_6 \rightarrow C_3H_6O + e$	10.9($\pm$0.7)	not obsd.	--
$O^- + 1\text{-}C_4H_8 \rightarrow 1\text{-}C_4H_8O + e$	18.5($\pm$1.3)	not obsd.	--
$O^- + 2\text{-}C_4H_8 \rightarrow 2\text{-}C_4H_8O + e$	20.4($\pm$1.5)	not obsd.	--
$O^- + C_2H_2 \rightarrow C_2H_2O + e$	17.0($\pm$0.8)	--	$13.0^{d.}$
$O^- - CH_3COCH_3\ C_3H_6O_2 + e$	11.4	--	--

a.   Reference reaction
b.   Ferguson, (1970).
c.   Moruzzi, Ekin, Jr., and Phelps, (1968)
d.   Parkes, (1973)

may be subject to large errors when the SF_6 scavenger technique
is used, because as shown by Mauer and Schulz (1973), the energy
distributions of the detached electrons produced in the initial
collision with the target gas shift to slightly higher energies
as the energy of the incident ion increases. The efficiency of the
SF_6 scavenging reaction is therefore expected to change somewhat
with increasing incident ion energy. Comparisons of our data for
these reactions with energy dependence data obtained by the flowing afterglow-drift tube technique show reasonably good agreement,
however. Such comparisons were presented by Fehsenfeld at an
earlier session of the conference and appear in another section of
this volume.

ACKNOWLEDGMENTS

    The author wishes to acknowledge the substantial contributions
to this research made by his colleagues, Drs. B.M. Hughes, C.
Lifshitz, R.P. Clow and J.C. Haartz. Without their assistance,
helpful criticism, and stimulation, this work would not have been
possible.

REFERENCES

BAEDE, A.P.M. (1972). Physica (Utr.) $\underline{59}$, 541.

BAILEY, T.L. and MAHADEVAN, P., (1970). J. Chem. Phys. $\underline{52}$, 179.

BARDSLEY, J.N. (1969). J. Chem. Phys. $\underline{51}$, 3384.

BENNETT, J.E., MILE, B. and THOMAS, A. (1965). Trans. Faraday Soc $\underline{61}$, 2357.

BERKOWITZ, J. CHUPKA, W.A., and GUTMAN, D. (1971). J. Chem. Phys. $\underline{55}$, 2733.

BETHE, H.A. (1937). Rev. Mod. Phys. $\underline{9}$, 69.

BOHME, D.K. and YOUNG, L.B. (1970). J. Am. Chem. Soc. $\underline{49}$, 3301.

BOHME, D.K., LEE-RUFF, E. and YOUNG, L.B. (1972). J. Am. Chem. Soc. $\underline{94}$, 5153.

CALCOTE, H.F. (1972). Ions in flames. "Ion-Molecule Reactions" (Ed. J.L. Franklin). Plenum Press, New York.

CELOTTA, R., BENNETT, R., HALL, J., SIEGEL, M.W. and LEVINE, J. (1970). Bull. Am. Phys. Soc. $\underline{15}$, 1515.

CELOTTA, R., BENNETT, R., HALL, J., LEVINE, J. and SIEGEL, M.W. (1971). Bull. Am. Phys. Soc. $\underline{16}$, 212.

CHANTRY, P.J. and SCHULZ, G.J. (1967). Phys. Rev. $\underline{156}$, 134.

CHANTRY, P.J. (1969). J. Chem. Phys. $\underline{51}$, 3369.

CHANTRY, P.J. (1971). J. Chem. Phys. $\underline{55}$, 2746.

CHUPKA, W.A., BERKOWITZ, J. and GUTMAN, D. (1971). J. Chem. Phys. $\underline{55}$, 2724 (1971).

COMPTON, R.N. CHRISTOPHOROU, L.G., HURST, G.S., and REINHARDT, P.W. (1966). J. Chem. Phys. $\underline{45}$, 4634.

COMPTON, R.N. (1974). Private communication.

COOPER, C.D. and COMPTON, R.N. (1972). Chem. Phys. Letters $\underline{14}$, 29.

CRAGGS, J.D. (1971). Negative ions. "Tenth International Conference on Phenomena in Ionized Gases. Invited Papers". Donald Parsons & Co., Ltd., Oxford.

DECORPO, J.J. and FRANKLIN, J.L. (1971). J. Chem. Phys $\underline{54}$, 1885.

DUNKIN, D.B., FEHSENFELD, F.C. and FERGUSON, E.E. (1972). Chem. Phys. Lett. $\underline{15}$, 257.

DURUP, J. (1971). Comments on the paper by D. Vogt, p. 225. "Advances in Mass Spectrometry. Vol 5". (Ed. A. Quayle). The Institute of Petroleum, London.

FEHSENFELD, F.C., FERGUSON, E.E. and SCHMELTEKOPF, A.L. (1966). J. Chem. Phys. $\underline{45}$, 1844.

FEHSENFELD, F.C. (1970). J. Chem. Phys. $\underline{53}$, 2000.

FELDMANN, D. (1970). Z. Naturforsch. $\underline{A25}$, 621.

FERGUSON, E.E. (1969). Can. J. Chem. $\underline{47}$, 1815.

FERGUSON, E.E. (1970). Accts. Chem. Res $\underline{3}$, 402.

FERGUSON, E.E. (1972). Flowing Afterglow Studies. "Ion-Molecule Reactions" (Ed. J.L. Franklin). Plenum Press, New York.

FUTRELL, J.H. and TIERNAN, T.O. (1972). Tandem mass spectrometric studies of ion-molecule reactions. "Ion-Molecule Reactions" (Ed. J.L. Franklin). Plenum Press, New York.

HARTMANN, K.O. and Hisatsune, I.C., (1966). J. Chem. Phys. $\underline{44}$, 1913.

HELBING, R.K.B. and ROTHE, E.W. (1969). J. Chem. Phys. 51, 1607.

HOTOP, H. PATTERSON, T.A. and LINEBERGER, W.C., (1974). High
    Resolution Photodetachment Studies of Negative Ions.
    "Advances in Mass Spectrometry. Vol. 6", (Ed. A.R. West).
    Applied Science Publishers, Ltd., Barking, Essex.

HUGHES, B.M., LIFSHITZ, C. and TIERNAN, T.O. (1973). J. Chem. Phys.
    59, 3162.

KEBARLE, P. (1972). Higher order reaction-ion clusters and ion
    solvation. "Ion-Molecule Reactions" (Ed. J.L. Franklin).
    Plenum Press, New York.

KRAUS, K., MULLER-DUSPING, W. and NEURT, H. (1961). Z. Naturforsch
    A16, 1385.

KRAUSS, M. and NEUMANN, D. (1972). Chem. Phys. Letters 14, 26.

LACMANN, K. and HERSCHBACH, D.R. (1970) . Chem. Phys. Lett. 6, 106.

LACMANN, K. and HERSCHBACH, D.R. (1971). Bull. Am. Phys. Soc. 16,
    213.LEFFERT, C.B., JACKSON W.M., and ROTHE, E.W. (L973).
    J. Chem. Phys. 58, 5801.

LIFSHITZ, C., HUGHES, B.M., and TIERNAN, T.O. (1970). Chem.
    Phys. Letters 7, 469.

LIFSHITZ, C., TIERNAN, T.O. and HUGHES, B.M. (1973). J. Chem.
    Phys. 59, 3182.

MARX, R., MAUCLAIRE, G., FEHSENFELD, F.C., DUNKIN, D.B. and
    FERGUSON, E.E., (1973). J. Chem. Phys. 58, 3267.

MAUER, J.L. and SCHULZ, G.J., (1973). Phys. Rev. A47, 593.

MILLIGAN, D.E., JACOX, M.E., and GUILLORY, W.A. (1970). J. Chem.
    Phys. 52, 3864.

MOISEIWITSCH, B. (1965). Adv. At. Mol. Phys. 1, 61.

MORUZZI, J.L. and DAKIN, J.T. (1968). J. Chem. Phys. 49, 5000.

MORUZZI, J.L., EKIN, JR., J.W., and PHELPS, A.V. (1968). J. Chem.
    Phys. 48, 3070.

NAFF, W.T., COOPER, C.D., and COMPTON, R.N. (1968). J. Chem. Phys.
    49, 2784.

NALLEY, S.J., COMPTON, R.N., SCHWEINLER, H.C. and ANDERSON, V.E.
    (1973). J. Chem. Phys 59, 4125.

PAGE, F.M. and GOODE, G.C. (1969). "Negative Ions and the Magnet-
    ron", Wiley, New York.

PARKES, D.A. (1972). J.C.S. Faraday I, 68, 2103.

PARKES, D.A. (1973). J.C.S. Faraday I, 69, 613.

PAULSON, J.F. (1966). Some negative ion reactions in simple gases.
    "Ion-Molecule Reactions in the Gas Phase". (Ed. R.F. Gould).
    American Chemical Society, Washington.

PAULSON, J.F. (1970a). J. Chem. Phys. 52, 959.

PAULSON, J.F. (1970b). J. Chem. Phys. 52, 963.

PAUSLON, J.F. (1972). Negative ion-neutral reactions. "Ion-
    Molecule Reactions" (Ed. J.L. Franklin). Plenum Press, New
    York.

PECHUKAS, P. and LIGHT J.C. (1965). J. Chem. Phys. 42, 3281

PHELPS, A.V. (1969). Can. J. Chem. 47, 1783.

RUTHERFORD, J.A. and TURNER, B.R. (1967). J. Geophys. Res. 72, 3795.

SCHAEFER, J. and HENIS, J.M.S. (1968). J. Chem. Phys. <u>49</u>, 5377.
SIEGEL, M.W., CELOTTA, R.J., HALL, J.L. LEVINE, J. and BENNETT, R.A. (1972). Phys. Rev <u>A6</u>, 607.
SPENCE, D. and SCHULZ, G.J. (1969). Phys. Rev. <u>188</u>, 280.
STOCKDALE, J.A.D., COMPTON, R.N. and REINHARDT, P.W. (1969). Phys. Rev. <u>184</u>, 81.
THYNNE, J.C.J. (1972). Negative ion studies using a time-of-flight mass spectrometer. "Dynamic Mass Spectrometry. Vol. 3" (Ed. D. Price). Heyden & Son Ltd., London.
TIERNAN, T.O. and MARCOTTE, R.E. (1970). J. Chem. Phys. <u>53</u>, 2107.
TIERNAN, T.O., HUGHES, B.M. and LIFSHITZ, C. (1971). J. Chem. Phys. <u>55</u>, 5694.
TRUHLAR, D.G. (1969). J. Chem. Phys. <u>51</u>, 4617.
VOGT, D. (1971). On the reaction mechanism of some negative ion molecule reactions of low energies (some eV–200eV). "Advances in Mass Spectrometry. Vol. 5", (Ed. A. Quayle). The Institute of Petroleum, London.
WARNECK, P. (1969). Chem. Phys. Lett. <u>3</u>, 532.

ASSOCIATIVE DETACHMENT

Fred. C. Fehsenfeld

Aeronomy Laboratory, Environmental Research Laboratories
National Oceanic and Atmospheric Administration
Boulder, Colorado 80302 U.S.A.

INTRODUCTION

Over the past ten years there has been considerable interest
in the associative detachment reaction. In the case of low interaction energies this interest has been fed by the potential importance of this type reaction in controlling deionization in the
earth's upper atmosphere, in gas lasers, and in certain types of
energy conversion devices.

It is the purpose of this article to review the low energy
experimental data which has been obtained and indicate the present
understanding of these results. In the first section the general
features of the associative detachment reaction will be described.
In the ensuing sections the results will be used to further illum-
inate the nature of the reaction. This discussion will begin with
the simplest collisions between atomic negative ions and atoms and
proceed to the more complicated reactions in which the compound
negative ions formed by the collision are polyatomic.

GENERAL DISCUSSION

The associative detachment reaction is mechanistically represented by the equation

$$A^- + B \rightarrow (AB)^* + e \qquad (1)$$

where A^- may be an atomic or molecular ion and B may be an atom or
molecule. The asterisk on the molecular reaction product, AB,
indicates that this molecule may be formed in an excited, and in

some instances, dissociating state. In this reaction the energy
necessary to remove the electron from the negative ion is abstract-
ed from the excess energy resulting from the formation of the chemical bond between A and B and from the energy carried by A⁻ and B
into the collision.

The interaction between A⁻ and B leading to detachment is
illustrated in Fig. 1. This diagram indicates a hypothetical potential energy diagram for a line of centers approach between A⁻
and B (solid line) and the corresponding diagram for a similar
approach between A and B. The internuclear separation is plotted
along the ordinate while the potential energy of systems is plotted
along the abscissa. If the potential energy $V(R)$ between A⁻ and B
and between A and B is spherically symmetric then the curve shown
in Fig. 1 would be unique. If this were not the case then this
curve would represent a slice taken through the potential surfaces
associated with the A⁻-B, A-B interactions.

At large internuclear separations the potential energy for the
interaction of A⁻ with B is displaced below the potential energy
for the interaction of A and B by the electron affinity of A,
EA(A). As A⁻ approaches B the curve splits into a manifold of
curves representing the various states of the negative ion AB⁻ that
this combination of A⁻ and B can produce. For the purposes of
discussion we have indicated three possible curves which this
interaction might produce.

The curve labeled I is associated with a repulsive state of
AB⁻. If A⁻ and B approach along this curve at low energy they will
not reach the crossing at (1) between I and the AB ground state
potential curve. In this case there will be scattering with no
reaction. At large collision energies A⁻ and B will be able to
penetrate more closely along I, finally approaching inside (1)
and crossing the attractive curve of AB. When this happens AB⁻
will be unstable against autodetachment. The molecule formed by
this autodetachment would be left in a high vibrational level or
perhaps even be unstable if the detachment occurred at internuclear
separations which could yield an AB* in the predissociating continuum above the dissociation limit of AB. As the collision
energy between A⁻ and B increases still further additional crossings between other states of AB and I may occur, indicated by (2)
opening other channels for detachment. An autodetachment of AB⁻
into a repulsive state of AB such as indicated in this figure
would yield the dissociated products A, B, and e.

A second curve listed as II corresponds to an excited state
of AB⁻. This state is drawn to be attractive to relatively small
internuclear separations, i.e. within the crossing labeled (3). In
this case, even at low energies, A⁻ and B may enter the autodetaching region.

When the molecule AB forms a stable negative ion in its ground state there will exist a potential curve of type III. If A⁻ and B approach along this curve, the minimum of which lies below that of ground state curve for AB, associative detachment may occur in one of two ways. If the equilibrium internuclear separation of AB⁻ in its ground state is significantly different from that of AB in its ground state then the potential curve for AB⁻ may intersect that of AB. Such a crossing is indicated by (4). In this case, after the crossing the AB⁻ is unstable against autodetachment. If, however, the internuclear spacing of AB⁻ at equilibrium is similar to that for AB and the shape of the potential wells of AB⁻ and AB are similar there may be no crossing of the potential curve of AB⁻ and AB. This case is indicated by the dashed line in curve III. Associative detachment may occur in this case only when the colliding energy coincides with a nuclear-energy state of AB so that a radiationless transition from the continuum nuclear state of AB⁻ to the bound nuclear state of AB can eject an electron without a large alteration of the positions and moments of the nuclei. This latter mechanism is not considered very probable.

The mechanistic description of this type of collision has recently been carried out in a number of articles (Bardsley, 1967; Herzenberg, 1967; Chen, 1967; Dalgarno, 1967; Chen, 1968; Brown, 1969; Mizuno, 1969, 1971). The approach taken by these authors is basically the same. The electronic states of the interacting system A⁻ + B are stable at large internuclear separations and change adiabatically as the internuclear separation decreases. At these separations each state is described by a real potential energy, $V(R)$. However, when a crossing between a AB⁻ potential surface and those of AB occurs, and the electronic state of AB⁻ becomes unstable with respect to autodetachment, the attenuation of A⁻ by detachment can be taken into account by defining an interaction potential energy between A⁻ and B which is complex,

$$W_i(R) = V_i(R) - \tfrac{1}{2}i\,\Gamma(R) \qquad i = 1, 2 \ldots n \qquad (2)$$

where $V_i(R)$ is the electronic interaction potential for AB⁻ which determines the relative motion of A⁻ with respect to B and is indicated, for example, by curves I, II, and III in Fig. 1. $\Gamma_i(R)$ is the resonance width of the compound AB⁻ state formed in the autodetaching region and gives the lifetime, τ_i of $(AB^{-*})_i$ against autodetachment

$$\tau_i = \frac{\hbar}{\Gamma_i} \qquad (3)$$

Each of the potential surfaces indicated in Fig. 1 for the collision of A⁻ and B would have its characteristic interaction potential $W_i(R)$. The relative probability that the collision would take place along a particular potential surface would be equal to the relative statistical weight of that surface to the combined statistical weights of all the possible states formed by the

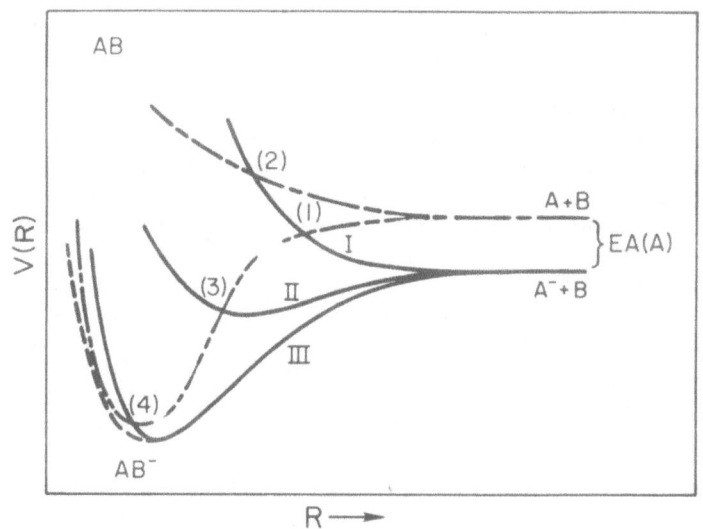

Figure 1. Hypothetical potential curves for the reaction $A^- + B \rightarrow$ $AB^* + e$. The internuclear separation of the reactants is plotted along the abscissa while the potential energy is plotted along the ordinate. The potential curves for A^- and B are shown as the solid lines, I, II and III. The potential curves for the inter- action of A and B are the alternating long and short dashes. The crossings between the negative ion curves and the neutral curves are indicated by the numbers (1), (2), (3) and (4).

combination of the negative ion and the neutral. Calculation of numerical values for the detachment cross section would, in general, require a detailed knowledge of $V_i(R)$ and $\Gamma_i(R)$ for all relevant internuclear separations.

The autodetaching states of the negative molecular ion complex- es with which we are presently concerned are formed by negative ion-neutral collisions. These states theoretically are also formed by collisions between electrons and the neutral molecule AB. Since the same compound state may be formed by the two processes recip- rocity relations exist which connect the cross sections for the two processes (O'Malley, 1966). A considerable body of experimental results exist for electron scattering by a wide variety of molecules. One is tempted to turn to this information as a source for both $V_i(R)$ and $\Gamma_i(R)$ listed in Eq. (2). In practice, however, it is not possible with any degree of certainty to predict the lifetime of an autodetaching complex formed by an ion-molecule collision from the lifetime associated with a compound state produced by electron impact on the ground state of the associated neutral. The electron scattering experiments do yield valuable information concerning these intermediate compound states. These experiments are able to locate the resonance in terms of energy with respect to the ground state of the neutral (and in some instances with respect to electronically or vibrationally excited states). Moreover, the experimental results classify the resonances according to process (i.e. shape resonance I and II, Feshbach resonance, etc.) and thereby indicate an order of magnitude range of lifetime one might expect for this compound state formed by ion-molecule collision. In this context it is interesting to note that a very wide resonance, 1 eV, would correspond from Eq. (3) to a lifetime of 6.6×10^{-16} sec ($\sim 10^{-15}$ sec) while a very narrow resonance, in electron scattering, 1 millielectron volt, would correspond to a lifetime of approximately 10^{-12} sec. The former lifetime is considerably shorter than a vibrational period of most simple molecules while the latter is considerably longer.

In certain instances a detailed knowledge of $\Gamma_i(R)$, or for that matter $V_i(R)$, in the autodetaching region is not required. If Γ_i is so large for all states that each time A^- and B enter the autodetaching region electron emission will occur, then the detachment cross section becomes independent of Γ_i and is determined solely by the maximum impact parameter for which A^- and B may reach the autodetaching region.

Under these circumstances at low energies in a semi-classical approach the detachment cross section will be determined by the fraction of collisions that find attractive surfaces into the autodetaching region multiplied by the total orbiting collision cross section σ_c, which is, in turn, determined by the long range neutral forces, (charge-dipole, charge-induced dipole, etc.) (Su, 1973). At higher energies (usually less than 1 eV) the geometric cross section defined for a particular state by the internuclear separation, R_i, within which AB^- is unstable against autodetachment becomes larger than the orbiting cross section and the detachment cross section will approach the statistically weighted geometric cross section of those states for which the autodetaching region is energetically accessible, i.e.

$$\sigma_d = \sum_i g_i \pi R_i^2 \qquad (4)$$

where g_i is the statistical weight in accessible states.

REACTIONS INVOLVING TWO ATOMS

    With these general remarks concerning the systematics of the reaction, we now move on to actual systems and specific results. The simplest of all associative detachment reactions is the one between the halogen negative ions and atomic hydrogen. Three such reactions involving F⁻, Cl⁻, and I⁻ have been studied and the results are listed in Table I. The reaction of Cl⁻ with H will serve to illustrate these reactions. In the case of the reaction between the halogen negative ions and atomic hydrogen (E. E. Ferguson, 1969; E. E. Ferguson, 1968; F. C. Fehsenfeld, 1973; C. J. Howard, 1974) only one molecular state, the $^2\Sigma$, can be formed and charge transfer from the halogen to H is highly endothermic. In Fig. 2 a hypothetical potential curve has been drawn for the Cl⁻-H interaction. The bond dissociation energy for HCl is 0.7 eV larger than the electron affinity of Cl so the associative detachment reaction is nominally exothermic by this amount. As Cl⁻ approaches H at large internuclear separations the charge-induced dipole interaction will produce a potential surface which is initially attractive. This curve would be expected to become repulsive at smaller internuclear separations as the s-state electron distributions of

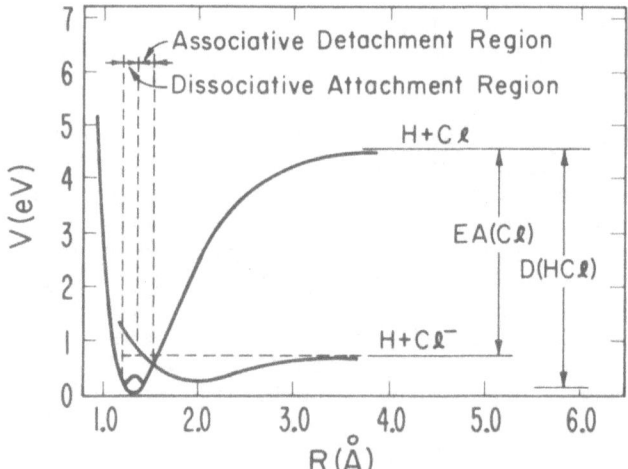

Figure 2. Hypothetical potential energy diagram for the interaction of Cl⁻ with H. The potential energy is plotted along the ordinate while the internuclear separation is plotted along the abscissa.

TABLE I

Rate constants for reactions between halogen negative ions and
atomic hydrogen measured at 300°K.

Reaction	Rate (cm^3/sec)
(1) $F^- + H \rightarrow HF + e$	1.6(-9)
(2) $Cl^- + H \rightarrow HCl + e$	9.6(-10)
(3) $I^- + H \rightarrow HI + e$	< 6 (-11)

Cl^- and H begin to overlap. Experimental evidence of this repulsive character can be inferred from observations of the dissociative attachment of electrons to HCl forming Cl^- which occurs at low
electron energies near the energetic threshold for Cl^- production.
This attachment must involve ground state Cl^- and H (D. C. Frost
and C. A. McDowell, 1958; L. G. Christophorou, 1968, 1971). As it
is represented here, HCl^- may be a stable weakly bound negative
ion with an equilibrium internuclear separation considerably larger
than HCl.

Of significance to the present discussion, the studies of low
energy electrons impacting on HCl near the threshold for the production of Cl^- indicate that the lifetime of the HCl^- compound
state so formed against autodetachment is 3.7×10^{-15} sec while the
lifetime of the state against dissociation into Cl^- and H is $3.4 \times
10^{-15}$ sec. (Christophorou, 1968). These lifetimes are quite short,
much shorter than the vibrational period of HCl which is $1.1 \times
10^{-14}$ sec (Herzberg, 1950).

At thermal energies the associative detachment reaction is
fast (Howard, 1974). This rate constant is approximately 50% of
the orbiting collision rate constant for Cl^- with H which is $1.96 \times
10^{-9}$ cm^3/sec, assuming the polarizability of H to be 0.666 $Å^3$.
That the reaction occurs for a large fraction of the total collisions
at thermal energies requires that the $HCl^-(^2\Sigma)$ potential curve be
attractive into the autodetaching region.

More recently the reaction has been studied as a function of
ion kinetic energy (McFarland, 1974). Preliminary results of this

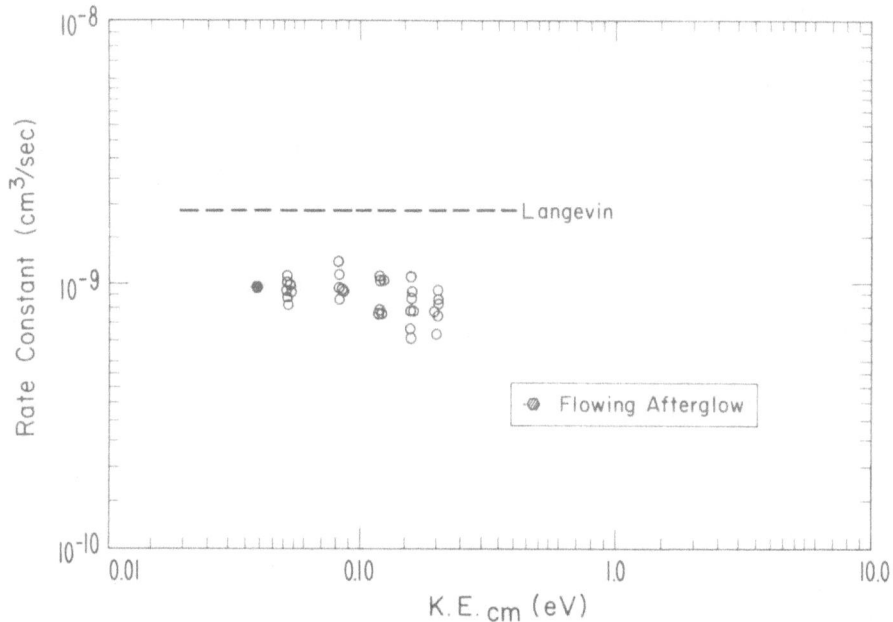

Figure 3. The measured reaction rate constant for the reaction of
  Cl⁻ with atomic hydrogen. The rate constant is plotted along
  the ordinate while the ion kinetic energy in the center of mass
  is plotted along the abscissa. The orbiting (Langevin) rate
  constant is shown as the dashed line

study are shown in Fig. 3 and indicate that the reaction rate constant declines slowly with increasing ion kinetic energy. These
results are compared with the energy independent orbiting rate
constant shown as the dashed line in Fig. 3. An obvious explanation for the measured decline in the detachment rate is that the
lifetime of the HCl against autodetachment is comparable to the
lifetime of the ion complex in the autodetaching region at lowest
energies and that the complex lifetime decreases with increasing
ion kinetic energy. This would in turn indicate that either (1)
the effective autodetaching lifetime of the compound state formed
by collision of H with Cl⁻ is somewhat larger than the autodetaching lifetime of the compound state produced by the collision of
electrons with HCl and/or (2) alternatively, that the lifetime of
the H-Cl⁻ reaction complex in the autodetaching region is less
than the lifetime of the compound HCl⁻ state produced by HCl-elec-
tron collision against dissociation.

    In a like fashion for reaction (1), the large rate constant

clearly implies that the $^2\Sigma$ ground state of HF$^-$ is attractive from large internuclear separation into the autodetaching region, i.e., until it crosses the HF ground state curve. This agrees with two theoretically calculated HF$^-$ curves, (A. W. Weiss, 1970; W. M. Hartmann, 1970) and disagrees with two others (Michels, 1968; V. Bondybey, 1972). Hence, the present result makes a clear judgement between those discordant calculations.

Finally, there was no observable reaction at thermal energies between I$^-$ and H, $k_3 < 0.6$ x 10^{-10} cm^3/sec. There is little if any difference in $D_0(HI)$ and EA(I), consequently, the reaction is almost thermoneutral. The present results would suggest a significant potential barrier (> 0.1 eV) may exist which prevents I$^-$ and H from forming an autodetaching complex.

The reaction between H$^-$ and H involving, as it does, only the three electron H$^-$-H system and the two electron H-H system, is amenable to a fairly exact theoretical treatment which has been described in a number of articles (Bardsley, 1967; Herzenberg, 1967; Chen, 1967; J. Mizuno, 1969,1970). The H$^-$-H interaction is somewhat more complicated than the interaction of halogen negative ions with atomic hydrogen. Since homonuclear atoms are involved two states may be formed by the combination of H$^-$ and H in their ground states. In addition, another reaction channel, resonant charge exchange, is open at all energies.

The potential curves for the H$^-$-H interaction are shown in Fig. 4. These curves were semi-empirically determined by Chen (1968). A similar set of curves were calculated somewhat earlier by Bardsley (1966), applying a variational technique. In this figure two plots are given, one for the real part of the interaction potential, $V(R)$, and the other for the corresponding resonance width $\Gamma(R)$ which yields the imaginary portion of the interaction potential.

The collision between H$^-$ and H gives rise to two potential surfaces, a $^2\Sigma_u^+$ surface which is attractive into the autodetaching region, and a $^2\Sigma_g^+$ surface which becomes repulsive before the autodetaching region is reached. In both states inside the autodetaching region Γ is very large, (of the order of volts). This means that the lifetime of the H$_2^-$ formed in either configuration is very short, much shorter in fact than a vibrational period of H$_2$.

Under these circumstances there is agreement that in the low energy region (KE$_{cm} \leq 10$ eV), all the H$^-$-H collisions which pass into the autodetaching region will result in an associative detachment. The calculations of the associative detachment cross section for the H$^-$-H reaction then are reduced to that determination. In

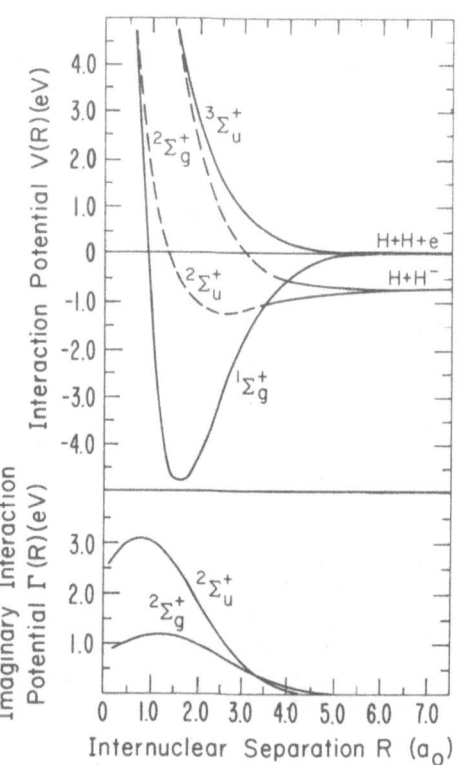

Figure 4. Plots of V(R) and Γ(R) vs R for the interaction of H⁻
with H. These curves are taken from the paper of Chen
(1968). The V(R) potential curves for the $^3\Sigma_u^+$ and $^1\Sigma_g^+$
are also shown.

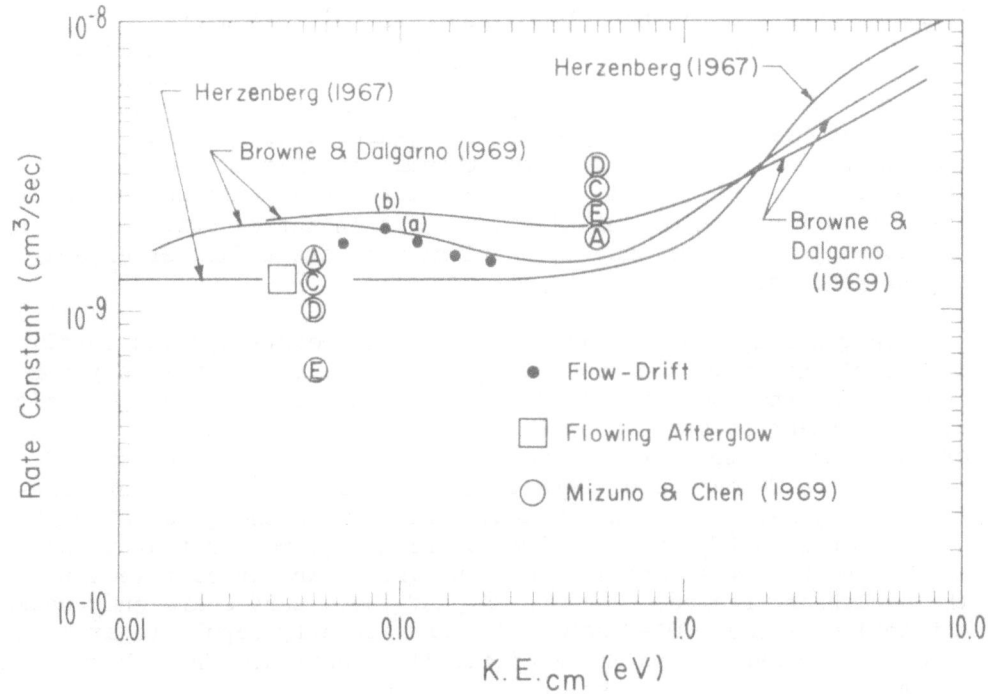

Figure 5. Calculated and measured rate constants for the associative detachment reaction between H⁻ and H. Included are the calculated values of Herzenberg (Herzenberg, 1967) and Brown and Dalgarno (Brown, 1969) shown as the designated solid lines. The results of Mizuno and Chen (Mizuno, 1969) are shown as various labeled open circles, ◯ etc. and are taken from tabulations for the associative detachment reaction cross section at 0.05 eV and 0.5 eV using various calculated values of V(R) and Γ(R). The designation, A, C, D, and E are the same designation used by Mizuno and Chen (Mizuno, 1969, p. 187). The room temperature flowing afterglow results (Schmeltekopf, 1967) are shown as the open box, ☐ while recent flow-drift results (McFarland, 1974) are shown as the closed circles, ●.

Fig. 5 the calculated and measured rate constant for the associative detachment reaction is plotted as a function of kinetic energy between 0.01 eV and 10 eV. The curve labeled Herzenberg (1967) is a semiclassical calculation based on the potential curve of Bardsley, et al (1966) and was done under the simplifing assumption that the survival probability of H⁻-H in the autodetaching region is zero. The curves labeled Brown and Dalgarno (a) and (b) (Brown and Dalgarno, 1969) are semiclassical calculation using V(R) and Γ(R) from Bardsley et al. (1966) in case (a) and

those of Chen and Peacher (1968) for case (b). Finally
Mizuno (1971), using a quantum mechanical treatment calculated
a series of cross sections tabulated at an H$^-$-H collision energies
of 0.005 eV, 0.05 eV, and 0.5 eV. In contrast to theory there
are, as yet, very little experimental results available for this
reaction. The open square is the thermal flowing afterglow data
(Schmeltekopf, 1967). The closed circles are preliminary results
taken as a function of ion kinetic energy in a flow-drift apparatus.
Any final judgment of the calculations must await further experimental results.

    In general, the associative detachment reaction, even involv-
ing atomic negative ions reacting with atoms, is considerably more
complicated than the case considered to this point. The aeronomic-
ally important reaction between O$^-$ and O is a good example of one
of these more complicated systems. The interaction of O$^-$ with O
may generate 24 separate states. The associative detachment reaction rate constant for this reaction at thermal energies is roughly
1/3 of Langevin (Fehsenfeld, 1966), which requires that at least
1/3 of these curves, statistically weighted, are attractive into
the detachment region. Because many of these curves are only weakly
attractive at large O$^-$-O separation and strongly repulsive at small
internuclear separation it is difficult to estimate from theoretical
arguments how many of these curves will be attractive into the autodetaching region. The attractive potential energy curves produced
by the interaction of O$^-$ with O are shown in Fig. 6 as solid lines.
In these curves the ground X $^2\Pi_g$ state curve of O_2^- is taken from
the results of recent photodetachment studies (Celotta, 1972) while
the excited state curves are from very recent correlation calculations of Michels (Michels, 1974). The preliminary results of
Michels indicate that about 40% of the collisions will proceed
along curves which are attractive into the autodetaching region.

    The lowest energy state O_2^- is the X $^2\Pi_g$ state. This state
yields a stable O_2^- with an adiabatic detachment energy of 0.43
eV. This potential curve passes through a minimum at 1.34 Å and
then rises, cutting the ground state $3\Sigma_g^-$ curve of O_2 (Linder,
1971). At this point the O_2^- becomes unstable against autodetachment. However, the autodetaching lifetimes of this state near the
crossing are quite long with various estimates ranging from about
10^{-10} sec. (Herzenberg, 1969) to 10^{-12} sec. (Goans, 1974). Autodetachment lifetimes this long would not seem to favor strong detachment from this state in a collision lifetime. Collisions
proceeding through the excited states of O_2^- would seem to be more
favorable for autodetachment. The dissociative attachment of
electrons to hot $O_2(^3\Sigma_g^-)$ to form O$^-$ (Henderson, 1969) has been interpreted as proceeding through the $^2\Pi_u$ state of O_2^- (O'Malley,
1967). O'Malley's interpretation also indicates that this curve
is attractive into the autodetaching region, that the lifetime

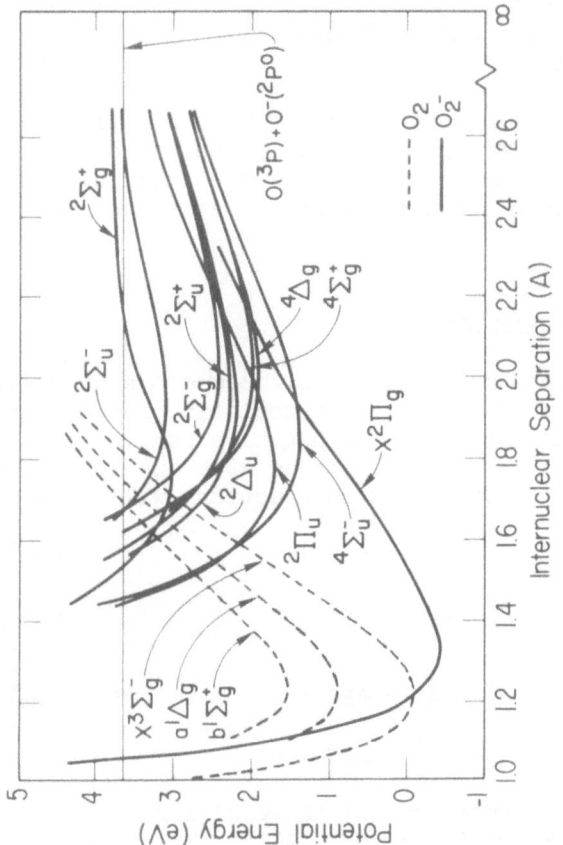

Figure 6. Potential energy curves for the interaction of O^- with O are shown as solid lines. The potential energy curves for the $^3\Sigma_g^-$, $^1\Delta_g$, and the $^1\Sigma_g^+$ states of O_2 are shown as the dashed lines.

against autodetachment is short, $< 10^{-13}$ sec for internuclear separations less than 1.5 Å, decreasing with decreasing internuclear separation, and that autodetachment from this state produces the $^1\Delta_g$ amd $^{3}\Sigma_g{}^-$ ground state. O'Malley's contention that autodetachment of the O_2^- ($^2\Pi_u$) state decays into the $O_2(^1\Delta_g)$ state has been confirmed recently by Barrow (Barrow, 1973) who found that the decay into $O_2(^1\Delta_g)$ to be about equal to that of $O_2(^3\Sigma_g{}^-)$. Other experiments (Wong, 1973) have observed strong vibrational excitation of the ground state $O_2(^3\Sigma_g{}^-)$ from the decay of $O_2^-(^2\Sigma_u{}^-)$ and $O_2^-(^4\Sigma_u{}^-)$ and indicate that these curves which are shown in Fig. 6 to be attractive into the autodetaching region may be short lived against autodetachment.

In a like fashion the reaction of O^- with N is fast indicating again that at least 30% of these statically weighted potential surfaces are attractive into the autodetaching region. At present, there is even less known about the compound states of NO than of O_2.

REACTIONS INVOLVING THREE ATOMS

When the reaction complex contains three atoms or more, additional constraints are placed on the associative detachment reaction. In this section the interaction of a negative ion and neutral will be considered in which one particle is an atom and the other is a diatomic molecule. Two types of such reactions will be designate: (1) addition, i.e.

$$A^- + BC \rightarrow ABC + e \qquad\qquad (4a)$$

or

$$A + BC^- \rightarrow ABC + e \qquad\qquad (4b)$$

(2) insertion, i.e.

$$B^- + AC \rightarrow ABC + e \qquad\qquad (5a)$$

or

$$B + AC^- \rightarrow ABC + e. \qquad\qquad (5b)$$

Several detachment reactions of the addition type, 4a and 4b, have been studied at 300°K (Ferguson, 1967). These results are listed in Table II. In the addition reaction the atom, A, is placed nearer one atom or the other of the diatomic molecule. Unless the reaction is highly exothermic, this position should correspond to that of A in the ground state of the triatomic neutral which is produced upon detachment. Most of these reactions which thus far have been studied are fast. Also, for negative ion reactions, they are reasonably exothermic.

All of the factors which are important to the associative detachment reaction between atomic negative ions and atoms are still

TABLE II

Rate constants at 300°K

	Reaction	Rate (cm^3/sec)	Exothermicity (eV)
1.	$H^- + CO \rightarrow HCO + e$	~5(-11)	0.5
2.	$H^- + NO \rightarrow HNO + e$	4.6(-10)	≤ 1.3
3.	$H^- + O_2 \rightarrow HO_2 + e$	1.2(-9)	1.3
4.	$C^- + CO \rightarrow C_2O + e$	4.1(-10)	2.0
5.	$O^- + CO \rightarrow CO_2 + e$	6.5(-10)	4.0
6.	$O^- + NO \rightarrow NO_2 + e$	2.5(-10)	1.6
7.	$O^- + N_2 \rightarrow N_2O + e$	< 1(-12)	0.2
8.	$O^- + O_2(^1\Delta) \rightarrow O_3 + e$	3(-10)	0.5
9.	$OH^- + H \rightarrow H_2O + e$	1.4(-9)	3.2
10.	$OH^- + O \rightarrow HO_2 + e$	2(-10)	1.0
11.	$CN^- + H \rightarrow HCN + e$	1.3(-9)	1.5
12.	$O_2^- + O \rightarrow O_3 + e$	~ 2(-10)	0.5

significant. There are additional constrainst which must now be considered, however. The bond length of the diatomic molecule may be different from the corresponding bond length in the triatomic molecule. This would mean that the triatomic neutral might be formed with a separation of BC associated with a vibrationally excited state of ABC.

A separate type of geometric constraint arises because the configuration of these triatomic molecules (and polyatomic molecules as well) in their lowest energy configuration depends on the number of valence electrons (Ferguson, 1967). Theoretical considerations concerning this observation have been made by Mulliken (1942) and Walsh (1953) and are often referred to as "Walsh rules". In the associative detachment reaction the lowest energy state for the compound state will correspond to a triatomic molecule with (n+1)

valence electrons while the neutral molecule produced upon detachment will be a triatomic with n valence electrons. To demonstrate
how these constraints influence this type of reaction at thermal
energy the fast reaction between O^- and CO and the unobservably
slow reaction between O^- and N_2 will be considered (Fehsenfeld,
1966). The reaction between O^- and CO to form CO_2 is nominally
exothermic by 4.2 eV while the reaction between O^- and N_2 to form
N_2O is only exothermic by 0.2 eV.

    In the reaction between O^- and N_2 the N-N bond in the diatomic
is almost identical to the N-N bond length in N_2O. This small
adjustment in bond lengths would not represent a significant barrier. The same is true for the C-O bond length relative to that
bond length in CO_2. Turning to the geometrical difference between
N_2O^- and CO_2^- and their neutral counter parts, N_2O and CO_2, the
Walsh rules state that triatomic molecules (not involving H) containing 16 or less valence electrons are linear in their ground
state, examples being CO_2, N_2O, CON, C_2O, etc. With 17 electrons,
such molecules are bent in the ground state, e.g. $NO_2(134.1^\circ)$; and
with 18 electrons they are bent still more, e.g. $O_3(116.8^\circ)$ (Herzberg, 1966). The N_2O^- or CO_2^- have 17 valence electrons. In their
lowest energy configuration these molecules will possess a bond
angle of about 134° as does NO_2 while in their lowest energy configuration CO_2 and N_2O have 16 valence electrons and a bond angle of
180°. In this case the difference in bond angle would correspond
to about 1 eV of vibrational excitation in a bending mode. Similar
considerations hold for the autodetaching complexes involving all
three atom molecules although the differences in equalibrium bond
angle in the ground state configuration between the neutral and its
associated ion is the most extreme in the case under discussion.

    In those cases where the potential surface into the autodetach-
ing region is not strongly attractive these geometrical differences
offer the strongest constraint against reaction. When the surface
is not strongly attractive to reach the autodetaching region the
ions must follow the most attractive path which of course will
correspond to the ground state configuration of the negative ion.
This is, in fact, the situation in the interaction of O^- with CO
and N_2.

    For the reaction of O^- with CO this is not a serious problem,
however. This reaction is nominally exothermic by 4.2 eV. Thus
the necessary chemical energy is available to detach the electron
from O^- and leave the product CO_2 with 1 eV or more of vibrational
excitation. On the other hand the reaction of O^- with N_2 possesses
an adiabatic exothermicity of only 0.2 eV. This small exothermicity does not allow for the formation of N_2O in a geometrical
configuration most favorable for a closest approach between O^- and
N_2 and represents a strong barrier against the reaction of O^- with
N_2 at low ($300^\circ K$) energy.

At 300°K the measured rate for the reaction of O^- with CO is 2/3 of the Langevin rate. When O^- approaches CO in a linear configuration two states may be formed a $CO_2^-(^2\Pi_u)$ and a $CO_2^-(^2\Sigma_g^+)$. The latter state correlates to an excited state of CO_2^- and is almost certainly repulsive. The former state corresponds to the ground state of CO_2^-. This state splits into a $^2A_1$ and a $^2B_1$ at a $(O-C-O)^-$ bond angle of 140°. In either configuration these lower states are probably attractive into the autodetaching region. The lifetime of these compound states of CO_2^- are very short (Sanche, 1973), probably of the order of 10^{-15} sec. Consequently, if the O^- and CO approach along the attractive surface they may enter the autodetaching region and detachment will almost certainly occur. At low energies approach along the repulsive state will not reach the autodetaching region. Thus statistics indicate that the reaction should occur on about 2/3 of collisions at low energies as is observed.

One additional feature of the associative detachemtn reaction which is to be expected from this type of reaction was observed in the results of Mauer and Schulz (Mauer, 1973). In these beam results an energy spectrum of the ejected electrons was obtained. This spectrum indicated that a large fraction of the electrons were formed with very low kinetic energy. The exothermicity of the reaction would allow for about 3 eV of kinetic energy in the ejected electrons even allowing for a sizable (1.2 eV) energy requirement to compensate for the geometrical differences between CO_2^- and CO_2. The predominance of the very low energy electron is due to the very short lifetime of the CO_2^- in the autodetaching region. This means, in turn, that most of the exothermicity of the reaction goes into the vibrational excitation of CO_2. The associative detachment reaction will probably tend to leave the molecular product in highly excited vibrational states and emit low energy electrons.

It is possible to draw some conclusions about the associative detachment reaction from the energy dependence of the rate constant. Figure 7 shows the data for the reaction of O^- with CO as a function of ion-kinetic energy. This reaction has been studied as a function of ion-kinetic energy from 0.04 eV to 10.0 eV and at present there is relatively good agreement among all the measurements. The rate constant is observed to decrease from 0.04 eV to about 2 eV, at this point it passes through a minimum and rises. The decline can be associated with two effects. The lifetime of CO_2^- against detachment, though small, is not zero. At larger energies and hence smaller collision lifetime the CO_2^- may survive against autodetachment. Also, during short interaction times the O^--CO may not assume a proper geometry. That is, a O^--OC approach would be much less favorable than O^--CO. A certain rearrangement time is required in most collisions for preferred geometries to occur.

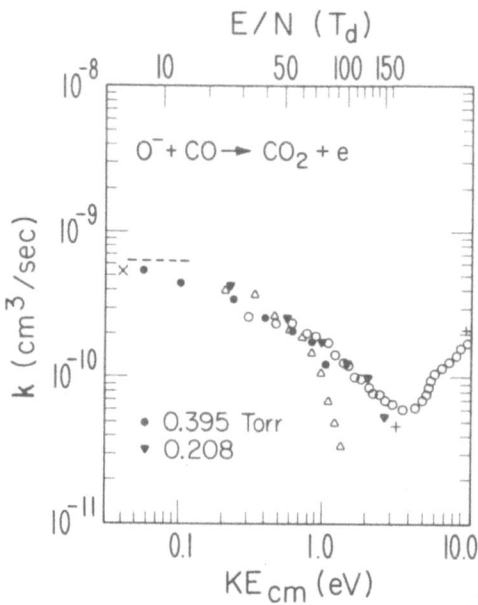

Figure 7. Rate constant for the reaction of O⁻ with CO as a function of ion kinetic energy in the center-of-mass. The X is the
thermal energy (296ºK) flowing afterglow measurement (Fehsenfeld,
1966); the dashed lines are the drift tube results of Moruzzi and
Phelps (Moruzzi, 1966); the closed circles, and triangles are the
flow drift experiment results (McFarland, 1973); the open circles
are the beam results of Comer and Schulz (Comer, 1974); the open
triangles are tandem mass spectrometer results (Tiernan, 1973);
and the pluses are ion beam results (Roche, 1969).

      Another detachment reaction which has been studied as a func-
tion of ion kinetic energy is O⁻ with NO. In this reaction the
rate constant falls rapidly with increasing kinetic energy due
probably to the fact that the lifetime of the NO_2^- against autodetachment is long. This results from the fact that the NO_2^-
represents an instance of a relatively long-lived Feshbach resonance in the compound state formation (Sanche, 1973).

Turning to the insertion type of reaction, 5a and 5b (Fehsen-feld, 1969), several rate constant determinations have been made for reactions of this type which are nominally exothermic. These rate constants are listed in Table III. In this case over half of the reactions of this type which were studied failed to have measureable associative detachment rate constants, indicating, not surprisingly that this is not a favorable type of reaction mech-anism. In this type of reaction, to assume the most stable config-uration the atomic species, B, must insert itself between the atoms of the diatomic molecule, AC. Thus the formation of the compound negative ion ABC⁻ will ordinarily involve a substantial increase in the AC internuclear distance with an associated activation barrier.

At present we feel the insertion reaction proceeds by the two step mechanism $B^- + AC \rightarrow (AB^- + C) \rightarrow ABC + e$. That is, the initial interaction between B^- and AC is an exothermic atom-ion interchange. However, the molecular negative ion is formed in close proximity to the atom C so that AB^- and C are either formed in or may enter the autodetaching region during the collision lifetime. According to this criterion then the insertion reaction requires an exothermic negative ion formation as an intermediate channel (McFarland, 1973). The reactions of O^- with H_2 and D_2 and the reaction of S^- with D_2 indicate the validity of this mechanism.

TABLE III

Rate constants measured at 300°K.

	Reaction	Rate (cm^3/sec)	Exothermicity (eV)
1.	$C^- + H_2 \rightarrow CH_2 + e$	$< 1(-13)$	3.3
2.	$C^- + O_2 \rightarrow CO_2 + e$	$< 5(-11)$	10.2
3.	$O^- + H_2 \rightarrow H_2O + e$	$7(-10)$	3.1
4.	$S^- + H_2 \rightarrow H_2S + e$	$< 1(-15)$	0.9
5.	$S^- + O_2 \rightarrow SO_2 + e$	$3(-11)$	3.8
6.	$OH^- + N \rightarrow HNO + e$	$< 1(-11)$	2.4
7.	$O_2^- + N \rightarrow NO_2 + e$	$4(-10)$	4.0

The reaction of O⁻ with H_2 and D_2 has been studied as a function of ion kinetic energy (McFarland, 1973). It is found that the reaction rate constant is independent of energy between 0.04 eV and 1 eV energy in the center-of-mass. There are two exothermic channels for the reaction, associative detachment and formation of OH⁻ (or OD⁻). The associative detachment, though dominant throughout the energy range studied, decreases slowly with increasing kinetic energy while the production of OH⁻ rises rapidly. According to the two step mechanism, there is initially a rapid, exothermic H atom abstraction by O⁻ from H_2 to form OH⁻ followed by the associative detachment between OH⁻ and H. The OH⁻ + H associative detachment reaction is independently known to be rapid (Fehsenfeld, 1973), i.e. to occur on essentially every collision. As the kinetic energy of the incident O⁻ is increased the lifetime of the intermediate reaction state, OH⁻ - H, will decrease which will favor the production of OH⁻ relative to associative detachment. In this same sense the substitution of D_2 for H_2 would be expected to favor the associative detachment channel since the lifetime of the intermediate reaction state, OD⁻ - D, will be increased by the reduced velocity of D relative to that of H. This is indeed the observation at all O⁻ kinetic energies studied (McFarland, 1974).

Finally the reaction of S⁻ with D_2 has been studied as a function of S⁻ kinetic energy. At thermal energy this reaction is immeasureably slow. With increasing S⁻ kinetic energy, however, the reaction becomes detectable. These results are shown in Fig. 8 (Tellinghuisen, 1974). The associative detachment reaction, if ground state D_2S is produced is exothermic by 0.9 eV. From Fig. 8 it can be seen that the reaction behaves somewhat as an endothermic reaction. The increase in the associative detachment would be compatible with an activation barrier of the order of 0.4 eV. The reaction channel for the production of SD⁻ is endothermic by 0.68 eV. In fact, the rapid increase observed in the production of SD⁻ is consistent with an activation energy of this amount. The similar behavior of the exothermic associative detachment channel and the endothermic ion molecule reaction channel in this reaction again suggests the two step mechanism proposed earlier.

REATIONS INVOLVING MORE THAN THREE ATOMS

As molecules become more complicated the associative detachment reaction seems to be an increasingly less favorable type of reaction scheme. There are several reasons for this trend. (1) With more complicated molecules, molecular rearrangement or charge transfer is, from the standpoint of exothermicity, usually the most favorable channel. With few exceptions the associative detachment is large only when there are no other ion-molecule exit channels of larger exothermicity. (2) As the number of atoms in the compound state

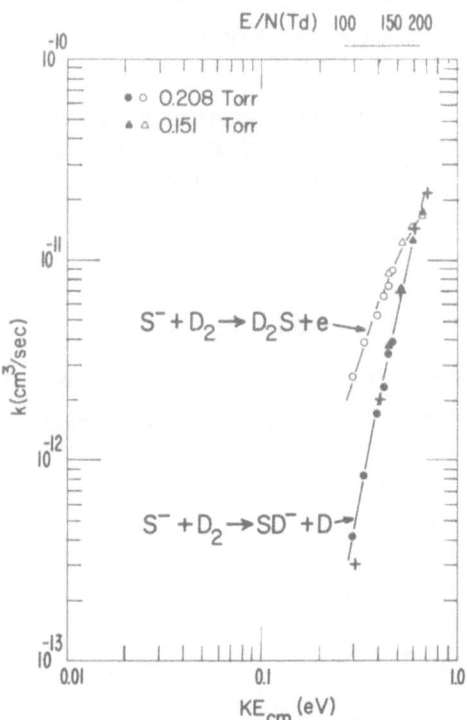

Figure 8. Reaction rate constants for the interaction of S^- with D_2 plotted as a function of ion kinetic energy in the center-of-mass system. Open circles and triangles are the experimental data for the associative detachment channel while the closed circles and triangles are the ion-molecule reaction producing SD^-. The pluses, +, indicate a theoretical fit to the production of SD^- using a reasonable ion velocity distribution and assuming the reaction has a rate constant of 1×10^{-10} cm^3/sec above a 0.68 eV threshold and zero below.

becomes larger the lifetime of this state against autoionization
becomes larger, assuming delocalization of the charge in the compound state. (3) As the electron affinity of the reactant negative
ion increases the number of possible chemical combinations which can
produce associative detachment is reduced. (4) Finally, when both
reactants are molecular considerable bond rearrangement will usually
be required to realize the maximum energy possible from the bond
formation. This in turn implies large steric hinderances and activation barriers.

In several cases associative detachment reactions have been
studied for reacting systems involving four atoms or more. These
reactions are listed in Table IV. Although the reaction is exothermic, the addition of one water of hydration greatly reduces
the reactivity of Cl^- with atomic hydrogen, i.e. by over an order

TABLE IV

The rate constants for several associative detachment
reactions involving 4 reactant atoms or more.

	Reactions	Rate cm^3/sec
1.	$O^- + C_2H_4 \rightarrow C_2H_4O + e$	4.05(-10)
	$\rightarrow C_2H_2^- + H_2O$	1.9 (-10)
2.	$O^- + C_2H_2 \rightarrow C_2H_2O + e$	1.3 (-9)
	$\rightarrow C_2H^- + OH$	8 (-10)
	$\rightarrow C_2OH^- + H$	8 (-11)
3.	$Cl^- \cdot H_2O + H \rightarrow HCl + H_2O + e$	< 8 (-11)
4.	$OH^-(H_2O)_2 + H \rightarrow 3H_2O + e$	~ 3 (-10)
5.	$OH^-(H_2O)_3 + H \rightarrow 4H_2O + e$	~ 2 (-10)
6.	$SO_3^- + H_2O \rightarrow H_2SO_4 + e$	< 1 (-12)
7.	$O^- + H_2O \rightarrow H_2O_2 + e$	< 1 (-12)
8.	$O^- + SO_2 \rightarrow SO_3 + e$	7 (-10)
9.	$C^- + N_2O \rightarrow CO + N_2 + e$	9 (-10)

of magnitude. This large reduction in reactivity may be due to an increase in the autodetachment lifetime of the $HCl(H_2O)^{-*}$ complex or to a repulsive barrier which prevents the formation of an auto-detaching complex. Unfortunately little is known concerning the lifetime of such a complex against autodetachment but generally the lifetime increases with increasing molecular complexity. The addition of a second water of hydration to the Cl^- renders the ΔH for the reaction positive, $\Delta H \cong 0.3$ eV. Under these conditions one would not expect a significant reaction and none is observed.

The reactions of $OH^-(H_2O)_2$ and $OH^-(H_2O)_3$ with H are much slower than that of OH^- with H. Because of the strong bonding between OH^- and water molecules, the exothermicity of these reactions are considerably reduced from that of OH^-. Nevertheless, for all reactions of $OH^- \cdot (H_2O)_n$, n = 0, 2, 3, with atomic hydrogen, ΔH is negative. To account for the reduction in reaction rate produced by the addition of two or three water molecules, similar arguments to those presented in the Cl^- reactions discussion are applicable.

SO_3^- was not observed to react with H_2O to form H_2SO_4 (Fehsenfeld, 1974). This reaction will be endothermic if the electron affinity of SO_3 is greater than 0.9 eV which it almost certainly is. Not surprisingly, the associative detachment reaction between O^- and H_2O to form H_2O_2 is very slow. Energetically the reaction is almost thermoneutral. However, there is probably a significant activation barrier ($\gtrsim 10$ kcal/mole) involved in forming H_2O_2 in its lowest energy configuration (H-O-O-H) by a collision between O^- and H_2O. Instead, O^- is observed to associate with H_2O, i.e. to form $O^- \cdot H_2O$ (or perhaps $H_2O_2^-$).

The reaction of C^- with N_2O (Fehsenfeld, 1970) is rapid with detachment occurring on nearly every collision. The reaction is probably not associative detachment in the sense that it has been used to this point by virtue of the fact that two product neutrals are produced, either CO and N_2 or CN and NO. In addition, the production of CN^- by reaction of C^- with N_2O is exothermic but was not observed. This tends to suggest that, in fact, N_2 and CO are probably the neutral products of the reaction. Neither N_2 or CO form stable negative ions. They do form short lived negative ion compound states which are responsible for the very strong vibrational excitation observed in N_2 and CO. These compound states are probably the intermediate by-products of the C^- reaction with N_2O.

REFERENCES

Bardsley, J. N., A. Herzenberg, and F. Mandl, Vibrational excitation
    and dissociative attachment in the scattering of electrons by
    hydrogen molecules, Proc. Phy. Soc. 89, 305, 321 (1966).

Bardsley, J. N., Electron detachment and charge transfer in H-H⁻
    collisions, Proc. Phy. Soc., 91, 300 (1967).

Barrow, P. D., Dissociative attachment from $O_2(a^1\Delta_g)$ state, J. Chem.
    Phys., 59, 4922 (1973).

Bondybey, V., P. K. Pearson and H. F. Schaefer, Theoretical Poten-
    tial energy curves for OH, HF⁺, HF, HF⁻, NeH⁺, NeH*, J. Chem.
    Phys. 57, 1123 (1972).

Browne, J. C., and A. Dalgarno, Detachment in collisions of H and
    H⁻, J. Phys. B. (Atom. Molec. Phys.) 2, 885 (1969).

Celotta, R. J., R. A. Bennett, J. L. Hall, M. N. Siegel, J. Levine,
    Molecular photodetachment spectroscopy II. Phys. Rev. A, 6,
    631 (1972).

Chen, J. C. Y., Theory of atomic collisions with negative ions:
    Associative Detachment, Phys. Rev. 156, 12 (1967).

Chen, J. C. Y. and J. L. Peacher, Interaction potential between
    the ground states of H and H⁻, Phys. Rev. 167, 30 (1968).

Christophorou, L. G., R. N. Compton, and H. W. Dickson, Dissocia-
    tive electron attachment to hydrogen halides and their deuter-
    ated analogues, J. Chem. Phys. 48, 1949 (1968).

Christophorou, L. G., Atomic and molecular radiation physics, Wiley
    Interscience, New York (1971).

Comer, J. and G. J. Schulz, Measurement of electron detachment cross
    sections from O⁻, J. Phys. B, Atomic and Molec. Phys. 7,
    L249 (1974).

Dalgarno, A. and J. C. Browne, The associate detachment of H and H⁻,
    Astrophys. J. 49, 231 (1967).

Fehsenfeld, F. C., E. E. Ferguson and A. L. Schmeltekopf, Thermal
    energy associative detachment reactions of negative ions,
    J. Chem. Phys. 45, 1844 (1966).

Fehsenfeld, F. C., A. L. Schmeltekopf, H. I. Schiff, and E. E.
    Ferguson, Laboratory measurements of negative ion reactions
    of atmospheric interest, Planet. Space Sci. 15, 373 (1967).

Fehsenfeld, F. C. and E. E. Ferguson, Model for the associative-
    detachment reaction of the insertion type, J. Chem. Phys. 51,
    3512 (1969).

Fehsenfeld, F. C. and E. E. Ferguson, Thermal energy reactions of
    C⁻ with O_2, N_2O, CO, and CO_2, J. Chem. Phys. 53, 2614 (1970).

Fehsenfeld, F. C., C. J. Howard, and E. E. Ferguson, Thermal energy
    reactions of negative ions with H atoms in the gas phase,
    J. Chem. Phys. 58, 5841 (1973).

Fehsenfeld, F. C. and E. E. Ferguson, Laboratory studies of negative
    ion reactions with atmospheric trace constituents, J. Chem.
    Phys. Oct. 1 (1974).

Ferguson, E. E., F. C. Fehsenfeld, and A. L. Schmeltekopf, Geometrical considerations for negative ion processes, 47, 3085 (1967).

Ferguson, E. E., Thermal energy ion-molecule reactions, Advances in Electronics and Electron Physics, 24, Academic Press, Inc. New York, 1, (1968).

Ferguson, E. E., F. C. Fehsenfeld, and A. L. Schmeltekopf, Ion-molecule reaction rates measured in a discharge afterglow, Advances in Chemistry 80, 83 (1969).

Frost, D. C. and C. A. McDowell, Electron capture processes in the hydrogen halides, J. Chem. Phys. 29, 503 (1958).

Goans, R. E. and L. G. Christophorou, Attachment of slow (< 1 eV) electrons to O_2 in very high pressures of nitrogen, ethylene, and ethane, J. Chem. Phys. 60, 1036 (1974).

Hartmann, W. M., T. L. Gilbert, K. A. Kaiser and A. C. Wahl, H-F$^-$ potential interaction and local mode frequencies for an interstitial hydrogen atom in alkaline earth fluorides, Phys. Rev. B, 2, 1140 (1970).

Henderson, W. R., W. L. Fite, and R. T. Brackmann, Dissociative attachment of electrons to hot oxygen, Phys. Rev. 183, 157 (1969).

Herzberg, G., Spectra of Diatomic Molecules, D. Van. Nostrand Co., New York (1950).

Herzberg, G., Electronic spectra and electronic structure of polyatomic molecules, Van Nostrand Co., New York (1966).

Herzenberg, A., Electron-detachment in slow collisions between atoms and negative ions, Phys. Rev. 160, 80 (1967).

Herzenberg, A., Attachment of slow electrons to oxygen molecules, J. Chem. Phys. 51, 4942 (1969).

Howard, C. J., F. C. Fehsenfeld, and M. McFarland, Negative ion-molecule reactions with atomic hydrogen in the gas phase at 296°K, J. Chem. Phys. 60, 5086 (1974).

Linder, F., and H. Schmidt, Z. Naturforsch. A. 26, 1617 (1971).

Mauer, J. L. and G. J. Schulz, Associative detachment of O$^-$ with CO, H_2, and O_2, Phys. Rev. A, 7, 593 (1973).

McFarland, M., D. L. Albritton, F. C. Fehsenfeld, E. E. Ferguson, and A. L. Schmeltekopf, Flow-drift technique for ion mobility and ion-molecule reaction rate constant measurements. III. The negative ion reactions of O$^-$ with H_2, D_2, CO, and NO, J. Chem. Phys. 59 6629 (1973).

McFarland, M., F. C. Fehsenfeld, C. J. Howard, and D. L. Albritton, Kinetic energy dependence of the associative detachment reactions of negative ions with atomic hydrogen, unpublished results.

Michels, H. H., R. E. Harris, and J. C. Browne, Electron detachment in H$^-$ + F and H + F$^-$ collisions, J. Chem. Phys. 48, 2821 (1968).

Michels, H. H., Correlation calculations of O_2^- potential energy surfaces, private communication (1974).

Mizuno, J. and J. C. Y. Chen, Theoretical investigation of complex potentials for atomic collisions I. Numerical solution of model (H⁻, H) collisions, Phys. Rev. 187, 167 (1969).

Mizuno, J. and J. C. Y. Chen, Theoretical investigation of complex potentials for atomic collisions II. Semi-classical description of atomic collisions, Phys. Rev. A., 4, 1500 (1971).

Moruzzi, J. L. and A. V. Phelps, Survey of negative ion-molecule reactions in O_2, CO_2, H_2O, CO, and mixtures of these gases at high pressure, J. Chem. Phys. 45, 4617 (1966).

Mulliken, R. S., Electronic structures and spectra of triatomic oxide molecules, Rev. Mod Phys. 14, 204 (1942).

O'Malley, T. F., Theory of dissociative attachment, Phys. Rev. 150, 14 (1966).

O'Malley, T. F., Calculation of dissociative attachment in hot O_2, Phys. Rev. 155, 59 (1967).

Roche, A. E. and C. C. Goodyear, Electron detachment from negative oxygen ions at beam energies in the range 3 to 100 eV, J. Phys. B, 2, 191 (1969).

Sanche, L. and G. J. Schulz, Electron transmission spectroscopy: Resonances in triatomic molecules and hydrocarbons, J. Chem. Phys. 58, 479 (1973).

Schmeltekopf, A. L., F. C. Fehsenfeld, and E. E. Ferguson, Laboratory measurements of the rate constant for H⁻ + H → H_2 + e, Astrophys. J. 148, L155 (1967).

Su, T. and M. T. Bowers, Theory of ion-polar molecule collisions, J. Chem. Phys. 58, 3027 (1973).

Tellinghuisen, J., M. McFarland, D. L. Albritton, F. C. Fehsenfeld and W. Lindinger, The reaction of S⁻ with D_2, unpublished results.

Tiernan, T. O., C. Lifschitz, and J. C. Haartz, unpublished results. Private communication.

Walsh, A. D., The electronic orbital shapes and spectra of poly-atomic molecules, J. Chem. Soc. 2260 (1953).

Weiss, A. W. and M. Krauss, Bound state calculation of scattering resonance energies, J. Chem. Phys. 52, 4363 (1970).

Wong, S. F., M. J. W. Boness, G. J. Schulz, Vibrational excitation of O_2 by electron impact above 4 eV, Phys. Rev. Letters 31, 969 (1973).

REACTION MECHANISMS OF ORGANIC AND INORGANIC

IONS IN THE GAS PHASE[*]

J. L. Beauchamp

Arthur Amos Noyes Laboratory of Chemical Physics[**]
California Institute of Technology
Pasadena, California 91109

INTRODUCTION

Progress is being made in the study of fairly complex chemical reactions under conditions which permit the results of bimolecular encounters to be observed and characterized. Under such conditions it is possible to make meaningful statements concerning the identity and initial interaction of the reactants, the nature of reaction intermediates, and the sequence of bond formation, rearrangement and breaking that occurs in proceeding from reactants to products. This is the case for both neutral-neutral reactions and ion-neutral reactions, and the future remains bright in terms of what remains to be learned from such studies. The day is approaching, for example, when it may be possible to isolate single polypeptide molecules and carry out sequential degradation reactions in the gas phase using selective bimolecular reactions. Such a claim is rendered closer to fact than fiction by recent improvements in the techniques for intact ionization of biological samples and new developments in techniques for the storage and detection of ions. It is now feasible to store and continuously detect a single charged particle for several days (Wineland, Ekstrom and Dehmelt, 1973). Furthermore, it should be clear from the contents of this paper that the knowledge required to select specific reagents for the cleavage of peptide bonds is being rapidly accumulated. It is the goal of this presentation to summarize our observations of the past several years regarding

---

[*] Work supported in part by the U.S. Atomic Energy Commission.
[**] Contribution No. 4938.

general classes of ion-molecule processes, and to formulate rules
for predicting reactivity based on a knowledge of ion structures and
charge distributions, simplified energy surfaces associated with
ion-neutral encounters, and ion and neutral thermochemical properties. Processes to be surveyed include acid and base induced
elimination reactions, processes which involve nucleophilic attack
at coordinately saturated carbon, processes such as esterification
reactions which result from nucleophilic attack at substituted carbonium ions and a range of processes involving reactions of metal
complexes. Many of the processes to be discussed have familiar
counterparts in solution chemistry, and indeed it is a comparison
of gas phase and solution chemistry which provides insight into
the moderating effect which solvents have on chemical properties
and reactivity in condensed phases. In this paper attention is
largely focussed on the reactions of closed shell systems in which
all valence electrons are paired. Reactant ions to be considered
are thus fragments of dissociative ionization processes or the
products of ion-molecule reactions. Experimental studies reported
herein utilize the techniques of ion cyclotron resonance spectroscopy, which have been described in detail elsewhere (Beauchamp,
1971; McMahon and Beauchamp, 1972; Blint, McMahon and
Beauchamp, 1974).

ION THERMOCHEMICAL PROPERTIES

A knowledge of ion thermochemical properties is essential to
interpret and predict ion-molecule reactions. Electron and proton
transfer reactions are by far the most widely studied ionic processes. Studies of equilibria involving proton transfer reactions
between n-donor bases permits a quantitative assessment of the
stability of the ionic species involved, which include the conjugate
acids of neutral bases (for example, reaction 1) and the conjugate
bases of neutral acids (for example, reaction 2). These processes
are quantified by the proton affinities of the bases, defined in
equation 3. Data for a range of bases are summarized in Table I.

$$NH_4^+ + CH_3NH_2 \rightarrow [H_3N \cdot \cdot \overset{+}{H} \cdot \cdot NH_2CH_3] \rightarrow CH_3NH_3^+ + NH_3 \quad (1)$$

$$CH_3O^- + HF \quad \rightarrow [CH_3O^- \cdot \cdot H^+ \cdot \cdot F^-] \rightarrow F^- + CH_3OH \quad (2)$$

$$B + H^+ \quad \rightarrow BH^+ \quad PA(B) = -\Delta H \quad (3)$$

The data in Table I details the energetics of proton solvation
for the interaction of a single neutral or ionic base with a proton.
Also of importance for the discussion of many processes are the
energetic changes associated with the second solvation step, i.e.,
the interaction of an n-donor base with an acidic proton in a
neutral or ionic species. Reactions 4 and 5 exemplify these processes, which involve the formation of a strong hydrogen bond.

TABLE I

Gas Phase Proton Affinities[a]

Species	PA(B)	Species	PA(B)	Species	PA(B)	Species	PA(B)
He	42	CH_3F	151	CH_3NH_2	216	H^-	399
Ne	48	C_2H_5F	163	$C_2H_5NH_2$	219	F^-	370
Ar	90	CH_3Cl	160	$(CH_3)_2CHNH_2$	221	Cl^-	333
Kr	101	C_2H_5Cl	167	$(CH_3)_3CNH_2$	223	Br^-	323
Xe	114	CH_3Br	163	$(CH_3)_2NH$	222	I^-	313
HF	114	C_2H_5Br	170	$(CH_3)_3N$	227	OH^-	390
HCl	140	CH_3I	170	HCN	170	SH^-	350
HBr	141	C_2H_5I	175	CH_3CN	188	SeH^-	339
HI	145	CH_3OH	182	quinuclidine	234	NH_2^-	402
H_2O	165	C_2H_5OH	188	$HCON(CH_3)_2$	214	PH_2^-	369
H_2S	169	$(CH_3)_2CHOH$	195	$CH_3CON(CH_3)_2$	218	AsH_2^-	360
H_2Se	170	$(CH_3)_3COH$	206	CH_3PH_2	207	HCO_2^-	342
NH_3	207	$(CH_3)_2O$	191	$(CH_3)_2PH$	219	$CH_3CO_2^-$	345
PH_3	188	$(C_2H_5)_2O$	203	$(CH_3)_3P$	228	$CF_3CO_2^-$	320
AsH_3	175	CH_2O	170	$(CH_3)_3As$	216	CH_3O^-	377
CH_4	128	CH_3CHO	186	CH_3SH	187	$C_2H_5O^-$	375
H_2	101	$(CH_3)_2CO$	195	$(CH_3)_2S$	197	$(CH_3)_2CHO^-$	374
N_2	114	HCO_2H	180	HC≡CH	152	$(CH_3)_3CO^-$	373
O_2	101	CH_3CO_2H	190	$CH_2=CH_2$	160	$C_6H_5CO_2^-$	336
CO	143	$CH_3CO_2CH_3$	198	$CH_2=CHCH_3$	179	HC≡C⁻	373
CO_2	127	$CH_3CO_2C_2H_5$	203	$CH_2=C(CH_3)_2$	195	$CH_3C≡C^-$	376

a All data in kcal/mole compiled from a literature survey. For leading references see Beauchamp (1971); Henderson, Holtz, Beauchamp and Taft (1972); Beauchamp and Caserio (1972); Hiraoka, Yamdagni and Kebarle (1973); McIver and Miller (1974); and the review by Bohme in this volume.

$$H_3O^+ + H_2O \rightarrow \quad \overset{H}{\underset{H}{\diagup}}O \cdot\cdot\overset{+}{H}\cdot\cdot O\overset{H}{\underset{H}{\diagdown}} \tag{4}$$

$$OH^- + H_2O \rightarrow HO^- \cdot\cdot\overset{+}{H}\cdot\cdot OH^- \tag{5}$$

The terminology is derived from the fact that such interactions are much stronger than is the case for hydrogen bond formation involving neutrals. Appropriate thermochemical data for such processes are summarized in Table II. More detailed considerations of the energetics of ion solvation have been recently reviewed by Kebarle (Kebarle, 1972).

TABLE II

Hydrogen-Bond Strengths in Proton Bound Dimers[a]

Base (B_1)	Base (B_2)	$D(B_1-\overset{+}{H}B_2)$	$D(B_1\overset{+}{H}-B_2)$
H_2O	H_2O	32	32
CH_3OH	CH_3OH	33	33
$(CH_3)_2O$	$(CH_3)_2O$	31	31
NH_3	NH_3	25	25
CH_3NH_2	CH_3NH_2	22	22
NH_3	CH_3NH_2	21	32
PH_3	PH_3	12	12
OH^-	OH^-	24	24
F^-	F^-	30	30
Cl^-	Cl^-	24	24
F^-	OH^-	23	44
F^-	CH_3O^-	24	40
F^-	Cl^-	50	14
Cl^-	OH^-	13	70
Cl^-	CH_3O^-	14	66

a All data in kcal/mole from Ridge and Beauchamp (1974); Yamdagni and Kebarle (1973); Grimsrud and Kebarle (1973) and Long and Franklin (1974).

While reactions 1 and 2 relate to relative stabilities of n-donor bases with regard to a particular reference acid (a proton in the case illustrated), the process can be generalized as indicated in reaction 6 for arbitrary acidic (A) and basic species (B_1 and B_2). Reversals in base strength occur with different

$$AB_1 + B_2 \rightleftharpoons AB_2 + B_1 \tag{6}$$

$$A_1B + A_2 \rightleftharpoons A_2B + A_1 \tag{7}$$

reference acids. Several are seen in comparing the data in Table I to that presented in Table III, where methyl cations are considered as the reference acid. Such reversals have significant implications with respect to general theories of acid-base interactions.

TABLE III

Binding Energies of Methyl Cations to n-Donor Bases[a]

Base	$D(CH_3^+-B)$	Base	$D(CH_3^+-B)$
Kr	21	H_2O	67
Xe	43	H_2S	82
Zn	45	NH_3	106
Cd	45	N_2	49
Hg	38	CO	75
HF	36	CH_3F	44
HCl	51	CH_3OH	82
HBr	56	CH_3NH_2	117
HI	67	CH_2O	74

a All data in kcal/mole. Calculated from data in Table I and Holtz, Beauchamp and Woodgate (1970).

In contrast to the information derived from studies of the generalized process 6, it is of interest to consider the alternative in which the relative binding energies of two acids (A_1 and A_2) to a reference base (B) are determined, reaction 7 being the generalized process. Reaction 8 is a specific example which permits the relative binding energy of two Lewis acids (the fluoromethyl and

$$CHF_2^+ + CH_2F_2 \rightleftharpoons CH_2F^+ + CHF_3 \tag{8}$$

difluoromethyl cations) to the reference base F^- to be determined (Blint, McMahon and Beauchamp, 1974; McMahon, Blint, Ridge and Beauchamp, 1972; Dawson, Henderson, O'Malley and Jennings, 1973). The energetic changes associated with processes such as reaction 8 can be conveniently characterized by the heterolytic bond dissociation energy, $D(R^+-X^-)$, defined in equation 9.

$$RX \rightarrow R^+ + X^- \qquad \Delta H = D(R^+-X^-) \tag{9}$$

Data for several reference bases are summarized in Table IV. Again it is noteworthy that reversals in acid strength occur for different reference bases. It is thus hazardous to infer intrinsic stability of ions from the ordering determined for a particular reference base.

TABLE IV

Heterolytic Bond Dissociation Energies[a]

R^+	$D(R^+-H^-)$	$D(R^+-F^-)$
CH_3^+	312	256
CH_2F^+	290	247
CHF_2^+	284	247
CF_3^+	292	253
$C_2H_5^+$	273	220
CH_3CHF^+	262	221
$CH_3CF_2^+$	259	222
$(CH_3)_2CH^+$	250	200
$(CH_3)_2CF^+$	244	204
$(CH_3)_3C^+$	235	182
$C_2H_3^+$	287	254
$C_6H_5^+$	283	233
$CH_2=CHCH_2^+$	255	202
$CH_2=NH_2^+$	216	- -
CH_2OH^+	253	- -
CH_3CO^+	233	205

a All data in kcal/mole from Blint, McMahon and Beauchamp (1974) and Ridge, McMahon, Park and Beauchamp (1974).

In analogy with the further solvation of reference acids with bases, it is possible to quantify the solvation of an acid base complex by an acid. The binding of CH_3^+ to CH_3F is an example of such processes, several of which are included in the data in Table III.

ION-MOLECULE INTERACTIONS

The interaction of closed shell ions with neutral molecules may be broadly classified according to the range of interaction energies, defined by the enthalpy difference between the separated reactants and the energized intermediate or reaction complex (equation 10). In the absence of radiative processes or interaction

$$A^+ + B \rightarrow [AB^+]^* \qquad E^* = \Delta H_f(A^+) + \Delta H_f(B) - \Delta H_f(AB^+) \qquad (10)$$

with a third body, the internal excitation, E^*, in the intermediate $(AB^+)^*$ is retained as internal energy and the species formed may be regarded as being chemically activated. If the internal excitation is sufficient the intermediate may decompose to energetically accessible products other than the reactants. It is convenient to consider three classes of interaction energies, termed weak (0-10 kcal/mole), intermediate (10-40 kcal/mole) and strong (above 40 kcal/mole). Weak interactions are usually those which involve no specific chemical bonding of the reactant ion to the reactant neutral. The intermolecular potential is described by a long range attractive part, such as determined by the various ion-neutral multipole interactions, and a short range repulsion at distances generally large compared with those associated with bonding interactions. The analogy in the case of neutrals are van der Waals complexes. An example is the interaction of CH_5^+ with CH_4, illustrated schematically in Figure 1a.

The intermediate range of interaction energies includes the majority of ion-neutral collision partners in which a strong hydrogen bond can be formed between a labile proton of an acidic reactant and an n-donor site of a reactant base. A widely studied example in this case involves the interaction of H_3O^+ with H_2O (reaction 4, illustrated schematically in Figure 1b). Theoretical calculations indicate a stable linear O-H-O configuration in which the proton moves in a potential well with a relatively flat bottom at the calculated equilibrium O-O separation of 2.38 Å (Kollman and Allen, 1970; Kraemer and Diercksen, 1970). The force constant for O-H-O bending is low, indicating considerable freedom in the O-H-O bending motion for a chemically activated species resulting from hydrogen bond formation. There are a wide range of chemical reactions which are a direct consequence of strong hydrogen bond formation, the energetics of which are summarized in Table II.

Chemical bond formation is responsible for the third class of interactions. For example, the interaction of CH_3^+ with NH_3 (Figure 1c) involves C-N bond formation, leading to protonated methyl amine containing an internal excitation of 106 kcal/mole (Table III).

In examples more complex than those shown in Figure 1, multiple bond formation or isomerization during formation of reaction intermediates may lead to formation of highly excited species whose internal excitation may be accurately assessed with thermochemical data such as that summarized in the previous section.

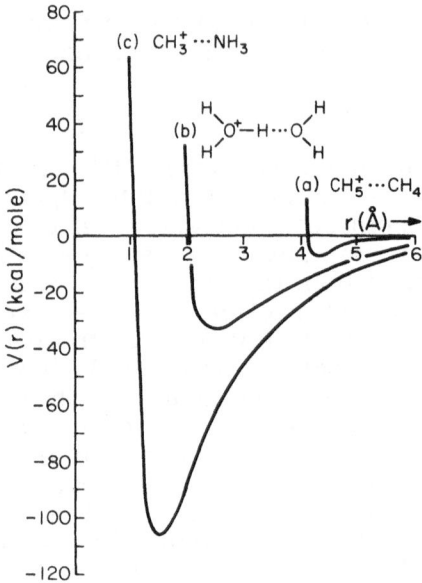

Fig. 1. Schematic potential energy curves for ion-neutral interactions. The separation is taken to be the spacing of the
two heavy atoms.

ORGANIC REACTION MECHANISMS

Weak Interactions

In the case of weak interactions it is expected that the chemical
activation energy available to the intermediate complex (Figure 1a,
0-10 kcal/mole) will be insufficient to cause chemical change.
The only chemistry which is characteristic of systems of this type
are simple atomic ion transfer reactions which occur in accordance
with the thermochemistry of the reaction.   Even thermoneutral
proton transfers may be fast, however, consistent with little or
no activation energy in the formation and dissociation of reaction
intermediates (Beauchamp, McMahon and Miasek, 1973).

Intermediate Interactions and the Chemical Consequences of
Strong Hydrogen Bonding

Reactions 1 and 2 are examples of exothermic proton transfer
reactions which involve the formation of a strong hydrogen bond
between the reactants. Appropriate reaction coordinate diagrams
are indicated in Figure 2, in which the relative energies of the
reactants, intermediates, and products are shown. For the

Proton Transfer Reactions

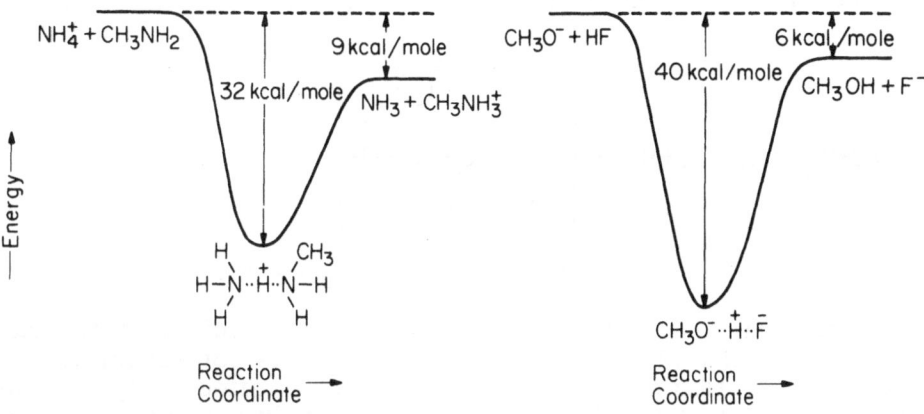

Fig. 2.   Reaction coordinate diagrams illustrating the energetics of proton transfer reactions.

examples cited there are no reactions which can compete with simple proton transfer. In somewhat more complex systems, however, other processes compete with proton transfer under appropriate conditions. The chemical activation energy provided by strong hydrogen bond formation can result in elimination reactions (induced by acids and bases), plays a central role in processes which involve nucleophilic attack at coordinately saturated carbon, and may trigger a wide range of decomposition, rearrangement and isomerization reactions.

Acid-Induced Elimination Reactions. The enthalpy change for 1, 2-elimination of HX (reaction 11) from a series of substituted hydrocarbons RX are summarized in Table V. Activation energies

$$\text{(11)}$$

for four-center elimination reactions occurring under homogeneous conditions in the gas phase are generally considerably in excess of the enthalpy change for the process (Benson, 1968). Homogeneous catalysis of some elimination reactions can be effected by the action of acidic neutrals. For example, addition of HBr to tert-butyl alcohol reduces the activation energy for elimination of H_2O from 57 to 30 kcal/mole (Maccoll and Stimson, 1966). The explanation for this behavior undoubtedly involves interaction of the acidic proton of the hydrogen halide with the basic oxygen, possibly involving the transition state shown in reaction 12. The conjugate

$$HBr + (CH_3)_3COH \rightarrow \left[\begin{array}{c} H-Br \\ H-O \quad H \\ C-C-H \\ CH_3 \quad CH_3 \quad H \end{array} \right]^*$$

$$\rightarrow (CH_3)_2C{=}CH_2 + H_2O + HBr \tag{12}$$

acids of neutral n-donor bases are much more acidic than neutral acids such as HBr (Table I). Consequently it is not surprising that the dehydration of tertiary butyl alcohol is effected by proton-ated acetone and protonated tert-butyl alcohol in a single bimolec-ular encounter without an apparent activation energy (Beauchamp, 1969). A reaction coordinate diagram for the latter process (reaction 13), which is formally an E2 elimination reaction, is shown in Figure 3. These reactions are envisioned as involving

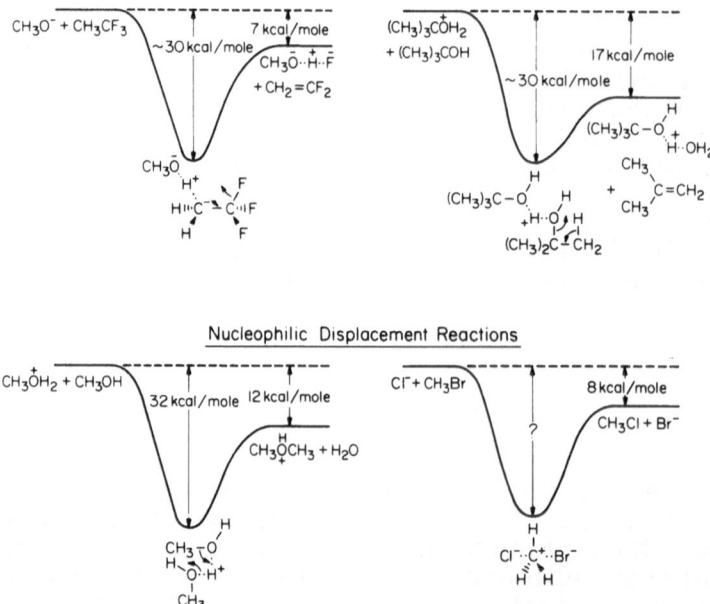

Fig. 3. Reaction coordinate diagrams illustrating the energetics of several reactions discussed in the text.

$$(CH_3)_3COH_2^+ + (CH_3)_3COH \rightarrow \begin{bmatrix} (CH_3)_3C-O \overset{H}{\underset{\cdot\cdot}{\diagdown}} \\ \overset{+}{H}\cdot\cdot O \overset{H}{\underset{\diagup}{\diagup}} H \\ (CH_3)_2 \overset{\mid}{C} \overset{\diagup}{\underset{}{=}} CH_2 \end{bmatrix}^*$$

$$\rightarrow (CH_3)_3COH\cdot\cdot\overset{+}{H}\cdot\cdot OH_2 + (CH_3)_2C=CH_2 \qquad (13)$$

the association of the acidic proton with the n-donor site of a reactant molecule, forming a chemically activated intermediate which may in part decompose by elimination of HX, with the overall process being exothermic by the difference in the enthalpy change for removal of HX from RX and the binding energy of HX to the reactant acid. The internal excitation of the intermediate in reaction 13 is high, being on the order of 30 kcal/mole, with a minimum critical energy for the dehydration process of about 13 kcal/mole. Thus an intrinsic activation energy (which would appear as a hump in the exit channel of the reaction coordinate diagram shown for process 13 in Figure 3) for the process may be masked. Interestingly, only the dehydration reactions of secondary and tertiary alcohols have been reported (Beauchamp and Caserio, 1972). Dehydration of ethanol has not been observed, possibly due to an intrinsic activation energy for the process in excess of the chemical activation energy available when various acids such as protonated acetaldehyde or protonated ethanol interact with ethanol. Although hydrogen bonding is not involved, it is of interest to note that Lewis acids can effect elimination reactions by a mechanism similar to that invoked for process 13. Thus isopropyl alcohol is dehydrated by CH_3^+ and NO^+ dehydrohalogenates 2-fluoropropane (Williamson and Beauchamp, 1974).

Base Induced Elimination Reactions. The interaction of the conjugate base of a neutral acid with another neutral acid (reaction 2) leads to the formation of a chemically activated intermediate with internal excitation comparable to that observed for the association of two neutral n-donor bases via a labile proton. Compare, for example, the reaction coordinate diagrams for reactions 1 and 2 in Figure 2. This led to the expectation, following observation of acid induced elimination reactions, that the type of strong hydrogen bonding illustrated in the generalized process 14 would form a chemically activated intermediate which might decompose as indicated. Base induced elimination reactions, which are formally E2 reactions, are indeed a prominent feature of the chemistry resulting from the interaction of strong bases with

$$Y^- + \overset{H}{\underset{/}{\diagup}}C-C\overset{X}{\underset{\diagdown}{\diagdown}} \rightarrow \begin{bmatrix} Y^- \\ \overset{\cdot\cdot}{H^+} \\ \underset{/}{\overset{}{C}} \overset{X}{\underset{\diagdown}{\diagdown}} \\ \overset{}{C}\overset{}{C} \end{bmatrix}^* \rightarrow Y^-\cdot\cdot H\cdot\cdot X^- + \overset{}{\underset{}{\diagup\diagdown}}$$

$$(14)$$

fluoroalkanes (Ridge and Beauchamp, 1974a, b). A reaction co-ordinate diagram for the elimination of HF from 1, 1, 1-trifluoro-ethane (reaction 15) is shown in Figure 3. The enthalpy change for elimination of HF from 1, 1, 1-trifluoroethane is 34 kcal/mole

$$CH_3O^- + CH_3CF_3 \rightarrow \left[\begin{array}{c} CH_3O^- \\ \cdot \cdot H^+ \\ H-C \underset{H}{\overset{H}{\rightleftharpoons}} C \overset{F}{\underset{F}{\rightarrow}} F \end{array} \right]^*$$

$$\rightarrow CH_3O^- \cdot \cdot \overset{+}{H} \cdot \cdot F^- + CH_2{=}CF_2 \qquad (15)$$

(Table V). The binding energy of HF to CH_3O^- is 40 kcal/mole, thus rendering the overall process 15 exothermic by 6 kcal/mole. From the trends observed in the bond strength measurements for anionic dimers (Yamdagni and Kebarle, 1971), the internal exci-tation in the intermediate for reaction 15 is estimated to be 30 kcal/mole. As is the case for acid induced elimination reactions, the elimination process 14 is only one of the reaction channels observed for the interaction of strong bases with alkyl halides in the gas phase. Also observed are the product ion X^- and carbanion generated by proton transfer to the reactant base. The factors which determine the product distribution have been elucidated by examining the interaction of a series of reactant bases of varying size and base strength with variously substituted fluoroethane

TABLE V

Enthalpy Changes for 1, 2-Elimination of HX from RX[a]

RX	HX	Enthalpy Change for Reaction
CH_3CH_2F	HF	8
CH_3CHF_2	HF	25
CH_3CF_3	HF	34
CH_3CH_2Cl	HCl	17
CH_3CCl_3	HCl	13
CH_3CH_2OH	H_2O	11
$(CH_3)_2CHOH$	H_2O	12
$(CH_3)_3COH$	H_2O	13

a All data in kcal/mole (Ridge and Beauchamp, 1974 and Beauchamp and Caserio, 1972).

substrates (Ridge and Beauchamp, 1974a, b and Sullivan and Beauchamp, 1974). Although base-induced elimination reactions are observed in other systems, including dehydrochlorination of 1, 1, 1-trichloroethane and β-chloroethanol, they are most prominent in the case of the fluoroethanes.

Nucleophilic Displacement Reactions. Two simple rules are useful in predicting the occurrence of a wide variety of gas phase nucleophilic displacement reactions occurring at coordinately saturated carbon (reaction 16 where X and Y are n-donor bases): (1) the reaction must be exoergic (i.e., the products must be

$$X + RY^+ \rightarrow Y + RX^+ \tag{16}$$

energetically accessible, a general requirement for ion-molecule reactions); and (2) no facile proton transfer can occur from the ionic substrate to the nucleophile (Holtz, Beauchamp and Woodgate, 1970). The displacement of HCl by H_2O serves to exemplify these considerations. The displacement of HF by HCl (reaction 17) was

$$HCl + CH_3FH^+ \rightarrow CH_3ClH^+ + HF \tag{17}$$

one of the first nucleophilic displacement reactions to be characterized as such (Holtz, Beauchamp and Woodgate, 1970). Formally reaction 17 is an example of the general process 6 in which the reference acid CH_3^+ is transferred from the base HF to HCl. Thus the data of Table III can be used to predict the order in which nucleophilic displacements occur (decreasing methyl cation affinity): $NH_3 > H_2S > CO > CH_2O > HI \cong H_2O > HBr > HCl > N_2 > HF$. Thus while HCl displaces HF, it is predicted that H_2O will displace HCl. Attempts to verify this led to the results shown in Figure 4. In a mixture of H_2O with CH_3Cl, reaction 18 is not observed; instead the simple proton transfer reaction 19 occurs.

$$CH_3ClH^+ + H_2O \longrightarrow$$

$$\xrightarrow{\;\;X\;\;} CH_3OH_2^+ + HCl \tag{18}$$

$$\longrightarrow H_3O^+ + CH_3Cl \tag{19}$$

However, in a mixture of C_2H_5Cl and H_2O, the nucleophilic displacement reaction (20) is observed to the exclusion of proton

$$C_2H_5ClH^+ + H_2O \rightarrow C_2H_5OH_2^+ + HCl \tag{20}$$

transfer (Figure 5). Reasonable representations of the reaction intermediates for processes 18 and 20 are indicated by structures I and II in which the labile proton is shared between the two n-donor bases. Quite significantly, the proton affinity of H_2O lies between the proton affinity of CH_3Cl and C_2H_5Cl (Table 1). Thus,

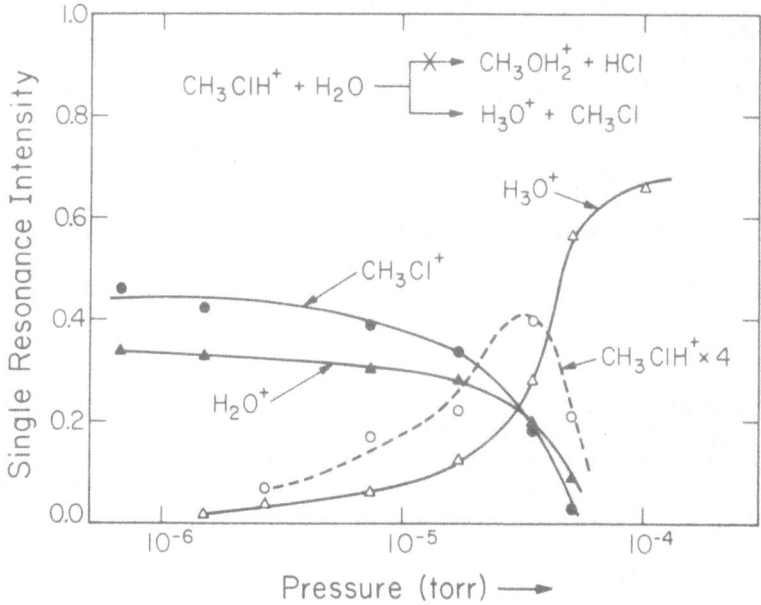

Fig. 4. Variation of ion abundance with total pressure for a 2:1 mixture of H_2O and CH_3Cl.

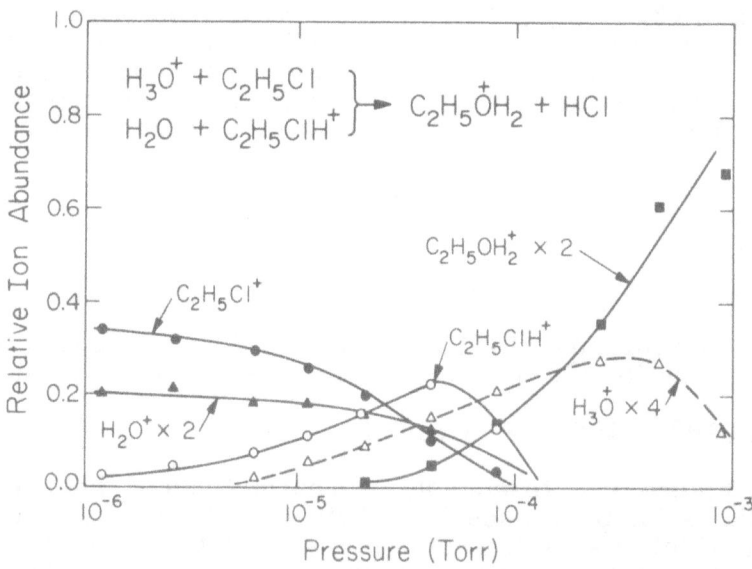

Fig. 5. Variation of ion abundance with total pressure for a 1:1 mixture of C_2H_5Cl and H_2O.

in I the proton is transferred to H_2O, destroying the nucleophilicity

$$CH_3-Cl$$

$$\overset{\cdot\cdot}{H-O}\cdot\cdot\overset{+}{H}$$
$$|$$
$$H$$

$$\text{I}$$

$$CH_3-CH_2-Cl$$

$$H-\overset{\cdot\cdot}{O}\cdot\cdot\overset{+}{H}$$
$$|$$
$$H$$

$$\text{II}$$

of water and inhibiting the displacement reaction. As a result reaction 18 does not occur, even though it is calculated to be exothermic. In II however, the proton remains bound to chlorine and HCl is displaced by H_2O.

The proposed rules, in conjunction with proton affinity data (Table I), make it possible to predict the outcome of a wide variety of encounters in which nucleophilic displacement reactions are a possible result. For example, halonium ion formation is a prominent feature of the gas phase ion chemistry of the alkyl halides (Blint, McMahon and Beauchamp, 1974 and Beauchamp, Holtz, Woodgate and Patt, 1972). In ethyl fluoride and methyl chloride alone, reactions 21 and 22 lead to formation of the diethyl fluoronium and dimethyl chloronium ions, respectively.

$$C_2H_5FH^+ + C_2H_5F \rightarrow (C_2H_5)_2F^+ + HF \tag{21}$$

$$CH_3ClH^+ + CH_3Cl \rightarrow (CH_3)_2Cl^+ + HCl \tag{22}$$

In a mixture of methyl chloride and ethyl fluoride, the only cross halonium ion observed is the methyl ethyl chloronium ion (reaction 23). These observations are consistent with ethyl fluoride being a

$$\left. \begin{array}{l} CH_3ClH^+ + C_2H_5F \\ \\ C_2H_5FH^+ + CH_3Cl \end{array} \right\} \rightarrow \left[\begin{array}{c} CH_3-CH_2-\overset{\cdot\cdot}{F} \\ Cl\cdot\cdot\overset{+}{H} \\ | \\ CH_3 \end{array} \right]^*$$

$$\rightarrow \overset{+}{C_2H_5ClCH_3} + HF \tag{23}$$

stronger base than methyl chloride in the gas phase. Other interesting examples of ions produced in nucleophilic displacement reactions include the methyl diazonium ion $(CH_3N_2^+)$, acyl cation (CH_3CO^+) and methyl xenonium ion (CH_3Xe^+) (Holtz and Beauchamp, 1971a, b).

A reaction coordinate diagram for a nucleophilic displacement reaction is shown for reaction 24 in Figure 3. The overall reaction is exothermic by 12 kcal/mole; the chemical activation

$$CH_3OH_2^+ + CH_3OH \rightarrow (CH_3)_2OH^+ + H_2O \qquad (24)$$

afforded the intermediate by strong hydrogen bonding is 32 kcal/mole. High pressure mass spectrometric techniques (Grimsrud and Kebarle, 1973) have recently been employed to estimate an activation energy for the dehydration of the proton bound dimer of methanol, arriving at a value of 30–35 kcal/mole. This is larger than the 20 kcal/mole endothermicity of the reaction, indicating an activation energy somewhat in excess of the endothermicity of the reaction.

The chemical activation afforded the reaction intermediate in nucleophilic displacement reactions is retained only while the hydrogen bond remains intact; for this reason it is proposed that the reaction proceeds through a four-centered front sided displacement mechanism. Another observation consistent with this proposal is that while methyl fluoride and methanol exhibit the discussed reaction, in methyl amine reaction 25 is not observed,

$$CH_3NH_3^+ + CH_3NH_2 \nrightarrow (CH_3)_2NH_2^+ + NH_3 \qquad (25)$$

even though it is exothermic by 11 kcal/mole. In contrast to methyl fluoride and methanol where an additional lone pair is available on the nucleophile to form a new bond to carbon, the single lone pair on nitrogen in reaction 25 is involved in the strong hydrogen bond. This would not seem to be a problem associated with back side attack in a nucleophilic displacement reaction. Alternatively, the internal excitation available to the reaction intermediate in process 25 may be less than an intrinsic activation energy for the process.

Evidence for back side attack in nucleophilic displacement reactions has recently been obtained by Brauman and co-workers (1974). They present evidence based on an analysis of neutral products for processes such as reaction 26 occurring via the

$$Cl^- + {>}C{-}Br \rightarrow \left[Cl^- \cdots C^+ \cdots Br^- \right] \rightarrow Cl{-}C{<} + Br^- \qquad (26)$$

indicated mechanistic pathway, formally an SN2 displacement reaction. A reaction coordinate diagram for this process is included in Figure 3. At present there is little data of use in attempting to assess the internal excitation available to the reaction intermediate. Attempts to prepare "captured" SN2 transition states have thus far been unsuccessful (Riveros, Breda and Blair, 1973). In contrast, the intermediates suggested for the other processes illustrated in Figure 3 can all be prepared under conditions where their internal energy is insufficient for decomposition to occur.

Strong Interactions

Strong interactions cover a wide variety of processes, ranging from the binding of CH_3^+ to NH_3 (Figure 1) to the interaction of a proton with quinuclidine. While the latter process is so exothermic (234 kcal/mole) that charge transfer and decomposition are the observed reaction channels, the former process or more generally the interaction of nucleophiles with carbonium ions comprises the most interesting group of reactions which can be classified under the heading of strong interactions. The reaction of CH_3^+ with NH_3 leading to the formation of protonated methylene immine (process 27) bears an intriguing analogy to the recombination of two methyl radicals leading to the formation of chemically activated ethane which can decompose to ethylene (process 28). While the intermediates and products of reactions 27 and 28

$$CH_3^+ + NH_3 \rightarrow [CH_3NH_3^+]^* \rightarrow CH_2=\overset{+}{N}H_2 + H_2 \qquad (27)$$

$$CH_3 + CH_3 \rightarrow [CH_3CH_3]^* \rightarrow C_2H_4 + H_2 \qquad (28)$$

are isoelectronic, the C—H bond in reaction 27 is formed heterolytically and the C—C bond in reaction 28 is formed homolytically. Heterolytic bond formation is a common feature of the reaction of n-donor and π-donor bases with carbonium ions.

A knowledge of heterolytic bond energies (Table IV) in conjunction with some simple rules can be used to correlate and predict the outcome of a variety of reactions which involve nucleophilic attack on substituted carbonium ions (Cook and Beauchamp, 1974). The following rules apply to the general process 29:

$$R_1^+ + R_2X \rightleftharpoons [R_1\overset{+}{X}R_2]^* \rightleftharpoons R_1X + R_2^+ \qquad (29)$$
$$\downarrow$$
$$\text{other products}$$

(1) the reaction must be exoergic, making the products energetically accessible; (2) if a facile proton transfer can occur as the ion and neutral reactants approach, it will precede rearrangement or condensation processes which involve more intimate interaction of the reactants; and (3) the outcome of reaction 29 in

which R_1^+ and R_2^+ are carbonium ions and X is an n-donor substituent depends on the relative RX heterolytic bond dissociation
energies. If $D(R_1^+-X^-) > D(R_2^+-X^-)$ then reaction 29 preferentially
proceeds to the right and R_2^+ is the major product ion observed.
If $D(R_1^+-X^-) \leq D(R_2^+-X^-)$, then a chemically activated intermediate
$[R_1XR_2^+]*$ is formed in which the internal excitation may be sufficient to permit rearrangement, disproportionation or elimination
reactions to compete with the simple bond cleavage to produce R_2^+
or regenerate R_1^+. It is important to note that the equality holds
when R_1^+ and R_2^+ are identical and the species $[RXR^+]*$ is a symmetrical "onium" ion.

Two interesting examples (Weiting, Staley and Beauchamp,
1974) of the general reaction 29 are illustrated in Figure 6. In a
mixture of 1, 2-dibromoethane and 1, 4-dibromobutane, reaction
30 is observed to proceed entirely to the right (Figure 6a). In

$$\rightarrow Br(CH_2)_2Br + \qquad \qquad (30)$$

reaction 31, 1-adamantyl bromide is observed to transfer Br^- to
$CH_3CH_2CO^+$, (Figure 6b). For a series of carbonium ions R^+, it
is possible in these experiments to establish relative carbonium

$$\rightarrow CH_3CH_2COBr + \qquad \qquad (31)$$

ion stabilities (as defined by R^+-Br^- heterolytic bond energies).

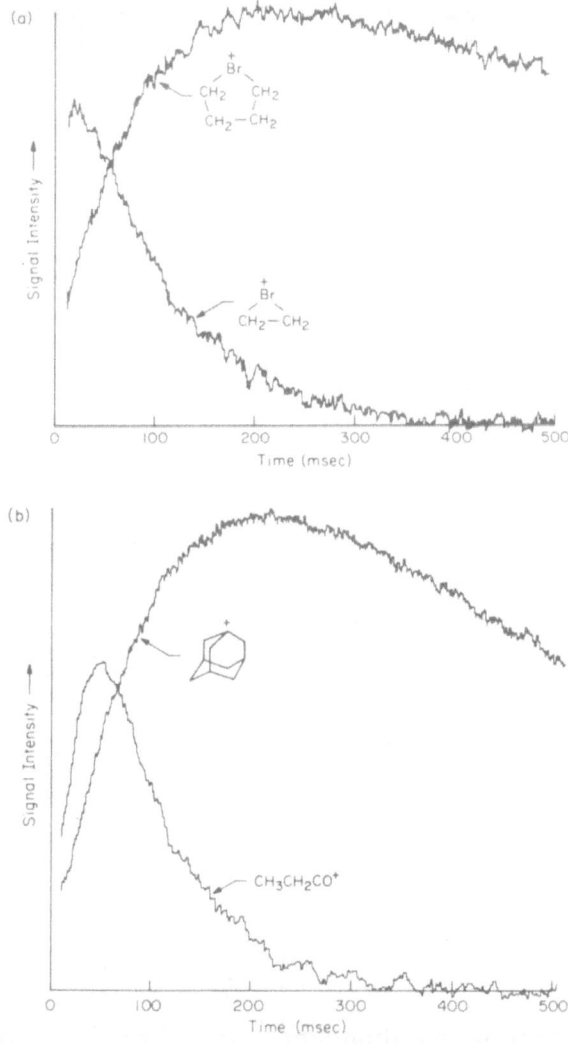

Fig. 6. Variation of ion abundance with time in two mixtures of
bromides. Relative carbonium ion stabilities in the gas
phase with respect to Br⁻ are determined from the direction of Br⁻ transfer. (a) 6:1 mixture of 1, 4-dibromo-
butane and 1, 2-dibromoethane at a total pressure of
2×10^{-6} torr: tetramethylene bromonium ion is more
stable than ethylene bromonium ion (reaction 30). (b) 1:1
mixture of adamantyl bromide and propionyl bromide at a
total pressure of 2×10^{-6} torr: adamantyl cation is more
stable than propionyl cation (reaction 31). Buildup of the
displayed species during the first 50 msec is due in part
to reaction by fragment ions (not shown) following the
initial 10 msec, 16 eV electron beam pulse.

The gas phase ion chemistry of 2-fluoropropane further exemplifies these considerations (Beauchamp and Park, 1974). The variation of ion abundance with time is shown in Figure 7 for

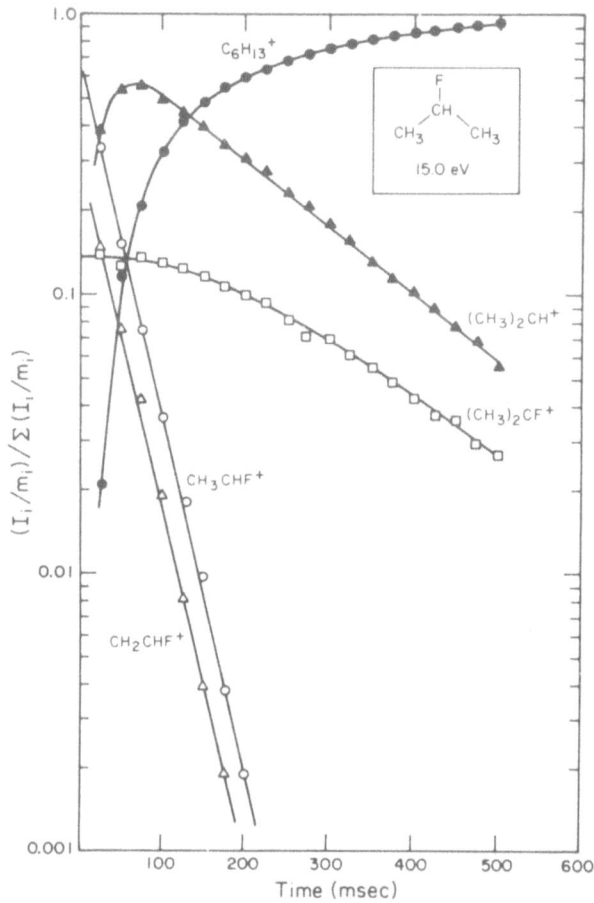

Fig. 7.   Variation of ion abundance with time for 2-fluoropropane at 6.2×10^{-7} torr ionized with a 6 msec electron beam pulse at 15.0 eV.

2-fluoropropane ionized by 15 eV electrons at a pressure of 7.0×10^{-7} torr. Fragmentation leads to the formation of CH_2CHF^+, CH_3CHF^+ and $(CH_3)_2CF^+$ as the principal reactant ions. All three of these ions reactant by exothermic F^- transfer to generate the isopropyl cation (reactions 32-34). The isopropyl cation reacts with 2-fluoropropane to form a diisopropyl fluoronium ion which decomposes by loss of propane (reaction 35) or by a more complex rearrangement process in which HF elimination leads to formation of a hexyl cation (reaction 36). Although reaction 35 has the

$$CH_2CHF^+ + (CH_3)_2CHF \rightarrow (CH_3)_2CH^+ + C_2H_3F_2 \tag{32}$$

$$CH_3CHF^+ + (CH_3)_2CHF \rightarrow (CH_3)_2CH^+ + CH_3CHF_2 \tag{33}$$

$$(CH_3)_2CF^+ + (CH_3)_2CHF \rightarrow (CH_3)_2CH^+ + (CH_3)_2CF_2 \tag{34}$$

$$(CH_3)_2CH^+ + (CH_3)_2CHF$$
$$\downarrow$$

$$\begin{bmatrix} CH_3 & & CH_3 \\ & \overset{+}{>}CH-F-CH< & \\ CH_3 & & CH_3 \end{bmatrix}^* \begin{array}{l} \nearrow (CH_3)_2CF^+ + (CH_3)_2CH_2 \qquad (35) \\ \\ \searrow C_6H_{13}^+ + HF \qquad\qquad\qquad (36) \end{array}$$

overall appearance of a hydride transfer process, it is unlikely that the reaction occurs directly; evidence for "onium" inter-mediates in related hydride transfer reactions has been presented elsewhere (Beauchamp, Holtz, Woodgate and Patt, 1972). It is of interest to note that reactions 34 and 35 taken together comprise a chain reaction, the overall process (reaction 37) resulting in the

$$(CH_3)_2CHF \rightarrow (CH_3)_2CH_2 + (CH_3)_2CF_2 \tag{37}$$

conversion of 2-fluoropropane to a mixture of 2, 2-difluoropropane and propane. Each step of the reaction is exothermic as a result of the reversal in R^+-X^- heterolytic bond energies for the iso-propyl and 2-fluoropropyl cations when the reference base is changed from F^- to H^-. While in the case indicated the chain reaction is terminated by the condensation process 36, it is pos-sible with the use of thermochemical data such as that presented in Table IV and the rules summarized above to predict a number of interesting chain reactions. Other chain reactions of this type have been reported which are not terminated under the conditions of the experiment (Blint, McMahon and Beauchamp, 1974).

Reactions of fluoromethyl cations with ROH (R = H, CH_3, and C_2H_5) can be generally written as indicated in processes 38 and 39 (Cook and Beauchamp, 1974). For R = H and CH_3, elimination of HF is observed (reaction 38). For R = C_2H_5, only reaction 39 is observed. From these results it can be inferred that $D(R^+\text{-}OH^-)$

$$CH_2F^+ + ROH \rightarrow \begin{bmatrix} F & H \\ | & | \\ CH_2-O-R \\ & + \end{bmatrix}^* \begin{array}{l} \nearrow CH_2=\overset{+}{O}-R + HF \qquad (38) \\ \\ \searrow CH_2FOH + R^+ \qquad\quad (39) \end{array}$$

decreases in the order $CH_3^+ > CH_2F^+ > C_2H_5^+$. This is the same order as observed for the reference bases F^- and H^-.

Esterification reactions (Cook and Beauchamp, 1974) are a particularly interesting class of reactions, the mechanism for which in part involves nucleophilic attack on substituted carbonium ions. Using deuterium and ^{18}O labelling, the mechanism of acetic acid esterification reactions in which the labile proton is initially bound to the acid is as indicated in reaction 40, with elimination of H_2O occurring from the reaction intermediate. However, if the the labile proton is initially bound to the alcohol, the reaction proceeds as in 41. Observation of the condensation product for

$$CH_3\text{-}C(\text{OH})(O\text{-}H)^+ + ROH \rightarrow \left[CH_3\text{-}C(\text{H-O-H})(O)(^+O\text{-}H)(R) \right]^* \rightarrow CH_3\text{-}C(O)(\overset{+}{O}\overset{-}{R})(H) + H_2O \qquad (40)$$

$$CH_3CO_2H + R\overset{+}{O}H_2 \rightarrow \left[CH_3\text{-}C(O\cdots\overset{+}{H}\cdots O\text{-}H\text{-}R)(O\text{-}H) \right]^*$$

$$R = CH_3, C_2H_5 \qquad\qquad R = (CH_3)_2CH,\ (CH_3)_3C$$

$$\left[CH_3\text{-}C(\text{H-O-H})(^+O\text{-}H)(R) \right]^* \qquad\qquad \left[CH_3\text{-}C(O\cdots\overset{+}{H})(O\text{-}H\text{-}R)(H) \right]^*$$

$$\downarrow \qquad\qquad\qquad\qquad\qquad\qquad \downarrow$$

$$CH_3\text{-}C(O)(\overset{+}{O}\text{-}R)(H) + H_2O \qquad\qquad CH_3\text{-}C(O)(\overset{+}{O}\text{-}R)(H) + H_2O$$

$$(41a) \qquad\qquad\qquad\qquad\qquad\qquad (41b)$$

reaction 40 suggests that the dihydroxymethylcarbonium ion is more stable than even the t-butyl cation. The relative proton affinities of acetic acid and ROH are $(CH_3)_3COH > (CH_3)_2CHOH > CH_3CO_2H > C_2H_5OH > CH_3OH$. Hence when the labile proton is initially bound to the alcohol (reaction 41) the mechanism conforms to the expectations for nucleophilic attack at coordinately saturated carbon. In the case of CH_3OH and C_2H_5OH the labile proton is transferred to the more basic acetic acid site and the reaction proceeds as in the case when the labile proton is initially bound to the acid.   This leaves open the question as to why the labile proton is not first transferred to the alcohol for ROH = $(CH_3)_2CHOH$ and $(CH_3)_3COH$ in process 40.   It appears that exothermic proton transfer reactions which can precede condensation reactions occur competitively only when they are facile in the sense of not involving a significant change in the charge distribution and structure of the donor ion.   Esterification reactions are thus far the only case in which this phenomenon has been observed, and can be attributed to the significant charge delocalization expected in protonated acids.

The reactions of π-donor bases (e.g., ethylene and benzene) with substituted carbonium ions are being examined in several laboratories and will likely prove to be of mechanistic interest. For example, addition of CD_3^+ to C_6H_6 results in the loss of H_2, HD and D_2 with product distribution indicated in reaction 42 (Blint, 1972).   This suggests that either hydrogen scrambling is fast compared to elimination of HD or that the intermediate has rearranged to yield an ionic product such as the tropyllium ion (cyclo-$C_7H_7^+$) other than the benzyl cation ($C_6H_5CH_2^+$).

$$CD_3^+ + C_6H_6 \rightarrow \left[\begin{array}{c} CD_3 \; H \\ \bigcirc_+ \end{array} \right]^* \quad \begin{array}{l} \xrightarrow{15\%} C_7H_6D^+ + D_2 \\ \xrightarrow{72\%} C_7H_5D_2^+ + HD \\ \xrightarrow{13\%} C_7H_4D_3^+ + H_2 \end{array} \qquad (42)$$

Addition of CH_2F^+ to C_6D_6 appears to proceed cleanly, however, the only observed product resulting from loss of DF (reaction 43)

$$CH_2F^+ + C_6D_6 \rightarrow \left[\begin{array}{c} F \\ D \; CH_2 \\ \bigcirc_+ \end{array} \right]^* \rightarrow C_6D_5CH_2^+ + DF \qquad (43)$$

In studying the photochemistry of benzyl cations, reaction 44 has been conveniently used as a "clean" source of difluorobenzyl cations whose photochemistry is quite different from the $C_7H_5F_2^+$ ion generated by electron impact from trifluoromethyl benzene (Freiser and Beauchamp, 1974).

$$CF_3^+ + C_6H_6 \rightarrow \left[\begin{array}{c} \overset{F}{\underset{H}{\diagdown}} CF_2 \\ \bigoplus_{+} \end{array} \right]^* \rightarrow C_6H_5CF_2^+ + HF \qquad (44)$$

A quite intriguing process involving nucleophilic attack on substituted carbonium ions is the recently reported reaction 45 of CF_3^+ with acetone which can be envisioned as a four-center elimination process (Eyler, Ausloos and Lias, 1974).

$$CF_3^+ + (CH_3)_2CO \rightarrow \left[\begin{array}{c} CF_2 \\ O \diagup \diagdown F \\ +C \\ CH_3 \quad CH_3 \end{array} \right]^*$$

$$\rightarrow (CH_3)_2CF^+ + CF_2O \qquad (45)$$

REACTION MECHANISMS OF INORGANIC IONS

The chemistry of organic ions in the gas phase has been investigated with increasing attention to detail for nearly two decades. It is only recently, however, that the real potential for studying the reactivity of transition metal complexes in the gas phase has been revealed (Foster and Beauchamp, 1971). Already it is clear that a variety of processes are amenable to study, including (1) formation of binuclear complexes, (2) ligand displacement reactions, (3) determination of ligand binding energies, (4) determination of the basicity of transition metal complexes, (5) examination of processes involving both electrophilic and nucleophilic attack on transition metal complexes, (6) generation and study of unusual π- and σ-bonded organometallic complexes and (7) studies of the photochemistry of transition metal complexes (including photodecomposition and photodetachment processes).

Examples of these studies are shown in Figures 8-10. In Figure 8 the effect of adding H_2O to $Fe(CO)_5$ is demonstrated. At 18 eV electron energy, all of the species $Fe(CO)_n^+$ (n = 1-5) are generated in addition to H_2O^+ (which reacts with H_2O to form H_3O^+ which in turn reacts by proton transfer to give $HFe(CO)_5^+$). As the H_2O pressure is increased, new peaks are observed which can be accounted for by sequential ligand displacement reactions such as 46 and 47. It is interesting to note that while $Fe(CO)^+$ exchanges

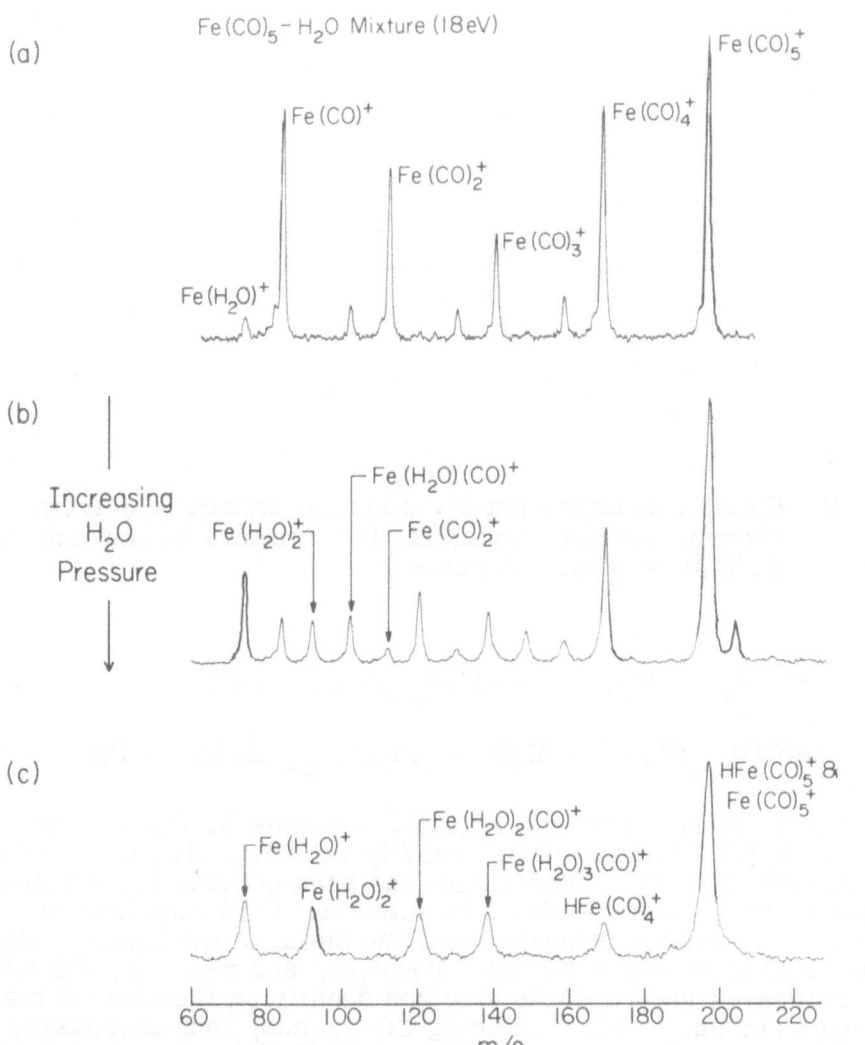

Fig. 8. Single resonance ion cyclotron resonance spectra showing displacement of CO by H_2O as the pressure of the latter is increased in a mixture of H_2O and $Fe(CO)_5$. Not shown is the ion H_3O^+ (formed from H_2O^+) which reacts with $Fe(CO)_5$ to form $HFe(CO)_5^+$.

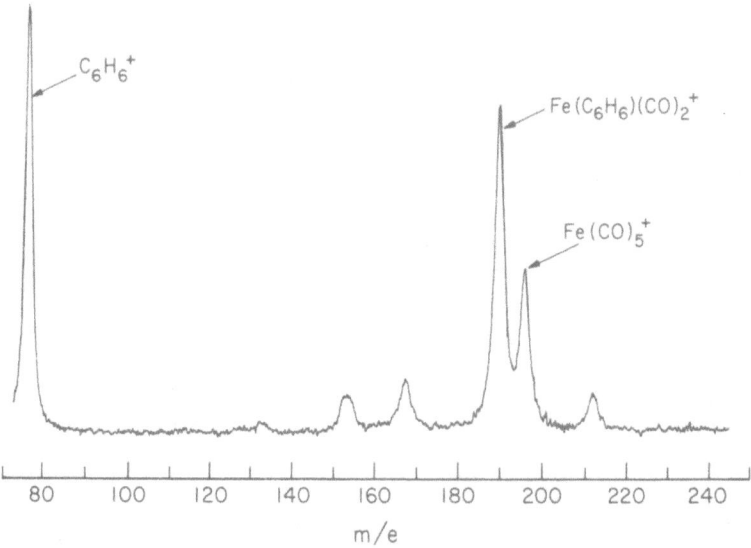

Fig. 9. Single resonance ion cyclotron resonance spectrum
showing the ions remaining in a mixture of benzene and
$Fe(CO)_5$ at high pressures.

$$Fe(CO)_n^+ + H_2O \rightarrow Fe(CO)_{n-1}(H_2O)^+ + CO \tag{46}$$

$$Fe(CO)_{n-1}(H_2O)^+ + H_2O \rightarrow Fe(CO)_{n-2}(H_2O)_2^+ + CO \tag{47}$$

completely to $Fe(H_2O)^+$ and $Fe(CO)_2^+$ becomes $Fe(H_2O)_2^+$, the
species $Fe(CO)_3^+$ and $Fe(CO)_4^+$ readily exchange all but one of the
CO ligands with H_2O. This effect has been observed in studies of
ligand displacement reactions in solution. CO bonds both by σ-
donation and π-back donation from the metal to the ligand. The
ligand H_2O is purely a σ-donor, however, and replacing CO with
H_2O increases the available electron density on the central metal
for back bonding to the remaining CO ligands, thus increasing the
metal ligand binding energy. The ion $Fe(CO)_5^+$ is inert against
substitution by H_2O. This species is a 17 electron complex (18
electrons around the central metal atom would be appropriate for
a rare gas configuration). With electron pair donors as ligands
Fe^+ prefers to form five-coordinate 17 electron complexes and
will not exceed the rare gas electron configuration. Similarly,
Fe^- forms four-coordinate 17 electron complexes with electron
pair donors. The reactivity of complexes of Fe^+ and Fe^- demonstrate this propensity. For example, in Figure 9, the most
abundant ion in a mixture of $Fe(CO)_5$ and benzene is the species

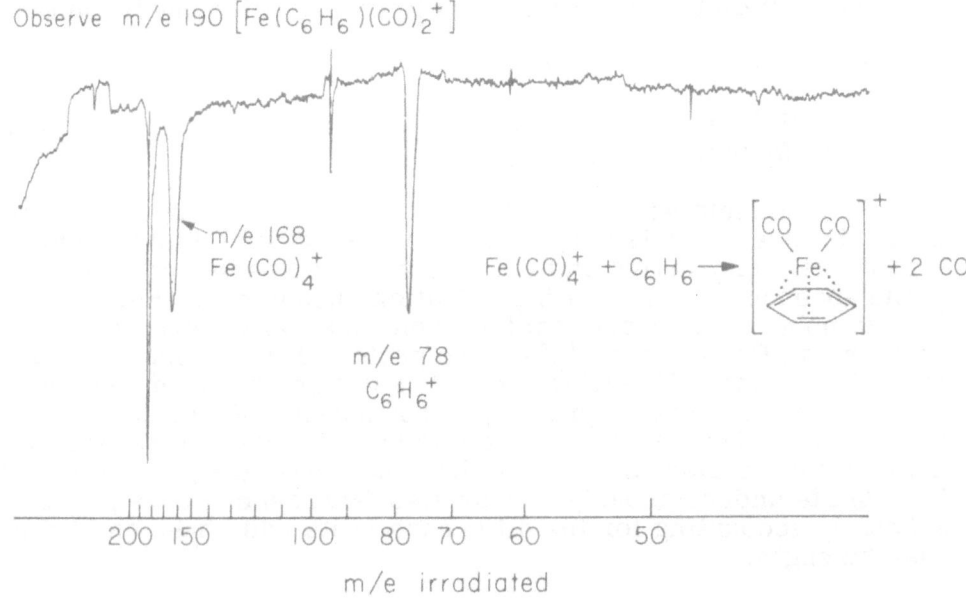

Fig. 10. Double resonance spectrum of the $Fe(C_6H_6)(CO)_2^+$ complex observed in Figure 9. The major contributions are from $Fe(CO)_4^+$ and $C_6H_6^+$.

$Fe(CO)_2(C_6H_6)^+$. The double resonance experiment in Figure 10 indicates that $Fe(CO)_4^+$ is the major precursor to this product ion (the contribution from $C_6H_6^+$ is probably via charge transfer to $Fe(CO)_5$, forming $Fe(CO)_4^+$). The indicated reaction 48 involves

$$Fe(CO)_4^+ + C_6H_6 \rightarrow \left[\begin{array}{c} CO \quad CO \\ Fe \\ \end{array} \right]^+ + 2CO \qquad (48)$$

displacement of two CO molecules by C_6H_6. Benzene thus acts as a tridendate ligand, forming the characteristic 17 electron five-coordinate complex.

The reactions of closed shell ions with $Fe(CO)_5$ leads to complexes with either 16 or 18 electrons around the central metal. For example, in a mixture of NF_3 and $Fe(CO)_5$, F^- formed from NF_3 by electron attachment interacts with $Fe(CO)_5$ in accordance with reaction 49 to displace two CO ligands. Other nucleophiles

$$F^- + Fe(CO)_5 \rightarrow FFe(CO)_3^- + 2DO \tag{49}$$

such as $C_2H_5O^-$ react similarly, the 4-coordinate 16 electron complex being formed exclusively.

Experiments with transition metal complexes are of particular interest since there is very little information available regarding the intrinsic properties and reactivity of the species considered, in the absence of complicating solvation phenomena. Ferrocene, for example, is an exceptionally strong base in the gas phase (Foster and Beauchamp, 1974), the proton affinity being comparable to methyl amine (217 kcal/mole). In solution, however, strongly acidic conditions are required for protonation. Spectroscopic studies have shown that the proton is bound to the central metal atom, which is also likely to be the case in the gas phase (structure III). Quite understandably, protonated ferrocene will be poorly solvated, accounting for the difference in its gas phase and solution base strength.

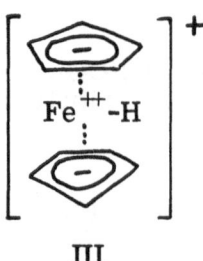

III

In addition to transition metal complexes there remains a wealth of chemistry to be explored in studying the ion chemistry of inorganic molecules. Some of these discoveries may have important consequences which extend beyond the intrinsic interest of the processes themselves. For example, as a result of its extremely efficient capture of thermal electrons, sulfur hexafluoride is widely used to chemically "remove" electrons from both gaseous and liquid radiolytic systems. In elucidating reaction mechanisms, the tacit assumption usually made is that SF_6^- is chemically inert. On the contrary, SF_6^- is highly reactive with a variety of acidic species which often are constituents of radiolytic systems (Foster and Beauchamp, 1974). For example, in a mixture of SF_6 and HBr (Figure 11), reactions 50-55 are observed. Br^- and $BrHBr^-$ are the final products in this system and do not react further. From the limiting slopes for the disappearance of the reactants, the total rate constants for reaction of SF_6^-, SF_5^- and $FHBr^-$ are 5.8, 3.5 and 2.4×10^{-10} cm^3 molecule^{-1} sec^{-1}, respectively. Reactions such as these can be described in terms

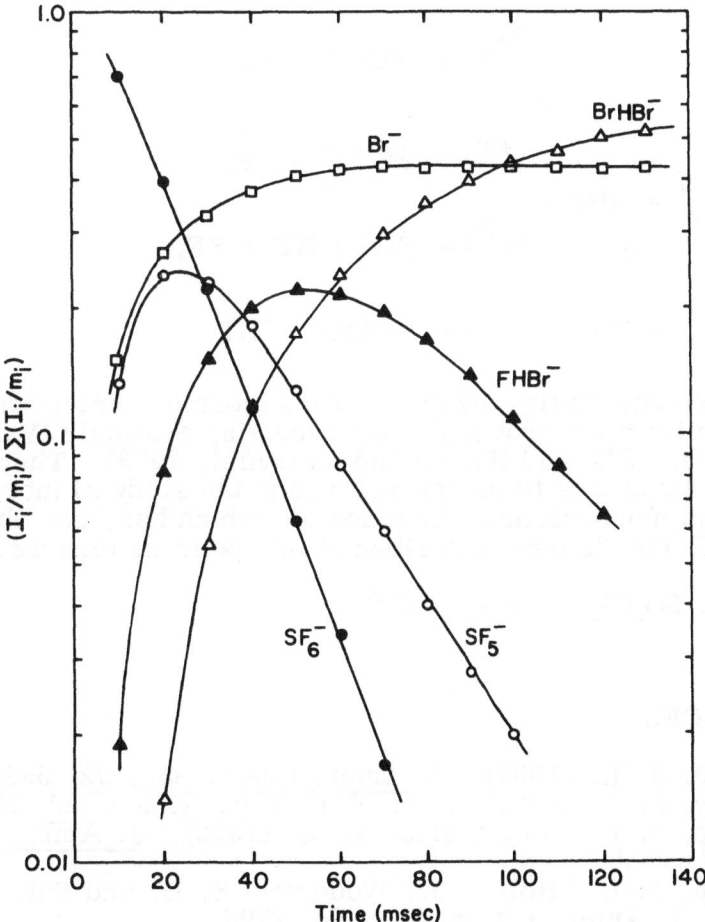

Fig. 11. Trapped-ion study of the negative ions in a 5:1 mixture
of HBr and SF_6 at 1.7×10^{-6} torr. SF_6^- is formed at
t = 0 by a 4 msec pulse of 70 volt electrons. Only Br^-
originally derived from SF_5^- is included; that formed
from dissociative electron attachment by HBr is
excluded.

$$SF_6^- + HBr \quad \begin{cases} \xrightarrow{4\%} Br^- + HF + SF_5 & (50) \\[2mm] \xrightarrow{74\%} SF_5^- + HF + Br & (51) \\[2mm] \xrightarrow{22\%} FHBr^- + SF_5 & (52) \end{cases}$$

$$SF_5^- + HBr \quad \begin{cases} \xrightarrow{48\%} FHBr^- + SF_4 & (53) \\[2mm] \xrightarrow{52\%} Br^- + HF + SF_4 & (54) \end{cases}$$

$$FHBr^- + HBr \longrightarrow BrHBr^- + HF \tag{55}$$

of acid-base chemistry and quantitative thermochemical data of use in interpreting such processes are becoming available (Yamdagni and Kebarle, 1971 and Haartz and McDaniel, 1973). There are still many surprises to be encountered in the study of inorganic ion reaction mechanisms. Reaction 56, which has been observed to be rapid, can be best described at this point as bizarre.

$$Cl^- + SO_2Cl_2 \rightarrow SO_2 + Cl_3^- \tag{56}$$

REFERENCES

Beauchamp, J. L. (1969). J. Amer. Chem. Soc. 92, 5925.

Beauchamp, J. L. (1971). Ann. Rev. Phys. Chem. 22, 527.

Beauchamp, J. L. and Caserio, M. C. (1972). J. Amer. Chem. Soc. 94, 2638.

Beauchamp, J. L., Holtz, D., Woodgate, S. D. and Patt, S. L. (1972). J. Amer. Chem. Soc. 94, 2798.

Beauchamp, J. L., McMahon, T. B. and Miasek, P. (1973). Advances in Mass Spectrometry, Vol. 6, Applied Science Publishers Ltd, England, in press.

Beauchamp, J. L. and Park, J. Y. (1974). To be published.

Benson, S. W. (1968). Thermochemical Kinetics. John Wiley and and Sons, New York, 75.

Blint, R. J. (1972). Ph.D. Thesis, California Institute of Technology.

Blint, R. J., McMahon, T. B. and Beauchamp, J. L. (1974). J. Amer. Chem. Soc. 96, 1269.

Cook, J. H. and Beauchamp, J. L. (1974). J. Amer. Chem. Soc. 96, in press.

Curphey, T. J., Santer, J. O., Rosenblum, M. and Richards, J. H. (1969). J. Amer. Chem. Soc. 82, 5249.

Dawson, J. H. J., Henderson, W. G., O'Malley, R. M. and Jennings, K. R. (1973). Int. J. Mass Spectrom. Ion Phys. 11, 61.

Eyler, J. R., Ausloos, P. and Lias, S. G. (1974). J. Amer. Chem. Soc. 96, 3673.

Foster, M. S. and Beauchamp, J. L. (1971). J. Amer. Chem. Soc. 93, 4924.

Foster, M. S. and Beauchamp, J. L. (1974a). J. Inorg. Chem. submitted for publication.

Foster, M. S. and Beauchamp, J. L. (1974b). To be published.

Freiser, B. S. and Beauchamp, J. L. (1974). To be published.

Grimsrud, E. P. and Kebarle, P. (1973). J. Amer. Chem. Soc. 95, 7939.

Haartz, J. C. and McDaniel, D. H. (1973). J. Amer. Chem. Soc. 95, 8562.

Henderson, W. G., Holtz, D., Beauchamp, J. L. and Taft, R. W. (1972). J. Amer. Chem. Soc. 94, 4728.

Hiraoka, K., Yamdagni, R. and Kebarle, P. (1973). J. Amer. Chem. Soc. 95, 6833.

Holtz, D. Beauchamp, J. L. and Woodgate, S. D. (1970). J. Amer. Chem. Soc. 92, 7484.

Holtz, D. and Beauchamp, J. L. (1971a). Nature Phys. Science 231, 204.

Holtz, D. and Beauchamp, J. L. (1971b). Science 173, 1237.

Kebarle, P. (1972). Higher-Order Reaction-Ion Clusters and Ion Solvation. Ion-Molecule Reactions (Ed. J. L. Franklin). Plenum Press, New York, 315.

Kollman, P. A. and Allen, L. C. (1970). J. Amer. Chem. Soc. 92, 6101.

Kraemer, W. P. and Diercksen, G. H. F. (1970). Chem. Phys. Lett. 5, 463.

Lieder, C. A. and Brauman, J. I. (1974). J. Amer. Chem. Soc. 96, 4028.

Long, J. W. and Franklin, J. L. (1974). J. Amer. Chem. Soc. 96, 2320.

Maccoll, A. and Stimson, V. R. (1966). J. Chem. Soc. 2836.

McIver, R. T. and Miller, J. S. (1974). J. Amer. Chem. Soc. 96, 4323.

McMahon, T. B. and Beauchamp, J. L. (1972). Rev. Sci. Instr. 43, 509.

McMahon, T. B., Blint, R. J., Ridge, D. P. and Beauchamp, J. L. (1972). J. Amer. Chem. Soc. 94, 8934.

Ridge, D. P. and Beauchamp, J. L. (1974a). J. Amer. Chem. Soc. 96, 637.

Ridge, D. P. and Beauchamp, J. L. (1974b). J. Amer. Chem. Soc. 96 3595.

Ridge, D. P., McMahon, T. B., Park, J. Y. and Beauchamp, J. L. (1974). To be published.

Riveros, J. M. Breda, A. C. and Blair, L. K. (1973). J. Amer. Chem. Soc. 95, 4066.

Sullivan, S. A. and Beauchamp, J. L. (1974). To be published.

Wieting, R. D., Staley, R. H. and Beauchamp, J. L. (1974). J. Amer. Chem. Soc. Submitted for publication.

Williamson, A. D. and Beauchamp, J. L. (1974). To be published.

Wineland, D., Ekstrom, P. and Dehmelt, H. (1973). Phys. Rev. Lett. 31, 1279.

Yamdagni, R. and Kebarle, P. (1971). J. Amer. Chem. Soc. 93, 7139.

Yamdagni, R. and Kebarle, P. (1973). J. Amer. Chem. Soc. 95, 3504.

ION-MOLECULE REACTIONS IN SILANE AND ORGANOSILANE SYSTEMS

F.W.Lampe[1,2]

Hahn-Meitner-Institut für Kernforschung Berlin GmbH,
Bereich Strahlenchemie, 1 Berlin 39, Germany

1. INTRODUCTION

Ion-molecule reactions in hydrocarbons and in hydrocarbon derivatives containing oxygen, nitrogen, sulfur, and halogens have been studied extensively (Franklin, 1972). The knowledge acquired from these investigations has been of considerable value, not only to physical chemists and chemical physists interested in reactions and molecular properties in the gaseous phase, but also to organic chemists in providing some new avenues of thought concerning the condensed phase chemistry of strong acid solutions of alkanes (Olah and Schlosberg, 1968). An impressive body of knowledge relative to ion-molecule reactions of hydrocarbons and their derivatives containing atoms of Groups V - VII now exists.

However, from the point of view of the inorganic and organo-metallic chemist, comparatively little is known about the ion-molecule reactions characteristic of compounds in which one or more carbon atoms of a hydrocarbon or hydrocarbon derivative are replaced by another atom or atoms of Group IVa of the periodic table, namely by silicon, germanium, tin, or lead. While structurally similar to hydrocarbons and hydrocarbon derivatives, the chemistry of these inorganic and organometallic compounds is generally quite different.

Present knowledge of ion-molecule reactions of such compounds is restricted almost entirely to silanes and organosilanes. The purpose of this paper is to review this body of knowledge, making comparisons wherever pertinent to the ion-molecule reaction chemistry of hydrocarbons.

445

2. MONOSILANE

From the point of view of the chemist, methane has played a very important role in the development of knowledge of ion-molecule reactions.The discoveries of the formation of CH_5^+ (Talroze and Lyubimova, 1952) and $C_2H_5^+$ (Schissler and Stevenson, 1956) in ionized methane by (1) and (2), viz.

$$CH_4^+ + CH_4 \longrightarrow CH_5^+ + CH_3 \qquad (1)$$

$$CH_3^+ + CH_4 \longrightarrow C_2H_5^+ + H_2 \qquad (2)$$

were central factors in calling attention to the rich chemistry of collision processes in ionized gases. Reactions (1) and (2) are by far the predominant ion-molecule reactions in ionized methane.

When the carbon atom of methane is replaced by silicon there results the simplest silane, namely monosilane, SiH_4. In Table I there are shown some comparative data for methane and monosilane that are pertinent to the gas phase ionic chemistry. From this table we conclude that while methane is quite stable relative to the elements that compose it, monosilane is unstable relative to its elements, all elements being of course in their standard states. Despite this difference in stability the bond dissociation energy of

TABLE I		
Comparison of Methane and Monosilane		
Property	CH_4	SiH_4
Standard Enthalpy of Formation, ΔH_f^0	-0.776 eV[a]	+0.36 eV[a]
Bond Dissociation Energy, $D(H_3M-H)$	4.5 eV[a]	3.6-4.1 eV[b]
Ionization Potential, $I_z(MH_4)$	12.7 eV[a]	11.7 eV[c]
Appearance Potential, $A(MH_3^+/MH_4)$	14.3 eV[a]	12.3 eV[b]
Appearance Potential, $A(MH_2^+/MH_4)$	15.2 eV[a]	11.9 eV[b]
Bond Dipole Moment, $\mu(M-H)$	$\overrightarrow{0.4}$ D[e]	$\overleftarrow{1.6}$ D[d]

a) Franklin, Dillard, Rosenstock, Herron, Drexl and Field, 1969
b) Potzinger and Lampe, 1969
c) Pullen, Carlson, Moddeman, Schweitzer, Bull and Grimm, 1970
d) Ball and McKean, 1962
e) Barrow, 1973
f) Potzinger, 1974

the Si-H bond in SiH_4 is at least 80 % of that of the C-H bond in CH_4. Less energy is required to ionize SiH_4 as compared to CH_4 and considerably less energy is required to form SiH_3^+ and SiH_2^+ from SiH_4 than is the case for CH_3^+ and CH_2^+ from CH_4. The bond dipole moment of the Si-H bond is of considerably greater magnitude than that of the C-H bond and is of opposite direction.

If one examines the mass spectrum of monosilane in the pressure range of 10^{-5} to 0.5 torr, one finds, in contrast to the case of methane, neither SiH_4^+ nor SiH_5^+ and therefore no reaction in monosilane analogous to (1). A rationalization for the virtual absence of SiH_4^+ and therefore the non-existence of (1) in mono- silane can be made on the basis of the data in Table I. Combinat- ion of the ionization and appearance potentials in Table I show that reactions (3) and (4), namely

$$SiH_4^+ \longrightarrow SiH_3^+ + H \tag{3}$$

$$SiH_4^+ \longrightarrow SiH_2^+ + H_2 \tag{4}$$

are endothermic by only 0.6 and 0.2 eV, respectively, while the corresponding processes in methane are endothermic by 1.6 and 2.5 eV, respectively. Thus SiH_4^+ formed from SiH_4 in a vertical Franck-Condon process may not be stable energetically with re- spect to SiH_2^+ and SiH_3^+ and thus SiH_4^+ is not present as a primary ion in the mass spectrum of SiH_4.

All primary ions of monosilane, namely Si^+, SiH^+, SiH_2^+, and SiH_3^+ react with SiH_4 (Henis, Stewart, Tripodi and Gaspar, 1972; Yu, Cheng, Kempter and Lampe, 1972). The predominant $(k > 10^{-10} \text{ cm}^3/\text{sec})$ secondary reactions are shown by (5) - (8). The primary ion

$$Si^+ + SiH_4 \longrightarrow Si_2H_2^+ + H_2 \tag{5}$$

$$SiH^+ + SiH_4 \longrightarrow Si_2H_3^+ + H_2 \tag{6}$$

$$SiH_2^+ + SiH_4 \longrightarrow Si_2H_4^+ + H_2 \tag{7}$$

$$SiH_2^+ + SiH_4 \longrightarrow SiH_3^+ + SiH_3 \tag{8}$$

SiH_3^+ does react with monosilane in a process analogous to (2) to form $Si_2H_5^+$ but this reaction is quite slow $(k \sim 10^{-11} \text{cm}^3/\text{sec})$.

The most rapid secondary reaction in monosilane is (8) and since SiH_2^+ is the major primary ion of monosilane and SiH_3^+ reacts only slowly with SiH_4 in bimolecular processes, increasing the pressure of monosilane in a mass-spectrometer ion-source results at first in an increase in the relative abundance of SiH_3^+. With further increases of pressure to ~0.1 torr, SiH_2^+ is sufficiently depleted and third-order processes become of sufficient importance for SiH_3^+ to be consumed by the ternary reaction (9), viz.

$$SiH_3^+ + 2SiH_4 \longrightarrow Si_2H_7^+ + SiH_4 \qquad (9)$$

The product of (9), namely $Si_2H_7^+$, attains a relative abundance of 35 % of the total ionization at a pressure of about 0.2 torr. Further increases in the pressure of monosilane in a mass spectrometer ion-source result in depletion of $Si_2H_7^+$ and formation of heavier ions, principally $Si_3H_8^+$, $Si_3H_9^+$, $Si_4H_{10}^+$, and $Si_4H_{11}^+$.

3. MONOSILANE-METHANE MIXTURES

In ionized monosilane-methane mixtures one finds, in addition to the ion-molecule reactions characteristic of the pure compounds, reactions between ions derived from methane with monosilane and vice-versa. These so-called cross-reactions have been identified and the cross sections measured using a tandem mass spectrometer (Cheng, Yu and Lampe, 1973). The relative cross-sections for formation of the product ions from the various reactant pairs are shown in Table II in which the identities of the product ions are given by the m/e (mass-to-charge) ratios.

TABLE II
RELATIVE CROSS-SECTIONS IN SiH_4-CH_4 MIXTURES[a) c)]

Reactant ion	Reactant molecule	Relative cross section for formation of secondary ion at given m/e								
		28	29	30	31	41	42	43	44	45
C^+	SiH_4	5.2	7.8	13.0	54.6	2.7[b]	7.1[b]			
CH^+	SiH_4	5.0	9.8	9.1	42.5		18.6	13.4		
CH_2^+	SiH_4	2.3	7.3	2.3	47.2		2.5	15.1		
CH_3^+	SiH_4				52.5			1.7		1.6
CH_4^+	SiH_4			10.9	35.2				1.0	1.5
CH_5^+	SiH_4				49.5					
Si^+	CH_4							2.5		
SiH^+	CH_4							6.7	10.4	
SiH_2^+	CH_4								4.4	2.6
SiH_3^+	CH_4									0.5

a) Reactant ion energy of 1 eV (laboratory) unless otherwise indicated. b) Reactant ion energy of 1.5 eV (laboratory). c) To convert to units of $Å^2$ the data in table should be multiplied by 1.2 m/e (mass-to-change ratio) values

One notes immediately the dominance of reactions in which ions derived from methane produce the ion having m/e 31 upon collision with SiH_4. Thus the major reactions of all methane primary ions as well as of CH_5^+ with SiH_4 is a hydride transfer process, which for the primary ions C^+, CH^+, CH_2^+, and CH_3^+ may be written as in (10), viz.

$$CH_n^+ + SiH_4 \longrightarrow CH_{n+1}^+ + SiH_3^+; \quad n = 0, 1, 2, 3 \qquad (10)$$

In the case of hydride ion transfer to CH_4^+ and CH_5^+, the reaction is most likely dissociative and can be written as in (11). Other experiments (Cheng and Lampe, 1972) showed

$$CH_m^+ + SiH_4 \longrightarrow CH_{m-1}^+ + H_2 + SiH_3^+; \quad (n = 4, 5) \qquad (11)$$

that $C_2H_5^+$, the other major product of ion-molecule reactions in pure methane, also reacts with monosilane to yield almost exclusively SiH_3^+. Since all primary and secondary ions derived from methane react predominantly with monosilane to yield SiH_3^+ and since the major ion-molecule reaction in pure monosilane is the formation of SiH_3^+ via (8), it follows that, over the entire range of composition, increasing the pressure of ionized monosilane-methane mixtures should result predominantly in the formation and subsequent reaction of SiH_3^+. This consequence of the domination of hydride ion transfer from monosilane has been confirmed by high-pressure mass spectrometry (Cheng, Yu and Lampe, 1973).

As is evident from Table II, and as shown by high-pressure mass spectrometry (Cheng, Yu and Lampe, 1973), only insignificant amounts, if indeed any, of SiH_5^+ are formed in monosilane-methane mixtures. The fact that SiH_5^+ is not observed in the high-pressure mass-spectrometric experiments up to 0.5 torr suggests that the formation of SiH_3^+ proceeds via a direct hydride ion transfer and not via protonation of SiH_4 followed by dissociation to SiH_3^+ and H_2. If the latter were true one would expect to see some collisional stabilization of the SiH_5^{+*} intermediate at the higher pressures.

This idea has been tested further by studying the isotopic composition of the silyl ion formed in the reaction of CD_4^+ and CD_5^+ with SiH_4 (Stewart, Henis and Gaspar, 1972; Cheng and Lampe, 1972). The alternatives are indicated in (12) - (15), viz.

$$[SiH_4D^{+*}] \xrightarrow{\quad M \quad} SiH_4D^+ + M \tag{12}$$

$$\rightarrow SiH_3^+ + HD \tag{13}$$

$$XD^+ + SiH_4$$

$$\rightarrow SiH_2D^+ + H_2 \tag{14}$$

$$\text{Direct} \quad \rightarrow SiH_3^+ + X + HD \tag{15}$$

It has already been mentioned that SiH_5^+ was not observed in the high-pressure mass-spectrometric experiment (CH_4-SiH_4 mixtures) up to 0.5 torr and this would seem to rule out the collisional stabilization shown by (12). At low pressures the isotopic composition of silyl ion is found   within experimental error to be 100 % SiH_3^+. Since statistical break-down of SiH_4D^{+*} would yield 60 % SiH_2D^+ and 40 % SiH_3^+, it would seem to be quite certain that the hydride transfer process yielding SiH_3^+ proceeds via a direct transfer of H^- from the monosilane to the attacking ion. In this vein it is instructive to point out that the bond dipole moment of the Si-H bond (cf. Table I) indicates that the hydrogenic species in monosilane are already partially hydridic in the isolated neutral molecule. The predominance of the direct hydride ion transfer would thus seem to be a natural consequence of the magnitude and direction of the polarity of the Si-H bond.

4. FORMATION OF SiH_5^+

Despite its apparent absence in pure monosilane and in monosilane-methane mixtures, SiH_5^+ does exist and has been found to be formed via the bimolecular reaction depicted by (16) - (19) (Cheng and Lampe, 1973a).

$$NH_2^+ + SiH_4 \longrightarrow NH + SiH_5^+ \tag{16}$$

$$C_2H_3^+ + SiH_4 \longrightarrow C_2H_2 + SiH_5^+ \tag{17}$$

$$C_2H_6^+ + SiH_4 \longrightarrow C_2H_5 + SiH_5^+ \tag{18}$$

$$C_3H_8^+ + SiH_4 \longrightarrow C_3H_7 + SiH_5^+ \tag{19}$$

The dependence of the reaction cross-sections of (16) - (19) on the relative kinetic energy of the collision partners indicates that (16) - (18) are exothermic or thermoneutral reactions while (19) is an endothermic process. Thus for (16) - (18), $\Delta H^\circ \lesssim 0$, while for (19) $\Delta H^\circ \gtrsim 0$. Using these two inequalities and known standard enthalpies of formation (Franklin, Dillard, Rosenstock, Herron, Drexl, and Field, 1969) one may derive upper and lower limits to the enthalpy of formation of SiH_5^+ or, equivalently, to the proton affinity of monosilane. One obtains the following:

$$217 < \Delta H_f^o(SiH_5^+) < 223 \text{ kcal/mole}$$

$$150 < P(SiH_4) < 156 \text{ kcal/mole}$$

where P denotes proton affinity and is defined as the exothermicity of the addition of a proton to monosilane. The proton affinity of monosilane is thus seen to be 25-30 kcal/mole higher than that of methane (Haney and Franklin, 1969). Hence monosilane is a considerably stronger base in the Bronsted sense than is methane. The relative energetics involved in the formation of SiH_3^+ from SiH_4 and CH_3^+ from CH_4, i.e. hydride ion abstraction from the parent molecule, indicate that monosilane is a stronger base in the Lewis sense than is methane by about 47 kcal/mole.

5. MONOSILANE-WATER MIXTURES

A direct correspondence has been shown to exist between the major ion-molecule reactions of methane, namely (1) and (2), and those characteristic of very strong acid solutions of methane (Olah and Schlosberg, 1968). On the other hand it has been known for more than half a century that monosilane reacts with water under alkaline or acidic conditions to yield, ultimately, hydrogen in amounts equal to twice that contained in the monosilane (Stock and Somieski, 1916; Stock, Somieski and Wintgen, 1917; Stock and Somieski, 1918). While the overall chemical conversion is thought to proceed via silanol and disiloxane intermediates, the necessity of having alkaline or acidic conditions suggests that ionic reactions are of importance.

Tandem mass spectrometric studies (Cheng and Lampe, 1973b) have shown that the predominant bimolecular ion-molecule reactions characteristic of ionized gaseous mixtures of monosilane and water are as shown by (20) - (28), viz.

$$OH^+ + SiH_4 \longrightarrow SiH_2^+ + H_2O + H \tag{20}$$

$$OH^+ + SiH_4 \longrightarrow SiH_3^+ + H_2O \tag{21}$$

$$OH^+ + SiH_4 \longrightarrow SiOH^+ + H_2 + H + H \tag{22}$$

$$H_2O^+ + SiH_4 \longrightarrow H_3O^+ + SiH_3 \tag{23}$$

$$H_2O^+ + SiH_4 \longrightarrow SiH_2^+ + H_2O + H_2 \tag{24}$$

$$H_2O^+ + SiH_4 \longrightarrow SiH_3^+ + H_2O + H \tag{25}$$

$$H_3O^{+*} + SiH_4 \longrightarrow SiH_3^+ + H_2O + H_2 \tag{26}$$

$$H_3O^{+*} + SiH_4 \longrightarrow SiH_5^+ + H_2O \qquad (27)$$

$$SiH_3^+ + H_2O \longrightarrow H_2SiOH^+ + H_2 \qquad (28)$$

All the above reactions with the exception of (28) appear to have rate constants of the order of 10^{-10} cm^3/sec. Detailed studies of the energy dependence of the cross-sections of (26) and (27) show that internally excited hydronium ions are involved at relative kinetic energies of the collision partners below 1.4 eV. At higher pressures (> 0.1 torr) the major reaction of SiH_3^+ with H_2O appears to be a collision-stabilized association to form $SiH_3OH_2^+$ (Mayer and Lampe, 1973).

It is of interest to point out here that (26) and (27) appear to proceed by independent mechanisms. That is to say, SiH_3^+ is formed predominantly by direct hydride ion transfer from SiH_4 and not via protonation of SiH_4 followed by dissociation to SiH_3^+ and H_2. Similar to the reaction of CD_5^+ with SiH_4 already mentioned, collision of D_3O^+ with SiH_4 yields SiH_4D^+ and SiH_3^+ but no significant amounts of SiH_2D^+, whereas statistical decomposition of SiH_4D^+ would yield 60 % SiH_2D^+ and 40 % SiH_3^+.

Since, as already pointed out, gaseous monosilane is a considerably stronger base than methane in both the Bronsted and Lewis concepts, it is suggested that a carry-over to the solution phase of this relative base strength is the reason that monosilane can be decomposed in aqueous acids while very much stronger acids are required to induce the decomposition of methane. By analogy with the ionic mechanism describing the behavior of methane in very strong acids (Olah and Schlosberg, 1968) and on the basis of the above ion-molecule reactions it is proposed that the acid hydrolysis of monosilane occurs as shown by (29) - (31). Silanol is

$$SiH_4 \quad \xrightleftharpoons{H_3O^+} \quad [SiH_5^+] \qquad (29)$$

$$SiH_4 \quad \xrightarrow{H_3O^+} \quad SiH_3^+ + H_2 \qquad (30)$$

$$SiH_3^+ + H_2O \rightleftharpoons [SiH_3OH_2^+] \underset{+H^+}{\overset{-H^+}{\rightleftharpoons}} SiH_3OH \qquad (31)$$

unstable, having never been isolated, and it is proposed that further decomposition proceeds by unknown mechanism to produce disiloxane, silicic acid and hydrogen. The stoichiometry is given by (32) and (33), viz.

$$SiH_3OH + SiH_3OH \longrightarrow SiH_3OSiH_3 + H_2O \qquad (32)$$

$$SiH_3OSiH_3 + 5H_2O \longrightarrow 2H_2SiO_3 + 6H_2 \qquad (33)$$

6. MONOSILANE-ETHYLENE MIXTURES

The monosilane-ethylene system is a very rich one from the point of view of the variety of ion-molecule reactions that occur (Mayer and Lampe, 1974a). The major bimolecular reactions observed by tandem mass spectrometry for each primary reactant ion and for a collision energy of 0.7 eV are as shown in (34)-(40), viz.

$$C_2H_2^+ + SiH_4 \longrightarrow SiH_3^+ + C_2H_3 \qquad (34)$$

$$C_2H_3^+ + SiH_4 \longrightarrow SiH_3^+ + C_2H_4 \qquad (35)$$

$$C_2H_4^+ + SiH_4 \longrightarrow SiH_3^+ + C_2H_5 \qquad (36)$$

$$Si^+ + C_2H_4 \longrightarrow SiC_2H_3^+ + H \qquad (37)$$

$$SiH^+ + C_2H_4 \longrightarrow SiC_2H_3^+ + H_2 \qquad (38)$$

$$SiH_2^+ + C_2H_4 \longrightarrow SiCH_3^+ + CH_3 \qquad (39)$$

$$SiH_3^+ + C_2H_4 \longrightarrow SiC_2H_7^+ \qquad (40)$$

It is clear from the above reactions that, as in all systems involving monosilane yet studied, the predominant reaction involving monosilane molecules is that of hydride ion transfer from monosilane to the attacking ion.

The most interesting reaction involving ions derived from monosilane with ethylene is (40) because it represents a simple association of the reactants. The pressure in the collision chamber of the tandem mass spectrometer in which this reaction was observed was too low (10^{-3} torr) for a stabilizing collision to have occurred and so (40) is a true bimolecular process. The transit time of the apparatus is of the order of 10^{-5} seconds and so the association product $SiC_2H_7^+$ must have a lifetime of at least this long. The observation of persistent collision complexes is rather rare in investigations of ion-molecule reactions, such prior reports having referred to studies in single-source mass spectrometers and mostly to complexes having more vibrational degrees of freedom than $SiC_2H_7^+$ (Pottie and Hamill, 1959; Henglein, 1962; Henglein, Jacobs and Muccini, 1963). As expected, the cross-section for (40) decreases very sharply with increasing energy and becomes zero at a collision energy of 1.3 eV.

In the absence of stabilizing collisions the reversion of an association complex to the original reactants is always energetically feasible, as demanded by the conservation of energy. A loosely bound complex in which the reactants retain essentially their original identities could not be expected to have a sufficiently long lifetime, relative to reversion to reactants to survive a time as long as 10^{-5} seconds (Johnston, 1966). However, if a chemical transformation occurs in which the loosely bound complex is transformed into an isomeric species (intimate collision complex) whose dissociation to reactants has a large barrier, sufficient lifetime for detection in a tandem mass spectrometer may result. With this in mind, the suggested mechanism is as shown by (41). The loosely bound complex, which most of the time

$$SiH_3^+ + CH_2{=}CH_2 \rightleftharpoons \left[\begin{array}{c} CH_2{-}{-}{-}CH_2 \\ SiH_3^+ \end{array} \right] \rightleftharpoons CH_3CH_2SiH_2^{+*}$$

(41)

reverts to reactants, is shown in brackets. The occasional formation of vibrationally excited ethylsilyl ion yields a species having a potential well relative to reactants that is presumably deep enough to provide a lifetime of greater than 10^{-5} seconds.

High-pressure studies of monosilane-ethylene mixtures in the pressure range of 0.01 - 0.3 torr (Mayer and Lampe, 1974a) confirm the observation of (40) as a bimolecular reaction. Moreover, such studies suggest that the $SiC_2H_7^+$ species adds successively two more molecules of ethylene as the pressure is raised, as shown by (42) and (43). That there appears to be no further addition

$$CH_3CH_2SiH_2^+ + C_2H_4 \rightleftharpoons \left[\begin{array}{c} CH_2{-}{-}{-}CH_2 \\ SiH_2^+ \\ | \\ CH_2 \\ | \\ CH_3 \end{array} \right] \rightleftharpoons (C_2H_5)_2SiH^+ \quad (42)$$

$$(C_2H_5)_2SiH^+ + C_2H_4 \rightleftharpoons \left[\begin{array}{c} CH_2{-}{-}{-}CH_2 \\ SiH^+ \\ C_2H_5 \quad C_2H_5 \end{array} \right] \rightleftharpoons (C_2H_5)_3Si^+ \quad (43)$$

of ethylene is considered as confirming evidence that ethylsilyl ions are involved in reactions analogous to (41).

7. KINEMATICS OF HYDROGEN TRANSFER REACTIONS IN MONOSILANE

As mentioned earlier the major ion-molecule reaction characteristic of ionized monosilane is (8), viz.

$$SiH_2^+ + SiH_4 \longrightarrow SiH_3^+ + SiH_3 \qquad (8)$$

The net process can be described as an H-atom or as an H^--ion transfer from SiH_4 to the attacking SiH_2^+ ion.

In view of the tendency of SiH_4 to transfer hydride ions to ions from hydrocarbons and water, one might imagine apriori that (8) is another example of hydride transfer from SiH_4. Substitution of SiD_4 for SiH_4 in the collision chamber of a tandem mass spectrometer (Mayer and Lampe, 1974b), giving the reactant pair $SiH_2^+ + SiD_4$, yields all three possible ions SiD_3^+, $SiHD_2^+$ and SiH_2D^+ whose relative intensities depend on collision energy as shown in Table III.

TABLE III

RELATIVE INTENSITIES OF SiD_3^+, SiD_2H^+, and $SiDH_2^+$ FROM COLLISIONS OF SiH_2^+ WITH SiD_4

Product	Collision Energy, eV					Statistical Prediction
	0.54	1.09	2.18	3.27	4.36	
SiD_3^+	100	98	100	100	100	100
SiD_2H^+	52	40	16	4	-	300
$SiDH_2^+$	100	100	75	48	64	100

The rapid decrease in the relative intensity of $SiHD_2^+$ with increasing collision energy and the persistence of SiD_3^+ and SiH_2D^+ suggest that $SiHD_2^+$ is formed principally via a complex $Si_2H_2D_4^+$, whereas most of SiD_3^+ and SiH_2D^+ are produced via direct or impulsive mechanisms. Examination of the kinetic energy distributions of the product ions for various collision energies confirms this conclusion.

It is found that the SiH_2D^+ ion is formed with appreciable kinetic energy in the forward direction. At low collision energies (1.5 eV) the center of the kinetic energy distribution lies very close to that calculated for dissociation of a complex $Si_2H_2D_4^+$ to the observed products, namely, close to the energy given by (44). The kinetic energy of the SiH_2D^+ ion increases

$$E_{SiH_2D^+} \text{ (Complex)} = \frac{M_{SiH_2^+} M_{SiH_2D^+}}{(M_{SiH_2^+} + M_{SiD_4})^2} E_{SiH_2^+} =$$

$$= 0.22 \, E_{SiH_2^+} \tag{44}$$

with increasing collision energy, with the center of the distribution, at high collision energies (5.0 eV), approaching closely to the energy calculated for SiH_2D^+ by the spectator stripping model (Henglein, Lacmann and Jacobs, 1965), namely close to energy given by (45). One may thus conclude that the mechanism

$$E_{SiH_2D^+} \text{ (stripping)} = \frac{M_{SiH_2^+}}{M_{SiH_2D^+}} E_{SiH_2^+} = 0.94 \, E_{SiH_2^+} \tag{45}$$

describing the formation of SiH_2D^+ undergoes a transition from something close to complex formation to something close to spectator stripping. At all energies so far studied the actual mechanism lies intermediate to these two idealized models, the expected product energies of which are given by (44) and (45). Such transitions have been observed before for hydrogen (deuterium) atom transfer reactions (Ding, Henglein, Hyatt and Lacmann, 1968).

In contrast to SiH_2D^+, the SiD_3^+ ion is formed with very little kinetic energy in the forward direction during SiH_2^+-SiD_4 collisons. The apparent center of the kinetic energy distribution of SiD_3^+ is found to be near 0.5 eV and the true value is probably lower. However, more importantly from the mechanistic point of view, the center, and indeed the position of the entire distribution, is independent of the collision energy or, equivalently, of the kinetic energy of the SiH_2^+ reactant ions. Thus formation of SiD_3^+ corresponds to neither of the model mechanisms discussed so far, since both models, represented by (44) and (45), demand that the kinetic energy of the product depend on the kinetic energy of the reactant ion.

The simplest explanation of this fact is that, at all energies, formation of most of the SiD_3^+ product occurs via a D^--ion stripping process similar to the D-atom stripping process described by (45). The difference is that in D-atom stripping the particle detected has a kinetic energy centered around $(M_i/M_i+2)E_i$, where M_i is the mass of the attacking ion and E_i is its kinetic energy. In idealized D^--ion stripping the particle detected is the so-called "spectator" and it receives non of the kinetic energy of the reactant ion, regardless of what that energy is. The two stripping processes discussed are represented by (46) and (47), viz.

$$\overrightarrow{SiH}_2^+ + \underset{\underset{SiD_3}{|}}{D} \longrightarrow \overrightarrow{SiH_2D}^+ + SiD_3 \quad \text{(D-Atom Stripping)} \tag{46}$$

$$\overrightarrow{SiH}_2^+ + \underset{\underset{\delta+SiD_3}{|}}{D}^{\delta-} \longrightarrow \overrightarrow{SiH_2D} + SiD_3^+ \quad \text{(D}^-\text{-Ion Stripping)} \tag{47}$$

in which the arrow over a molecular entity indicates which entities have kinetic energy in the laboratory frame of reference. It is believed that the D^--ion stripping is especially probable in silanes because of the magnitude and direction of polarity of the Si-H bond (cf. Table I). That is to say, in neutral silanes the hydrogen atoms bound to silicon have a significant partial hydridic character even before the approach of an attacking positive ion.

The $SiHD_2^+$ ion formed in SiH_2^+-SiD_4 collisions exhibits a broad distribution that is centered on the energy calculated if this ion were formed by dissociation of a complex $Si_2H_2D_4^+$. This is consistent with the date of Table III.

It is found that the dominant reaction of SiH_3^+ with SiD_4 is also a D^--ion stripping reaction in which the SiD_3^+ product energy is independent of reactant ion energy. Further, in methylsilane the dominant processes have been found to be hydride ion transfer processes from Si-H bonds to the attacking ions which follow H^--ion stripping (Mayer and Lampe, 1974c). It is suggested that this is a general feature of ionic attack on silanes that is made possible by the magnitude and direction of polarity of the Si-H bond. While kinetic energy measurements have not been made on SiH_3^+ ions formed via attack of ions derived from hydrocarbons and water on monosilane, the predominance of H^--transfer in these reactions has been established and it is likely that these reactions also follow a hydride ion stripping mechanism.

REFERENCES

Ball, D.F. and McKean, D.C. (1962). Spectrochim. Acta 18, 1019
Barrow, G.M. (1973). Physical Chemistry, McGraw-Hill Book Company, 3rd Edition, New York, p.407
Cheng, T.M.H. and Lampe, F.W. (1972). Unpublished results
Cheng, T.M.H. and Lampe, F.W. (1973a). Chem. Phys. Letts. 19, 532

Cheng, T. M. H. and Lampe, F. W. (1973b). J. Phys. Chem. 77, 2841
Cheng, T. M. H., Yu, T-Y. and Lampe, F. W. (1973), J. Phys. Chem.
        77, 2587
Ding, A., Henglein, A., Hyatt, D. and Lacmann, K (1968), Z. Natur-
        forsch. 23a, 2090
Franklin, J. L. (1972) Ion-Molecule Reactions (Ed. J. L. Franklin)
        Plenum Press, New York
Franklin, J. L., Dillard, J. G., Rosenstock, H. M., Herron, J. T.,
        Drexl, K. and Field, F. W. (1969) Nat. Stand. Ref. Data
        Sec., Nat. Bur. Stand. (U.S.) 76
Haney, M. A. and Franklin, J. L. (1969), J. Phys. Chem. 73, 4328
Henglein, A., Lacmann, K. and Jacobs, G. (1965), Ber. Bunsen-
        Ges. phys. Chem. 69, 279
Henis, J. M. S., Stewart, G. W., Tripodi, M. K. and Gaspar, P. P.
        (1972). J. Chem. Phys. 57, 389
Johnston, H. S. (1966). Gas-Phase Reaction Rate Theory, The Ro-
        nald Press Company, New York, pp. 280 ff.
Mayer, T. M. and Lampe, F. W. (1973). Unpublished results
Mayer, T. M. and Lampe, F. W. (1974a). J. Phys. Chem. (submitted)
Mayer, T. M. and Lampe, F. W. (1974b). J. Phys. Chem. (in press)
Mayer, T. M. and Lampe, F. W. (1974c). J. Phys. Chem. (in press)
Olah, G. A. and Schlosberg, R. H. (1968). J. Am. Chem. Soc. 90, 2726
Potzinger, P. (1974). Private Communication
Potzinger, P. and Lampe, F. W. (1969). J. Phys. Chem. 73, 3912
Pullen, B. P., Carlson, T. A., Moddeman, W. E., Schweitzer, G. K.,
        Bull, W. E. and Grimm, F. A. (1970). J. Chem. Phys. 53,
        768
Schissler, D. O. and Stevenson, D. P. (1956). J. Chem. Phys. 24, 926
Stewart, G. W., Henis, J. M. S. and Gaspar, P. P. (1972). J. Chem.
        Phys. 57, 2247
Stock, A. and Somieski, C. (1916). Ber. 49, 111
Stock, A. and Somieski, C. (1918). Ber. 51, 989
Stock, A., Somieski, C. and Wintgen, R. (1917). Ber. 50, 1754
Talroze, V. L. and Lyubimova, A. K. (1952). Dokl. Akad. Nauk SSSR
        86, 909
Yu, T-Y., Cheng, T. M. H., Kempter, V. and Lampe, F. W. (1972).
        J. Phys. Chem. 76, 3321

FOOTNOTES

1) U. S. Senior Scientist Awardee of Alexander von Humboldt-Stif-
        tung 1973-74
2) On leave from Department of Chemistry, The Pennsylvania
        State University, University Park, Pennsylvania 16 801, USA

THERMOCHEMICAL INFORMATION FROM GAS PHASE ION EQUILIBRIA

P. Kebarle

Chemistry Department, University of Alberta

EDMONTON, Alberta, Canada T6G 2G2

a.   Introduction

Results from ion equilibria measurements in solution recorded
in the form of dissociation constants, instability constants, sol-
ubility products and electrochemical reduction potentials represent a mainstay of information for the chemist. The thermodynamic
concepts and theory underlying most of these measurements were
obtained in the early nineteen hundreds and a large store of data
was obtained quite early. Similarly valuable information was
potentially available from measurement of ion equilibria in the
gas phase. For example since the ions are not surrounded by solvent molecules, measured gas phase acidities or basicities would
reveal the intrinsic molecular properties determining acidity and
basicity. In spite of such exciting possibilities systematic gas
phase ion equilibria measurements were not started until quite
recently. The reason for this relatively late development is
quite simple. Ions in aqueous solution are stable, one can wait
for equilibrium to be reached. Ions in the gas phase are unstable,
positive ions recombine with negative ions and both positive and
negative ions become discharged on the walls of the apparatus.
Therefore special techniques and instrumentation are required for
gas phase ion equilibria measurements. These techniques grew with
the development of the field of ion molecule reactions, and
particularly the study of ions at higher pressures. The present
work will give a brief description of the types of ion equilibria
subject to measurement and the development of the ion equilibrium
method. The thermodynamic basis of the measurements and their
close relationship with the kinetics of the equilibria will be
considered also. Attendant instrumental requirements and agree-

ment between different methods will be examined briefly. Finally
a brief summary of the results on ion-clusters and ion-solvent
molecule interactions obtained in the author's laboratory will be
given. This presentation follows closely the talk given at the
conference in Biarritz. The reader should be aware that this is
not a review article. Thus while the references given are intended to serve as a guide to the literature they are not comprehensive and do not give a balanced view. Well known aspects of ion
equilibria measurements have not been discussed. On the other
hand some general aspects, not available in the literature, have
been considered. This is true for section e: One legged thermodynamics, f: Relationship between the equilibrium and bracketing
techniques and secion h: High pressure mass spectrometry and
stationary afterglow, which contain largely original material.

b.   Examples of ion equilibria and thermodynamic equations

A typical ion molecule reaction which reaches equilibrium
is the proton transfer reaction (1)

$$NH_4^+ + CH_3NH_2 = NH_3 + CH_3NH_3^+ \qquad (1)$$

If the equilibrium concentrations of the neutrals and ions can be
determined then one can evaluate the equilibrium constant K_1 from
expression (2).

$$K_1 = \frac{[CH_3NH_3^+][NH_3]}{[NH_4^+][CH_3NH_2]} \qquad (2)$$

Furthermore equation (3) allows one to determine the free

$$-RT\ln K = \Delta G_1^\circ \qquad (3)$$

energy change ΔG_1° and Van't Hoff plots of the temperature dependence of K_1, measured at different temperatures, lead to the
evaluation of ΔH_1° and ΔS_1°.

On principle very many classes of ion-molecule reactions can
reach equilibrium and provide information of interest. A list of
such reactions is given in Table I.

TABLE I

Types of ion molecule reactions whose equilibria have been measured

Reaction type	Thermochemical information that can be obtained
Charge Transfer	
$C_2F_4^+ + C_2HF_3 = C_2F_4 + C_2HF_3^+$ (a) $\Delta G = 0.02$ e.v.	Difference between ionization potentials of molecules

$$NO_2^- + Cl_2 = NO_2 + Cl_2^- \quad (b)$$

Differences between electron affinities

Proton transfer

$$CH_3OH_2^+ + (CH_3)_2O =$$
$$CH_3OH + (CH_3)_2OH^+ \quad (c)$$

Proton affinity differences, i.e. difference between gas phase basicities

$$CH_3COO^- + CH_2ClCOOH =$$
$$CH_3COOH + CH_2ClCOO^- \quad (d)$$

Difference in gas phase acidities of acids AH also $D(A-H) - EA(A)$ i.e. difference between bond dissociation energy and electron affinity

Hydride ion transfer

$$t-C_4H_9^+ + i-C_5H_{12} =$$
$$C_5H_{11}^+ + i-C_4H_{10} \quad (e)$$

Carbonium ion stabilities $D(R-H) + I_p(R)$

Halide ion transfer

$$R_1^+ + R_2X = R_1X + R_2^+ \quad (f)$$

Carbonium ion stabilities $D(R-H) + I_p(R)$

Clustering = adduct formation

$$O_2^+ + O_2 = O_4^+ \quad (g)$$

Stability of clusters i.e. $D(O_2^+ - O_2)$, $D(H_3O^+ - H_2O)$

$$K^+(H_2O)_{n-1} + H_2O = K^+(H_2O)_n \quad (h)$$

Ion solvent molecule interactions in the gas phase

$$H^+(H_2O)_{n-1} + H_2O = H^+(H_2O)_n$$

$$Cl^-(CH_3OH)_{n-1} + CH_3OH = \quad (h)$$
$$Cl^-(CH_3OH)_n$$

Hydrogen bonding of protic solvents to positive and negative ions, etc.

$$CH_5^+ + CH_3 = C_2H_9^+ \quad (i)$$

Clustering exchange

$$Cl^-(HOH) + CH_3OH = Cl^-(CH_3OH) + HOH \quad (h)$$

(a) Bowers, M.T. (1974)   (b) Ferguson E. E. (1974)

(c)   Hiraoka, K. (1974)   (d)   Yamdagni, R. (1973)
(e)   Solomon, J. J. (1974)   (f)   McMahon, T. B. (1972)
(g)   Conway, D. C. (1970)   (h)   Kebarle, P. (1972a), (1972b)
                                        (1974)
                                  (i)   Hiraoka, K. (1975)

Evidently a large variety of reactions are amenable to equilibrium
measurements yielding thermochemical information on ionization
potentials, electron affinities, bond dissociation energies,
proton affinities, hydrogen bonding to ions etc. As will be seen
in the next section not all reaction types have received equal investigation or proven equally useful.

c.   The development of equilibria measurements

      It is interesting to note that the early ion equilibria
measurements were largely incidental. The researchers observing
the equilibria had not consciously set out to measure an equilibrium but when observing it recognized it and experimentally
exploited it. The first such measurement was done by Bill Chupka,
Chupka, W. A. (1959), who was mass spectrometrically studying the
species effusing from a Knudsen cell containing alkali halides.
The heated salts evaporated neutral species like $(KCl)_n$ and also
ions K^+, KCl_2^- etc. Amongst the ions Chupka found K^+OH_2 formed by
the presence of water vapour. He took the opportunity and studied
the equilibrium (3) obtaining an equilibrium constant which was

$$K^+(H_2O) = K^+ + H_2O \qquad\qquad (3)$$

verified by later more comprehensive measurements, Searles, S. K.
(1969). Probably a thorough search of the literature will reveal
other early and independent equilibria observations. The present
author, when studying ions in alpha particle irradiated gases at
high pressure, accidentally observed the proton hydrate equilibria
(4), Kebarle P. (1963).

$$H^+(H_2O)_{n-1} + H_2O = H^+(H_2O)_n \qquad\qquad (4)$$

      Systematic studies of ion equilibria and their temperature
dependence were initiated somewhat later independently in two
laboratories. Yang and Conway, (1964), studied the equilibria
(5) in a high pressure mass spectrometer with a tritiated foil as

$$O_2^+ + O_2 = O_4^+ \qquad\qquad (5)$$

ionizing medium, while Kebarle and Hogg, (1965), made a study of
the proton hydrate equilibria (4).

      Another independently conceived study was that of Pack and

Phelps, Pack J. L. (1966), who utilized a drift tube in which negative ions were obtained by the pulsed creation of photoelectrons.
This very elegant investigation of the equilibrium (6) gave the

$$O_2^- = O_2 + e \qquad\qquad (6)$$

first accurate electron affinity of O_2.

In the years between 1965-72 measurements in Alberta were
expanded to include a large variety of systems. Attention was
also paid to the development of apparatus specially suited for
the measurement of ion equilibria and the kinetics of equilibrium
reactions. All equilibria studied involved clustering reactions
(see Table I). The results provided information on the energetics
of ion-molecule complexes in the gas phase, strong hydrogen bonding to positive and negative ions, differences between protic and
aprotic solvents etc. Kebarle P. (1972a,b).

The field of ion molecule equilibria experienced a very
important expansion with the initiation of proton transfer equilibria measurements by trapped ion cyclotron techniques, McIver
(1971), Bowers, M. T. (1971), Henderson, W. H. (1972) and high
pressure mass spectrometry, Briggs, J.P. (1972). These measurements permitted accurate determinations of proton affinities and
gas phase acidities of literally hundreds of organic bases and
acids. As was mentioned earlier these results are of particular
interest since they reveal the intrinsic molecular effects in
acid-base reactions. A discussion of proton transfer reactions
is to be found in the Chapter by Diether Bohme. A combination of
proton transfer and clustering equilibria allows one to study the
gradual change of acidity or basicity with the gradual addition
of individual solvent molecule to the ionic acid or base, Hiraoka
K. (1974). In this manner gas phase ion equilibria provide not
only information on the intrinsic properties of acids and bases
but also throw a bridge to the behaviour in solution.

Recently, the very promising group of hydride transfer reactions (see Table I) was studied by Solomon and Field, Solomon
(1974). These reactions allow the accurate determination of the
heats of formation of carbonium ions. Carbonium ion energetics
can also be determined from halide ion transfer reactions (Table I).
These reactions were investigated earlier by Beauchamp's group,
McMahon T. B. (1972). They are particularly well suited for the
determination of fluoro substituted carbonium ions.

Charge transfer equilibria involving positive ions (Table I)
have received little study. Probably these reactions will be of
use in isolated cases where the conventional methods for determination of molecular ionization potentials experience difficulties.

Charge transfer equilibria involving negative ions provide electron affinities of molecular species. (Table I). Again the method has been used only in special cases.

The great promise of ion molecule reaction equilibria has led to a considerable increase in the number of research groups involved in such measurements. Table II gives a summary of the present day practitioners classified by the type of instrument used.

TABLE II

Groups and Instrumental Methods Involved in Ion Equilibria

Measurements

High pressure				Low pressure
0.5 - 30 torr				$p < 10^{-3}$ torr
flow	no flow, high pressure mass spec		drift tubes	ion cyclotron resonance
flowing afterglow	pulsed	continuous	pulsed	pulsed trapped ion
Bohme[a]	Field[c]	Conway[e]	Castleman[l]	Aue[p]
Ferguson[b]			Franklin[m]	
Fehsenfeld[b]	Kebarle[d]	Franklin[f]	McKnight[n]	Beauchamp[q]
		Futrell[g]	Phelps[o]	Bowers[p]
		Herman[h]		McIver[r]
		Tiernan[k]		

Recent representative work:

a	Bohme D.K. (1973)	k	Chang C. (1974)
b	Fehsenfeld F.C. (1973)	l	Tang I. N. (1972)
c	J. J. Solomon (1974)	m	Long J.W. (1974)
d	Cunningham A. (1972)	n	McKnight (1972)
e	Conway D.C. (1970)	o	Pack J. L. (1965)
f	Cheng S.L. (1972)	p	Bowers M.J. (1971)
g	Arshadi, M. (1974)	q	McMahon T.B. (1972b)
h	Wincel H. (1973)	r	McIver R.T. (1970)

d.   Thermodynamic considerations

We will examine the thermodynamic requirements for the study
of gas phase ion equilibria using the proton transfer reaction
(6) as an example.

$$A_1H + A_2^- = A_1^- + A_2H \qquad\qquad (6)$$

The following conditions must be present:

1. $A_1H + A_2^-$ and A_1^- + A_2H must be in thermal equilibrium with
   their common surroundings, (carrier gas or walls of the vessel).
It is not necessary that the reaction complex be in thermal equilibrium with the surroundings.

2.   The path reactively coupling the equilibrium must be faster
than all other processes affecting the concentrations of the reactants.

3.   Sufficient time must be given to the system to reach equilibrium.

Since the neutrals AH are generally stable and present at
vastly higher concentrations than the ions one may generally assume
that they are in thermal equilibrium with their surroundings. However, care must be taken to ensure that the ions are thermal.
Ions produced by electron impact are generally excited. Such
excitation can be removed by the addition of a third gas at concentrations two to three orders higher than the concentration of
the neutral reactants. Excitation may be also induced by the reaction itself. Thus if the proton transfer reaction (6) is exothermic A_1^- will contain excess energy. In the absence of a third
gas A_1^- entering the reverse reaction may not be thermal.

Non thermal ion distributions can be also due to the presence
of imposed electric fields or space charge effects. The presence
of fields can be avoided by working as nearly as possible without
the imposition of external fields and at charge densities considerably lower than 10^6 ions per cm^3.

It is not very difficult to meet also the second condition requiring the equilibrium reactions to be faster than other processes.
Competitive processes destroying ions are either ion recombination
or discharge of ions to the wall. Ion recombination is slow at low
charge densities, discharge to the wall is slowed down by a third
gas if one works at high pressures. At low pressures (ICR) ion
trapping techniques must be applied.

The third condition of sufficient time for achievement of

equilibirium is relatively easily met. Ion molecule reactions are
fast and it is possible to have an equilibrium establish itself in
tens to hundreds of microseconds. Important is either to be able
to directly observe the approach to equilibrium with time or if
that is instrumentally not possible, than to know the average ion
residence time in the reaction chamber and know the approximate
values of the rate constants involved in the reaction so as to be
able to assess whether the conditions selected are realistic. An
understanding of the kinetics of the reaction system and their
temperature dependence is therefore extremely desirable. A
discussion of the temperature effects of ion molecule reactions
can be found in the chapter by R. Porter. Some additional information is given in the next section which examines some re-
lationships between equilibria and rate constants.

e.   One legged thermodynamics, the temperature dependence of ion
      molecule reactions

      Ion molecule reaction equilibria have a peculiarity which
should be discussed. Very many (bimolecular) exothermic ion
molecule reactions proceed with rate constants which can be predicted either by the Langevin-Stevenson or the ADO theory (see
chapter by M. T. Bowers). These rate constants depend on polarizability and dipole moment and are in the neighbourhood of 10^{-9}
molecules^{-1} cm^3 sec^{-1}. We may take reaction (6) as an example
for the discussion.

$$A_1^- + A_2H \xrightleftharpoons[k_r]{k_f} A_1H + A_2^- \tag{6}$$

The equilibrium constant K can be expressed by eq. (7). Assuming

$$K = \frac{k_f}{k_r} \tag{7}$$

that (6) is (appreciably) exothermic we may expect that k_f will
be close to 10^{-9}. Furthermore we expect k_f to show only a very
minimal temperature dependence as predicted by the Stevenson or
ADO equations. Therefore k_f does not contain any specific information relating to the internal parameters of the reaction
system. Yet K through the equation: $RT\ln K = \Delta G°$ must express the
free energy change and thus must contain information on the
internal energy and entropy of the reaction system.   We conclude that this information should enter K through k_r. This means
that the properties of the system are described only through the
reverse leg of the reaction. This somewhat puzzling situation
deserves further examination. The results of such an examination

are of interest not only to ion equilibria but also improve our understanding of the temperature dependence of ion molecule re- §
actions. We will assume that (6) proceeds via an intermediate §
$A_1HA_2^-$. One can then write the Lindeman type mechanism (8) where

$$A_1^- + A_2H \xrightleftharpoons[b]{c} (A_1HA_2^-)^* \xrightarrow{d} A_1H + A_2^- \qquad (8)$$

$$\xrightarrow[M]{s} (A_1HA_2)^-$$

c stands for complex formation, b for back decomposition of the complex, d for dissociation of the complex along the reactive channel and s for stabilization of the complex by the third body M. A steady state treatment for $(A_1HA_2^-)^*$ leads to equation (9) for the overall forward rate constant k_f.

$$k_f = \frac{k_c k_d}{k_d + k_b + k_s[M]} \qquad (9)$$

A schematic representation for the potential energy of the exo-thermic reaction (6) is given in Figure 1.

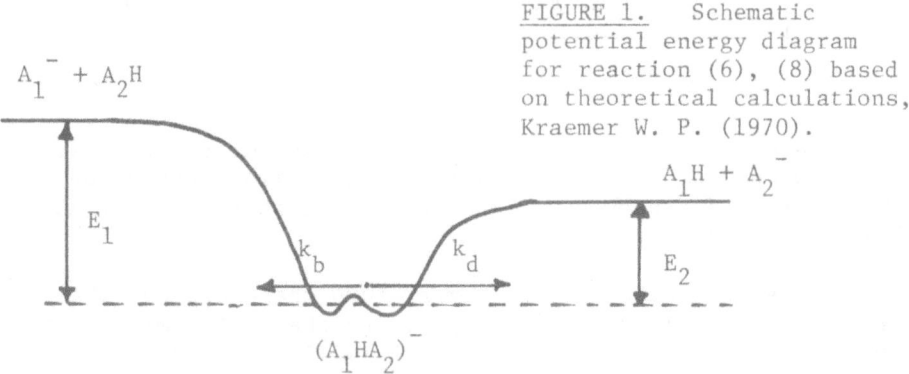

FIGURE 1. Schematic potential energy diagram for reaction (6), (8) based on theoretical calculations, Kraemer W. P. (1970).

---

§ At temperatures which are not very high (T < 800°K) many if not most ion molecule reactions proceed via relatively long lived re-action complexes. The excess energy contained in the complex con-sequent to its formation may be quite large. For example the formation of $NH_4^+ \cdot NH_2CH_3$ from NH_4^+ and CH_3NH_2 releases 32 kcal/ mol, Yamdagni (1973a). At high pressure the complex can be col-lisionally stabilized and observed mass spectrometrically. It is these stabilized complexes which are the reactants in ion cluster equilibria.

Since the reaction is exothermic we expect at low temperature $k_d \gg$
k_b. At low pressures $k_d \gg k_s[M]$ such that equation (9) reduces to
(10). k_c is the Langevin or ADO type rate constant which is of

$$k_f = k_c \approx 10^{-9} \text{ cm}^3 \text{ molec}^{-1} \text{ sec}^{-1} \qquad (10)$$

the order 10^{-9} cm^3 molec^{-1} sec^{-1}. Thus we expect in this temperature and pressure region

$$K_{eq} = \frac{k_c}{k_r} \approx \frac{10^{-9}}{k_r} \qquad \text{and since } K = e^{-\frac{\Delta G}{RT}}$$

$$e^{-\frac{\Delta G}{RT}} = \frac{10^{-9}}{k_r}$$

$$k_r \approx 10^{-9} e^{-\frac{\Delta G}{RT}} = 10^{-9} e^{-\frac{\Delta S_r}{R}} \cdot e^{-\frac{\Delta H_r}{RT}} \qquad (11)$$

ΔH_r and ΔS_r are the enthalpy and entropy changes for the reverse
reaction. We notice that eq. (11) predicts a simple Arrhenius
type temperature dependence for (strongly) endothermic bimolecular ion molecule reactions.

Evidently at some very high temperature the energy of the
activated complex will be large compared to the exothermicity of
the (forward) reaction. At this point k_b will become competitive
with k_d (see Fig. 1). For such a situation k_f will be given by
eq. (12)

$$k_f = \frac{k_c k_d}{k_d + k_b} \qquad (12)$$

A simple model calculation based on the RRK equation can be
made in order to obtain a rough idea, for a given exothermicity,
at what temperature will k_b become competitive with k_d. The RRK

expression is given by eq. (13), Robinson, P. J. (1972). E^* is the

$$k \propto \left(\frac{E^* - E^\circ}{E^*} \right)^n \tag{13}$$

total energy contained in the chemically activated complex. E° is the minimum energy required for the dissociation. n is a measure of the internal degrees of freedom of the complex. Assuming that the thermal energy in the complex can be simply expressed by cpT where cp is the sum of the heat capacities of the reactants we can write for k_d and k_b eq. (14) and (15), where E_1 and E_2

$$k_d \propto \left(\frac{E_1 - E_2 + cpT}{E_1 + cpT} \right)^n \tag{14}$$

$$k_b \propto \left(\frac{cpT}{E_1 + cpT} \right)^n \tag{15}$$

are the energies indicated in Figure 1. The ratio of k_d/k_b is then given by (16)

$$k_d/k_b = \left(\frac{E_1 - E_2 + cpT}{cpT} \right)^n \tag{16}$$

Since $E_1 - E_2$ is the exothermicity of the reaction the expression inside the brackets depends only on the exothermicity and the heat capacity of the educts. Table III gives numerical results for the k_d/k_b ratio calculated for three different temperatures for two exothermicities $E_1 - E_2$ = 10 kcal/mole and 2 kcal/mole. cp = 30 cal/degree was assumed for the heat capacity. n was assumed equal to 8. (Since the RRK treatment is only a rough approximation n is essentially a parameter corresponding to roughly half the normal vibrations of the complex. Thus n = 8 should represent a small to medium sized polyatomic molecule). The values for k_f given in the table were evaluated by substituting the calculated ratio for k_d/k_b into eq. 12 and assuming that $k_c = 10^{-9}$ cm^3 molec^{-1} sec^{-1}. While the results contained in Table III originate from an admittedly primitive treatment they are instructive. We see that for reactions with exothermicities of 10 kcal and higher, k_f will be equal to k_c and thus have only a very small temperature dependence. The equilibrium constant will decrease with temperature only because k_r increases

TABLE III

Values for k_b/k_d and k_f predicted by simple RRK treatment

Temp °K	Exothermicity = 10 kcal/mole		Exothermicity = 2 kcal/mole	
	k_d/k_b	k_f	k_d/k_b	k_f
300	$\left(\frac{19}{9}\right)^8 = 394$	10^{-9}	$\left(\frac{11}{9}\right)^8 = 4.98$	0.83×10^{-9}
600	$\left(\frac{28}{18}\right)^8 = 34.3$	0.97×10^{-9}	$\left(\frac{20}{18}\right)^8 = 2.32$	0.7×10^{-9}
1000	$\left(\frac{40}{30}\right)^8 = 9.9$	0.908×10^{-9}	$\left(\frac{32}{30}\right)^8 = 1.67$	0.6×10^{-9}

with temperature. Reactions with low exothermicities (i.e. 3 - 0 kcal) will have a k_f which is less than k_c and which decreases with temperature. k will therefore decrease with temperature because of the decrease of k_f and increase of k_r.

Little experimental work on the temperature dependence of k_f and k_r of bimolecular reactions has been done so far, therefore the predictions   the above treatment have not yet been experimentally verified.

f.   Relationship between the equilibrium method and the "occurrence-non occurrence" or bracketing method

A large amount of extremely useful thermochemical data particularly proton affinities and gas phase acidities have been obtained by the method where one observes the occurrence or non occurrence of a given reaction, Brauman J.I. (1970), Beauchamp, J.L. (1971), Long J.W. (1973). The method was initiated by Talroze, Talroze V.L. (1962). A discussion of the assumption has been given by Long and Franklin, Long J.W. (1974). In the early mass spectrometers with which ion molecule reactions were studied a reaction was observed only if it was fast, i.e. if it had a near Langevin rate constant. Thus if a reaction was observed to occur it had no activation energy which meant that it is exothermic (see preceeding section). The technique is most successful when the same type of reaction i.e. proton transfer is studied in a series of compounds. Thus if one observes that B_1H^+ proton transfers to B_2 then it is concluded that the proton affinity $PA(B_1) < PA(B_2)$. Such an experiment is typically done by working with a mixture of B_1 and B_2. The observations amount to determining that k_f the forward rate constant for reaction (17) is larger than k_r, the reverse rate constant.

$$B_1H^+ + B_2 = B_1 + B_2H^+ \qquad (17)$$

This means that the equilibrium constant for reaction (17) is bigger than unity.

$$K = k_f/k_r > 1 \qquad (18)$$

Furthermore substituting (18) into the free energy relationship (3) one obtains (19).

$$\Delta G^\circ = RT\ln K \quad (3) \qquad \boxed{\Delta G^\circ < 0} \qquad (19)$$

Thus the "occurrence non occurrence" technique establishes the sign of the _free_ energy change and not the enthalpy change as is often assumed or implied.

Obviously all the precautions regarding thermalization of the reactants, necessary for equilibrium meausrements apply equally to the "occurrence non occurrence" technique. In fact this technique can be considered as a qualitative form of ion equilibria.

g.   Instrumental - the pulsed high pressure mass spectrometer

Space limitations do not permit a detailed discussion of all types of apparatus utilized for ion equilibria measurements. We propose to describe here briefly the pulsed high pressure mass spectrometer used in our laboratory and then in the following two sections consider some features of other instruments.

The apparatus used in Alberta has been described previously, Cunningham A.J. (1972). Shown in Figure 2 is the version presently in use. Ionization is obtained with a pulsed 2 kev electron beam. The separation of the electron filament from the ion source permits better ion source temperature control particularly at low temperatures, and leads to longer filament life. The ion source contains the reactant gases at some 1-10 torr. The gas flows in and out of the ion source. This maintains constant reactant concentration ratio and prevents buildup of impurities. The electrons beam is pulsed on for some 10 μsec and is off for several msec.

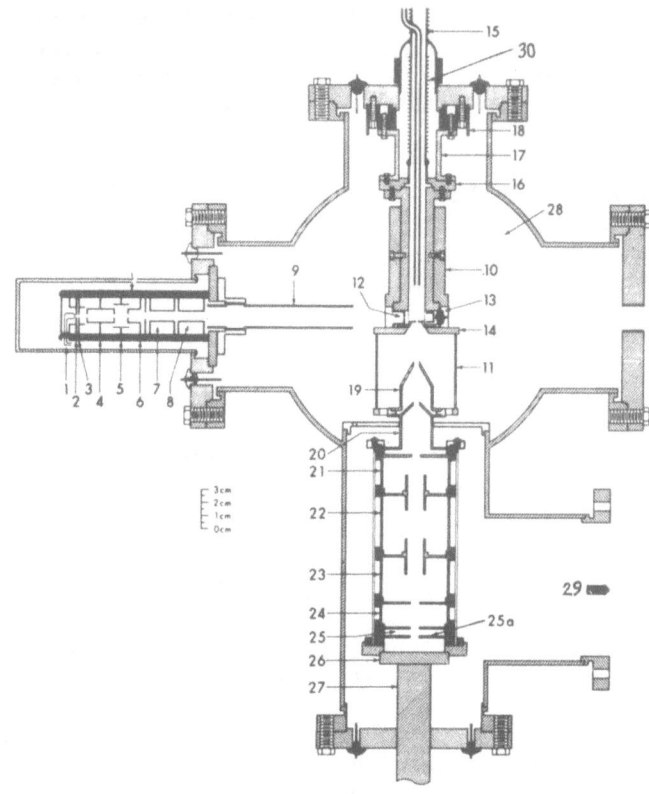

1-9 electron gun
10 ion source
11 shielding cage
12 electron entrance
19-26 accelerating
      electrodes
27 to mass analysis
28 to 6" pump
29 to 4" pump
30 gas flow system

FIGURE 2. APPARATUS

Some 10^6 electrons per pulse produce charge density of about 10^6 ions/cc. The ion source does not contain imposed electric fields. Some of the ions diffuse to the vicinity of the sampling leak and escape into the evacuated region where they are captured by electric fields, accelerated and subjected to magnetic or quadru- pole mass analysis. Ion detection is obtained with an electron multiplier followed by a counting system and a multiscaler. The sweep of the multiscaler is synchronized with the start of the electric pulse. In this manner the time dependence of the ions after the ionizing pulse is obtained. The time profile relates to processes in the ion source, since the ion time of flight in the low pressure analyzer section is only several μsec while the ions are observed over milliseconds. The ion source can be heated or cooled allowing measurements to be made from −180° to +680°C.

Negative ion time profiles observed in the moist oxygen, J.D. Payzant (1972), are shown in Figure (3a). The reaction sequence $O_2^- \longrightarrow O_2^- \cdot H_2O \longrightarrow O_2^- (H_2O)_n$ is easily followed by observing the successive maxima.

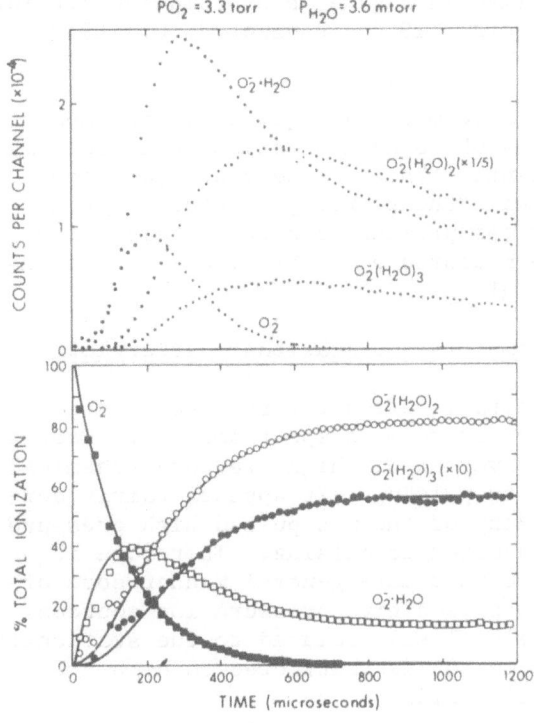

FIGURE 3a

Negative ions observed in moist oxygen

FIGURE 3b

Same as above but normalized

The kinetics and approach to equilibrium are more easily observed after normalization of the ion intensities (Fig. 3b). The "normalization" is obtained by dividing the ion count of a given ion at time t by the total ion count at time t. The achievement of equilibrium between $O_2^-(H_2O)$, $O_2^-(H_2O)_2$ and $O_2^-(H_2O)_3$ is clearly visible in Figure 3b. The concentrations of the reactants are such that the rates of the equilibrium reactions are not very much faster than the diffusion rates of the ions to the wall. Thus the condition "much faster equilibrium rates than ion loss rates" set out in section d would appear to be violated. However, measurements made with increased oxygen pressure and water concentration where the equilibrium rates are faster and the ion loss rates slower are found to lead to equilibrium constants in agreement with the results of figure 3. This agreement must be due to the similarity of the diffusion coefficients of ions like $X^-(H_2O)_n$ and $X^-(H_2O)_{n+1}$. This similarity relaxes the requirement that the equilibrium rates should be much faster than the ion loss rates.

Measurements of the reaction rate constants is also possible

on basis of normalized plots like in Figure 3b by computer fitting
the observed ion profiles, Good A. (1970), Payzant J. D. (1972),
Payzant J.D. (1973).

In general, when conditions permit, the equilibria are measured
in the presence of a buffer gas whose concentration is 20-50 times
higher than that of the reactants. Under these conditions the ions
suffer some 20-50 collisions with the buffer gas before meeting
neutral reactant. The buffer gas pressure can be increased from
1-10 torr. Invariance of the measured equilibrium constant means
that the ions are truely thermal.

h. High pressure mass spectrometery and stationary afterglow

The apparatus used in Alberta is often referred to in the
literature as a "pulsed high pressure mass spectrometer". Recently
Field, Solomon J.J. (1974), converted his high pressure (chemical
ionization) mass spectrometer to pulsing. It appears fairly certain that in the near future many of the non pulsed high pressure
mass spectrometers will be converted to pulsing. Therefore it
is of some interest to describe in a more general manner some of
the conditions existing in the pulsed high pressure ion sources
and to show that they are really closely related to the stationary
afterglow apparatus. The discussion is restricted to high pressure
ion sources using no electrical fields.

Time profiles of the total positive and negative ions observed
with the Alberta instrument are shown in Figures 4 and 5. The
system in Figure 4 (moist methane) does not contain substances
which capture electrons rapidly. In this system initially (0-0.4
msec) only positive ions are observed. Their decay in this initial
stage is more rapid than later. After ~0.4 msec the positive current suddenly drops by about a factor of two. Coincident with this
is the appearance of negative ions. After this transition period
positive and negative ions decay more slowly with equal or nearly
equal slope. Nearly identical behaviour, but on a longer time
scale has been observed by Puckett and Lineberger (1970) in their
stationary afterglow apparatus. This apparatus consists of a
large vessel (18 x 36") in which ionization is obtained with a
pulsed krypton lamp. The ions escaping from a leak located in the
wall of the vessel are observed mass spectrometrically. Conditions
in such apparatus where the ionization is nearly uniform are more
tractable and Lineberger and Puckett using an analysis by Oskam
were able to explain the phenomena responsible for the observed
peculiar ion decay. Initially the much lighter electrons diffuse
faster to the wall. The resulting charge imbalance causes the
buildup of a self field which leads to positive ion-electron ambipolar diffusion and trapping of the negative ions i.e. zero diffusion to the walls for the negative ions. The gradual reduction

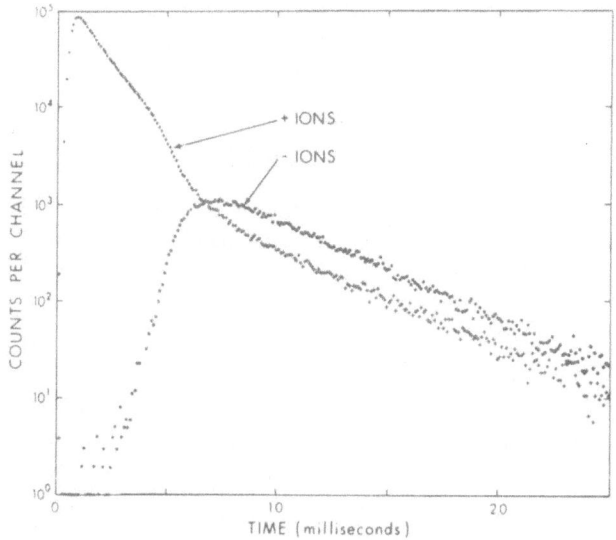

FIGURE 4

Observed total ion decay in absence of fast electron capture agents

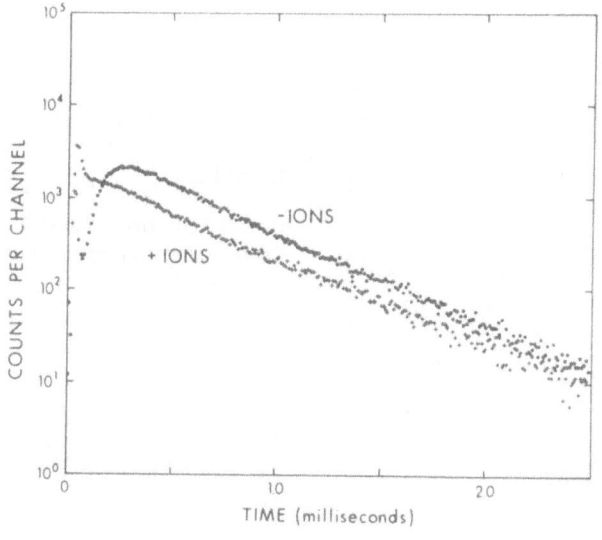

FIGURE 5

Observed total ion decay in presence of dissociative electron capture

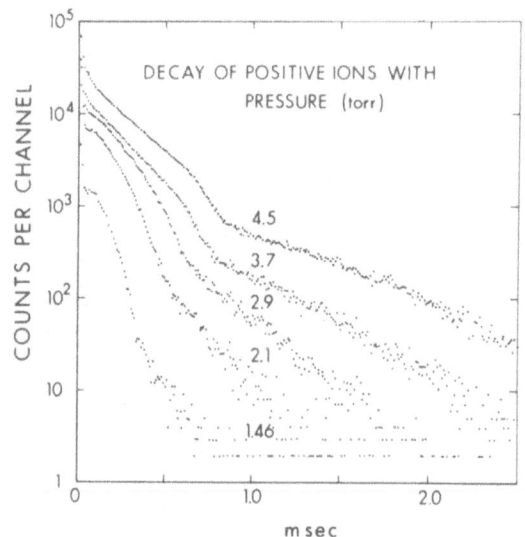

<u>FIGURE 6</u>

Decay of positive
ions with pressure

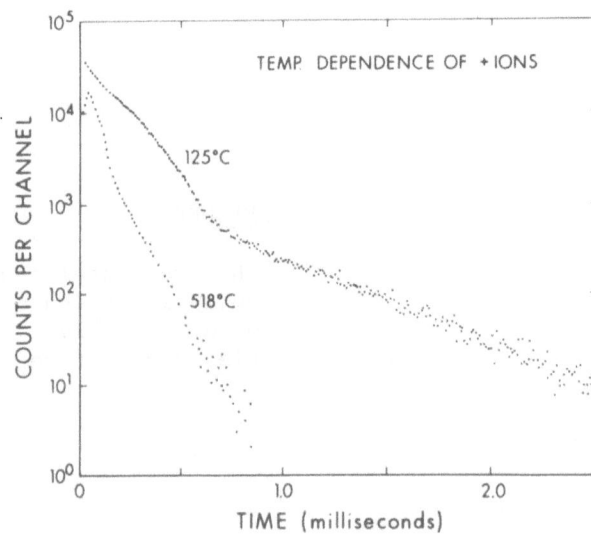

<u>FIGURE 7</u>

Decay of positive
ions with temperature

of the electron density by diffusion and electron capture leads at
a given point to a collapse of the self field. This is the transition point where the negative ions can also diffuse to the wall.
After the transition, slower positive-negative ion ambipolar diffusion followed by free diffusion prevails.

An experimental confirmation of this analysis is given in
Figure 5 obtained under conditions otherwise identical to those
in Figure 4 but with the addition of a small amount of CCl_4 which
rapidly captures electrons by the reaction: $e + CCl_4 = Cl^- + CCl_3$.
Rapid conversion of the electrons to negative ions eliminates
the positive ion-electron ambipolar stage. No negative ion trapping is observed and the decay is by slow positive-negative ion
ambipolar diffusion. This condition of slow and uniform ion
diffusion and no self field appears best suited to equilibrium
measurements. Therefore when conditions permit measurements in
Alberta are made in the presence of an electron capture agent.

Figure 6 shows positive ion decay at different pressures while
Figure 7 shows the effect of temperature. More rapid diffusion at
low pressure leads to much faster decay. The effect of temperature
is even more dramatic. The decrease of ion signal at 500°C is
several times faster than at 100°C. These results are instructive not only for pulsed but also for not pulsed high pressure
mass spectrometers. Thus even though at room temperature the
ion lifetimes may be long enough for the ions to reach equilibrium,
at higher temperatures the shortened lifetimes may not permit
establishment of equilibrium. The situation becomes even more
serious when clustering equilibria are involved. Since clustering reactions have negative temperature coefficients, Cunningham
A. J. (1973) (see also Chapter by Porter this volume), the
approach to equilibrium requires a longer time at higher temperature.

In high pressure ion sources without pulsing particularly when
high ionizing currents are used (electron currents in the
microamp range) one may expect that the ion diffusion rates will
be considerably faster and thus the ion lifetimes considerably
shorter than those in pulsed system. This will be a consequence
of the selffield created by the fast electron diffusion. Since
(secondary) electrons are created continuously appreciable self
fields can be expected. In addition to reducing the ion lifetimes
these fields may lead to non thermal ion distributions even in the
absence of repeller or other external electrical fields.

k.   Flowing afterglow, drift tubes, ion sampling errors,
                   ion cyclotron resonance

The flowing afterglow method which can produce very good equilibrium measurements will not be discussed here since the chapters

by Ferguson and particularly Bohme deal with this topic.

Drift tubes incorporating mass spectrometric sampling of the last stage can produce good equilibrium measurements provided the last stage is at near zero drift field. Successful drift tube apparatus in which negative ions are produced by photoelectrons has been described by Pack and Phelps, Pack J. L. (1966).

Drift apparatus is particularly useful for the measurement of clustering equilibria of ions which can be produced only by thermionic emission i.e. alkali ions, Hg^+, Bi^+ etc. The most rudimentary form of such apparatus consists of a two stage drift tube. The first stage contains the thermionic filament and is operated at appreciable drift field. The second stage, the equilibration chamber, is thermally isolated from the filament and operated at near zero drift field. Experiments on the clustering of water around alkali ions, Dzidic, I. (1970), Cs^+, McKnight L. G,(1972) and Hg^+ and Bi^+, I. N. Tang (1972), (1974), have been reported.

High pressure mass spectrometers, afterglow and drift tube methods obtain ion detection by passing part of the reactant gas through an exit leak into an evacuated mass analysis region. This method of sampling can lead to errors. The problem is a complex one and a complete discussion cannot be given here. Probably the most serious errors can occur in the measurement of clustering equilibria. If ion-molecule collisions occur in the region outside the sampling leak, cluster dissociation may be induced, since the ions have acquired some kinetic energy from the electric fields that are present. This dissociation leads to the formation of lower clusters. The error is particularly serious under conditions where the equilibrium constant $K_{n,n-1}$, and the ion ratio I_n/I_{n-1} is very large, say $I_n/I_{n-1} = 10^4$. In such a case even only a small degree of the dissociation of higher clusters leading to the formation of non equilibrium I_{n-1} changes the observed ion ratio very significantly. Obviously measurements with an expected high I_n/I_{n-1} ratio should be avoided. Collision outside the ion source can be practically eliminated by working at near molecular flow conditions at the ion exit leak and having an uncluttered electrode geometry. In order not to experience a too large decrease of ion intensity the Alberta apparatus utilizes long and narrow ion exit leaks (10μ x 2000μ). Molecular flow is obtained up to quite high pressures since the narrow dimension (10 μ) determines the upper pressure limit for molecular flow i.e. mean free path >10 μ. On the other hand sufficient gas flow to maintain adequate ion intensity is obtained by the extensive length of the leak.

Ion cyclotron resonance with a pulsed electron beam and a trapped

ion cell, McIver R.T. (1970), McMahon T.B. (1972b), is the only
apparatus with which ion equilibria can be measured in situ. The
absence of sampling problems makes this method extremely valuable .
The presence of trapping fields, the very low pressures utilized
in ICR (p ≈ 10^{-5} torr), and the absence of a third gas justify some
concern regarding the thermal nature of the ionic reactant. The
ions created not at the bottom of the trapping well will have
excess kinetic energy. However in the pulsed experiments the ions
are observed in a time period during which hundreds of collisions
occur. Therefore the excess kinetic energy becomes gradually
dissipated as the ions relax towards the centre of the cell. The
question regarding excitation induced by the exothermicity of the
reaction is somewhat more subtle. As discussed in section d,
ions like A_2^- formed in an exothermic reaction like (6) might

$$A_1^- + A_2H = A_2^- + A_1H \qquad\qquad (6)$$

have non thermal distribution when reacting in the reverse direction. The self proton transfer reaction:

$$A_2^- + A_2H = A_2H + A_2^- \qquad\qquad (20)$$

is probably a very efficient mode by which excess energy in the
A_2^- can be dissipated. Generally ICR measurements are done only
for systems where the free energy change is small. The ratio:
weaker to stronger acid i.e. $[A_1H]/[A_2H]$ utilized is not very much
bigger than unity. If A_2^- was non thermal, decrease of the ratio
$[A_1H]/[A_2H]$ in a series of measurements should lead to significantly
increased thermalization of A_2^- by reaction (20). This should lead
to a change in the observed equilibrium constant. Since the
equilibrium constants are found to be invariant with such concentration change we must conclude that the reactants are thermal.

1. Internal consistency and comparison of results obtained by
                    different methods

    Data obtained from ion equilibria must be consistent with data
obtained by other physical methods or theoretical calculations. A
review of such comparisons goes beyond the scope of the present
paper. However it can be stated that on the whole the agreement
has been good where such comparisons have been possible. However,
the ion equilibrium method is rapidly providing data in many areas
where no competitive experimental methods exist. Therefore
internal consistency checks and agreement between the results
from different groups are important criteria to the reliability
of the method.

The consistency of equilibrium data can be often checked through thermodynamic cycles. Shown in Figure 8 are free energy differences observed in substituted benzoic acids acidity measurements (Yamdagni R, 1974). If we consider reaction (6) it is obvious

$$A_1^- + A_2H = A_1H + A_2^- \qquad (6)$$

that apart from the direct measurement of K_6, ΔG_6 could be obtained by many other pathways involving equilibria with other intermediate acids. The ΔG_6° obtained by these multiple pathways should be the same. Acids at which more than one arrow terminates (Figure 8) are measured by more than one pathway. In general the consistency observed with such multiple pathways has been good.

Consistency checks through thermodynamic cycles is also possible for some clustering equilibria. Figure 9 shows results for ΔG measurements involving protonated clusters containing water W, dimethyl ether E and mixtures thereof. The results were obtained in three independent series of experiments involving pure water, pure ether and water and ether mixtures, Hiraoka K. (1974). The ΔG values shown in Figure 9 permit the evaluation via cycles of the proton affinity difference PA(dimethyl ether) - PA(water) = 23 kcal/mole. This is close to the result obtained by proton affinity measurements involving proton transfer reactions, Long. J. (1973).

In general agreement between equilibrium data obtained in different laboratories and by different methods has also been good. Table IV shows comparisons of proton transfer equilibrium data and Table V shows comparison of cluster equilibria data.

m. Brief summary of results from ion equilibria measured in Alberta

Earlier work has been summarized in three review articles P. Kebarle (1972a), P. Kebarle (1972b) and P. Kebarle (1974). Since the appearance of the last review article work on the basicity of amines and proton induced cyclization of diamines has been reported, Yamdagni (1973a). The acidities of aliphatic, Yamdagni (1973b), Yamdagni (1973c), aromatic, Yamdagni (1974b) and carbon acids, McMahon T.B. (1974) were also determined.

The clustering of the proton with methanol, dimethyl ether and mixtures of water, methanol, diethyl ether was excamined, Grimsrud (1973), Hiraoka (1974), as well as the proton in water and ammonia mixtures. Payzant (1973). An investigation on (ClHCl)$^-$ and Cl$^-$(HCl) Yamdagni (1974c) and O$_2^-$, Cl$^-$ clustered with H$_2$O, CH$_3$OH and CH$_3$CN were also reported on, Yamdagni (1973d).

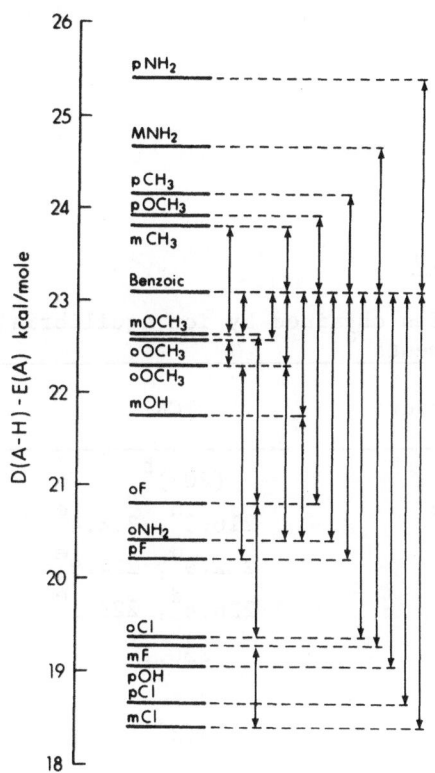

FIGURE 8

Gas phase acidity differences
of substituted benzoic acids

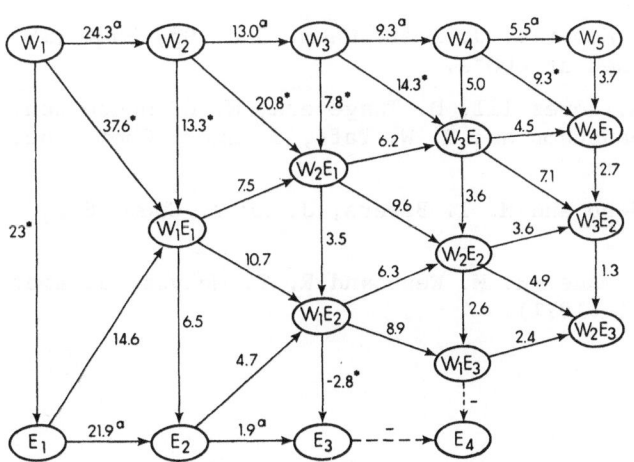

FIGURE 9

Thermo dynamic
cycles involving
clusters containing
the proton and W =
water, E = di-
methyl ether
molecules

TABLE IV

Comparison of some proton affinities obtained by ion equilibria measurements[a]

	high pressure mass spec	ICR
NH_3	$(207)^b$	$(207)^b$
CH_3NH_2	217.8^c	216.3^d, 218.4^e
$(CH_3)_2NH$	225.3^c	222.9^d, 224.9^e
$(CH_3)_3N$	230.3^c	226.6^d, 229.1^e
Piperidine	231.5^c	230.1^f
Pyridine	225.6^c	225^d
Aniline	215.9^c	216^d

a. All values in kcal/mole

b. Reference value, M. A. Haney and J. L. Franklin, J. Chem. Phys. 50, 2029 (1969).

c. R. Yamdagni and P. Kebarle, J. Amer. Chem. Soc. 95, 3504 (1973) measurements obtained at 600°K.

d. E. M. Arnett, F. M. Jones III, M. Taagepera, W. G. Henderson, D. Holtz, J. L. Beauchamp and R. W. Taft, J. Amer. Chem. Soc. 94, 4724 (1972).

e. D. H. Aue, H. M. Webb and M. T. Bowers, J. Amer. Chem. Soc. 94, 4726 (1972).

f. M. T. Bowers, D. H. Aue, H. M. Webb and R. T. McIver, J. Amer. Chem. Soc. 93, 4314 (1971).

TABLE V

Comparison of data obtained for clustering equilibria

Reaction	High pressure mass spec.	Flowing Afterglow
$NH_4^+(NH_3)_2 + NH_3 = NH_4^+(NH_3)_3$	6.1^a, 6.0^b	6.5^c
$NH_4^+(NH_3)_3 + NH_3 = NH_4^+(NH_4)_4$	3.7^a, 3.4^b	3.4^c
$OH^-(H_2O)_2 + H_2O = OH^-(H_2O)_2$	7.7^e	8.2^d
$Cl^- + H_2O = NO_2^-(H_2O)$	8.0^f	8.0^d
$NO_3^- + H_2O = NO_3^-(H_2O)$	6.8^f	6.8^d

a. Payzant, J. D., Cunningham, A. J., and Kebarle, P, Can. J. Chem. 51, 3242 (1973).

b. Arshadi, M. R. and Futrell J. H., J. Phys. Chem. 78, 1482 (1974).

c. Fehsenfeld, F.C. and Ferguson E.E., J. Chem. Phys. 59, 6272 (1973).

d. Fehsenfeld, F. C. and Ferguson, E. E. J. Chem. Phys. (to be published).

e. Arshadi, M. R., Yamdagni, R. and Kebarle, P. J. Phys. Chem. 74, 1475 (1970).

f. Payzant, J. D., Yamdagni, R., and P. Kebarle, Can. J. Chem. 49, 3308 (1971).

Several useful generalizations can be made on basis of the ion-molecule cluster and the gas phase acidity-basicity studies.

Negative ions and protic molecules

The stability of one molecule-ion clusters formed from negative ions and protic molecules HR like HOH, CH_3OH, HCl, increases with the gas phase acidity of HR. Thus increasing binding energies are observed between Cl^- and HR = HOH, CH_3OH, C_2H_5OH, CH_3COOH, HCOOH. The order of HR shown above is the same as the gas phase acidity order of the compounds HR.

The above rule is generally valid also for the addition of the second molecule but changes gradually occur such that water becomes the best solvent at high cluster numbers. This is due to the bulkiness of groups like CH_3 or C_2H_5 which become a hindrance when more molecules are present in the cluster. Furthermore such groups cannot hydrogen bond.

Positive ions and bases like OH_2, NH_3 etc.

The interactions between <u>spherical positive</u> ions like K^+ and various polar bases follow approximately the gas phase basicities (or proton affinities) of the clustering molecules. However, the effect is small i.e. large differences in basicity lead to small differences in the clustering interactions. Thus NH_3, H_2O, CH_3OH and CH_3OCH_3 all form clusters with K^+ which have very similar stability (Davidson 1974).

Positive onium ions RH^+

Positive onium ions RH^+ like: H_3O^+, NH_4^+, $CH_3NH_3^+$ form clusters with the bases B whose stability increases with the acidity of RH^+ (i.e. decreases with the proton affinity of R). The stability of a given ion RH^+ clustering with different bases B increases with the gas phase basicity of B. Thus $NH_4^+ . NH_3$ is more strongly bonded than $NH_4^+ . H_2O$. In large clusters the trends change, ultimately water becomes the favoured clustering molecule.

Positive and negative ions and dipolar aprotic molecules

Compounds like CH_3CN, $(CH_3)_2SO$, $(CH_3)_2SO_2$ which have large permanent dipoles but no acidic hydrogens capable of hydrogen bonding are called dipolar aprotic molecules. The interactions of such molecules with positive ions are stronger than with negative ions. This is due to the charge distribution of the dipole. The negative pole is relatively concentrated and accessible for interaction with a positive ion. The positive charge of the dipole is quite difusely distributed over several atoms. This puts the positive pole inside the molecule and makes the distance between it and a negative ion large, which results in a weak interaction with negative ions. Thus the detailed charge distribution in the molecule is of considerable importance in polar aprotic solvents.

Inner-outer shell transitions

When the molecules involved in clustering can hydrogen bond the transition from an inner to an outer shell does not lead to a very large fall off in stability. Large drop off in outer shell stability is observed when the molecules involved cannot hydrogen bond.

References

Note: References given in the text have been identified with the name of the first author only.

Arshadi, M.R. and Futrell J. H. J. Phys.Chem. 78, 1482 (1974).

Beauchamp J.L. Amer. Rev. Phys. Chem. 22, 527 (1971).

Bohme D.K., Hemsworth R.S., Rundle, H.W. and Schiff H. I. J. Chem. Phys. 58, 3504 (1973).

Bowers M.T., Aue, D. H., Webb, H. M., McIver Jr. R.T., J. Amer. Chem. Soc. 33, 4314 (1971).

Bowers M.T. paper presented at 22nd Annual Conference for Mass Spectrometry, Philadelphia 1974.

Briggs, J. P., Yamdagni R., Kebarle P. J. Amer. Chem. Soc., 94, 5128 (1972).

Brauman, J.I. and Blair L. K., J. Amer. Chem. Soc. 92, 5968 (1970).

Chang C. and Tiernan T.O. paper presented at 22nd Annual Meeting on Mass Spectrometry, Philadelphia (1972).

Chang S.L.and Franklin J.L., J. Amer. Chem. Soc., 94, 6347 (1972).

Chupka W.A. J. Chem. Phys. 30, 458 (1959).

Cunningham A.J., Payzant J.D., Kebarle P. J. Amer. Chem. Soc., 94, 7627 (1972).

Conway, D.C. and Janik G.S., J. Chem. Phys. 53, 1859 (1970).

Davidson W. and Kebarle P. to be published (1974).

Dzidic I. and Kebarle P., J. Phys. Chem. 74, 1466 (1970).

Fehsenfeld F.C. and Ferguson E.E. J. Chem. Phys. 59, 6272 (1973).

Ferguson E.E. paper presented at 22nd Annual Conference on Mass Spectrometry, Philadelphia (1974).

Good A., Durden, D.A. and Kebarle P., J. Chem. Phys. 52, 222 (1970).

Grimsrud E. and Kebarle P., J. Amer. Chem. Soc. 95, 7939 (1973).

Haney M.A. and Franklin J. L. J. Phys. Chem. 73, 4328 (1969).

Henderson W.H. Taagerpera D, Oltz R., McIver R.T., Beauchamp, J.L. and Taft R.W., J. Amer. Chem. Soc. 94, 4728 (1972).

Hiraoka K., Grimsrud E.P. and Kebarle P., J. Amer. Chem. Soc. <u>96</u>, 3359 (1974).

Hiraoka K. and Kebarle P., J. Chem. Phys. to be published.

Kebarle P. and Godbole E.W. J. Chem. Phys. <u>39</u>, 1131 (1963).

Kebarle P. and Hogg A.M., J. Chem. Phys. <u>42</u>, 798 (1965).

Kebarle P. "Ions and ion solvent molecule interactions in the gas phase", chapter in Ions and ion pairs in organic reactions Vol. 1 Edited by M. Szwarc, Wiley, New York (1972a).

Kebarle P. "Higher order reaction-ion clusters and ion solvation" in Ion-molecule reactions Edited by J. L. Franklin, Plenum Press New York (1972b).

Kebarle P. Gas phase ion equilibria and ion solvation, Chapter in Modern Aspects of Electrochemistry Vo.1. 9 B. E. Conway and J.O. M. Bockris Editors, Plenum Press (1974).

McKnight L.G. and Sawina J.M. J. Chem. Phys. <u>57</u>, 5156 (1972).

Kraemer W.P. and Diercksen G.H.F. Chem. Phys. Letters <u>5</u>, 463 (1970).

Long, J.W. and Munson Burnaby, J. Amer. Chem. Soc. <u>95</u>, 2427 (1973).

Long, J. W. and Franklin J. L. J. Amer. Chem. Soc. <u>96</u>, 2320 (1974).

McMahon T.B., Blint R.J., Ridge D. P. and Beauchamp J. L. J. Amer. Chem. Soc. <u>94</u>, 8934 (1972a).

McMahon T.B. and Beauchamp J.L. Rev. Sci. Instr. <u>43</u>, 509 (1972b).

McMahon T.B. and Kebarle P. J. Amer. Chem. Soc. <u>96</u>   (1974).

McIver R.T. Rev. Sci. Instr. <u>41</u>, 555 (1970).

Pack, J. L. and Phelps A.V. J. Chem. Phys. <u>44</u>, 1870 (1966).

Payzant J.D. and Kebarle P. J. Chem. Phys. <u>56</u>, 3482 (1972a).

Payzant J.D. Cunningham A.J. and Kebarle P. Can. J. Chem., <u>51</u>, 3242 (1973b).

Payzant J.D. Cunningham A.J. and Kebarle P. J. Chem. Phys. <u>59</u>, 5615 (1973c)

Puckett L. J. and Lineberge W.C. Phys. Rev. <u>6</u>, 1635 (1970).

Robinson P.J. and Holbrook K. A. "Unimolecular Reactions" Wiley-Interscience London (1972).

Solomon J.J. Meot-Ner M. and Field F. H. J. Amer. Chem. Soc. 96, 3727 (1974).

Searles S.K. and Kebarle P. Can. J. Chem. 47, 2619 (1969).

Talroze V.L. Pure and Appl. Chemistry 5, 455 (1962).

Tang I.N. and Castleman Jr. A. W. J. Chem. Phys. 57, 3638 (1972).

Tang I.N. and Castleman Jr. A.W. J. Chem. Phys. 60, 3981 (1974).

Wincel H. and Herman J.A. J.C.S. Faraday Trans. 69, 1797 (1973).

Yamdagni R. and Kebarle P. J. Amer. Chem. Soc. 95, 3504 (1973a)

Yamdagni R. and Kebarle P. J. Amer. Chem. Soc. 95, 4050 (1973b).

Yamdagni R. and Kebarle P. J. Amer. Chem. Soc. 95, 6833 (1973c).

Yamdagni R. Payzant J. and Kebarle P. Can. J. Chem., 51, 2507 (1973d).

Yamdagni R., McMahon T.B. and Kebarle P. . Amer. Chem. Soc. 96, 4035 (1974a).

Yamdagni R. and Kebarle P. Can. J. Chem., 52, 861 (1974b).

Yamdagni R. and Kebarle P. Can. J. Chem., 52, 2449 (1974c).

Yang, J. H. and Conway D.C. J. Chem. Phys. 40, 1729 (1964).

THE KINETICS AND ENERGETICS OF PROTON TRANSFER

D. K. Bohme

Department of Chemistry and C.R.E.S.S., York University

4700 Keele Street, Downsview, Ontario M3J 1P3, Canada

INTRODUCTION

At the time of the last international exposé on proton transfer (P.T.) reactions in 1968 (Volpi, 1970) information concerning these processes was still scanty. However, the fundamental significance of P.T. reactions in the overall regime of ion-molecule (I.M.) reaction kinetics had been recognized and the crucial role of such reactions in a variety of charged chemical environments was beginning to be appreciated (see for example Ausloos and Lias, 1965) and in certain instances even exploited (Munson and Field, 1966). Since 1968, as part of the general information explosion in the chemical physics of I.M. reactions, there has become available considerable new experimental and theoretical insight into the nature of protonated species and their reactions, viz. the transfer of a proton from an ion to a molecule:

$$XH^+ + Y \underset{\leftarrow}{\rightarrow} YH^+ + X \tag{1}$$

and the transfer of a proton from a molecule to an ion:

$$YH + X^- \underset{\leftarrow}{\rightarrow} XH + Y^- \tag{2}$$

This lecture will focus on those aspects of P.T. which have emerged out of recent experimental studies and which in the opinion of this author, have proven to be most attractive and rewarding to the experimental kineticist.

COMPARISONS WITH CLASSICAL THEORIES

Laboratory measurements of rate coefficients have indicated that >95% of all (>150) exothermic P.T. reactions observed to date proceed extremely rapidly at thermal energies, viz. 10^{-10} cm^3 molecule^{-1} sec^{-1} $\lesssim$ k$_{300°K}$ $\lesssim$ 10^{-8} cm^3 molecule^{-1} sec^{-1}, essentially in the absence of energies and entropies of activation. It can be inferred from these experimental findings that the transfer of a proton, when energetically allowed, often proceeds on nearly every collision, viz. that the reaction rate coefficient is likely to be very nearly equal to the capture rate coefficient for most exo-thermic P.T. reactions (at least those involving small molecular systems). Consequently P.T. reactions should represent a group of I.M. reactions highly suited for the systematic testing of current theories of I.M. reactions. For example, comparisons of measured reaction rate coefficients, k$_{exp}$, with predicted capture (collision) rate coefficients, k$_c$, should either provide insight into any natural limitations of the current theories of I.M. reactions or (upon the assumption that theory adequately predicts the collision rate) allow the identification of the possible presence of energies or entropies of activation (Schiff and Bohme, 1974).

For a system having a Maxwell-Boltzmann (M.B.) energy distri-bution, current classical theories of ion-molecule interactions predict a collision or capture rate coefficient given by

$$k_c = 2\Pi e \left(\frac{\alpha}{\mu}\right)^{\frac{1}{2}} + C\left(\frac{2\Pi e\mu_D}{\mu}\right)\left(\frac{2\mu}{\Pi kT}\right)^{\frac{1}{2}} \qquad (3)$$

where e is the charge on the ion, μ the reduced mass of the colli-dants, α is the polarizability and μ_D is the permanent dipole moment of the neutral molecule. Equation 3 is composed of the familiar Langevin term (Gioumousis and Stevenson, 1958) and a 'correction' term reflecting the ion-permanent dipole interaction. In the Average-Dipole-Orientation (A.D.O.) theory, C is a measure of the extent to which the dipole is orientated w.r.t. the direction of the approaching ion (Su and Bowers, 1973a). For C=1, Equation 3 reduces to the 'locked-dipole' limit (Moran and Hamill, 1963; Gupta et al, 1967).

For a series of reactions involving a fixed substrate, the three classical theories define three straight lines on a plot of k vs. $\mu^{-\frac{1}{2}}$. Figure 1 shows a comparison of reaction rate coeffi-cients measured at 300°K with the flowing afterglow (F.A.) technique (Hemsworth et al, 1974) with capture rate coefficients predicted by the three theories for proton transfer to NH$_3$(α=2.16Å^3, μ_D=1.47D). The reactions appear to proceed with unit efficiency (k$_{exp}$/k$_c$=1) when comparison is made with the A.D.O. theory which predicts the correct dependence of k on μ and moreover has provided a reasonable value for C in this case (C=0.222). The 'locked-dipole' model clearly overestimates the effect of the dipole.

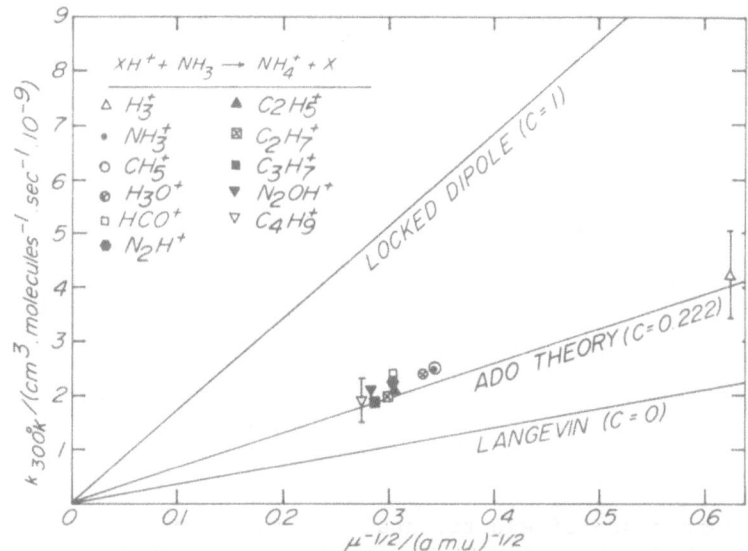

FIG. 1 A comparison of experimental reaction rate coefficients
for P.T. with NH_3 (the solid bars represent an estimated accuracy
of ±25%) with collision rate coefficients predicted by classical
theories.

Figure 2 shows a comparison of rate coefficients measured in
a F.A. (Betowski et al, 1974) with the capture rate coefficients
predicted by the classical theories for P.T. reactions with H_2O
(α=1.45Å^3, μ_D=1.84D). The A.D.O. model again appears to be the
most realistic. The low experimental value of the rate coefficient
for the P.T. reaction of H_2O with H^- may reflect the presence of a
small activation energy barrier for this reaction.

Measured rate coefficients (Payzant et al, 1974) for P.T.
reactions with HCN (α=2.59Å^3, μ_D=2.98D) are compared with theoretical capture rate coefficients in Figure 3. The A.D.O. capture
rate coefficients again reflect most accurately the observed
reaction rate coefficients except perhaps at low μ where a positive
deviation from the A.D.O. line is evident. The reaction of HCN
with C^- is anomalously slow.

Several generalities have emerged from comparisons of experiment with theory such as those given above and others not presented
here (see for example Su and Bowers, 1973b). The rate coefficients
for P.T. reactions involving polar molecules are in reasonable
accord with the A.D.O. theory. We can conclude that the A.D.O.
treatment provides a more realistic model for ion-molecule collisions than the 'locked-dipole' approach - it appears that the
rotation of the neutral molecule at 300°K is sufficient to prevent

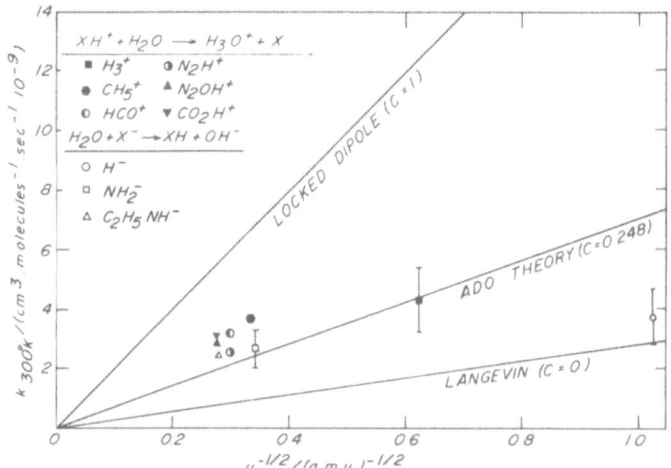

FIG. 2. A comparison of experimental reaction rate coefficients for P.T. with H_2O (the solid bars represent an estimated accuracy of ±30%) with collision rate coefficients predicted by classical theories.

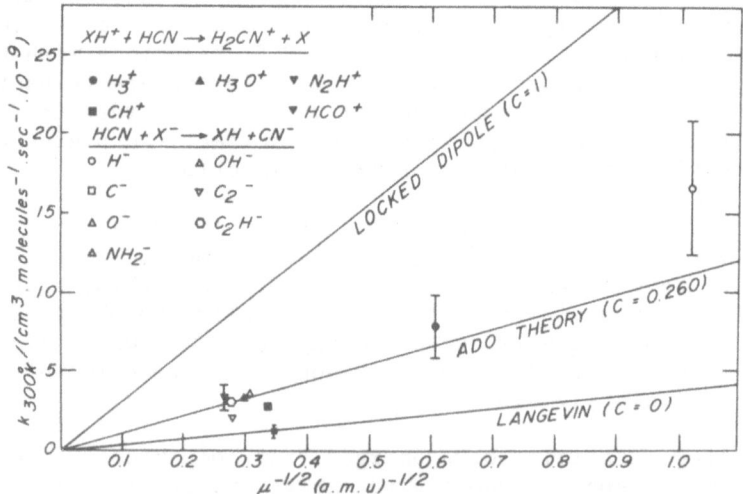

FIG. 3. A comparison of experimental reaction rate coefficients for P.T. with HCN (the solid bars represent an estimated accuracy of ±25%) with collision rate coefficients predicted by classical theories.

a total alignment of the dipole in the electric field of the ion. There is an indication that the current A.D.O. model underestimates the effect of the dipole for reactions involving molecules with large permanent dipoles and small reduced masses of the collidants. Refinements of the A.D.O. theory are currently in progress (Su and Bowers, 1974). The dearth of systematic studies of P.T. reactions involving non-polar molecules has discouraged a similar assessment of the Langevin model although the data available to date indicates that this model predicts quite accurately (within ca. 20%) the observed rate coefficients for these P.T. reactions (Hemsworth et al, 1974). Only a few P.T. reactions have been identified for which $k_{exp}/k_{L,ADO} \lesssim 0.5$, viz.

$$H_2O + H^- \rightarrow H_2 + OH^- + 8 \text{ kcal mole}^{-1}, \; k_{exp}/k_{ADO} = 0.5 \qquad (4)$$

$$HCN + C^- \rightarrow CH + CN^- + 15 \text{ kcal mole}^{-1}, k_{exp}/k_{ADO} = 0.3 \qquad (5)$$

and the reactions (Bohme et al, 1973a, and unpublished results)

$$H_2 + NH_2^- \rightarrow NH_3 + H^- + 3 \text{ kcal mole}^{-1}, \; k_{exp}/k_L = 0.01 \qquad (6a)$$

$$CH_3NH^- + H_2 \rightarrow H^- + CH_3NH_2 + 3 \text{ kcal mole}^{-1}, \; k_{exp}/k_L = 0.1 \; (6b)$$

One could ascribe the relatively low efficiency of these reactions to the presence of a small activation energy barrier. Curiously, however, the exothermicities of all of these reactions are relatively small, $\Delta H^o_{298} \lesssim 15$ kcal mole^{-1}. This raises questions concerning the influence of excess energy in the form of reaction exothermicity on the reaction rate coefficient for P.T. reactions.

INFLUENCE OF EXCESS ENERGY

P.T. reactions provide an, as yet rare, opportunity to explore the effect of excess energy upon the rates, product channel distributions, and kinematics for a series of I.M. reactions different in exothermicity but otherwise quite similar and generally not subject to restrictive activation energy or entropy barriers - most P.T. reactions are probably characterized by a basin in the potential energy vs. reaction diagram (see Figure 4). We can consider three contributions to the excess energy (see Figure 4): the C.M. translational energy of the reactants, E_t^R, the total internal excitation energy of the reactants, E_i^R, and the reaction energy, ΔE^o_0 or ΔH^o_0, which in the case of P.T. is defined by the difference in the proton affinity, $\Delta P.A._{0^oK}$, of either the neutral species (Reaction 1) or the negative ions (Reaction 2). Meaningful experimental measurements of the influence of excess energy on the magnitude of P.T. rate coefficients have become available only within the last year or so. Quantitative studies of the influence of

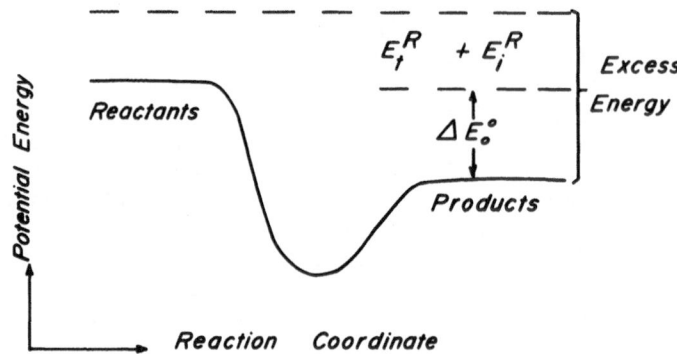

FIG. 4. Hypothetical reaction coordinate – potential energy
diagram for P.T. reactions.

excess energy on product distributions have also recently been
initiated (Bowers et al, 1973) whereas knowledge concerning the
influence on dynamics awaits the further exploitation of crossed
beam studies (Vestal et al, 1974).

Figure 5 shows the only experimental results (Hemsworth et al,
1973) presently available for the variation of the P.T. rate coefficient with temperature, i.e. as E_t^R and E_i^R are increased concurr-
ently. The P.T. reaction

$$N_2OH^+ + CO \rightarrow HCO^+ + N_2O \tag{7}$$

is observed to proceed at ~60% of the Langevin collision rate (in
this case the correction for the permanent dipole of CO is negligible) and shows Langevin behaviour in that it is independent of
temperature from 280 to 500°K. Further studies of this type are
obviously desirable.

A number of experimental results have recently become available
which provide insight into the effect of internal energy, E_i^R, alone
on P.T. rates. The tandem mass spectrometer of Smith and Futrell
(1974) has provided information on the influence of vibrational
excitation of D_3^+ on the P.T. reaction with NH_3 (see Figure 6).
The P.T. rate coefficient approaches the M.B. limit as D_3^+ is
increasingly relaxed by collisions with D_2. A similar behaviour
has been observed by Ryan (1974) for the P.T. between H_3^+ and N_2

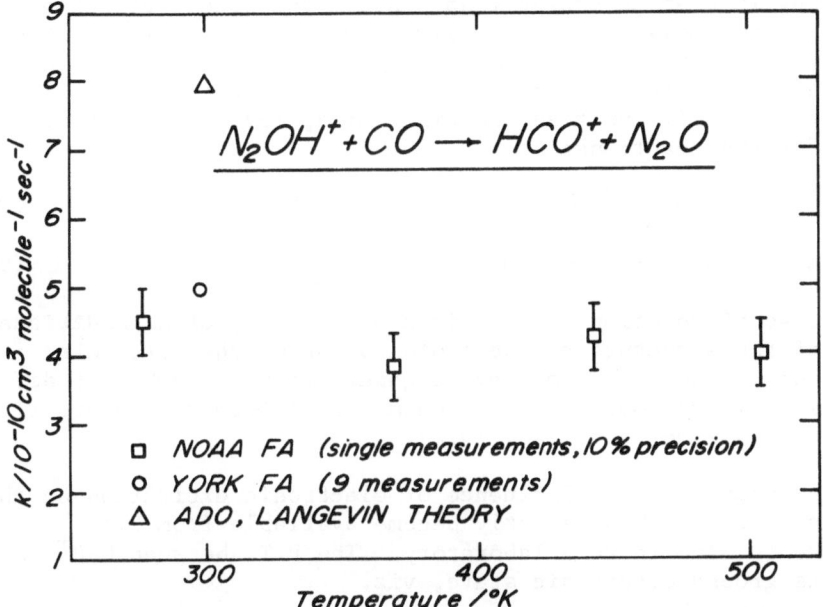

FIG. 5. Temperature dependence of the P.T. reaction rate coefficient observed for the P.T. reaction of N_2OH^+ with CO.

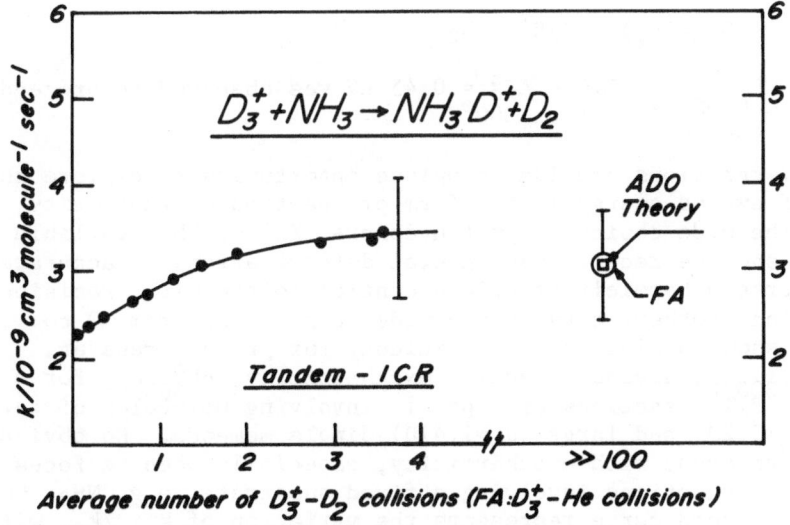

FIG. 6. Influence of internal excitation on the P.T. reaction of D_3^+ with NH_3.

proceeding in a space charge trap. Again the rate coefficient
extrapolates nicely to the M.B. limit as determined with a flowing
afterglow. A similar influence of internal excitation on the P.T.
rate coefficient has also been identified (Chupka and Russell, 1968;
Sieck et al, 1971; Huntress and Pinizzotto, 1973; Su and Bowers,
1973c) for the reactions:

$$NH_3^+ + NH_3 \rightarrow NH_4^+ + NH_2 \tag{8}$$

$$C_4H_9^+ + NH_3 \rightarrow NH_4^+ + C_4H_8 \tag{9}$$

In the case of reaction 9 there is a possibility of an additional
effect of the structure of the proton donor on the P.T. rate
coefficient. The $C_4H_9^+$ ion may be generated in a number of different ways and often more than one geometrical isomer is energetically
accessible (Ausloos and Lias, 1972).

One example of the influence of electronic excitation on the
efficiency of P.T. has recently become available from F.A.
measurements made in this laboratory. The P.T. between N_2H^+ and
O_2 in its ground electronic state, viz.

$$N_2H^+ + O_2(X^3\Sigma_g^-) \rightarrow O_2H^+ + N_2 \tag{10a}$$

for which $E_{excess} = -\Delta E_o^o = -0.55$ eV was observed to be highly
inefficient, $k_{exp}/k_c \leq 10^{-3}$ at $300°K$. On the other hand the
electronic excitation of O_2 was observed to promote the transfer
of the proton, i.e. the reaction

$$N_2H^+ + O_2(a^1\Delta_g) \rightarrow O_2H^+ + N_2 \tag{10b}$$

for which $E_{excess} = E_i^R - \Delta E_o^o = 0.45$ eV was observed to proceed
rapidly, $k_{exp}/k_c \gtrsim 10^{-1}$ at $300°K$.

P.T. reactions provide an unique opportunity to explore the
effect of excess energy in the form of reaction exothermicity, ΔE_o^o,
alone. The wide choice of proton donors (XH^+ or YH) available in
practice and the recent experimental determinations of accurate
values for exothermicities afford control (often with precision)
on reaction exothermicity over a wide range, viz. from ∿0 to ∿5 eV.
Figure 7 shows a plot of the efficiency for proton transfer,
$(k_{exp}/k_c)_{300°K}$, against reaction exothermicity, $\Delta H_{300°K}^o$, for a
number of P.T. reactions of type 1 involving molecules of zero,
small ($\mu_D \sim 0.2D$) and large ($\mu_D = 1.47D$) dipole moments. No obvious
trend is apparent with exothermicity, especially when we focus on
a series of reactions involving a fixed substrate, e.g. NH_3, CO or
CH_4. The dashed curve represents the variation of k_{exp}/k_{LD} with
$-\Delta H_{300°K}^o$ observed (Solka and Harrison, 1973) at ion energies = 0.43
V (lab) for P.T. reactions of the type

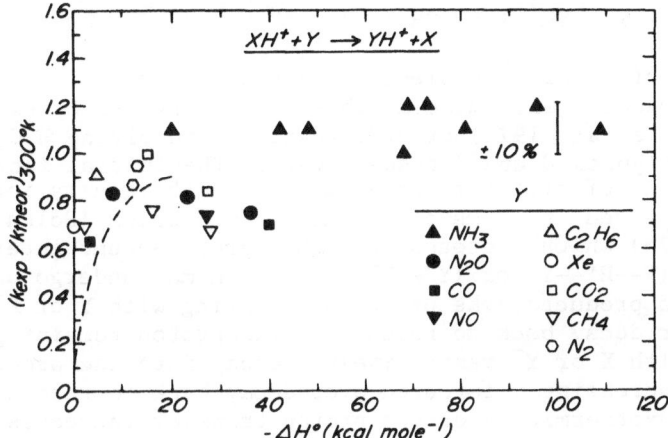

FIG. 7. Influence of reaction exothermicity on the probability of P.T.

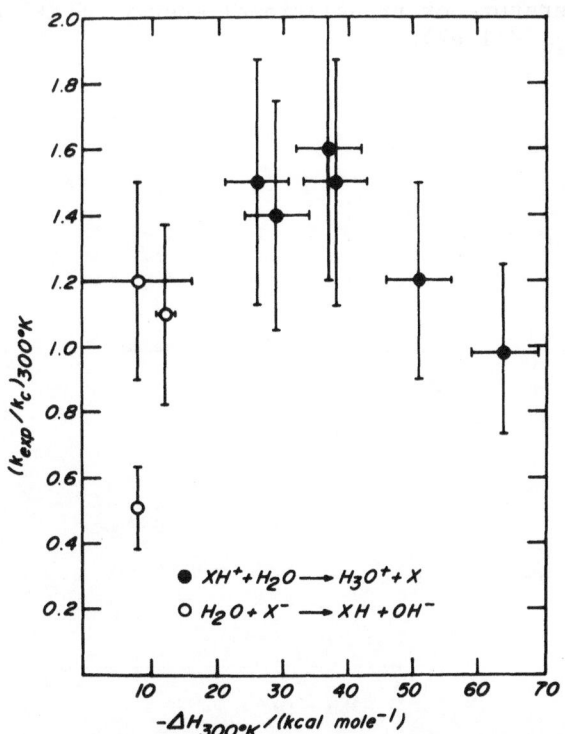

FIG. 8. Influence of reaction exothermicity on the probability of P.T. with H_2O.

$$CH_3SH_2^+ + Y \rightarrow YH^+ + CH_3SH \qquad (11)$$

for a series of polar molecules, Y. A similar trend is apparent
from recent measurements made in this laboratory (Betowski et al,
1974; Payzant et al, 1974) of P.T. reactions involving H_2O and HCN
as shown in Figures 8 and 9 respectively. The initial increase in
the probability of proton transfer as the exothermicity increases
may be interpreted in terms of a reaction mechanism (Solka and
Harrison, 1973) which proceeds through a proton-bound intermediate
of the type $(X\text{--}H^+\text{--}Y)$ or $(Y^-\text{--}H^+\text{--}X^-)$ which may undergo unimolecular decay to products, the proton associating with Y or X^- res-
pectively, or decay back to reactants, the proton remaining
associated with X or Y^- respectively. Decay into the product
channel is increasingly favoured over decay back to reactants as
the overall exothermicity of the proton transfer increases. It
will be recalled from the previous section that the low probability
of several slightly exothermic P.T. reactions may also be interpreted in terms of the possible presence of a small activation
energy barrier. Further insight into the nature of the effect
responsible for the relatively low probability of these reactions
would be provided from a study of the rate coefficients as a
function of temperature or translational energy of XH^+ or X^- at
low energies $(E_{lab} < 1 \text{ eV})$.

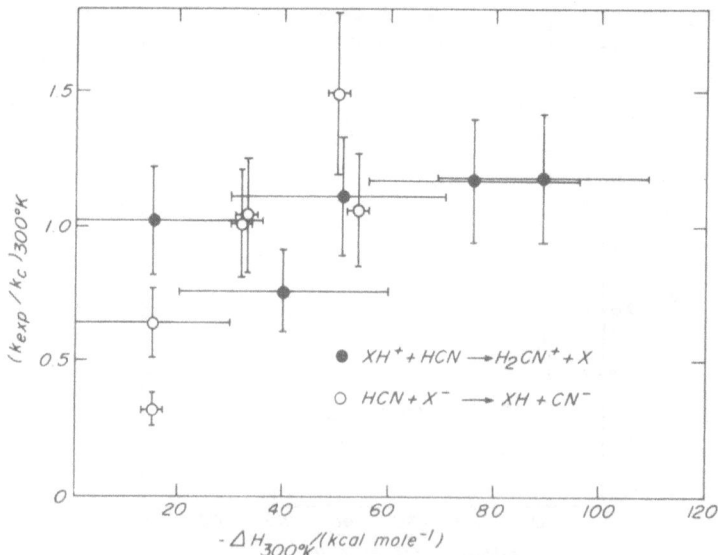

FIG. 9.   Influence of reaction exothermicity on the probability
of P.T. with HCN.

DERIVATION OF FUNDAMENTAL MOLECULAR PROPERTIES

Experimental determinations of the reaction exothermicities for many P.T. reactions ($\Delta H^o_{300} \lesssim 0.5$ eV) have recently become accessible to equilibrium studies. The fundamental measurement involves the determination of the equilibrium constant, K, either from the ratio of concentrations of products and reactants measured at equilibrium, e.g. for reaction 1:

$$K = \left(\frac{[YH^+][X]}{[XH^+][Y]} \right)_{equ_m} , \qquad (12)$$

or from the ratio of rate coefficients for the forward and reverse directions measured for MB energy distributions:

$$K = \left(\frac{k_f}{k_r} \right)_{MB} . \qquad (13)$$

The measurement of K at various temperatures can completely specify the energetics of the P.T. reaction:

the standard free energy change $\Delta G^o_T = -RT \ln K_T$,
the standard enthalpy change $\Delta H^o_T = -R(d \ln K_T / d(1/T))$,
and the standard entropy change $\Delta S^o_T = (\Delta H^o_T - \Delta G^o_T)/T$.

Numerous equilibrium studies of both P.T. reactions 1 and 2 have been reported in the past few years. Both inorganic (Bohme et al, 1973b; Hemsworth et al, 1973; Fennelly et al, 1973; McIver and Eyler, 1971; Bohme et al, 1973a) and organic (see for example Bowers et al, 1971 and Yamdagni and Kebarle, 1973a - amines; Bohme et al, 1971 and McIver and Miller, 1974 - alcohols; Yamdagni and Kebarle, 1973b - carboxylic acids) systems have been investigated. As well as providing values for fundamental molecular properties, such studies have proven useful in the experimental assessment of ab initio molecular orbital calculations and in the thermodynamic analysis of the transfer of protonated species from the gas phase to solution (see for example, Henderson et al, 1972).

The algorithms shown in Figure 10 summarizes the fundamental thermochemical properties of atoms and molecules which may be derived once the exothermicity, ΔH^o_T, of a P.T. reaction has been established (Schiff and Bohme, 1974). For P.T. reactions 1 knowledge of ΔH^o_T together with other relevant thermochemical information provides values for the difference in the heats of formation of the protonated species YH^+ and XH^+ or neutrals X and Y:

$$\Delta H^o_T = \Delta H^o_{f,T}(YH^+) - \Delta H^o_{f,T}(XH^+) + \Delta H^o_{f,T}(X) - \Delta H^o_{f,T}(Y), \qquad (14a)$$

the difference in the proton affinities of X and Y:

$$\Delta H^o_T = PA_T(X) - PA_T(Y), \qquad (14b)$$

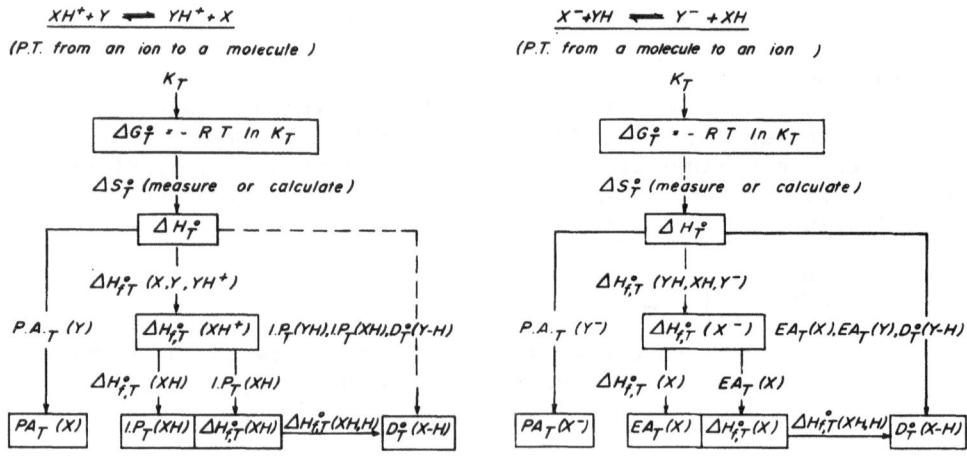

FIG. 10. Derivation of fundamental thermochemical properties of atoms and molecules.

and the difference in the ionization potentials or bond energies of YH and XH (although YH and XH are often unstable species):

$$\Delta H_T^O = 1P_T(YH) - 1P_T(XH) + D_T^O(X-H) - D_T^O(Y-H) . \qquad (14c)$$

To obtain absolute values reference must always be made to values established with some other technique. For example, Figure 11 shows the differences in proton affinities determined for a number of simple atoms and molecules using the F.A. technique. Absolute proton affinities become available only when an absolute value has been adopted for the proton affinity of one of these species (Schiff and Bohme, 1974). For P.T. reactions 2 knowledge of ΔH_T^O provides values for the difference in the heats of formation of the protonated species YH and XH or the negative ions X^- and Y^-:

$$\Delta H_T^O = \Delta H_{f,T}^O(XH) - \Delta H_{f,T}^O(YH) + \Delta H_{f,T}^O(Y^-) - \Delta H_{f,T}^O(X^-), \qquad (15a)$$

the difference in the proton affinities of Y^- and X^-:

$$\Delta H_T^O = PA_T(Y^-) - PA_T(X^-), \qquad (15b)$$

and the difference in the electron affinities of X and Y or the bond energies of YH and XH:

$$\Delta H_T^O = EA_T(X) - EA_T(Y) + D_T^O(YH) - D_T^O(XH) . \qquad (15c)$$

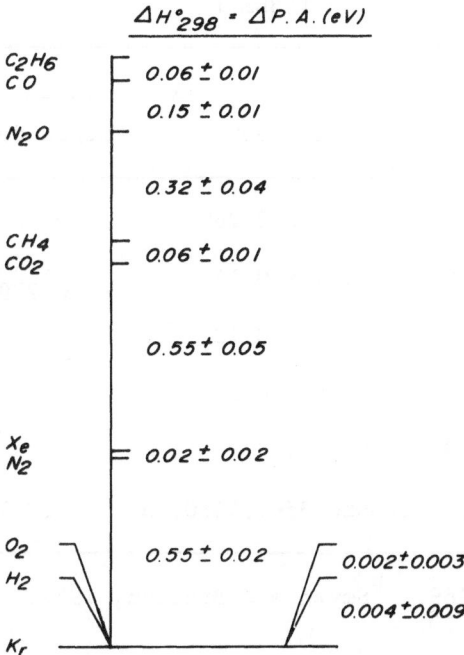

FIG. 11. Differences in the proton affinities of atoms and molecules as determined from equilibrium studies performed with a flowing afterglow.

Table I summarizes the standard heats of formation and electron affinities of several representative species determined from equilibrium studies of reactions 2 (Bohme et al, 1973a; Mackay et al, 1974; Bohme et al, 1974). The general agreement with values determined with other techniques, viz. photoionization and photodetachment, provides a measure of the confidence one may assign to this approach which yields values for $\Delta H^o_{f,T}(X^-)$ reliable to within a few kcal mole^{-1} and values for EA(X) reliable to within a few tenths of an eV. This approach is particularly attractive in the evaluation of electron affinities of polyatomic species for which otherwise more accurate methods, e.g. photodetachment, can provide only upper or lower bounds. When reliable values for the electron affinities of X^- and Y^- are available from other techniques, equilibrium studies of reactions 2 can provide reliable values for bond dissociation energies. For example, equilibrium studies (Bohme et al, 1973a; and unpublished results; McIver and Eyler, 1971) of reaction 6a and the reaction

$$SH^- + HCN \rightarrow CN^- + H_2S \qquad (16)$$

TABLE I

ION	$\Delta H^o_{f,298}(X^-)$ kcal mole^{-1}	$EA_{298}(X)$/eV molecule^{-1}	
		This work	Other determinations
CN^-	16.7 ± 3.6	3.79 ± 0.26	3.82 ± 0.02[a]
NH_2^-	25.4 ± 0.9	0.64 ± 0.17	0.744 ± 0.022[b] 0.779 ± 0.037[c]
CH_3NH^-	30.5 ± 1.2	0.57 ± 0.10	
CH_3O^-	-27.9 ± 1.1	1.36 ± 0.09	
$CHCO^-$	3.72 ± 2.31	1.43 ± 0.14	
C_2H^-	71.5 ± 1.1	$1.86\pm0.35-2.54\pm0.13$	$\leq3.73\pm0.05$[d]

[a]Berkowitz et al, 1969. [b]Smyth and Brauman, 1972. [c]Celotta et al, 1974. [d]Feldman, 1970.

have provided values for $D^o_o(NH_2-H) = 106.0 \pm 1.1$ kcal mole^{-1} and $D^o_o(H-CN) = 124 \pm 2$ kcal mole^{-1} which are believed to be more reliable than the currently accepted values (Darwent, 1970) of 103 ± 2 and 127 ± 5, respectively.

The experimental determination of ΔS^o_T from a measurement of K at various temperatures can provide hitherto unavailable information about the geometries of protonated molecules. For example, equilibrium studies (Hemsworth et al, 1973) of P.T. reaction 7 and the P.T. reaction

$$CO_2H^+ + CH_4 \rightleftarrows CH_5^+ + CO_2 \tag{17}$$

have provided values for the following differences in entropies:

$$S^o_{298}(CO_2H^+) - S^o_{298}(CH_5^+) = +5.2 \pm 0.6 \text{ e.u.} \tag{18a}$$

$$S^o_{298}(N_2OH^+) - S^o_{298}(HCO^+) = +7.1 \pm 1.1 \text{ e.u.} \tag{18b}$$

The application of statistical thermodynamics to the optimum ground-state geometries of HCO^+ and CH_5^+ determined from molecular orbital calculations allows the determination of absolute entropies for these ions and from equations 18a and 18b the absolute entropies $S^o_{298}(CO_2H^+) = 55.6 \pm 0.6$ e.u. and $S^o_{298}(N_2OH^+) = 54.9 \pm 1.1$ e.u. These values are greater than the entropies of N_2OH^+ and

CO_2H^+ calculated assuming linear geometries by amounts larger than the estimated errors. This may be taken as evidence that the ground-state geometries of N_2OH^+ and CO_2H^+ are non-linear. On the basis of the isolectronic principle we expect the H-atoms to be at an angle to the line of centres of the heavy atoms - for the isolectronic (22 el) neutral species N_3H, $<N_3-H = 113°$.

The previous section was intended to demonstrate that, in addition to completely specifying the energetics of P.T. reactions, equilibrium measurements of P.T. reactions 1 and 2 can provide a valuable technique (often superseding other established methods) for the determination of fundamental molecular parameters including heats of formation, proton affinities, electron affinities, bond dissociation energies, and geometries of ions. Both in this connection, as well as in the other areas of the kinetics and energetics of P.T. reactions briefly considered here, we can look forward to further developments of significance and interest as P.T. reactions continue to be actively investigated.

REFERENCES

Ausloos, P. and Lias, S. G. (1965). *Disc. Faraday Soc.* 39, 36.
Ausloos, P. and Lias, S. G. (1972). Structure and Reactivity of Hydrocarbon Ions. *Ion-Molecule Reactions* (Ed. J. L. Franklin). Plenum Press, New York.
Berkowitz, J., Chupka, W. A. and Walter, T. A. (1969). *J. Chem. Phys.* 50, 1497.
Betowski, D., Payzant, J. D., Mackay, G. I. and Bohme, D. K. (1974) *Chem. Phys. Letters*, submitted for publication.
Bohme, D. K., Lee-Ruff, E. and Young, L. B. (1971). *J. Am. Chem. Soc.* 93, 4608.
Bohme, D. K., Hemsworth, R. S. and Rundle, H. W. (1973a). *J. Chem. Phys.* 59, 77.
Bohme, D. K., Hemsworth, R. S., Rundle, H. W. and Schiff, H. I. (1973b). *J. Chem. Phys.* 58, 3504.
Bohme, D. K., Mackay, G. I., Schiff, H. I. and Hemsworth, R. S. (1974). *J. Chem. Phys.*, to be published.
Bowers, M. T., Ave, D. H., Webb, H. W. and McIver, R. T. (1971). *J. Amer. Chem. Soc.* 93, 4314.
Bowers, M. T., Chesnavich, W. J. and Huntress, W. T. (1973). *Int. J. Mass Spectrom. Ion Phys.* 12, 357.
Celotta, R. J., Bennett, R. A. and Hall, J. L. (1974). *J. Chem. Phys.* 60, 1740.
Chupka, W. A. and Russell, M. E. (1968). *J. Chem. Phys.* 48, 1527.
Darwent, B. de B. (1970). Nat. Stand. Ref. Data Ser., Nat. Bur. Stand. (U.S.), 31.
Feldman, D. (1970). *Z. Naturforsch.* 25a, 621.
Fennelly, P. F., Hemsworth, R. S., Schiff, H. I. and Bohme, D. K. (1973). *J. Chem. Phys.* 59, 6405.

Gioumousis, G. and Stevenson, D. P. (1958). *J. Chem. Phys.* <u>29</u>, 294.

Gupta, S. K., Jones, E. G., Harrison, A. G. and Myher, J. J. (1967). *Can. J. Chem.* <u>45</u>, 3107.

Hemsworth, R. S., Payzant, J. D., Schiff, H. I. and Bohme, D. K. (1974). *Chem. Phys. Letters* <u>26</u>, 417.

Henderson, W. G., Beauchamp, J. L., Holtz, D. and Taft, R. W. (1972). *J. Amer. Chem. Soc.* <u>94</u>, 4724.

Huntress, W. T. and Pinizzotto, R. F. (1973). *J. Chem. Phys.* <u>59</u>, 4742.

Mackay, G. I., Payzant, J. D., Hemsworth, R. S., Schiff, H. I. and Bohme, D. K. (1974). Twenty-Second Annual Conference on Mass Spectrometry and Allied Topics, Philadelphia.

McIver, R. T. and Eyler, J. R. (1971). *J. Amer. Chem. Soc.* <u>93</u>, 6334.

McIver, R. T. and Miller, J. S. (1974). *J. Amer. Chem. Soc.* <u>96</u>, 0000.

Moran, T. F. and Hamill, W. H. (1963). *J. Chem. Phys.* <u>39</u>, 1431.

Munson, M. S. B. and Field, F. H. (1965). *J. Amer. Chem. Soc.* <u>88</u>, 1621.

Payzant, J. D., Betowski, D., Schiff, H. I. and Bohme, D. K. (1974). *Chem. Phys. Letters*, to be submitted for publication.

Ryan, P. (1974). Private communication.

Schiff, H. I. and Bohme, D. K. (1974). *Intern. J. Mass Spectrom. Ion Phys.*, in press.

Sieck, L. W., Hellner, L. and Gorden, R. (1971). *Chem. Phys. Letters* <u>10</u>, 502.

Smith, D. L. and Futrell, J. H. (1974). *Chem. Phys. Letters* <u>24</u>, 611.

Smyth, K. C. and Brauman, J. I. (1972). *J. Chem. Phys.* <u>56</u>, 4620.

Solka, B. H. and Harrison, A. G. (1973). Twenty-First Annual Conference on Mass Spectrometry and Allied Topics, San Francisco.

Su, T. and Bowers, M. T. (1973a). *Intern. J. Mass Spectrom. Ion Phys.* <u>12</u>, 347.

Su, T. and Bowers, M. T. (1973b). *J. Amer. Chem. Soc.* <u>95</u>, 7609.

Su, T. and Bowers, M. T. (1973c). *J. Amer. Chem. Soc.* <u>95</u>, 7611.

Su, T. and Bowers, M. T. (1974). *J. Chem. Phys. and Int. J. Mass Spectrom. Ion Phys.*, to be published.

Vestal, M. L., Blakley, C. R., Ryan, P. and Futrell, J. H. (1974). Twenty-Second Annual Conference on Mass Spectrometry and Allied Topics, Philadelphia.

Volpi, G. G. (1970). Gas-Phase Proton-Transfer Reactions. *Molecular Beams and Reaction Kinetics* (Ed. Ch. Schlier). Academic Press, New York.

Yamdagni, R. and Kebarle, P. (1973a). *J. Amer. Chem. Soc.* <u>95</u>, 3504.

Yamdagni, R. and Kebarle, P. (1973b). *J. Amer. Chem. Soc.* <u>95</u>, 4050.

CHEMICAL IONIZATION MASS SPECTROMETRY: ANALYTICAL APPLICATIONS

OF ION-MOLECULE REACTIONS

Burnaby Munson

Department of Chemistry, University of Delaware

Newark, Delaware 19711

Chemical ionization mass spectrometry (CIMS) has grown tremendously in the past few years since the technique was developed (Munson and Field, 1966a). Several reviews have been written on CIMS which explain the technique and discuss reactions which have been observed with different reactant ions and with different classes of compounds (Field, 1968a; Field, 1968b; Munson, 1971; Arsenault, 1972; Field, 1972).

In order to explain a CI mass spectrum of a known compound, or what is more difficult, to deduce the structure of an unknown compound from a CI mass spectrum, good chemical intuition is needed. However, much data are needed to develop this chemical intuition. The molecules which are being analyzed by CIMS are generally high molecular weight species of significant molecular complexity. The molecules which are being studied in most ion-molecule reaction studies are relatively simple species. Consequently generalizations about rates of reaction, sites of attack, energy transfer, and mechanisms of decomposition must be made which can be extrapolated to these complex molecules. Data on ion-molecule reactions in complex systems would also be very useful.

There are two general needs in CIMS. There is a need for rather non-specific reactant systems, systems that are general ones for the detection of all compounds with high and approximately equal sensitivities. In addition, there is a need for highly specific systems, those which will react with only a few or perhaps only one class of compounds. Data on rate constants for reactions and the effects of changes in the structures or

functional groups of the neutral species on these rate constants
are needed. Similarly, data are needed on the effects of variations in reactant ions on the rate constants of reaction with
the same molecules.

Thermochemical data for gaseous ions are very important in
CI studies. The occurrence or non-occurrence of many reactions
under CI conditions can be explained from the exothermicity or
endothermicity of the reactions. The selectivity of proton
transfer reactions depends on the base strengths of the compounds
to be analyzed and the acid strengths of the reactant ions.
Hydride transfer reactions can sometimes be predicted from the
differences in hydride affinities of the reactant and product
ions. Some correlation of thermochemical properties with molecular structure is necessary since reliable experimental data
for complex gaseous molecules and ions are not available.

In polyfunctional compounds it is useful to know if one
group is strongly preferred over others as a site of attack. If
so, one might be able to design derivatives of polar compounds
which greatly enhance, for example, the abundance of $(M + H)^+$
ions. Consequently, we are interested in the details of the ion-
molecule collisions that give the $(M + H)^+$ species as well as the
mechanisms of the subsequent decompositions of these ions. Data
obtained from model compounds can be used to predict the reactions
in complex species.

The extent of fragmentation of the protonated species,
$(M + H)^+$, depends on the exothermicity of the proton transfer
reaction. Addition ions like $(M + C_2H_5)^+$ from CH_4, $(M + 57)^+$
from $i-C_4H_{10}$, and $(M + NO)^+$ from NO are useful to determine the
molecular weights of compounds. The relative abundances of these
ions depend on both the experimental conditions and on the
structures of the molecules to be analyzed. Consequently, we
need information on collisional stabilization of excited ions
and relative rates of dissociation and deactivation.

It is also true that some fundamental data, perhaps only of
a qualitative or semi-quantitative nature, can be obtained from
CIMS studies. Estimates or limits for gaseous ionic heats of
formation and approximate or relative values for rate constants
can sometimes be inferred from CI spectra.

Most CIMS studies have been done with mass spectrometers
which have been modified or designed to operate at high pressures.
High pressure is not essential, however, and CI studies have been
made with ion cyclotron resonance (ICR) techniques (Bursey, Elwood,
Hoffman, Lehman, and Tesarek, 1970; McIver, Jr., Miller, and
Ledford, 1974). The present discussion will be limited to high

pressure studies of positive ions.

Methane was the first CI reagent gas and is still in common use. It is a non-specific system and should produce significant ionization for any compound to be analyzed. The reactions in methane have been well studied and the ionic distributions in "pure" methane at high mass spectrometric pressures have been well characterized (Field and Munson, 1965; Haynes and Kebarle, 1966). The non-reactivity of CH_5^+ and $C_2H_5^+$ with CH_4 is indicated by the invariance in the abundances of these ions for pressures above 0.2 torr. Upper limits for rate constants for reaction of CH_5^+ and $C_2H_5^+$ with methane were estimated by Field and Munson (1965) to be 10^{-12} cc/molecule-sec. From the work at higher pressures Haynes and Kebarle (1966) estimated upper limits of 10^{-16} cc/molecule-sec for these reactions.

Because of the lack of reaction of CH_5^+ and $C_2H_5^+$, as shown by the constancy in the abundances of these ions at high pressures, it is relatively easy to detect reactions of these ions with other compounds by studying the pressure dependence of ionic abundances in mixtures of constant composition (Munson, 1965). From studies of mixtures of methane with simple compounds (Munson and Field, 1965a; 1965b; 1969), it was possible to say that CH_5^+ and $C_2H_5^+$ reacted with many compounds with rate constants of approximately 10^{-9} cc/molecule-sec for these exothermic reactions. The rate constants for reaction of CH_5^+ appeared to be larger than rate constants for $C_2H_5^+$, but the effect was less than a factor of two.

The absence of any XeH^+ ions in the CH_4 CI mass spectrum of Xe (Munson and Field, 1965b), however, indicated the absence of fast proton transfer from CH_5^+ to Xe. The slowness of this reaction (essentially undetectable in these experiments) was interpreted to mean that the proton transfer reaction was endothermic and that Xe was a weaker base than CH_4. That is, the proton affinity of Xe, PA(Xe), is less than the proton affinity of CH_4.

$$CH_5^+ + Xe \quad \longrightarrow\!\!\!\!\!\times\!\!\!- \quad XeH^+ + CH_4 \tag{1}$$

$$\Delta H = \Delta H_f(XeH^+) + \Delta H_f(CH_4) - \Delta H_f(CH_5^+) - \Delta H_f(Xe) \tag{2}$$

$$\backsimeq PA(CH_4) - PA(Xe) \tag{3}$$

$PA(X) \equiv D(X - H^+)$ is the energy of the following reaction

$$XH^+ \rightarrow X + H^+ \tag{4}$$

$$PA(X) = D(X - H^+) = \Delta H = \Delta H_f(X) + \Delta H_f(H^+) - \Delta H_f(XH^+) \tag{5}$$

This conclusion is in disagreement with the dissociation energy of XeH^+ from scattering experiments reported by Ding in this volume but agrees with the results on PA(Xe) reported elsewhere in this volume by Bohme from flowing afterglow experiments.

Let us now consider the general question of sensitivity in CI studies and how changes in the structures of the neutral molecules effect the sensitivity. One measure of relative sensitivity for two different compounds is the ratio of ion currents (or peak heights) from the two compounds under as nearly identical experimental conditions as possible. The generalized reactions are the following,

$$CH_5^+ + M \xrightarrow{\quad k_{17} \quad} \Sigma I_i^+ (+ \ldots) \tag{6}$$

$$C_2H_5^+ + M \xrightarrow{\quad k_{29} \quad} \Sigma I_i^+ (+ \ldots) \tag{7}$$

where M is the compound to be analyzed and ΣI_i^+ represents the sum of ion currents of all ions produced by reactions with M. If the extent of conversion of CH_5^+ and $C_2H_5^+$ to product ions, ΣM_i^+, is small then

$$\frac{d(\Sigma I_i^+)}{dt} = \{k_{17}[CH_5^+] + k_{29}[C_2H_5^+]\}[M] \cong \frac{\Delta(\Sigma I_i^+)}{\Delta t} = \frac{\Sigma I_i^+}{\Delta t} \tag{8}$$

and, therefore,

$$\Sigma I_i^+ \cong \Delta t\{k_{17}[CH_5^+] + k_{29}[C_2H_5^+]\}[M] \tag{9}$$

That is, the ion current from the sample ions is proportional to the rate constants for reaction of CH_5^+ and $C_2H_5^+$ with the added compound at fixed reaction time and concentrations.

In one of the early papers on CIMS (Munson and Field, 1966b), estimates were made of relative rates of CH_5^+ and $C_2H_5^+$ with a homologous series of aliphatic esters. In these experiments it was not possible to make a direct measurement of a very small sample pressure with a methane pressure of one torr. Consequently, it was necessary to estimate the relative pressures in the source of equal volumes of liquid introduced through a molecular leak.

Since the densities of the esters are approximately the same (Timmermans, 1950), the number of moles introduced for a fixed volume of liquid is inversely proportional to the molecular weight of the ester. The rate of flow of sample through the molecular leak decreases with increasing molecular weight. The product $M^{3/2}\Sigma I_i^+$ in Table I, then refers to the total ion current for the products of reaction of CH_5^+ and $C_2H_5^+$ with the esters, corrected to the same gaseous concentration of ester in the reaction chamber. There is virtually no change in this quantity; therefore, there should be little change in rate constants within this homologous series.

Table I

Relative Sensitivities of Propionate Esters

$$CH_3CH_2COOR$$

R	ΣI_i^+	$M^{3/2}\Sigma I_i^+$
Methyl	8558	8.56×10^3
n-Propyl	6413	9.75×10^3
n-Butyl	4693	8.45×10^3
n-Pentyl	4034	8.47×10^3
Cyclohexyl	5006	11.8×10^3
n-Heptyl	3597	9.80×10^3

CH_4 CI; Munson, M. S. B. and Field, F. H. (1966). J. Amer. Chem. Soc. **88**, 4337.

Few data on rate constants for proton transfer reactions are available for a homologous series of compounds for comparison. Bowers has reported elsewhere in this volume on rate constants for CH_5^+ with a series of alkyl chlorides and has observed only a small increase in rate constant: 2.6×10^{-9} cc/molecule-sec for CH_3Cl and 3.3×10^{-9} cc/molecule-sec for $C_5H_{11}Cl$. Similarly, the data reported in this volume by Cacace on reactions of HeH^+ showed no significant variation in rate constant with molecular structure.

Systematic comparisons of sensitivities among different classes of compounds with methane have not, to my knowledge, been reported. Qualitative observations, however, indicate no drastic changes in sensitivity among different classes of compounds.

Some of the recent measurements on rate constants for CH_5^+ with polar compounds are listed in Table II. The variations with dipole moment are measurably significant; however, these variations are generally less than a factor of two. For the present purposes, the rate constants may be considered equal.

Table II

Rate Constants for Reaction of CH_5^+ With Polar Compounds

M	$k, 10^{-9} \dfrac{cc}{molecule\text{-}sec}$	$\alpha, 10^{-25} \dfrac{cm^3}{molecule}$	μ, D [f]
CH_3COCH_3	3.2 ± 0.3 [a]	63 [d]	2.86
	5.3 ± 0.2 [b]		
CH_3CH_2CHO	2.7 ± 0.2 [a]	(61) [e]	2.75
$CH_2{=}CH{-}CH_2{-}OH$	1.5 ± 0.1 [a]	(60) [e]	1.64
CD_3CDO	4.2 ± 0.4 [b]	(43) [e]	2.71
CH_3CN	4.9 ± 0.2 [b]	43 [d]	3.96
CD_3OCD_3	3.0 ± 0.2 [b]	52 [d]	1.30
NH_3	2.3 [c]	23 [d]	1.47
$(CH_3)_2NH$	2.2 [c]	(59) [e]	1.03

[a] Ridge, D. P.(1973). Thesis, Cal. Inst. Tech.

[b] Blair, A. S. and Harrison, A. G. (1973). Can. J. Chem. 51, 1645.

[c] Su, Timothy and Bowers, Michael T. (1973). Int. J. Mass Spectrom. and Ion Physics 12, 347.

[d] Landolt, H. H. and Bornstein, R. (1950). Zahlenwerte und Functionen, Part 3, 6th Edition, Springer-Verlag, Berlin.

[e] Estimated.

[f] McLellan, A. L. (1963). Tables of Experimental Dipole Moments, W. H. Freeman and Co., San Francisco.

One may estimate limiting values of rate constants from the molecular polarizabilities (Gioumousis and Stevenson, 1958) or the locked dipole model (Moran and Hamill, 1963; Gupta, Jones, Harrison and Myher, 1967). Table III shows calculations for three ketones of widely different structures and molecular weights. Polarizability and dipole moment data for the 17-keto androstane are only estimates and such data are generally not available for complex molecules. However, these approximate calculations do show that very large changes in structure or molecular weights will not alter rate constants by more than a factor of two. Similar conclusions would be reached from calculations using the average dipole orientation (ADO) theory (Su and Bowers, 1973).

Table III

Rate Constants for Reaction of CH_5^+ with Ketones

	k_P	k_D
I.	1.6	6.9
II.	2.0	7.0
III.	3.3	8.1

Calculated rate constants, 10^{-9} cm^3/molecule-sec; k_P = polarizability only; k_D = locked dipole.

There is little precise data available for the effect of changing the reagent gas on the sensitivity for a given compound. Qualitative observations indicate roughly comparable sensitivities for different reagent gases. Whitney, Klemann, and Field (1971) reported similar sensitivities for polyamines using N_2, CH_4, and i-C_4H_{10} as reagent gases. Michnowicz and Munson (1972a) observed no significant decrease in additive ion current from valerophenone with increasing amounts of water in CH_4. They concluded that proton transfer from CH_5^+, $C_2H_5^+$, and H_3O^+ occurred with approximately equal rate constants. Recent experiments in our laboratory on N_2/NO, CH_4/NH_3, and CH_4/$(CH_3)_2$O mixtures show effects of less than a factor of two on the ion currents from the samples as the reactant ions change from N_2^+ and N_3^+ to NO^+, CH_5^+ and $C_2H_5^+$ to NH_4^+, and CH_5^+ and $C_2H_5^+$ to $(CH_3)_2OH^+$.

Bohme has reported on rate constants for simple proton transfer reactions elsewhere in this volume. His data indicate that the rate constants show the expected small decrease with increasing reduced mass and show no obvious variation with reaction exothermicity or structure.

If the CI reactions are reasonably exothermic, we may expect that the rate constants for reaction and hence the sensitivities will show no major changes with changes in the structures of

either the reactant ions or the molecules. Significant effects
on rate constants and hence the selectivity of reactant ions
appear to occur only when a reaction becomes endothermic as a
result of changes in either the reactant ion or neutral molecule.

Almost without exception the CH_4 CI spectra of organic
compounds contain $(M + H)^+$, $(M - H)^{+4}$, and associated fragment ions
as their major species and essentially no molecular ions, M^+. The
CH_4 CI spectra of aromatic hydrocarbons contain M^+ ions at relative
abundances of 2-3% (Munson and Field, 1967). There are, however,
a few compounds that are exceptions to this generalization.

Table IV

CH_4 CI Spectra of Organometallic Compounds

% Base Peak

	$(M + H)^+$	M^+	$(M + H - CO)^+$
$C_4H_6Fe(CO)_3$[a]	100	---	50
$Cr(CO)_6$[a]	100	20	4
$C_6H_6Cr(CO)_3$[b]	100	34	3
$(C_5H_5)_2Co$[a]	6	100	
$(C_6H_6)_2Cr$[b]	8	100	

[a]Hunt, D. F., Russell, J. W. and Torian, R. L. (1972). J.
Organometal. Chem. 43, 175.

[b]Anderson, W. P., Hsu, N., Stanger, Jr., C. W. and Munson, B.
(1974). J. Organometal. Chem. 69, 249.

Table IV shows partial CH_4 CI spectra of a few organometallic
compounds. Butadieneiron tricarbonyl shows a "normal" CH_4 CI mass
spectrum: $(M + H)^+$ and fragment ions. The spectra of chromium
hexacarbonyl and benzenechromium tricarbonyl have significantly
larger amounts of M^+ ions than the spectrum of butadieneiron tricarbonyl. This difference may be attributed to charge exchange
reactions of the $C_2H_5^+$ ion since $IP[Cr(CO)_6] = 8.0$ eV (Franklin,
Dillard, Rosenstock, Herron, Draxl, and Field, 1969),
$IP[C_6H_6Cr(CO)_3] = 7.4$ eV (Pignataro and Lossing, 1967), and
$IP(C_2H_5) = 8.4$ eV (Lossing and Semeluk, 1970).

For cobaltocene and dibenzenechromium, however, the dominant ion is the molecular ion and only small amounts of $(M + H)^+$ ions are observed in the spectra. The basicities of the two compounds are certainly much greater than the basicity of CH_4, and the proton affinities are probably greater than 180 kcal/mole. Consequently, rapid proton transfer from CH_5^+ to these two compounds is expected to occur and it will be strongly exothermic. We have proposed a two step process for the formation of these ions in dibenzenechromium (Anderson, Hsu, Stanger, and Munson, 1974) and the process is equally applicable to cobaltocene:

$$CH_5^+ + (C_6H_6)_2Cr \rightarrow (C_6H_6)_2CrH^+* + CH_4 \qquad (10)$$

$$(C_6H_6)_2CrH^+* \rightarrow (C_6H_6)_2Cr^+ + H \qquad (11)$$

The proton transfer reaction, (10), is vigorously exothermic; consequently, some decomposition of these excited ions is expected if a low energy decomposition path is available. Because of the low ionization potentials of the molecules, $IP[(C_6H_6)_2Cr] = 5.7$ eV (Pignataro and Lossing, 1967) and $IP[(C_5H_5)_2Co] = 6.1$ eV (Pignataro and Lossing, 1967; Muller and D'Or, 1967), the lowest energy decomposition path for the $(M + H)^+$ ions involves the loss of H rather than another fragment.

This reaction is very similar to the rapid reactions of H_3O^+ with alkali metals that have been observed in flames and are discussed by Sugden elsewhere in this volume.

$$H_3O^+ + M \rightarrow M^+ \ (+ H_2O + H) \qquad (12)$$

From the calculated value for $D(Li^+ - H)$ of 3 kcal/mole (Browne, 1964), one may determine that Li is a strong base, $Pa(Li) = 192$ kcal/mole. Proton transfer is vigorously exothermic since $PA(H_2O) = 165$ kcal/mole (Long and Munson, 1973). Because of the low dissociation energy of LiH^+ it is possible that reaction (12) also occurs in two steps and no LiH^+ ions are detected in the flames because of their rapid dissociation:

$$Li + H_3O^+ \rightarrow LiH^+* + H_2O \qquad (13)$$

$$LiH^+* \rightarrow Li^+ + H \qquad (14)$$

Interpretations of CI mass spectra of polyfunctional compounds based on the reactions of monofunctional compounds are complicated by the possibility of intramolecular interactions. Table V shows an illustration of such intramolecular interactions in the CH_4 CI spectra of some aliphatic diols:

Table V

Intramolecular Hydrogen Bonding

CH_4 CI Spectra

Compound	$(M + H)^+$, % Additive Ionization
$CH_3-(CH_2)_9-OH$	0^a
$HO-(CH_2)_6-OH$	11
$HO-(CH_2)_8-OH$	11
$HO-(CH_2)_{10}-OH$	$15, 14^b$
$HO-(CH_2)_{12}-OH$	17
$HO-(CH_2)_{22}-OH$	28^b
$HO-(CH_2)_{34}-OH$	10^b
$HO-(CH_2)_{46}-OH$	5^b

[a] Munson, M. S. B. and Field, F. H. (1966). J. Amer. Chem. Soc. 88, 2621.

[b] Dzidic, I. and McCloskey, J. A. (1971). J. Amer. Chem. Soc. 93, 4955.

Aliphatic alcohols show essentially no $(M + H)^+$ ions in their spectra as indicated by the CH_4 CI mass spectrum of 1-decanol, but aliphatic diols do, even when the two hydroxy groups are separated by several carbon atoms. The lower abundance of $(M + H)^+$ ions for the 46-carbon diol is probably the result of the higher temperature necessary for vaporization.

Either these molecules are internally hydrogen bonded in the gas phase or the two ends come together in a time short compared with H_2O elimination from the protonated hydroxyl group. For these flexible compounds there is no obvious effect of ring size on stability of the internally di-solvated proton. If, however, the hydroxyl groups are rigidly separated, as on opposite sides of the steroidal nucleus in the androstane-3,17-diols, essentially no internal hydrogen bonding is possible and very few $(M + H)^+$ ions are observed (Michnowicz and Munson, 1972b).

These internally di-solvated protons are very similar to the

solvated protons discussed by Kebarle elsewhere in this volume.
Similar internal hydrogen bonding effects were noticed by Morton
and Beauchamp (1972) for C_2-C_6 diethers and also for diamines
(Aue, Webb, and Bowers, 1973; Yamdagni and Kebarle, 1973).

Such internal hydrogen bonding effects are not observed for
aliphatic dibromides, since the CH_4 CI mass spectra of 1,6-di-
bromohexane and 1,4-dibromocyclohexane contain no $(M + H)^+$ ions
of any significance.

One of the everpresent impurities in our system is water,
and the rapid proton transfer reactions of CH_5^+ and $C_2H_5^+$ to form
H_3O^+ (Munson and Field, 1965b) can be both a nuisance and an
advantage. Because water is readily detected as H_3O^+ in the high
pressure mass spectrum of CH_4, one must take care to purify the
methane and bake out the inlet lines to the mass spectrometer.
Without such care the relative amount of H_3O^+ will be comparable
to the amounts of CH_5^+ and $C_2H_5^+$ and the spectra will contain
product ions produced by all three species.

It has been discussed earlier in this paper that the rate
constants for proton transfer are not strongly dependent upon
reaction exothermicity. However, the extent of fragmentation, or
dissociative proton transfer, does depend critically on the exothermicity of the proton transfer reactions. Michnowicz and
Munson (1972a) noted significant decreases in the extent of
fragmentation of valerophenone as increasing amounts of water
changed the reactant ions from predominantly CH_5^+ and $C_2H_5^+$ to
predominantly H_3O^+. It has also been noted that there is a
significant decrease in the extent of fragmentation as the
reagent gases are changed in the order H_2, CH_4, C_2H_6, C_3H_8,
i-C_4H_{10} (Michnowicz and Munson, 1970; Munson, 1971).

For charge exchange reagent gases, the extent of fragmentation increases with increasing recombination energy of the
reactant ions (Einolf and Munson, 1972). Recent work on mixtures
shows major enhancement of M^+ ions and decreases in the extent of
fragmentation in N_2/NO mixtures as the concentration of NO is
increased and the ions are converted from N_2^+, N_3^+, and N_4^+ to
NO^+ (Jelus, Munson, and Fenselau, 1974b).

The effects of changing the acid strength of the reactant
ions can be demonstrated easily from the spectra of 5α-androstan-
17-one (compound III of Table III) obtained with CH_4 and with
mixtures of CH_4 containing small amounts of NH_3. The CH_4 CI
spectrum of the ketone contains as the two major species,
$(M + H)^+$, 28% of ion current, and $(M + H - H_2O)^+$, 38% of ion
current (Michnowicz and Munson, 1974). As increasing amounts of
NH_3 are added to CH_4, the reactant ions change from CH_5^+ and

$C_2H_5^+$ to NH_4^+ and higher solvated protons. For a reagent gas mixture of 2% NH_3 in CH_4, the predominant reactant ion is NH_4^+ and the major ion in the spectrum of the ketones is $(M + NH_4)^+$ and the abundance of $(M + H)^+$ ions has been reduced and most of the $(M + H - H_2O)^+$ ions have been eliminated.

The proton affinity of NH_3 is about 207 kcal/mole and the proton affinities of CH_4 and C_2H_4 are 127 and 159 kcal/mole (Long and Munson, 1973). Recent data on proton affinities of ketones (Jelus, 1972) suggest that the proton affinity of the 17-keto androstane is about 203 kcal/mole, slightly below that for NH_3. Consequently, proton transfer should be endothermic and slow unless some excitation energy remains in the NH_4^+ ions formed from CH_5^+ and $C_2H_5^+$. Solvation of NH_4^+ by the polar ketone is a reaction that one might reasonably expect.

If one uses NH_3 as a reagent gas, then, proton transfer is expected only to very basic compounds and non-polar or weakly solvating molecules will not be detected. Work has been done recently to show that one may estimate relative proton affinities of complex molecules from their CI spectra with strongly basic reagent gases. Dzidic and McCloskey (1972) observed striking differences in the ratios of $(M + H)^+$ and $(M + NH_4)^+$ ions for two isomeric steroidal ketones. These differences are sufficiently pronounced that one may with confidence say that the proton affinity of the conjugated ketone, V, is higher than that of the non-conjugated ketone, IV. The differences in these ratios are sufficiently large that one could confidently distinguish between the isomers using the NH_3 CI spectra.

IV   V

$[MH]^+$ 5% Rel. int. $[MH]^+$ 100% Rel. int.
$[MNH_4]^+$ 100% Rel. int. $[MNH_4]^+$ 18% Rel. int.

Dzidic (1972) has also used the CI spectra obtained with ammonia, methylamine, dimethylamine, and trimethylamine to obtain the order of basicities of several complex amines.

Another interesting and useful application of selective ionization with a weak acid is the work of Hunt, McEwen, and Upham (1971) using ND_3 as a reagent gas. The primary amine, 2,6-dimethyl-aniline gives only a two species spectrum with NH_3: $(M + H)^+$ and $(M + NH_4)^+$:

$$(15)$$

With ND_3 as the reagent gas, instead of $(M + D)^+$ and $(M + ND_4)^+$ ions at m/e = 123 and 142, ions are observed at m/e = 125 and 145. The hydrogens bonded to the nitrogen have been replaced by deuterium. One can use this technique to differentiate among primary, secondary, and tertiary amines.

In efforts to find selective systems for different types of compounds, many reagent gases have been tried, either as mixtures or as pure compounds. A partial list includes the following, H_2, CH_4, C_2H_6, C_3H_8, $i-C_4H_{10}$, $n-C_6H_{14}$, $n-C_8H_{18}$, N_2, Ar, NO, CO, H_2O, NH_3, CH_3OCH_3, CH_3COCH_3, CH_3CHO, $Si(CH_3)_4$, CF_4, CH_3NH_2, $(CH_3)_2NH$, $(CH_3)_3N$, $CH_2OH-CH_2NH_2$ (ethanolamine), and $CH_2NH_2-CH_2NH_2$ (ethylene-diamine). The most extensively used low energy reagent gases are $i-C_4H_{10}$ and NO, either pure or in N_2/NO mixtures.

Table VI gives proton and hydride affinities associated with the ions of several common reagent gases. The proton affinities were defined earlier in this paper and hydride affinities are defined in a similar manner:

$$XH \rightarrow X^+ + H^- \qquad\qquad (16)$$

$$H^-A(X^+) \equiv D(X^+ - H^-) \equiv \Delta H = \Delta H_f(X^+) + \Delta H_f(H^-) - \Delta H_f(XH) \quad (17)$$

A large proton affinity of a neutral molecule implies a strong conjugate base and hence a weak acid; a large hydride affinity implies a strong acid.

Table VII compares the CH_4 and isobutane CI spectra of cis 1,2-cyclohexanediol. These data provide a specific example of the lesser extent of fragmentation that is generally observed with isobutane than with methane. The differences are consistent with a generally lower heat of reaction associated with proton transfer, hydride transfer, or even hydroxide transfer reactions of $t-C_4H_9^+$ compared with CH_5^+ and $C_2H_5^+$. These data also provide another

example of internal hydrogen bonding since both the methane
spectra and the isobutane spectra of aliphatic mono-alcohols
contain essentially no $(M + H)^+$ ions (Field, 1970).

The absence of abundant $(M + H)^+$ ions in the isobutane CI
spectra of alcohols (Field, 1970) is somewhat surprising. The

Table VI

Proton and Hydride Ion Affinities

PA(B), kcal/mole	BH^+	$H^-A(BH^+)$, kcal/mole
100	H_3^+	---
129	CH_5^+	---
160	$C_2H_5^+$	272
165	H_3O^+	---
179	$C_3H_7^+$	248
195	$t-C_4H_9^+$	232
207	NH_4^+	---

Table VII

Comparison of CH_4 and $i-C_4H_{10}$ Spectra

cis-1,2-Cyclohexanediol

m/e	Species	$I_i/\Sigma I_i$	
		CH_4	$i-C_4H_{10}$
81	$(M + H - 2H_2O)^+$	.15	.08
97	$(M - H - H_2O)^+$	.03	.003
99	$(M + H - H_2O)^+$	.61	.46
115	$(M - H)^+$	.03	.02
117	$(M + H)^+$	.04	.23

major ion in the $i-C_4H_{10}$ CI spectrum of cyclohexanol is the $(M - OH)^+$ ion, which Field (1970) reports as being 66% of the ionization at 180°C. Direct data on the heat of formation of the cyclohexyl ion and the proton affinity of cyclohexanol are not available. However, from the data on the proton affinities of alcohols (Long and Munson, 1973) and heats of formation of other hydrocarbon ions (Franklin, Dillard, Rosenstock, Herron, Draxl, and Field, 1969), one may estimate the thermochemical data needed for the following reactions:

$$t-C_4H_9^+ + c-C_6H_{11}OH \rightarrow c-C_6H_{11}OH_2^+ + i-C_4H_8 \quad \Delta H \cong -2 \text{ kcal} \quad (18)$$

$$t-C_4H_9^+ + c-C_6H_{11}OH \rightarrow c-C_6H_{11}^+ + t-C_4H_9OH \quad \Delta H \cong +5 \text{ kcal} \quad (19)$$

Both reactions may perhaps be considered as thermoneutral within the error of the estimated thermochemical data. However, the proton transfer reaction is not expected to be sufficiently exothermal that dissociation of $c-C_6H_{11}OH_2^+$ would be very rapid. It is unlikely, therefore, that the $C_6H_{11}^+$ ions are formed by dissociative proton transfer.

Because the reactions of $t-C_4H_9^+$ with $c-C_6H_{11}OH$ appear to be thermoneutral, it appeared worthwhile to see if there was an activation energy. There is a strong temperature dependence of the relative abundances of the ions in the isobutane CI spectrum of cyclohexanol, an increase in the extent of fragmentation with increasing temperature. The ratios of the three major ions in the spectrum, $(M + 57)^+/(M + H)^+/(M - OH)^+$, are 1.0/.40/1 at 70° and .41/.21/1 at 110°.

More extensive temperature studies than these have been reported for the isobutane CI mass spectra of t-amyl and benzyl-acetate. Activation energies of 12 kcal/mole were noted for the decompositions of the protonated acetates (Field, 1969, Laurie and Field, 1972).

Although one normally considers that product ions in CI spectra are produced from rapid, exothermic reactions, these data suggest that caution must be applied to isobutane data, particularly if one is concerned with thermochemical inferences that may be drawn.

One of the purposes of using low energy, selective reactant ions is to distinguish between molecules that differ only slightly in properties. An example of this is shown in Table VIII which compares the isobutane CI spectra of cis- and trans-1,4-cyclo-hexanediol at 30°. Because of the very marked temperature effects just mentioned for the isobutane spectra of cyclohexanol, comparisons will be significant only if made under very similar conditions.

Table VIII

i-C_4H_{10} CI Spectra of cis- and trans-1,4-Cyclohexanediol

$$I_i / \Sigma I_i$$

m/e	Species	cis	trans
81	$(M + H - 2H_2O)^+$	.11	.12
99	$(M + H - H_2O)^+$	.07	.31
117	$(M + H)^+$	.58	.18
173	$(M + C_4H_9)^+$	.07	.13

The differences are sufficiently large that one could easily distinguish between the two compounds. Comparisons made at other temperatures show similarly obvious differences. If one assumes that either ring inversion is rapid or that the molecules maintain their lowest energy configuration, then in the boat configuration the hydroxyl groups of the cis-isomer are sufficiently close together to allow internal hydrogen bonding for $(M + H)^+$ ions. No such favorable configuration exists for the trans isomers. These spectra are consistent with the spectra of the aliphatic diols: large abundance of $(M + H)^+$ ions when internal hydrogen bonding is possible.

The isobutane CI spectra of the cis- and trans-1,2-cyclo-hexanediols were indistinguishable. However, in molecular models, the distances between the hydroxyl groups are about the same for these two isomers; so large differences are not expected. Longevialle, Milne, and Fales (1973) observed no internal hydrogen bonding in the isobutane CI spectra of some 1,3-dihydroxy steroids.

Another low energy reactant ion that has been used is NO^+ (Einolf and Munson, 1972; Hunt and Ryan, 1972; Jelus, Munson and Fenselau, 1974a; 1974b). One of the major advantages of NO^+ as a reactant ion for quantitation or confirmation of molecular weight is the observation that there are very few ions in the spectra.

Three probable reactions of NO^+ are illustrated for 2-propanol. The occurrence of charge exchange is determined by the recombination energy of NO^+ and the ionization potential of the sample. IP(NO) = 9.3 ev (Franklin, Dillard, Rosenstock, Herron, Draxl, and Field, 1969) but it has been suggested that the recombination energy is somewhat lower (Einolf and Munson, 1972).

$$\text{NO}^+ + \text{CH}_3\text{CHOHCH}_3 \Bigg\langle \begin{array}{l} \text{CH}_3\text{CHOHCH}_3{}^+ + \text{NO} \quad \Delta H = +20 \text{ kcal} \quad (20a) \\[1em] \text{CH}_3\text{COHCH}_3{}^+ + \text{HNO} \quad \Delta H = -33 \text{ kcal} \quad (20b) \\[1em] \text{CH}_3\text{CHCH}_3{}^+ + \text{HNO}_2 \quad \Delta H = +3 \text{ kcal} \quad (20c) \end{array}$$

$\text{H}^-\text{A}(\text{NO}^+) = 244$ kcal/mole; hence NO^+ is a stronger Lewis acid than $\text{t-C}_4\text{H}_9{}^+$ and most secondary alkyl ions. H^- transfer is a relatively common reaction. Abstraction of groups other than OH^- as indicated in (20c) is possible and the NO spectra of some pesticides suggest that Cl^- transfer reactions occur rapidly.

An additional reaction path not mentioned above is association to give $(\text{M} + \text{NO})^+$ ions. The relative amount of $(\text{M} + \text{NO})^+$ ions depends strongly on the existence of decomposition paths. For simple ketones, charge exchange is endothermic and the only significant ion is $(\text{M} + \text{NO})^+$. For substituted benzenes, the ratio $(\text{M} + \text{NO})^+/(\text{M}^+)$ is strongly dependent on the ionization potential of the substituted benzene. This ratio increases with increasing ionization potential of the sample. Table VIII shows the striking effect of ionization potential on the relative amounts of M^+ and $(\text{M} + \text{NO})^+$ ions using a 10% NO in N_2 mixture. These observations are essentially the same as those reported earlier for pure NO (Einolf and Munson, 1972). Ionization potentials are not available for DDT and Methoxychlor, but if one assumes similar substituent effects on ionization potentials of these compounds as are noted for the simple substituted benzenes, the marked differences in the NO spectra are easily explained.

Hunt and Ryan (1972) have reported some unusual ionic reactions with NO^+, for example the formation of $(\text{M} - 2 + 30)^+$ ions in the spectra of primary and secondary alcohols. We have also observed these ionic products.

In some of our experiments with low concentrations of NO in N_2 we observed very marked increases in the extent of fragmentation and large decreases in the $(\text{M} + \text{NO})^+/\text{M}^+$ ratio with increasing temperature. Part of this temperature effect is the result of the increased dissociation of $(\text{M} + \text{NO})^+$ ions with increasing temperature, but a significant amount of this temperature effect resulted from a decrease in the $\text{NO}^+/\text{N}_2{}^+$ ratio with increasing temperature. We interpret these results to mean that the charge exchange reaction,

$$\text{N}_2{}^+ + \text{NO} \rightarrow \text{NO}^+ + \text{N}_2 \tag{21}$$

has a negative temperature coefficient. Such negative temperature coefficients are not uncommon and are discussed elsewhere in this volume by Ferguson. However, studies to determine a negative temperature coefficient for reaction (21) have not been reported.

Table VIII

Effect of Ionization Potential[a]

Species	IP[b]	$(M + NO)^+/M^+$
$C_6H_5OCH_3$	8.2	0.05
C_6H_5Cl	9.2	1.4
C_6H_5CN	9.7	45
DDT[c]		1.1
Methoxychlor[d]		<0.05

[a] 10% NO in N_2; $P \cong .6-.7$ torr; t = 80–120°C.

[b] Franklin, J. L. Dillard, J. G., Rosenstock, H. M., Herron, J. T. Draxl, K., and Field, F. H. (1969). NSRDS-NBS26.

[c] DDT = Cl—⟨O⟩—CH—⟨O⟩—Cl
       |
       CCl_3

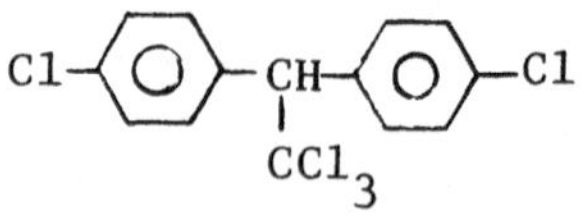

[d] Methoxychlor = CH_3O—⟨O⟩—CH—⟨O⟩—OCH_3
            |
            CCl_3

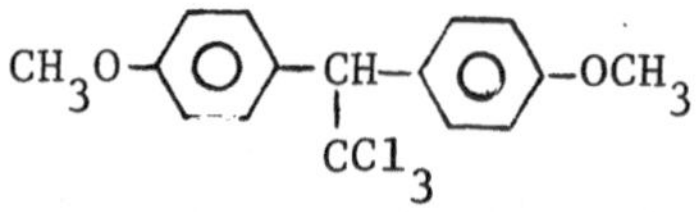

Acknowledgement

This work has been supported in part by the National Science Foundation, Grant No. GP20231, The author is grateful to Barbara Jelus, Douglas Ridge, Frank Hatch, and Norman Hsu for their advice and assistance in obtaining some of these data.

REFERENCES

Anderson, W. P., Hsu, N., Stanger, Jr., C. W., and Munson, B. (1974). J. Organometal. Chem. <u>69</u>, 249.

Arsenault, G. P. (1972). Chemical Ionization Mass Spectrometry. Biochemical Applications of Mass Spectrometry, (Ed. G. R. Waller), Wiley-Interscience, New York.

Aue, D. H., Webb, H. M., and Bowers, M. T. (1973). J. Amer. Chem. Soc. 95, 2699.

Browne, J. C. (1964). J. Chem. Phys. 41, 3495.

Bursey, M. M., Elwood, T. A., Hoffman, M. K., Lehman, T. A., and Tesarek, J. M. (1970). Anal. Chem. 42, 1370.

Dzidic, I. (1972). J. Amer. Chem. Soc. 94, 8333.

Dzidic, I. and McCloskey, J. A. (1972). Org. Mass Spectrom. 6, 939.

Einolf, N. and Munson, B. (1972). Int. J. Mass Spectrom. Ion Phys. 9, 141.

Field, F. H. (1968a). Accounts Chem. Res. 1, 42.

Field, F. H. (1968b). Chemical Ionization Mass Spectrometry. Advances in Mass Spectrometry, Vol. 4. (Ed. E. Kendrick). Institute of Petroleum, London.

Field, F. H. (1969). J. Amer. Chem. Soc. 91, 2827.

Field, F. H. (1970). J. Amer. Chem. Soc. 92, 2672.

Field, F. H. (1972). Chemical Ionization Mass Spectrometry. (Ed. J. L. Franklin). Ion-Molecule Reactions, Vol. 1. Plenum Press, New York.

Field, F. H. and Munson, M. S. B. (1965). J. Amer. Chem. Soc. 87, 3289.

Franklin, J. L., Dillard, J. G., Rosenstock, H. M., Herron, J. T., Draxl, K., and Field, F. H. (1969). NSRDS-NBS26.

Gioumousis, G. and Stevenson, D. P. (1958). J. Chem. Phys. 29, 294.

Gupta, S. K., Jones, A. G., Harrison, A. G., and Myher, J. J. (1967). Can. J. Chem. 45, 3107.

Haynes, R. M. and Kebarle, P. (1966). J. Chem. Phys. 45, 3899.

Hunt, D. F., McEwen, C. N., and Upham, R. A. (1971). Tetrahedron Letters, 4539.

Hunt, D. F. and Ryan, J. F. (1972). Chem. Comm., 620.

Jelus, B. L. (1972). Ph.D. Thesis, Univ. of Cincinnatti.

Jelus, B., Munson, B., and Fenselau, C. (1974a). Anal. Chem. 46, 729.

Jelus, B. L., Munson, B., and Fenselau, C. (1974b). Biomed. Mass Spectrom. 1, 96.

Laurie, W. A. and Field, F. H. (1972). J. Phys. Chem. 76, 3917.

Long, J. and Munson, B. (1973). J. Amer. Chem. Soc. 95, 2427.

Longevialle, P., Milne, G. W. A., and Fales, H. M. (1973). J. Amer. Chem. Soc. 95, 6666.

Lossing, F. P. and Semeluk, G. P. (1970). Can. J. Chem. 48, 955.

McIver, Jr., R. T., Miller, J. S., and Ledford, E. B. (1974). Paper presented at ASMS Meeting, Phila.

Michnowicz, J. and Munson, B. (1970). Org. Mass Spectrom. 4, 481.

Michnowicz, J. and Munson, B. (1972a). Org. Mass Spectrom. 6, 283.

Michnowicz, J. and Munson, B. (1972b). Org. Mass Spectrom. 6, 765.

Michnowicz, J. and Munson, B. (1974). Org. Mass Spectrom. 8, 49.

Moran, T. F. and Hamill, W. H. (1963). J. Chem. Phys. 39, 1413.

Morton, T. H. and Beauchamp, J. L. (1972). J. Amer. Chem. Soc. 94, 3671.

Muller, J. and D'Or, L. (1967). J. Organometal. Chem. 10, 313.

Munson, M. S. B. (1965). J. Amer. Chem. Soc. 87, 2332.

Munson, B. (1971). Anal. Chem. 43, 28A.

Munson, M. S. B. and Field, F. H. (1965a). J. Amer. Chem. Soc. 87, 3294.

Munson, M. S. B. and Field, F. H. (1965b). J. Amer. Chem. Soc. 87, 3294.

Munson, M. S. B. and Field, F. H. (1966a). J. Amer. Chem. Soc. 88, 1621.

Munson, M. S. B. and Field, F. H. (1966b). J. Amer. Chem. Soc. 88, 4337.

Munson, M. S. B. and Field, F. H. (1967). J. Amer. Chem. Soc. 89, 1047.

Munson, M. S. B. and Field, F. H. (1969). J. Amer. Chem. Soc. 91, 3413.

Pignataro, S. and Lossing, F. P. (1967). J. Organometal. Chem. 10, 531.

Su, T. and Bowers, M. T. (1973). J. Chem. Phys. 58, 3027.

Timmermans, J. (1950). Physico-Chemical Constants of Pure Organic Compounds. Elsevier, New York.

Whitney, T. A., Klemann, L. P., and Field, F. H. (1971). Anal. Chem. 43, 1048.

Yamdagni, R. and Kebarle, P. (1973). J. Amer. Chem. Soc. 95, 3504.

β-DECAY OF TRITIATED MOLECULES AS A TOOL FOR STUDYING ION-MOLECULE REACTIONS.

Fulvio Cacace

Laboratorio di Chimica Nucleare del C.N.R., Rome

Universita' Di Roma, 00100 Rome, Italy

INTRODUCTION

Apart from its intrinsic fundamental interest, the study of gas-phase ion-molecule reactions is currently proving of great value to physical organic chemistry, providing unified and extremely simplified models for the correspondent ionic processes occurring in solution, and allowing a direct experimental test of the theoretical calculations concerning an increasing variety of organic ions.

In this particular application, mass spectrometry, the major experimental tool for the study of gaseous ions, presents certain limitations, especially if evaluated by solution-chemistry standards, since it provides scarce information, if any, on the structure and the stereochemistry of the ionic species, their isomeric composition, their neutral products, etc., which represent essential mechanistic data, generally accessible in solution.

Such limitations have promoted the exploration of new experimental approaches, other than, and largely complementary to mass spectrometry, suitable for the study of gas-phase ionic processes with the classical methods of physical organic chemistry (cfr. Ausloos, 1966, Cacace, 1970).

The present review describes the principles and the applications of a technique based on the β decay of covalently bound tritium atoms for the production of gaseous cations and the study of their reactivity in both gaseous and condensed systems.

THEORETICAL AND MASS SPECTROMETRIC BACKGROUND

The spontaneous decay of T atoms causes their transformation into ^{3}He$^+$ ions, following the emission of an antineutrino and a β-particle, whose energy ranges continuously up to ca. 18 keV, with a mean value of 5.6 keV:

$$^3_1\text{T} \xrightarrow[\text{12.26 y}]{\text{Half life}} {}^3_2\text{He}^+ + \beta^- + \bar{\nu}$$

The molecular consequences of the β decay have been the subject of extensive theoretical and mass spectrometric work, thoroughly reviewed by Wexler (1965). Referring first to _isolated_ T atoms, there are two major causes of decay-induced excitation, i.e. the momentum imparted to the daughter ^{3}He$^+$ ions by the emission of the β$^-$ and $\bar{\nu}$ particles, and the perturbation of their electron cloud due to the sudden increase of the nuclear charge ("shaking" effect).

The _maximum_ energy imparted to _isolated_ ^{3}He$^+$ ions is calculated (Edwards and Davies, 1948, Wu, 1955) around 3.6 eV, while over 80% of the daughter ions recoil with energies below 0.08 eV.

The "shaking" effect has been theoretically treated by evaluating the probability of each electronic transition from the ground state of T to either the ground state, or to the various excited states of ^{3}He$^+$, using hydrogen-like wave functions. The probability is computed from the square of the overlap integral:

$$P_{(n,1) \rightarrow (n',1')} = \left| \int \psi^*_{(n',1')} \psi_{(n,1)} d\tau \right|^2$$

where n,1 are the quantum numbers referring to the ground state of T, n',1' those referring to each orbital of the daughter ^{3}He$^+$, dτ is the volume element. The results (Migdal, 1941) indicate that most (70%) of the daughter ions are formed in their 1s ground state 25% in the 2s, 1.3% in the 3s, 0.45% in the 4s excited states, while 2.5% are doubly ionized ^{3}He^{++} ions.

Considering now the decay of covalently bound T atoms, only part of the recoil energy is available for bond rupture, as shown in the simple case of diatomic parent molecules by the expression derived by Monaham (1958):

$$E_i = \frac{p_\beta^2}{2M_1} \left(1 - \frac{M_1}{M_1 + M_2} \right)$$

where E_i is the energy available for bond rupture, p_β is the momentum of the β particle, M_1 and M_2 the masses of the radioactive and of the stable atom.

The electronic excitation of the daughter molecule due to the "shaking" effect has been rigorously treated by Wolfsberg (1956), whose equation requires the knowledge of the wave functions of vibrationally and rotationally excited states, and can be therefore used only for very simple molecules, like HT and T_2. Approximate models have been used for more complex molecules, in particular the "loose" model of Wexler and Hess (1958), which arbitrarily restricts the "shaking" perturbation to the bond electrons of the radioactive atom. The overall conclusion of the theoretical calculations is that a large fraction of the decay ions, ranging from 75 to 85%, are formed in their ground state, receiving only negligible recoil and/or electronic excitation.

Nevertheless, it should be underlined that the change of chemical identity (and therefore of binding ability) undergone by the T atom can represent, per se, an effective factor of molecular dissociation. Thus, the very weak C-He bond, formed from the decay of tritiated hydrocarbons, is expected to dissociate within a few vibrations, while the H-He$^+$ bond, formed from the decay of HT, is quite stable and is expected to survive dissociation.

The theoretical results have been verified by the experimental study of the decay-induced dissociation of isolated tritiated molecules with special instruments, the "charge" mass spectrometer (Snell et al., 1957, Wexler, 1959) and, more recently, a modified ICR spectrometer (Pobo et al., 1973). The mass spectrometric results can be summarized as follows:

i.   Over 90% of the T_2 (HT) decays lead to stable ^{3}HeT$^+$ (^{3}HeH$^+$) daughter ions, with less than 10% fragment ions, such as ^{3}He$^+$, ^{3}He^{++}, T$^+$(H$^+$).

ii.   The decay of tritiated hydrocarbons is invariably followed by the loss of neutral ^{3}He. A high fraction (ca. 80%) of the daughter organic ions, remarkably independent of their structure, survives further fragmentation during the long (10^{-4} to 10^{-5} sec) transit time through the spectrometer. Typical yields of stable daughter ions are listed in Table I.

iii. The organic ions which dissociate further undergo extensive fragmentation, different from that promoted by electron impact. Thus, according to Snell and Pleasonton (1958), the CH$_3$T decay yields, together with a 82.4% abundance of stable CH$_3^+$ ions, the following charged fragments: CH$_2^+$ (4.9%), CH$^+$ (4.0%), C$^+$ (4.9%), and H$^+$ (2.4%).

TABLE I

Yields of major daughter ions from the decay of tritiated hydrocarbons

Precursor	Daughter Ion	Yield	Ref.
CH_3T	CH_3^+	82%	Snell and Pleasonton, 1958
C_2H_5T	$C_2H_5^+$	78%	Wexler and Hess, 1958
C_6H_5T	$C_6H_5^+$	72%	Carlson, 1960
$o-CH_3-C_6H_4T$	$C_7H_7^+$	78%	Wexler et al., 1960
$m-CH_3-C_6H_4T$	$C_7H_7^+$	79%	" " " "
$p-CH_3-C_6H_4T$	$C_7H_7^+$	76%	" " " "
$C_6H_5-CH_2T$	$C_7H_7^+$	79%	" " " "
$c-C_4H_7T$	$C_4H_7^+$	ca. 80%	Pobo et al., 1973
$c-C_5H_9T$	$C_5H_9^+$	ca. 75%	" " "

As a whole, the mass spectrometric results largely support the theoretical conclusions on the following significant points:

i.   Extensive dissociation of the daughter ion occurs only as a result of the change of chemical identity of T, as shown by the loss of ^{3}He, invariably following the decay of tritiated hydrocarbons. On the other hand, the high binding energy of the He-H$^+$ pair, ca. 2 eV according to Wolniewicz (1965), is reflected by the high abundance of the ^{3}HeH$^+$ (^{3}HeT$^+$) daughter ions.

ii. The theoretical contention that most (over 80%) of the daughter ions are formed in their ground state is supported by the experimental evidence.

OUTLINE OF THE DECAY TECHNIQUE

The theoretical and mass spectrometric results discussed in the previous section show that ionic species of well-defined structure can be obtained in high yields from the decay of tritiated precursors. Furthermore, since the formation of the ions does not depend on external excitation, but rather on the spontaneous nuclear transformation of a constituent atom, the process is quite independent of the environment, and can be used in gaseous or condensed systems. These considerations provided the base of a new technique for the study of ion-molecule reactions (Cacace, 1964). A tracer amount of a precursor <u>containing at least two T atoms in equivalent</u> <u>positions of the same molecule,</u> is allowed to decay within a gaseous or liquid system. The decay of one of the radioactive atoms of a given molecule leads to the formation of a labeled daughter cation, whose reactions can be traced, and the final products identified, owing to the presence of the other radioactive atom(s), which acts as a label.

The reaction products can be directly determined by efficient analytical techniques, such as radio gas chromatography. If required, they can be also isolated and purified, following the addition of inactive carriers, by preparative gas chromatography and subjected to chemical degradation in order to ascertain the molecular distribution of T.

The necessity of using labeled daughter ions arises from their low rate of formation, since only about 0.5% of the tritiated precursor undergoes β decay in one month. Owing to the limitations posed to the specific activity of the sample by radiolytic problems (vide infra), the number of decay ions formed in the system within any reasonable period of time is too small to allow the detection of their products, unless some radioactive tracer is used. Entirely new synthetic and analytical techniques had to be developed for the preparation of chemically, radiochemically and isotopically pure precursors containing two or more T atoms in the same molecule. As an example, pure methane-T_4 has a specific activity exceeding 10^5 Ci per mol, and must be prepared in the carrier-free state, before dilution with CH_4 and gas chromatographic purification can be undertaken (Ciranni and Guarino, 1966).

The decay of one of the T atoms contained in the multilabeled molecules introduces into the system, together with tritiated daughter ions, β particles with a mean energy of 5.6 keV, which could promote the radiolysis of a significant fraction of the labeled precursor, yielding tritiated products indistinguishable from those formed from the reactions of the decay ions. It is therefore necessary to keep the concentration of the tritiated molecules sufficiently low to make the rate of their radiolysis insignificant in comparison with the rate of formation of the decay ions. It has been shown by kinetic considerations, repeatedly verified by blank experiments involving monotritiated precursors (where only radiation-induced processes are operative, the formation of labeled products via decay ions being a priori suppressed) that in most systems self-radiolysis becomes entirely insignificant if the specific activity of the sample does not exceed 0.1-0.5 mCi per mmol.

In the actual experiments, the specific activity of the gas can be further reduced by a factor of at least 10, without sacrificing the accuracy of the products analysis, by using sufficiently long decay periods (2 to 6 months).

ALKYL AND CYCLOALKYL CARBENIUM IONS FROM β DECAY

Several alkyl and cycloalkyl cations have been obtained from
the decay of appropriate multilabeled precursors, and their reactions studied over a wide range of pressures in hydrocarbon gases,
including CT_3^+ from methane-T_4 (Cacace et al., 1966), $C_2H_4T^+$ from
ethane-1,2-T_2 (Aliprandi et al., 1967), $C_3H_6T^+$ from propane-1,2-T_2
(Cacace et al., 1967), c-$C_4H_6T^+$ from c-C_4H_6-1,2-T_2 (Pobo et al.,
1973), c-$C_5H_8T^+$ from c-C_5H_8-1,2-T_2 (Babérnics and Cacace, 1971).

The studies have been extended to other gaseous substrates,
such as arenes (Nefedov et al., 1970), water vapor (Nefedov et al.,
1968a), hydrogen halides and alkyl haides (Nefedov et al., 1968b),
alcohols (Nefedov et al., 1968c), ammonia and amines (Nefedov et al.,
1971), ethers (Nefedov et al., 1973a), and olefines (Nefedov et al.,
1973b).

Detailed discussion of these studies is beyond the scope of
the present review. On one hand, in fact, the results confirm and
extend those obtained by mass spectrometric techniques in comparable systems. Thus, the kinetic analysis of the CT_3^+ attack on CH_4,
or on CH_4/C_3H_8 mixtures at 760 torr indicates that (i) the reaction
rates of CT_3^+ with CH_4 and C_3H_8 are comparable, (ii) the CT_3^+ attack on CH_4 yields ethyl ions, without detectable H^- abstraction
from CH_4, and (iii) the CT_3^+ reaction with C_3H_8 involves exclusively H^- abstraction, with no further intramolecular H atoms mixing.

These conclusions support, and extend to much higher pressures,
the mass spectrometric results on ion-molecule reactions in methane
reported by Lampe and Field (1959), Derwish et al., (1964), Field
and Munson (1965), Munson and Field (1965), and Haynes and Kebarle
(1966), as well as those obtained with their radiolytic technique
by Ausloos and Lias (1963), and by Ausloos and Gorden (1964). A
typical result of the decay technique, the actual isolation of CT_3H
by a gas chromatographic technique allowing the full resolution of
isotopic methanes (Bruner and Cartoni, 1965), has provided the direct experimental proof of the often postulated hydride-ion abstraction, $CT_3^+ + C_3H_8 \rightarrow CT_3H + C_3H_7$.

On the other hand, the results obtained by the decay technique
represent an extension to the gas phase of recent observations
concerning the formation and the reactivity of alkyl ions (Olah,
1973) dialkylhalonium ions (Olah et al., 1970), trialkyloxonium
ions (Olah et al., 1973b), and, in general, protonated heteroalipha-
tic compounds (Olah et al., 1970). Finally, the decay technique
can provide valuable information on the structure, in addition to
the reactivity, of gaseous carbenium ions. Thus, the isolation of
labeled cyclopentyl derivatives among the products from the c-$C_5H_8T_2$
decay provides convincing evidence for the occurrence of the cyclopentyl cation in the gas phase (Babérnics and Cacace, 1971).

CARBONIUM IONS FROM THE GAS-PHASE ATTACK OF HeT^+ ON ALKANES AND CYCLOALKANES

The unusual properties of HeT^+, formed with a 95% yield from the decay of T_2, have been exploited for the preparation of gaseous alkonium and cycloalkonium ions, and the study of their reactivity in gaseous and liquid systems.

HeT^+ is a stable molecular ion, characterized by a ΔH_f^0 value of ca. 320 kcal/mol^{-1}, and a bond distance of 0.765 Å. Owing to the extreme weakness of its conjugate base, the He atom having a PA of only 40 kcal/mol^{-1}, the HeT^+ ion is an exceedingly strong gaseous acid, which exothermically protonates (more properly, tritonates) all organic compounds, including very weak bases like methane. The technique used in these experiments involves the decay of a tracer activity of T_2 in a large excess of the gaseous substrate in any suitable range of temperatures and pressures, for periods up to several months, followed by the analysis of products with experimental techniques analogous to those outlined in the previous paragraphs. The HeT^+ has been used as a proton-donor to a variety of substrates, including methane, ethane, propane, n- and i-butane (Cacace et al., 1968a, 1968b), cyclopropane, cyclobutane, cyclopentane and cyclohexane (Cacace et al., 1969), cis- and trans-1,2-dimethylcyclopropane (Cacace et al., 1971a), bicyclo|2.1.0.|pentane, bicyclo|3.1.0|hexane and bicyclo|4.1.0.|heptane (Cacace et al. 1973), and meso-1,2-difluoro-1,2-dichloroethane (Cacace and Speranza, 1972).

The results of the decay technique are in general agreement with those of mass spectrometric and solution-chemistry studies on comparable systems. For instance, the gas-phase attack of HeT^+ on methane is fully comparable to the reactions of milder Brønsted acids, such as H_3^+ and D_3^+, investigated at much lower pressures in the mass spectrometer (Munson et al., 1963, Wexler, 1963, Aquilanti and Volpi, 1966). In particular, the thermoneutral proton transfer reaction $CH_4T^+ + CH_4 \rightarrow CH_3T + CH_5^+$ which has been observed by independent workers to occur in such different environments as the low-pressure (0.1 torr) ion source of a mass spectrometer and a "magic acid" ($HSO_3F \cdot SbF_5$) solution of methane (Olah and Schlosberg, 1968), seems to represent a quite general channel of the methonium ion reactivity.

There is no doubt, however, that the unique features of the decay technique proved most useful in problems concerning the structure of gaseous carbonium ions, which cannot be determined by other techniques. Thus, the isolation of cyclopropane-T (whose yield far exceeds that of propylene-T) from the attack of HeT^+ on cyclopropane at 760 torr provides very good evidence for the occurrence of the cyclopropanium cation as a gaseous ionic intermediate, with a lifetime exceeding 10^{-9}-10^{-10} sec.

The evidence for gaseous protonated cyclopropanes becomes com-pelling in the case of 1,2-dimethylcyclopropanium cations, owing to the complete lack of cis - trans isomerization following the HeT$^+$ attack on the geometric isomers of 1,2-dimethylcyclopropane.

These results are of considerable interest both to mass spec-trometry, in connection with the long-postulated intervention of cyclopropanium structures in the fragmentation of certain alkanes (Rylander and Meyerson, 1956), and to solution chemistry of carboca-tions, where protonated cyclopropanes play a prominent role (cfr. Olah and von Schleyer, 1972).

Finally, the decay technique has allowed the study of the stereo-chemical course of a gas-phase electrophilic attack at saturated carbon. The predominant retention of configuration in the tritio-deprotonation of meso-1,2-dichloro-1,2-difluoroethane by HeT$^+$ pro-vides direct evidence for a front attack of the electrophile to the asymmetric center, as postulated by most authors on theoretical grounds.

GASEOUS ARENIUM IONS FROM THE HeT$^+$ ATTACK ON AROMATICS

Electrophilic substitutions at aromatic carbon, one of the most classical subjects of physical organic chemistry, have been exten-sively investigated in the gas phase, using the HeT$^+$ decay ion as the electrophilic reagent, and a variety of gaseous aromatic sub-strates, including benzene, toluene and toluene-d$_8$ (Cacace and Caronna, 1967), halobenzenes (Cacace and Perez, 1971), anisole, ben-zonitrile, α-trifluorotoluene and t-butylbenzene (Cacace et al., 1971b).

In general, when a tracer activity of T$_2$ is allowed to decay in a gaseous aromatic substrate at several hundreds torr, the major reaction channel of the HeT$^+$ decay ion is tritiodeprotonation, leading to high yields of the tritiated aromatic substrate, with much lower yields of labeled fragmentation products, as illustrated in Table II.

A general outline of the mechanism suggested for tritiodepro-tonation involves the primary formation of excited arenium ions, followed by their partial collisional stabilization and by the loss of a proton to some base, including other substrate molecules, con-tained in the gas:

$$C_6H_5X \; + \; HeT^+ \longrightarrow He \; + \; (C_6H_5TX)^+_{exc}$$

$$(C_6H_5TX)^+_{exc} \xrightarrow{} \begin{array}{c} \nearrow \text{Fragments} \\[1ex] +M \\ \searrow \\ (C_6H_5TX)^+ \xrightarrow{-H^+} C_6H_4TX \\ >70\% \end{array}$$

TABLE II

Yields of tritiated products from the HeT$^+$ attack on gaseous aromatics at 760 torr

Substrate	Products				
Toluene	Toluene-T	:	60.0%	Benzene-T	: 3.8%
Toluene-d$_8$	Toluene-d$_7$-T	:	60.0%	Benzene-d$_5$-T	: 3.6%
Fluorobenzene	Fluorobenzene	:	67.0%	Benzene-T	:<0.2%
Chlorobenzene	Chlorobenzene-T	:	76.1%	Benzene-T	: 3.7%
Bromobenzene	Bromobenzene-T	:	69.3%	Benzene-T	: 8.1%
Anisole	Anisole-T	:	48.8%	Benzene-T	: 2.0%
α-Trifluorotoluene	α-Trifluorotoluene-T:		81.4%	Benzene-T	: 2.9%
t-Butylbenzene	t-Butylbenzene-T:		65.0%	Benzene-T	:25.0%

In general, the excited arenium ions are efficiently stabilized in the gaseous substrate at one atmosphere, with the notable exception of t-butylarenium ion, whose facile decomposition into benzene and the stable t-butyl cation:

$$(t-C_4H_9-C_6H_5T)^+ \xrightarrow{} C_6H_5T + t-C_4H_9^+$$

gives high yields of benzene-T, in close analogy with mass spectrometric (Munson and Field, 1967) and solution chemistry (Brouwer, 1968, Olah et al., 1972) observations on the extensive protodealkylation of t-butylbenzene.

Gaseous arenium ions have been observed both in mass spectrometric studies on the attach of H$_3^+$ (D$_3^+$), CH$_5^+$ and C$_2$H$_5^+$ on arenes (Field, 1967, Munson and Field, 1967, Aquilanti et al., 1968), and by n.m.r. spectrometry in cold solutions of superacids (cfr. Olah, 1973). The unique, and most useful feature of the decay technique in the specific application to gas-phase aromatic substitutions is

the possibility of measuring directly both the substrate and the
positional selectivity of the electrophilic attack.

The substrate selectivity is measured in competition experiments where two gaseous aromatics, in a known molar ratio, are allowed to compete for a tracer concentration of the HeT$^+$ electrophile.
The results, summarized in Table III, demonstrate that the HeT$^+$ cation displays in the gas phase a low, but significant substrate selectivity, which however reveals striking differences from the selectivity measured in solution, since all substrates, with the exception of α-trifluorotoluene, react faster than benzene, irrespec-
tive of the nature of the substituent groups.

TABLE III

Substrate selectivity of the gas-phase attack of HeT$^+$ on aromatics

Substituent group	Yields relative to benzene
CF$_3$	0.8
H	1.0
F	1.2
CH$_3$	1.5
Cl	1.6
Br	2.0
CN	2.2
OCH$_3$	3.0

The positional selectivity of HeT$^+$, or at least its lower
limit, can be deduced from the T distribution within the labeled
products. This can be measured with a stepwise substitution procedure, involving the replacement of the H atoms bound to each
molecular position with a suitable inactive group, and the determination of the correspondent decrease of the molar activity. The
results, illustrated in Table IV, reflect the typical electrophilic
character of the HeT$^+$ attack, whose distinctive feature is the relatively pronounced positional selectivity, in contrast with the low
and unusual substrate selectivity. This behaviour may have interesting implications on the problem concerning the long-range interactions between the HeT$^+$ cation and the polar substrate molecules,
especially in connection with the possible formation of an electrostatic reaction complex as the primary step of the gas-phase aromatic substitution (Cacace et al., 1972).

TABLE IV

Distribution of T in the products from the gas-phase HeT$^+$ attack on aromatics

Substrate	% T Activity			
	2 *ortho*	2 *meta*	*para*	3α
C_6H_5F	53.6	32.2	14.2	-
C_6H_5Cl	46.8	31.6	21.5	-
C_6H_5Br	37.4	38.6	23.8	-
$C_6H_5CH_3$	54.0	17.6	23.0	5.7
$C_6D_5CD_3$	50.0	19.0	23.0	8.1
$C_6H_5CF_3$	29.2	58.4	11.4	-
$C_6H_5OCH_3$	49.6	33.6	16.7	< 0.2

REACTIONS OF THE HeT$^+$ DECAY IONS WITH LIQUID ARENES

The study of the aromatic substitutions by the HeT$^+$ cation has been extended to the liquid phase by a technique based on the decay of a tracer activity of T_2 dissolved in the liquid substrate (Cacace and Caronna, 1972). This approach is of considerable interest since, in the author's knowledge, it is the only technique presently available to compare the reactivity in the gaseous and in the liquid phase of exactly the same ionic species, unaffected by solvation, ion-pairing, counterions, etc. even in the condensed phase.

The attack of HeT$^+$ on liquid benzene, toluene, and their mixtures, leads to quantitative yields of the correspondent tritiated arenes, according to a reaction scheme analogous to that outlined for the gas phase. The higher (100%) yields of labeled arenes, and the lack of fragmentation products, reflect the higher efficiency of collisional stabilization in the liquid phase. Essentially no excited arenium ions decompose in liquid arenes, despite the large exothermicity of the HeT$^+$ attack on benzene, ca. -110 kcal/mol^{-1}, and on toluene, ca. -100 kcal/mol^{-1}. Competition experiments have failed to detect any substrate selectivity of HeT$^+$, whose attack on benzene and toluene appears entirely indiscriminate. On the other hand, the intramolecular distribution of T within the labeled toluene formed:

Molecular position: 2 ortho: 2 meta: para: 3α :

Tritium content: 55.8%   17.2%   32.4%   4.5%

indicates the appreciable positional selectivity of the HeT^+ in the liquid phase. If the extreme electrophilic character of the unsolvated reagent, and the indiscriminate nature of its attack on the different substrates are taken into account, the observation of the positional selectivity of HeT^+ in the liquid phase is of interest. Indeed, any mechanism on which the individual positions of the aromatic ring compete for the electrophile in the rate-determining step seems to be ruled out, and the reactivity of HeT^+ can be best explained by the early formation of some kind of "electrophile-substrate adduct" or "encounter pair", kinetically if not structurally analogous to the π-complex advocated by Olah (1971), followed by a step (formation of σ-complex) leading to orientation of the substituent. The observation that the positional selectivity of HeT^+ is essentially the same in the gas phase and in the liquid phase is also of considerable interest to the theory of aromatic substitutions.

CONCLUSIONS

     The nuclear technique outlined in the present review represents a valuable tool for the study of ion-molecule interactions. When applied in conjunction with the appropriate mass spectrometric background to the study of gaseous ions, it provides otherwise unaccessible information on the molecular structure, the stereochemistry and the neutral products of the ionic species investigated, allowing the extension to the gas phase of the classical kinetic methods of physical organic chemistry.

     In the liquid systems, it affords the only tool so far available for producing ionic species of well-defined structure, unaffected by solvation, ion-pairing, counterions, etc., and for studying their reactivity by classical kinetic methods.

ACKNOWLEDGMENT

     The author is indebted to G. Ciranni for her invaluable help in the preparation of this contribution.

REFERENCES

Aliprandi, B., Cacace, F., and Guarino, A. (1967). J. Chem. Soc.
(B), 519.
Aquilanti, V., and Volpi, G. G. (1966). J. Chem. Phys. 44, 2307.
Aquilanti, V., Giardini, A., and Volpi, G. G. (1968). Trans. Fara-
day Soc., 64, 3282.
Ausloos, P. (1966). Ann. Rev. Phys. Chem. 17.
Ausloos, P., and Gorden, R., Jr. (1964). J. Chem. Phys., 41, 1278.
Ausloos, P., and Lias, S. G. (1963). J. Chem. Phys., 38, 2207.
Ausloos, P., Lias, S. G., and Scala, A. A. (1970). Ion-Molecule
Reactions in the Gas Phase, Amer. Chem. Soc., New York p. 264.
Babérnics, L., and Cacace, F. (1971). J. Chem. Soc. (B), 2313.
Brouwer, D. M. (1968). Recl. Trav. Chim. Pays-Bas, 87, 210.
Bruner, F., and Cartoni, G. P. (1965). J. Chromatog., 18, 390.
Cacace, F. (1964). Proceedings of the Conference on the Methods of
Preparing and Storing Marked Molecules, EURATOM, Bruxelles p. 719.
Cacace, F. (1970). Adv. Phys. Org. Chem., 8, 79.
Cacace, F., and Caronna, S. (1967). J. Amer. Chem. Soc., 89, 6848.
Cacace, F., and Caronna, S. (1972). J. Chem. Soc., Perkin II, 1604.
Cacace, F., and Perez, G. (1971). J. Chem. Soc. (B), 2085.
Cacace, F., and Speranza, M. (1972). J. Amer. Chem. Soc. 94, 4447.
Cacace, F., Caroselli, M., and Guarino, A. (1967). J. Amer. Chem.
Soc., 89, 4584.
Cacace, F., Cipollini, R., and Ciranni, G. (1968a). J. Amer. Chem.
Soc., 90, 1122.
Cacace, F., Cipollini, R., and Ciranni, G. (1971b). J. Chem. Soc.
(B), 2089.
Cacace, F., Cipollini, R., and Occhiucci, G. (1972). J. Chem. Soc.,
Perkin II, 84.
Cacace, F., Ciranni, G., and Guarino, A. (1966), J. Amer. Chem. Soc.
88, 2903.
Cacace, F., Guarino, A., and Possagno, E. (1969). J. Amer. Chem.
Soc., 91, 3131.
Cacace, F., Guarino, A., and Speranza, M. (1971a). J. Amer. Chem.
Soc., 93, 1088.
Cacace, F., Guarino, A., and Speranza, M. (1973). J. Chem. Soc., Per-
kin II, 66.
Cacace, F., Caroselli, M., Cipollini, R., and Ciranni, G. (1968b).
J. Amer. Chem. Soc., 90, 2222.
Ciranni, G., and Guarino, A. (1966). J. Labeled Comps., 2, 198.
Carlson, T. A. (1960). J. Chem. Phys., 32, 1234.
Derwish, G. A. W., Galli, A., Giardini, A., and Volpi, G. G. (1964).
J. Chem. Phys., 40, 5.
Edwards, R. R. and Davies, T. H. (1948). Nucleonics, 2, 44.
Field, F. H. (1967). J. Amer. Chem. Soc., 89, 5328.
Field, F. H., and Munson, M. S. B. (1965). J. Amer. Chem. Soc., 87,
3288.
Haynes, R. M., and Kebarle, P. (1966). J. Chem. Phys., 45, 3899.
Lampe, F. W., and Field, F. H. (1959). Tetrahedron, 7, 189.

Migdal, A. (1941). J. Phys. (USSR), 4, 449.

Monaham, J. E. (1958). Argonne National Laboratory Report, ANL-5911.

Munson, M. S. B., and Field, F. H. (1965). J. Amer. Chem. Soc., 87, 3294.

Munson, M. S. B., and Field, F. H. (1967). J. Amer. Chem. Soc., 89, 1047.

Munson, M. S. B., Field, F. H., and Franklin, J. L. (1963). J. Amer. Chem. Soc., 85, 3584.

Nefedov, V. D., Sinotova, E. N., and Akulov, G. P. (1968b). Radiokhimiya, 10, 609.

Nefedov, V. D., Sinotova, E. N., Akulov. G. P. (1971). Metod. Iztop. Indikatorov Nauch. - Issled. Prom. Proizvod, 346.

Nefedov, V. D., Sinotova, E. N., Akulov, G. P., and Korsakov, M. V. (1970). Zh. Org. Khim., 6, 1214.

Nefedov, V. D., Sinotova, E. N., Akulov, G. P., and Korsakov, M. V. (1973a). Zh. Org. Khim., 9, 629.

Nefedov, V. D., Sinotova, E. N., Akulov, G. P., and Korsakov, M. V. (1973b). Radiokhimiya, 15, 286.

Nefedov, V. D., Sinotova, E. N., Akulov, G. P., and Sass, V. P. (1968c). Radiokhimiya, 10, 761.

Nefedov, V. D., Sinotova, E. N., Akulov, G. P., and Syreischchikov, V. A. (1968a). Radiokhimiya, 10, 600.

Olah, G. A. (1971). Accounts Chem. Res., 4, 240.

Olah, G. A. (1973). Angew. Chem. Internat. Edit., 12, 173.

Olah, G. A., and Schlosberg, R. H. (1968). J. Amer. Chem. Soc. 90, 2726.

Olah, G. A., Olah, J. A., and Svoboda, J. J. (1973b). Synthesis, 490.

Olah, G. A., White, M. A., and O'Brien, D. H. (1970). Chem. Rev., 70, 561.

Olah, G. A., De Member, J. R., Mo, K. Y., Svoboda, J. J. Schilling, P., and Olah, J. A. (1973a). J. Amer. Chem. Soc. 96, 884.

Olah, G. A., Schlosberg, R. H., Porter, R. D., Mo, Y. K., Kelly, D. P., and Mateescu, G. D. (1972). J. Amer. Chem. Soc., 94, 2034.

Pobo, L. G., Wexler, S., and Caronna, S. (1973). Radiochimica Acta, 19, 5.

Rylander, P. N., and Meyerson, S. (1956). J. Amer. Chem. Soc. 78, 5799.

Snell, A. H., and Pleasonton, F. (1958). J. Phys. Chem. 62, 1377.

Snell, A. H., Pleasonton, F., and Leming, H. E. (1957). J. Inorg. Nuclear Chem. 5, 112.

Wexler, S. (1959). J. Inorg. Nuclear Chem. 10, 8.

Wexler, S. (1965). Actions Chimiques et Biologiques des Radiations (Ed. M. Haissinsky). VIII, 107. Masson, Paris.

Wexler, S., and Hess, D. C. (1958). J. Phys. Chem. 62, 1382.

Wexler, S., Anderson, G. R., and Singer, L. A. (1960). J. Chem. Phys., 32, 417.

Wolfsberg, M. (1956). J. Chem. Phys. 24, 24.

Wolniewicz, L. (1965). J. Chem. Phys. 49, 1087.

Wu, C. S. (1955). Beta and Gamma Spectroscopy. (Ed. K. Siegbahn) North Holland Publ. Co., Amsterdam.

ION-MOLECULE REACTIONS IN RADIATION CHEMISTRY

S. G. Lias

National Bureau of Standards

Washington, D. C. 20234, USA

HISTORICAL INTRODUCTION

Radiation chemistry is the study of the chemical effects resulting from the action of high energy radiation on matter. From the beginnings of the study of this complex subject in the 1890's, it was realized that ionization occurred in materials subjected to high energy irradiation, and it was even suspected, in an undefined way, that the chemical transformations which occurred in such samples might somehow come about as a result of this ionization. In 1912, when the first systematic studies of radiation-induced events were begun by S. C. Lind (Lind, 1961) and separately, by W. Mund (Mund, 1956), the concept of an elementary reaction between an ion and a molecule had not yet been conceived. These early workers postulated that the ions formed by the action of the radiation collected a cluster of neutral molecules about themselves, and that when the central ion underwent charge recombination, the energy released promoted chemical reactions among the clustered molecules, leading to product formation. The "cluster theory" was largely abandoned in 1936 when Eyring, Hirschfelder, and Taylor (Eyring et al, 1936) predicted theoretically that clustering should not be important in irradiated hydrogen. After this, it was generally held that parent or fragment ions formed in radiolytic systems would undergo charge neutralization to form excited species which would usually dissociate; reactions of the resulting free radicals were believed to be the most important product forming processes.

In the meantime, early observations of simple ion-molecule reactions had been made in the late 1920's (Hogness and Lunn, 1925; Hogness and Harkness, 1928; Barton and Bartlett, 1928) and experi-

ments carried out in gas discharge sources under pressure conditions
such that the ions could undergo many collisions also gave evidence
for reactions between ions and molecules as early as 1930 (Eisenhut
and Conrad, 1930). However, it was not until the extensive quantitative mass spectrometric studies of ion-molecule reactions carried
out by Talroze, by Stevenson, by Hamill and collaborators, and by
Field, Franklin, and Lampe appeared in the literature (Talroze and
Ljubimova, 1952; Stevenson and Schissler, 1955; Stevenson, 1963;
Lampe, Franklin, and Field, 1961; Durup, 1960) that radiation
chemists considered the possibility that such reactions might occur
under the conditions of their experiments. In 1957 Meisels, Hamill
and Williams (Meisels et al, 1957) investigated the gas phase radiolysis of methane and interpreted their results entirely in terms of
ion-molecule reactions, concluding that "yields can be virtually
completely explained without assuming any contribution by excitation".
After Field and Lampe (Field and Lampe, 1958) reported the occurrence
of hydride transfer reactions from alkanes to carbonium ions in 1958,
Futrell (Futrell, 1959; 1960) suggested that these reactions might
be the major product-forming processes in the radiolysis of alkanes.
(Of course, these early attempts to define the role of ion-molecule
reactions in the radiolysis were over-simplifications; it is now
known that the decomposition of neutral excited molecules is also
an important mode of product formation.)

During this period, it was fashionable for mass spectrometrists
to rationalize their reasons for investigating ion-molecule reactions
in terms of the probable importance of these reactions in systems
under high energy irradiation. Nevertheless, it was only in 1962
that firm experimental evidence was presented for the occurrence of
an ion-molecule reaction in a gas phase radiation chemistry experiment. It was shown in that study (Ausloos and Lias, 1962) that the
ethane formed in an equimolar mixture of C_3H_8 and C_3D_8 in the presence of a free radical scavenger consisted of approximately equal
amounts of C_2D_6, C_2D_5H, C_2H_5D, and C_2H_6 formed in the reactions such
as:

$$C_2D_5^+ + C_3H_8 \rightarrow C_2D_5H + C_3H_7^+ \tag{1}$$

By the mid-1960's, the understanding of the mechanisms occurring in
many gas phase radiolytic systems, particularly in hydrocarbons,
had become sophisticated enough that ion-molecule reactions were
actually investigated in radiolytic studies (Ausloos and Lias, 1967;
Ausloos, 1970), usually by analyzing the neutral products formed in
isotopically labeled systems. For instance, proton transfer from
CH_5^+ to higher alkanes was first reported in a radiolytic study
(Ausloos et al, 1963), and the H_2^- transfer reactions from alkanes
to olefin ions:

$$C_nH_{2n}^+ + C_mH_{2m+2} \rightarrow C_nH_{2n+2} + C_mH_{2m}^+ \tag{2}$$

as well as the H_2 transfer reaction from parent alkane ions to
olefins:

$$C_nH_{2n} + C_mH_{2m+2}{}^+ \rightarrow C_nH_{2n+2} + C_mH_{2m}{}^+ \tag{3}$$

were first studied in radiolysis experiments (Ausloos and Lias,
1967; Ausloos, 1970).

In the 1960's, the development of instruments which could
deliver a short (nanosecond to microsecond) high intensity pulse of
electrons brought about a shift of emphasis of the mainstream of
radiation chemical research (Matheson and Dorfman, 1969). In such
experiments, high concentrations of reactive intermediate species
can be produced, and their loss through reaction or charge recombination can be followed as a function of time using physical means of
detection such as optical spectroscopy or conductivity. As a result,
current radiation chemical research is concerned mainly with following the fate of the electron, the free radical kinetics, or
elucidating the energy deposition or charge neutralization processes.
Ion-molecule reactions are rarely investigated in such studies.
Fast ion-molecule reactions occur at times shorter than the shortest
observation time, especially in the liquid phase, and also, the
optical and conductivity methods generally used for observation are
a less discriminating means of identifying reactant or product ions
than mass analysis or product analysis.

DEFINITION AND DESCRIPTION OF RADIATION CHEMISTRY

Before beginning our discussion of information about ion-
molecule reactions derived from radiation chemistry experiments, it
is useful to examine briefly exactly what is meant by "radiation
chemistry", to see how the conditions of such experiments compare
with those encountered in other types of experiments in which ion-
molecule reactions are studied. Strictly speaking, the word "radiation" means only electro-magnetic radiation and the "high energy
radiation" used by radiation chemists therefore would be presumed
to mean only X-rays or gamma rays. Actually, X-rays or gamma rays
are not directly responsible for much of the chemical change which
is observed in irradiated materials. Such high energy quanta lose
energy to the medium through the photoelectric effect, the Compton
effect, or pair production - processes which ultimately result in
the release of energetic electrons throughout the irradiated material. These primary or secondary electrons, which have a wide spectrum of energies, are responsible for more than 99% of the ionization or excitation events occurring in an irradiated system. Almost
universally, scientists adopt a verbal shorthand which ignores the
strict dictionary definition of "radiation" and includes under the
name the high energy electrons released by gamma rays, as well as
other high energy charged particles released in radioactive decay or

generated by a pulsed accelerator. Thus, "radiation chemistry" is
the study of the chemical effects brought about by the interaction
of matter with high energy charged particles, even including as a
special case, the reaction chamber of any mass spectrometer in
which ionization is effect by a beam of high energy electrons or by
alpha particles. In actual usage, the term "radiation chemistry"
is usually restricted to experiments in which a gas, liquid, or
solid is irradiated in a closed system by X-rays, gamma rays, alpha
particles, or a beam of high energy electrons. Experimental results
and theoretical calculations indicate that the energy transferred
to the molecule is fairly insensitive to variations in the energy
of the ionizing electron as long as the energy is above about 2 I.P.
(Eyring and Wahrhaftig, 1961; Vestal, Wahrhaftig, and Johnston,
1962; Kebarle and Godbole, 1962; Monahan and Stanton, 1962). Actually, the energy distribution of ions formed in a radiation
chemistry system is comparable to that formed in a mass spectrometer
in which ionization is effected by electrons of 50-100 eV energy.

In the gas phase, irradiation of a sample by high energy radiation results in a macroscopically homogenous, microscopically
random, distribution of ions and electrons (as well as electronically excited molecules) throughout the sample. Under these conditions, the charged species will ultimately disappear through homogenous recombination in the body of the gas (although at dose rates
lower than about 10^{11} eV/cm^3-second, the ions may drift to the wall
where they will be neutralized). Thus, in the gas phase, the mean
lifetime of a charge is:

$$\tau = (I\alpha)^{-1/2} \tag{4}$$

where I is the number of ions formed per unit volume and unit time,
and α is the recombination rate coefficient. Most values of α
reported in the literature fall in the range 10^{-6}-10^{-5} cm^3/molecule-
second. Under typical gamma radiolysis conditions, I will be in
the range 10^8-10^{14} ions per cm^3 per second, so the lifetime of a
charge will be 10^{-4} to 10^{-1} second. The lifetime of a charge in
propane and in hydrogen at a typical gamma radiolysis dose rate of
10^{17} eV/g-second is shown in Figure 1 as a function of pressure.

In a gas phase pulse radiolysis experiment, dose rates are
much higher; values of I will range around 10^{16}-10^{22} ions per cm^3
per second, and ion lifetimes will be in the range 10^{-5} to 10^{-9}
second. The lifetime of a charge in propane is shown in Figure 1
as a function of pressure for a dose rate of 10^{26} eV/g-second. We
can expect that the overall ion-molecule reaction mechanism in a
system will depend in large part on the lifetime of the ions with
respect to charge neutralization. For example, at pressures of 1
torr to one atmosphere, the collision interval for an ion will be
about 10^{-8} to 10^{-11} second, as illustrated for propane in Figure 1.
As the curves shown in Figure 1 clearly demonstrate, under some

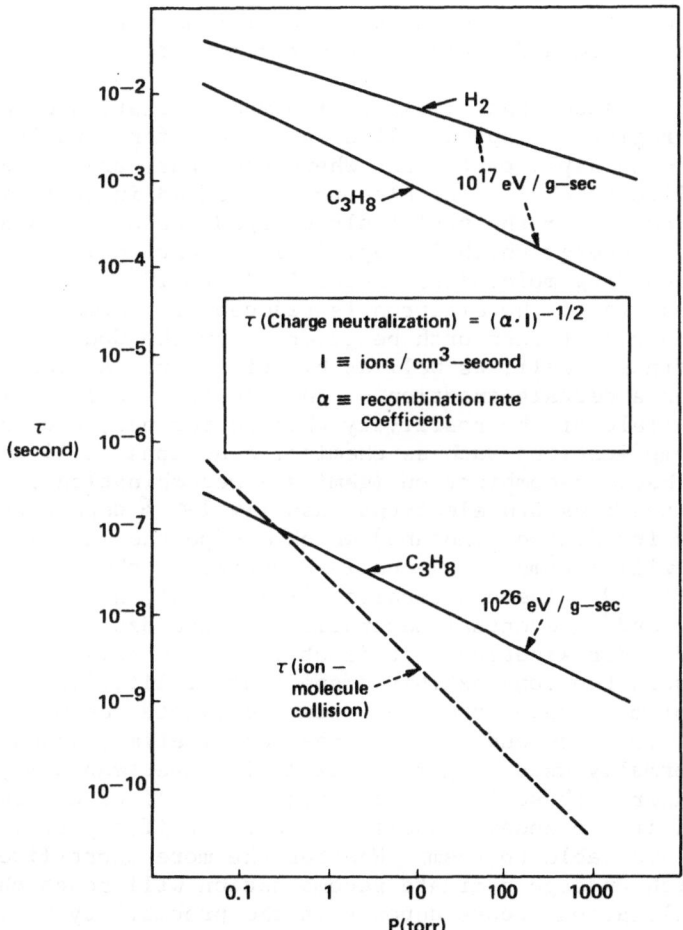

Figure 1. The lifetimes of charged species formed at pressures from 0.1 to 1000 torr in propane at a high dose rate (10^{26} eV/g-second) such as that obtained in a pulse radiolysis experiment, and in propane and hydrogen at the dose rate of a typical gamma radiolysis experiment (10^{17} eV/g-second). In propane, the absorption of 23.4 eV leads to the formation of one ion pair, while in hydrogen, 35.9 eV is required. The figure also shows the collision interval for ions colliding with propane over this pressure range.

conditions in a high dose rate pulsed experiment, charge recombination may actually compete with fast ion-molecule reactions.   In
the low dose rate gamma-ray irradiations, on the other hand, it is
obvious that the species carrying the charge will, on the average,
undergo many hundreds or thousands of collisions with the surrounding molecules before undergoing charge recombination.

     In the condensed phase, excitation and ionization events occur
in localized regions of space called spurs (or, for densely ionizing
radiation such as alpha particles, where the spurs are formed close
together, called tracks).   The electrons released in an ionization
process move away from the positively charged partner with a certain
initial kinetic energy which is rapidly lost through interactions
with the surrounding molecules.   About 10^{-14} to 10^{-13} second after
the ionization event, the electron is reduced to thermal energy,
and its motion will thenceforth be governed by the Coulombic field
in which it finds itself, as well as by diffusion.   At the time of
thermalization a certain fraction of the electrons will still be
found in the field of the positively charged partner, and in the
absence of complications such as chemical reactions, will return
and undergo charge recombination (geminate recombination).   The
remainder of the ions and electrons (usually $1-70\%$ depending on the
nature of the irradiated compound) which escape the mutually attractive forces, will become homogenously distributed throughout the
body of the liquid.   These are called "free ions", and their chemical reactions and/or eventual neutralization are described by homogenous second-order kinetics.   It is obvious from this description
that in a liquid the ions exhibit a spectrum of lifetimes covering
many orders of magnitude from the fastest geminate recombination
time of about 10^{-13} second, to the free ion lifetime, which at the
dose rates normally used in gamma radiolysis experiments may be 0.1
second or longer.   These long-lived free ions will of course have
ample opportunity to undergo chemical reactions if any reactive
pathways are available to them.   Whether the more short-lived positive ions which undergo geminate recombination will react chemically
before neutralization occurs depends on the probability that the ion
will encounter a molecule with which it can react before the return
of the electron, about 10^{-13} to 10^{-9} second after ionization.   Of
course, if an ion comes into being in the liquid surrounded by reactive parent molecules, it may have a very high probability of
reacting before the return of the electron.

THE INVESTIGATION OF ION-MOLECULE REACTIONS IN RADIATION CHEMISTRY
EXPERIMENTS

     It is our intention here to present a brief suvey of informa-
tion about ion-molecule reactions which has been derived from the
results of radiation chemistry investigations.   Since the great majority of studies of ion-molecule reactions are carried out in some type

of mass spectrometer, and all mass spectrometers devised thus far
operate only in the gas phase, it might be expected that the greatest value of radiation chemistry to ion-molecule reaction research
would be an examination of reaction mechanisms under conditions not
limited to any particular density range. Indeed, it is true that
an abundance of information about ion-molecule reactions at atmospheric pressures has been obtained in gas phase radiation chemistry
experiments. As we shall see, ion-molecule reactions in condensed
phases have also been investigated in radiolytic systems. However,
it must be admitted that the understanding of ion-molecule reaction
mechanisms in the liquid phase is still quite limited for most systems.

In principle, information about the kinetics and/or mechanism
of an ion-molecule reaction:

$$A^+ + B \rightarrow C^+ + D \tag{5}$$

can be obtained from a detection of some physical property of a
product or reactant, or from a chemical analysis of the products.
Most investigations of ion-molecule reactions rely on a determination of the charge to mass ratio of charged species A^+ and/or C^+
and the relative abundances of these charged species under varying
conditions. It is true that experiments have been carried out in
which the charged species present in gas phase systems under α-
particle irradiation were detected by ordinary mass spectrometric
techniques at pressures as high as one atmosphere (Kebarle et al,
1966). However, we shall here be concerned more with results from
studies utilizing conventional radiation chemistry techniques which
do not rely on mass spectrometric detection. In some studies, information has been derived from analysis of neutral molecules produced in ion-molecule reactions, while in others absorption spectroscopy or electron spin resonance have been used as detection techniques. It is obvious that a particular ion-molecule reaction will
not be amenable to study by these methods unless it results directly
or indirectly (through subsequent reaction) in the formation of a
product which can be uniquely traced to the reaction in question,
or unless one of the reactants or products possesses some unique
property (such as an absorption spectrum) which unambiguously
identifies it. The reason that ion-molecule reactions have not yet
been characterized in many liquid phase systems is that these conditions are not met.

INFORMATION ABOUT ION-MOLECULE REACTION RATES FROM RADIATION CHEM-
ISTRY EXPERIMENTS

Absolute Rate Coefficient Determinations

What is known about absolute rate coefficients of ion-molecule reactions in the liquid phase has been derived mainly from absorption spectroscopic measurements made in liquid phase pulse radiolysis experiments. For example, Dorfman and collaborators (Arai and Dorfman, 1964; Arai et al., 1967; Dorfman, 1970) followed the decay of optical absorption of aromatic anions generated in the pulse radiolysis of liquid solutions of aromatic compounds in various alcohols, and derived the rate coefficients for the reactions of aromatic anions with alcohols:

$$(Arene)^- + ROH \rightarrow (Arene)H + RO^- \tag{6}$$

shown in Table 1. As reported earlier from results obtained in gamma-irradiated glassy solutions of alcohols at $77^\circ K$ (Shida and Hamill, 1966), the rate of reaction of a particular anion increases with an increase in the liquid phase acidity of the reactant alcohol, showing that the proton transferred to the anion originates in the hydroxyl group of the alcohol. No measurements of the rate constants of these reactions in the gas phase have been made, but the values given in Table 1 are all very low compared to gas phase rate constants for similar reactions. Solvation of the anion and hydrogen bonding of the hydroxyl hydrogens means that a reorientation of the liquid structure must occur before or during reaction, thus slowing the reaction rate. Bohme and Young (Bohme and Young, 1970) have pointed out that the gas phase rate constant of the reaction:

$$O^- + CH_3OH \rightarrow OH^- + CH_3O \tag{7}$$

measured in a tandem mass spectrometer, is approximately three orders of magnitude higher than the value derived by Gall and Dorfman (Gall and Dorfman, 1969) from the pulse radiolysis of an aqueous solution containing methanol, clearly demonstrating that solvent effects do contribute to the slowing of a reaction rate of a particle transfer reaction, such as proton transfer reactions 6 and 7.

In the gas phase, rate constants for reaction of N_2^+ with various molecules have been determined by following the time depen-

Table 1. Liquid Phase Rate Coefficients of the Reaction:
$(Arene)^- + ROH \rightarrow (Arene)H + RO^-$

| | $k(cm^3/molecule\text{-}second) \times 10^{16}$ | | |
	Diphenylide	Anthracenide	p-Terphenylide
Methanol	1.15	1.35	0.006
Ethanol	0.3	0.27	0.003
1-Propanol	0.29	0.22	n.d.
2-Propanol	0.049	0.032	n.d.

Temperature = $298^\circ K$

Table 2. Rate Coefficients for Reactions of N_2^+ (Perner and Dryer, 1971)

	k_{Rn}(Pulse Radiolysis)	k_{Rn}(Literature)
$N_2^+ + N_2$	3.0×10^{-29} cm^6/mol^2-s	8×10^{-29} cm^6/mol^2-s [a]
$N_2^+ + NO$	4.2×10^{-10} cm^3/mol-s	3.3×10^{-10} cm^3/mol-s [b]
$N_2^+ + 1,3-C_4H_6$	13.4×10^{-10}	-
$N_2^+ + c-C_3H_6$	14.8×10^{-10}	-
$N_2^+ + NH_3$	18.3×10^{-10}	-
$N_2^+ + O_2$	$\leq 0.8 \times 10^{-10}$	0.5×10^{-10} [c]

[a] Good, Durden, and Kebarle, 1970.
[b] Fehsenfeld, 1970.
[c] Ferguson, et al, 1969.

dence of the 391.44 nm absorption of the N_2^+ species generated by a pulse of high energy electrons in pure N_2 and in mixtures of N_2 with several other compounds (Perner and Dreyer, 1971). At this writing, this is the only published study in which absorption spectroscopic measurements have been used to measure a gas phase rate coefficient. The rate constants derived from the results of that study are given in Table 2, where they are compared with rate constants derived for some of these same reactions by conventional mass spectrometric or flowing afterglow techniques (Good et al., 1970; Fehsenfeld, 1970; Ferguson, et al., 1969). The reasonably good agreement between the two sets of values verifies the validity of the radiolytic technique.

In another gas phase pulse radiolysis study (Lias et al., 1972), absolute rate coefficients for ion-molecule reactions have been derived from the absolute yields of the neutral reaction products taking advantage of the fact (illustrated in Figure 1) that at high dose rates, charge neutralization is actually in competition with ion-molecule reactions, and therefore, the observed yield of a given product of an ion-molecule reaction will be lower than that observed in the gamma radiolysis under conditions such that all the ions react. This method for determining rate constants is less direct than the spectroscopic technique described above, in that initially an accurate value for the rate coefficient of the competing charge recombination reaction:

$$A^+ + N^- \rightarrow \text{neutral products} \qquad (8)$$

(where N^- is an electron or negative ion) must be determined, and the accuracy of this determination (again based on the yield of a neutral product formed in an ion-molecule reaction in competition with neutralization) ultimately depends on knowing the rate coefficient for a competing standard reaction. Also, the system as a

whole must be well understood, i.e. the actual yield of A^+ initially
formed in the system must be known. The technique was applied to a
determination of the rate coefficients for the hydride transfer
reactions between the t-butyl ion and various alkane molecules:

$$(CD_3)_3C^+ + RH \rightarrow (CD_3)_3CH + R^+ \tag{9}$$

The results, shown in Table 3, are compared to rate constants derived for these same reactions in low dose rate gamma radiolysis
experiments, with excellent agreement.

Relative Rate Coefficient Determinations

Accurate relative rate constants for hydride transfer reactions between ethyl, sec-propyl, sec-butyl, and t-butyl ions and
the lower alkanes have been determined in experiments in which the
neutral reaction product formed in mixtures of fully deuterated
and non-deuterated alkanes were analyzed (Ausloos, 1970). The
reactant ion is generated in the presence of two reactive alkanes,
one of them serving as a standard reactant:

$$A^+ + RD \rightarrow AD + R^+ \tag{10}$$

$$A^+ + R'H \rightarrow AH + R'^+ \tag{11}$$

From a determination of the relative amounts of AD and AH in a
series of mixtures of known composition, accurate relative rate
coefficients are generated, and can be expressed in absolute rate
constant units if a determination of the absolute value of one of
the rate coefficients is available. The rate constants for the t-
butyl ion reactions listed in Table 3 from gamma radiolysis experi-

Table 3. Rate Constants for the Reaction: $t\text{-}C_4D_9^+ + RH \rightarrow i\text{-}C_4D_9H + R^+$ Derived from Pulse Radiolysis

RH	$k(cm^3/molecule\text{-}second) \times 10^{11}$	
	Pulse Radiolysis	Gamma Radiolysis
2,2-Dimethylbutane	0.0	0.0
3-Methylpentane	3.0[a]	2.7
	2.8[b]	
2-Methylpentane	4.2[a]	3.9
	4.25[b]	
2,3-Dimethylbutane	5.7[a]	5.7
	5.7[b]	
Methylcyclopentane	7.35[a]	7.4

[a] Competing process: $t\text{-}C_4D_9^+ + e$, $k = 1.92 \times 10^{-6}$.
[b] Competing process: $t\text{-}C_4D_9^+ + SF_6^-$, $k = 4.0 \times 10^{-7}$.
$\overline{T} = 300°K$

ments were determined in this way (Ausloos and Lias, 1970). Related
experiments with partially deuterated alkanes provided information
about the probabilities of reaction of a given ion at primary, secondary, or tertiary sites in the alkane molecules.   From these
studies (Ausloos, 1970; Ausloos and Lias, 1972; Lias and Ausloos,
1975) the following general conclusions emerge: (1) the probability of a reactive collision between an alkyl ion and an alkane is
less than unity unless reaction of the ion is very exothermic at
every site in the molecule; (2) for reactions of different ions
with the same molecule, it is seen that the probability of a reactive collision is greater, the more exothermic the reaction; (3) for
reactions of a given ion with a series of alkanes, the probability
of a reactive collision depends on the fraction of sites in the
molecule at which an exothermic reaction can occur; (4) when reaction is only marginally exothermic, Van der Waals repulsions between
the ion and parts of the molecule can contribute to a slowing of
the reaction rate.

Other Information About Ion-Molecule Reaction Rates From Radiolytic
Studies

        In many radiolytic studies, particularly those carried out in
the condensed phases, quantitative measurements of reaction rates
of ion-molecule reactions can not be made directly, but related
information is sometimes obtained from which inferences about the
rates of certain ion-molecule reactions can be made.

        For instance, it has been reported (Zador et al., 1973) that
in the liquid phase radiolysis of cyclohexane and methylcyclohexane
when tetramethyl-para-phenylenediamine (TMPD) or pyrene are added as
charge acceptors, the rate constants for formation of the TMPD or
pyrene cations are greater by at least an order of magnitude than
those which would result if the rate were controlled by the diffusion of the positive alkane ions to the charge acceptor molecules.
It was concluded that the most likely explanation for these high
rate constants is that the positive charge moves through the liquid
through a series of resonance charge transfer reactions between
alkane ions and neighboring alkane molecules.   However, in similar
experiments in cyclooctane, isooctane, cyclopentane, and n-hexane
much lower rate constants for charge transfer to the TMPD or pyrene
were observed, and it was concluded that resonance charge transfer
involving the parent alkane ion is not a general phenomenon.

        In Table 1 we showed some rate constant determinations made in
liquid phase radiolytic systems which indicated that solvent effects
cause a slowing of the rate of certain particle transfer reactions
in the liquid phase. The same conclusion has been reached in spectroscopic investigations of the liquid phase pulse radiolysis of
acetone (Robinson and Rodgers, 1973), where the absorption band

ascribed to the acetone cation (Shida and Hamill, 1966) decays very
slowly after an electron pulse ($\tau_{1/2} \sim 10^{-7}$ second). In the gas
phase the acetone cation reacts with acetone with a rate coefficient
of 6×10^{-10} (MacNeil and Futrell, 1972). However, the acetone system illustrates some of the difficulties associated with investigating ion-molecule reactions in the condensed phase using absorption
spectroscopy as a detection technique; in the solid phase, the absorption band ascribed to the acetone cation is only slightly affected by the presence of 10 mole percent of a charge acceptor, casting
some doubt on the identification of the band.

On the other hand, there is evidence for fast proton transfer
reactions in the liquid phase radiolysis of several polar compounds.
When an organic compound such as an alcohol, aldehyde, or ether is
irradiated in the presence of an aromatic compound, it is expected
that the parent ions of the major component of the mixture will
transfer charge to the aromatic solute forming aromatic ions which
will subsequently undergo charge neutralization to form a certain
yield of triplet excited aromatic molecules which can be detected
by their fluorescence. The fact that the yield of such triplet
excited species is very low in irradiated alcohols, aldehydes, and
ethers has been interpreted to mean that the parent ions of these
compounds react quickly with neighboring parent molecules to transfer a proton (Arai and Dorfman, 1965; Hayon, 1970; Hentz and Altmiller, 1971).

INFORMATION ABOUT IONIC REACTION MECHANISMS FROM RADIATION CHEMIS-
TRY STUDIES

In studies of ion-molecule reactions in hydrocarbon systems
carried out using analysis of neutral products in deuterium-labeled
or partial deuterium-labeled systems, unique information about the
details of ionic reaction mechanisms has been derived, which in
many cases could not have been obtained by other techniques.

For instance, although it is simple to determine mass spectrometrically, using partially deuterated reactant molecules, that
a t-butyl ion abstracts only a tertiary H(D) from isopentane:

$$t\text{-}C_4H_9^+ + (CH_3)_2CDCH_2CH_3 \rightarrow (CH_3)_3CD + C_5H_{11}^+ \tag{12}$$

only an experiment in which the neutral product is analyzed could
verify that the transferred particle ends up as a tertiary hydrogen
(deuterium) in the product isobutane, i.e., that the reaction is a
simple particle transfer which does not involve any shuffling of
H(D) species (Ausloos and Lias, 1970).

This is found to be generally true for many types of particle
transfer reactions in hydrocarbon systems. For instance, in the H_2

transfer reactions from alkane or cycloalkane parent ions to cyclo-
propane-d_6, the product propane always has the structure $CD_2HCD_2CD_2H$.
This indicates that in the radiolysis of c-C_6H_{12}-c-C_3D_6 mixtures,
for instance, the reaction complex of the H_2 transfer reaction can
be visualized as follows:

$$\longrightarrow C_6H_{10}^+ + CD_2HCD_2CD_2H$$

(13)

(Ausloos and Lias, 1967). Actually there are two reaction channels
between an alkane parent ion and an olefin or cyclopropane molecule,
H_2 transfer (such as reaction 13, for example) and H transfer:

$$c\text{-}C_6H_{12}^+ + c\text{-}C_3D_6 \rightarrow C_6H_{11}^+ + C_3D_6H \tag{14}$$

Extensive radiolytic investigations (Ausloos and Lias, 1965;
Doepker and Ausloos, 1965; Collin and Ausloos, 1975; Ausloos, 1970;
Lias and Ausloos, 1975) carried out in isotopically labeled alkane-
olefin mixtures have elucidated certain details of the mechanism of
these one- and two-particle transfer reactions (as well as of the
analogous one- and two-particle transfer reactions involving olefin
ions and alkane molecules). It is seen, for instance, that the
ratio of one-particle to two-particle transfer for a given alkane-
olefin pair is the same no matter whether the charge is initially
located on the alkane or the olefin; from this it is surmised that
in the complex, the charge always locates itself on the entity
with the lowest ionization potential. Secondly, if a one-particle
transfer reaction between the pair is not exothermic, no two-particle
transfer reactions are observed, even though these may be exothermic;
this leads to the conclusion that the two particle transfer reactions
occur in two clearly defined steps involving the transfer of one
particle, then, if the complex holds together, the transfer of a
second particle across the double bond of the olefinic species. This
conclusion is confirmed by a number of experiments in which it is
shown that the two-particle transfer reactions are extremely stereo-
specific, and the stereospecificity can be neatly explained if one
accepts that the first step of the reaction is the occurrence of an
exothermic one-particle transfer reaction. For example, in the
reaction between a propylene ion and isobutane labeled with deuterium at the tertiary position, the product propane consists of
$CH_3CH_2CH_2D$ but never CH_3CHDCH_3 or C_3H_8, indicating that the end CH_2

group of the propylene always attacks at the tertiary hydrogen (deuterium):

$$CH_3CHCH_2^+ + (CH_3)_3CD \rightarrow CH_3CH_2CH_2D + C_4H_8^+ \tag{15}$$

In fact, the competing one-particle transfer reaction always involves the tertiary D atom and results in the formation of a sec-propyl radical:

$$CH_3CHCH_2^+ + (CH_3)_3CD \rightarrow CH_3CHCH_2D + C_4H_9^+ \tag{16}$$

If the two-particle transfer reactions involved the simultaneous transfer of the two particles, one would not expect such specificity, since the heat of reaction would be unchanged if the positions of the transferred species in the final product were reversed. Thus the relative importance of one- to two-particle transfer for such reactant pairs is simply the probability that the reaction complex does nor does not fall apart after the transfer of the first particle.

The H_2^- and H_2 transfer reactions were the first specific ion-molecule reactions demonstrated to occur in the liquid and solid phase radiolysis of hydrocarbons (Scala et al., 1966; Ausloos, et al., 1967; Rzad and Schuler, 1968; Rzad et al., 1969; Van Ingen and Cramer, 1970; Collin and Ausloos, 1975). In the early experiments, 1-10% of $c-C_3D_6$ was added to various alkanes in the liquid phase, and the formation of the characteristic $CD_2HCD_2CD_2H$ product was observed. It is of interest that for the $c-C_6D_{12}^+-c-C_3H_6$ reactant pair, H_2S scavenging experiments have shown (Ausloos et al., 1967) that in both the liquid and gas phase, both D and D_2 transfer occur:

$$c-C_6D_{12}^+ + c-C_3H_6 \rightarrow CH_2DCH_2CH_2 + c-C_6D_{11}^+ \tag{17}$$

$$\rightarrow CH_2DCH_2CH_2D + C_6D_{10}^+ \tag{18}$$

In the condensed phase, the ratio k_{17}/k_{18} was equal to 2.6 ± 0.2 as compared to 5.5 in the gas phase. However, the liquid phase ratio takes into account only those product propyl radicals which are intercepted by H_2S. It is possible that geminate combination of the reactive products of reaction 17, or ineffective separation of the ion-molecule complex occurs at high densities, as evidenced by the formation of n-propylcyclohexane as a product in such mixtures (Rzad et al., 1969). If one accepts this as an alternate fate for the products of reaction 17, then liquid and gas phase values for the one-particle to two-particle transfer ratio are in the same range. In the gas phase, there is no change in one- to two-particle tranfer ratios observed mass spectrometrically at pressures of 10^{-4} to 10^{-2} torr (Sieck and Searles, 1970) or radiolytically at pressures of 1-760 torr (Collin and Ausloos, 1975).

    In view of the fact that there apparently may be no effect of
density on the ratio of one- to two-particle transfer for a given
reactant pair, it is of interest that when one observes only the H_2
or H_2^- transfer reactions, the relative rates of a series of reactions are much different in the liquid than in the gas phase. This
change may well represent a change in the probability of reaction
between a given pair in the liquid as compared to the gas phase.
It is of interest that for a given ion reacting with several different molecules, there is an apparent correlation between the efficiency of the two-particle transfer reaction in the liquid phase and
the enthalpy of the charge transfer reaction between the pair. As
is shown in Figure 2, the efficiency of the two-particle transfer

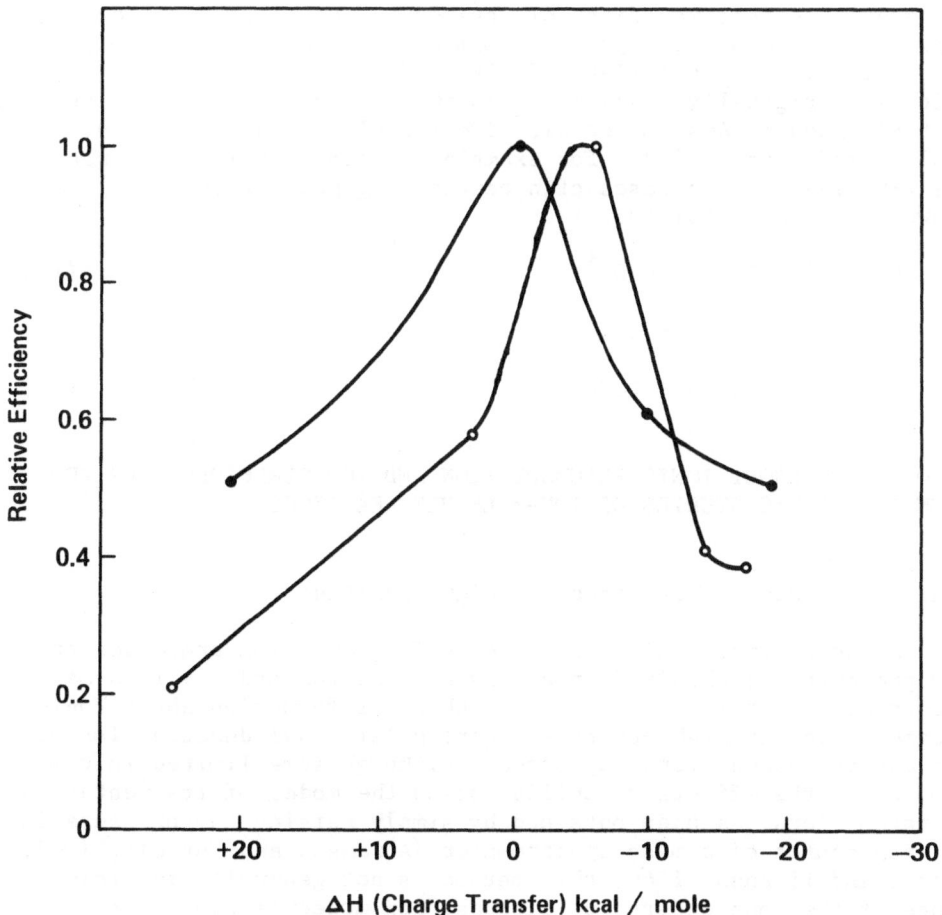

Figure 2.   Relative efficiencies of H_2-transfer in the liquid phase
for $i\text{-}C_5H_{12}^+$—olefin (o) and $c\text{-}C_5H_{10}^+$—olefin (●) reacting pairs.

reactions shows a maximum in the energy range where charge transfer is nearly thermoneutral. It has been suggested (Ausloos, et al., 1967) that the occurrence of resonance charge transfer may facilitate transfer of an H_2 species in the liquid phase. As pointed out by Durup (Durup, 1960) such an occurrence might lower the potential energy of interaction and thus increase the probability of reaction. On the other hand, it is conceivable that the differences between gas phase and liquid phase results are at least in part due to a change in the nature of the reactant ion because of clustering or solvation effects at high densities.

Gas phase radiolytic studies have also given information about the mechanisms of dissociative proton transfer reactions from CH_5^+ or H_3^+ to alkanes. (Actually, the occurrence of these proton transfer reactions was first discovered in radiolytic studies.) Analysis of the neutral product formed in the dissociation of a protonated isotopically-labeled alkane demonstrated that the proton which was originally transferred to the molecule ends up in the neutral product (Ausloos et al., 1963; Ausloos and Lias, 1964; Ausloos and Lias, 1965). For example, at atmospheric pressures, the C-C cleavage processes of n-pentane-d_{12} protonated by H_3^+ were shown to occur as follows.

$$C_5D_{12}H^+ \rightarrow CD_3H + C_4D_9^+ \tag{19}$$

$$\rightarrow C_2D_5H + C_3D_7^+ \tag{20}$$

$$\rightarrow C_3D_7H + C_2D_5^+ \tag{21}$$

INFORMATION ABOUT IONIC FRAGMENTATION AND ION STRUCTURES DERIVED FROM RADIOLYTIC STUDIES OF ION-MOLECULE REACTIONS

Ionic Fragmentation Processes at High Densities

Although information about ionic fragmentation processes at low pressures is obtained in a very straightforward way in a mass spectrometer, it is not so easy to obtain information about ionic fragmentation at high pressures where collisional deactivation of the excited parent ions may occur. Although some limited information about the effects of collisions on the modes of fragmentation of parent ions has been obtained by simply raising the pressure in the ion source of a mass spectrometer (Abramson and Futrell, 1967; Hughes and Tiernan, 1969) this method is not generally feasible, since if the ions undergo collisions, ion-molecule reactions may occur so that the observed abundances of the ions can not always be correlated to the yields of these ions formed in the initial decomposition step. Even in systems where the fragment ions are unreactive with the parent compound, this method of assessing the

effect of pressure on the modes of decomposition is not straightforward since the abundances of unreactive ions will be extremely
sensitive to the presence of reactive impurities.

Actually, the most reliable information currently available
about the effects of pressure on the modes of fragmentation of parent
ions has been obtained from analyses of products formed in ion-molecule reactions in radiolytic systems. The yields of these products
determined at several pressures give information about the variations
in the yields of the precursor ions with pressure. This straightforward approach has been particularly successful in studies of
saturated hydrocarbons where the fragment ions undergo fast reactions
with the parent compounds to form well characterized reaction products.

The yields of several major fragment ions have been determined
over a wide pressure range in the radiolysis of propane (Ausloos et
al., 1963). These results are shown in Figure 3. In agreement with
theoretical predictions (Vestal, 1968), an increase in pressure
results in the quenching of secondary decompositions such as:

$$(C_2H_5^+)^* \rightarrow C_2H_3^+ + H_2 \tag{22}$$

so that the yield of the $C_2H_5^+$ ion is enhanced and that of $C_2H_3^+$ is
reduced. If the reaction time is further diminished, the parent ion
decomposition process leading to the formation of $C_2H_5^+$ is quenched
as well, as shown by the decline in the yield of this ion at pres-

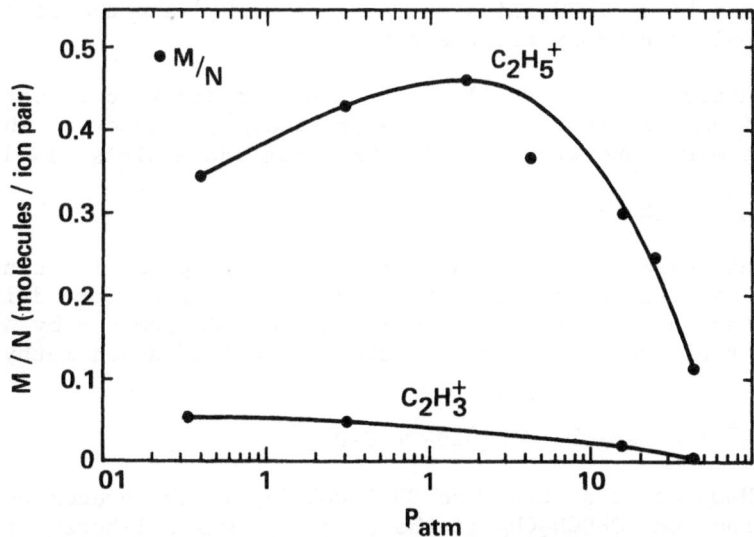

Figure 3. Ion pair yields of $C_2H_5^+$ and $C_2H_3^+$ ions, determined
through analysis of neutral reaction products in the radiolysis of
propane at pressures up to ~ 40 atmospheres.

sures of several atmospheres. This observation is consistent with the low yields of $C_2H_5^+$ (M/N$_+$ ~ 0.06) reported in the liquid phase radiolysis of propane (Koob and Kevan, 1968; Tanno and Shida, 1969).

More information about the decomposition of alkane ions in the liquid phase is summarized in Table 4. These results have all been obtained in studies in which the neutral products formed in ion-molecule reactions have been analyzed in liquid phase radiolysis experiments (Fujisaki, et al., 1971; Koob and Kevan, 1968; Tanno and Shida, 1969; Collin and Ausloos, 1971; Tanno et al., 1968). Branched alkanes dissociate more extensively at the high densities of liquid phase experiments than do n-alkanes because of the weaker C-C bond at the branching site. The short dissociative lifetime of highly branched alkanes is especially noticeable in the case of neopentane.

Ion Structures and Isomerization Processes

Much quantitative information about structures of organic ions—particularly hydrocarbon ions—has been obtained in studies utilizing the classical method of determining ionic structure, that is by analyzing the structure of a neutral reaction product formed under radiolysis conditions. For example, the structure of the butane product formed in reaction 9 shows that the precursor ion was a branched butyl ion, and the isotopic structure demonstrates that the reactant was a t-butyl rather than an isobutyl ion. The radiolytic studies of ionic structures have recently been reviewed (Ausloos and Lias, 1972; Lias and Ausloos, 1975) and therefore will be given only a cursory treatment here.

A particularly good illustration of a radiolytic investigation of ionic isomerization processes is provided by a study in which $C_4H_8^+$ ions were generated in ethylene (Lias and Ausloos, 1971):

$$C_2H_4^+ + C_2H_4 \rightarrow C_4H_8^+ \tag{23}$$

and by ionization of cyclobutane, methylcyclopropane, 1-butene, 2-butene, and isobutene. The structures of the ions formed in these systems under different conditions were determined by analysis of the neutral products formed in the presence of added methylcyclopentane-d$_{12}$:

$$C_4H_8^+ + c\text{-}C_5D_9CD_3 \rightarrow C_4H_8D_2 + C_5D_{11}^+ \tag{24}$$

(where $C_4H_8D_2$ has the structure $CH_3CHDCHDCH_3$ if the precursor was a 2-butene ion, $CH_2DCHDCH_2CH_3$ if the precursor was a 1-butene ion, and so forth), or by the analysis of neutral products formed in the presence of added charge acceptor such as dimethylamine:

Table 4. Fragmentation of Parent Alkane Ions in the Liquid Phase

FRAGMENT	$-C-C-C-$	$-C-C-C-C-$	$-C-C-C-$ (branched)	$-C-C-C-C-$ (branched)	$-C-C-C-C-$ (di-branched)
$C_2H_4^+$	0.037	≤0.05			
$C_2H_5^+$	0.070	≤0.05			0.025
$C_3H_6^+$		~0.005			
$C_3H_7^+$		~0.024	0.29		0.093
$C_4H_8^+$				0.2	
$C_4H_9^+$				≥0.6	0.03
Parent Ion: liquid — gas —	~0.9 0.10	~0.9 0.05	~0.7 0.02	~0.2 0.00	~0.8 ~0.01

Units: Fraction of ions

273° K

$$C_4H_8^+ + A \rightarrow C_4H_8 + A^+ \qquad\qquad\qquad\qquad\qquad\qquad (25)$$

(where the structure of the neutral C_4H_8 product corresponds to the structure of the precursor ion) (Lias and Ausloos, 1971). The higher energy ions initially generated in ethylene, cyclobutane, and methyl-cyclopropane rearrange to the more stable $1\text{-}C_4H_8^+$, $2\text{-}C_4H_8^+$, and iso-$C_4H_8^+$ structures, with the relative yields of the three final product ions depending on the energy with which the $C_4H_8^+$ species originated and on the pressure. The rearranging ions could be stabilized in the $1\text{-}C_4H_8^+$ structure only at high pressures. At low pressures only the more stable $2\text{-}C_4H_8^+$ and $i\text{-}C_4H_8^+$ structures were observed with the $2\text{-}C_4H_8^+$ structure predominating under all conditions and becoming more important relative to $i\text{-}C_4H_8^+$ as the pressure is raised, indicating that the rearrangement to the $i\text{-}C_4H_8^+$ structure proceeds with a lower rate and thus probably has a greater energy requirement.

CONCLUSION

Unfortunately, within the confines of this lecture, it is impossible to give a complete survey of the radiation chemistry studies which give quantitative or qualitative information about ion-molecule reactions. We have not even mentioned, for instance, the extensive work being carried out on the pulse radiolysis of aqueous solutions, which is leading to new insights about ionic processes in aqueous media.

At the present time, most of the studies which can shed light on ion-molecule reaction mechanisms at high densities (especially in the condensed phases) have been carried out using gamma rays or high energy electron beams to generate ions, and classical spectroscopic or chemical analysis techniques to follow the course of the ion-molecule reactions. Although as a result of these investigations, we now have an elementary understanding of the effects of density and of solvation on some types of ion-molecule process, further work is required before a complete picture emerges of ion-molecule reactions at high densities.

ACKNOWLEDGEMENT

Work supported in part by the U. S. Atomic Energy Commission. The author would also like to acknowledge the constructive criticisms of Dr. Peter Ausloos, which were a great help in the preparation of this manuscript.

REFERENCES

Abramson, F. P. and Futrell, J. H. (1967). J. Phys. Chem. 71, 3791.

Arai, S. and Dorfman, L. M. (1965). J. Phys. Chem. 69, 2239.

Arai, S. and Dorfman, L. M. (1964). J. Chem. Phys. 41, 2190.

Arai, S., Tremba, E. L., Brandon, J. R. and Dorfman, L. M. (1967). Can. J. Chem. 45, 1119.

Ausloos, P. (1970). Prog. Reaction Kin. V, 113.

Ausloos, P. and Lias, S. G. (1962). J. Chem. Phys. 36, 3163.

Ausloos, P. and Lias, S. G. (1964). J. Chem. Phys. 40, 3599.

Ausloos, P. and Lias, S. G. (1965). Disc. Faraday Soc. 39, 36.

Ausloos, P. and Lias, S. G. (1965). J. Chem. Phys. 43, 127.

Ausloos, P. and Lias, S. G. (1967). Actions Chim. et Biol. des Rad. 11, 1.

Ausloos, P. and Lias, S. G. (1970). J. Amer. Chem. Soc. 92, 5037.

Ausloos, P. and Lias, S. G. (1972). Ion-Molecule Reactions, J. L. Franklin, Ed., Plenum Press, New York.

Ausloos, P., Lias, S. G. and Gorden, Jr., R. (1963). J. Chem. Phys. 39, 3341.

Ausloos, P., Lias, S. G. and Sandoval, I. B. (1963). Disc. Faraday Soc. 36, 66.

Ausloos, P., Scala, A. A. and Lias, S. G. (1967). J. Amer. Chem. Soc. 89, 3677.

Barton, H. A. and Bartlett, J. H. (1928). Phys. Rev. 31, 822.

Bohme, D. K. and Young, L. B. (1970). J. Amer. Chem. Soc. 92, 7345.

Collin, G. J. and Ausloos, P. (1971). J. Amer. Chem. Soc. 93, 1336.

Collin, G. J. and Ausloos, P. (1975). Can. J. Chem., in press.

Doepker, R. D. and Ausloos, P. (1965). J. Chem. Phys. 42, 3746.

Dorfman, L. M. (1970). Acc. Chem. Res. 3, 224.

Durup, J. (1960). Les Reactions Entre Ions Positifs et Molecules en Phase Gaseuse, Gauthier-Villars, Paris.

Eisenhut, O. and Conrad, R. (1930). Z. Elektrochem. 36, 654.

Eyring, H., Hirschfelder, J. O. and Taylor, H. S. (1936). J. Chem. Phys. 4, 479.

Eyring, E. M. and Wahrhaftig, A. L. (1961). J. Chem. Phys. 34, 23.

Fehsenfeld, F. C., Dunkin, D. B., and Ferguson, E. E. (1970). Planet. Space Sci., 18, 1267.

Ferguson, E. E., Bohme, D. K., Fehsenfeld, F. C. and Dunkin, D. B., (1969). J. Chem. Phys. 50, 5039.

Field, F. H. and Lampe, F. W. (1958). J. Amer. Chem. Soc. 80, 5587.

Fujisaki, N., Shida, S., Hatano, Y. and Tanno, K. (1971). J. Phys. Chem. 75, 2854.

Futrell, J. H. (1959). J. Amer. Chem. Soc. 81, 5921.

Futrell, J. H. (1960). J. Phys. Chem. 64, 1634.

Gall, B. L. and Dorfman, L. M. (1969). J. Amer. Chem. Soc. 91, 2199.

Good, A., Durden, D. A. and Kebarle, P. (1970). J. Chem. Phys. 52, 212.

Hayon, E. (1970). J. Chem. Phys. 53, 2353.

Hentz, R. R. and Altmiller, H. G. (1971). J. Phys. Chem. 75, 2394.

Hogness, T. R. and Harkness, R. W. (1928). Phys. Rev. 32, 784.

Hogness, T. R. and Lund, E. G. (1925). Phys. Rev. 26, 44.

Hughes, B. M. and Tiernan, T. O. (1969). J. Chem. Phys. 51, 4373.

Kebarle, P. and Godbole, E. W. (1962). J. Chem. Phys. 36, 302.

Kebarle, P., Haynes, R. M. and Searles, S. (1966). Advan. Chem. Series 58, 210.

Koob, R. D. and Kevan, L. (1968). Trans. Faraday Soc. 64, 1.

Lampe, F. W., Franklin, J. L. and Field, F. H. (1961). Prog. Reaction Kin. 1, 69.

Lias, S. G. and Ausloos, P. (1971). NBS J. Res. 75A, 591.

Lias, S. G. and Ausloos, P. (1975). "Ion-Molecule Reactions and Their Role in Radiation Chemistry", in press.

Lias, S. G., Rebbert, R. E. and Ausloos, P. (1972). J. Chem. Phys. 57, 2080.

Lind, S. C. (1961). ACS Monograph Series, Reinhold Publishing Corp. New York.

MacNeil, K. A. G. and Futrell, J. H. (1972). J. Phys. Chem. 76, 409.

Matheson, M. S. and Dorfman, L. M. (1969). The MIT Press.

Meisels, G. G., Hamill, W. H. and Williams, Jr., R. R. (1957). J. Phys. Chem. 61, 1456.

Monahan, J. R. and Stanton, H. E. (1962). J. Chem. Phys. 37, 2654.

Mund, W. (1956). Actions Chim. et Biol. des Rad. 2, 1.

Perner, D. and Dreyer, J. W. (1971). Ber. Buns. Ges. fur Phys. Chem. 75, 420.

Robinson, A. J. and Rodgers, M. A. J. (1973). J. Chem. Soc. Faraday Trans. I. 69, 2036.

Rzad, S. J. and Schuler, R. H. (1968). J. Phys. Chem. 72, 228.

Rzad, S. J., Schuler, R. H. and Hummel, A. (1969). J. Chem. Phys. 51, 1369.

Scala, A. A., Lias, S. G. and Ausloos, P. (1966). J. Amer. Chem. Soc. 88, 5701.

Shida, T. and Hamill, W. H. (1966). J. Amer. Chem. Soc. 88, 3683.

Shida, T. and Hamill, W. H. (1966). J. Amer. Chem. Soc. 88, 3689.

Sieck, L. W. and Searles, S. K. (1970). J. Amer. Chem. Soc. 92, 2937.

Stevenson, D. P. (1963). Mass Spectrometry, Chapter 13, C. A. McDowell, Ed., McGraw Hill Book Co., New York.

Stevenson, D. P. and Schissler, D. O. (1955). J. Chem. Phys. 23, 1353.

Talroze, V. L. and Ljubimova, A. K. (1952). Dokl. Akad. Nauk. 86, 909.

Tanno, K. and Shida, S. (1969). Bull. Chem. Soc. Japan 42, 2128.

Tanno, K., Shida, S. and Miyazaki, T. (1968). J. Phys. Chem. 72, 3496.

Van Ingen, J. W. F. and Cramer, W. A. (1970). Trans. Faraday Soc. 66, 857.

Vestal, M., Wahrhaftig, A. L. and Johnston, W. H. (1962). ARL Technical Documentary Report 62-426.

Vestal, M. (1968). Fundamental Processes in Radiation Chemistry, P. Ausloos, Ed., Interscience, New York.

Zador, E., Warman, J. M. and Hummel, A. (1973). Chem. Phys. Letters 23, 363.

LUMINESCENCE FROM LOW ENERGY ION MOLECULE INTERACTION

R. MARX

Laboratoire de Résonance Electronique et Ionique

Université de Paris-Sud, Centre d'Orsay, France

I. - Introduction

Luminescence from ion molecule interaction is observed when products are formed in electronically excited states from which radiative transition to some lower state can occur. If the radiative lifetime is much shorter than that of any other relaxation process, spectral analysis of the luminescence gives direct informations on the energy states of the products and therefore on the energetics of the collision.

Different processes leading to excited products may occur :

1. Excitation of one of the products without chemical change :

$$I^+ + M \nearrow (I^+)^* + M$$
$$\searrow I + M^*$$

2. Ionisation of the neutral : $I^+ + M \longrightarrow I^+ + (M^+)^* + e$

3. Charge exchange with or without dissociation of the products:

$$I^+ + M \longrightarrow (M^+)^* + I$$

4. Ion molecule reactions : $I^+ + AB \longrightarrow (IA^+)^* + B$

With energetic ion beams endothermic as well as exothermic processes are observed indicating that at least part of the translational energy of the colliding particles is converted into electronic energy. Processes 1 and 2 become important only at high kinetic energies and we shall not discuss them here. In this energy range process 4 does not occur and excitation by process 3 is very similar to electron or photon excitation.

Specific excitation by charge exchange and ion molecule reactions appears only at low kinetic energies when the interaction time becomes long enough to allow not only electronic but also nuclear rearrangments.

Most of the luminescence works during the past decade have been focussed on the low energy behavior of ion molecule interaction to which this review will be restricted. Even though it is impossible in the allowed space to give a complete review of the field, so that a choice had to be made of a few rather simple systems involving at most triatomic particles.

II. - Experimental techniques

1 - Accelerated ions. Beam techniques

Detailed description of the experimental devices may be found in the original papers but the basic principles are very similar. The ions produced by electron impact, microwave discharges or plasma sources are extracted from the source by an accelerating voltage, focussed and mass analysed. Then the energy of the ion beam is ajusted to the desired value before crossing a molecular beam or entering a collision chamber filled with the neutral target gas. Light emission coming from the interaction zone is usually observed at right angles to the ion beam.

Recent improvements in getting low energy ion beams (down ᴗ2eV) of high enough intensity and in detecting low intensity light emissions have been of great help to lower the available energy limit (Brandt, D. (1973) ; Hughes, B. M. (1973)).

2 - Thermal energy ions

Thermal energy ions are usually studied in flowing afterglow tubes (Schmeltekopf, A. L. (1968); Albritton, D. L. (1973)). Recently a new technique using a pulsed ICR cell with an optical detection device (Marx, R. (1974)) has been adapted to study ion molecule reactions.

In flowing afterglows the ions are produced in a DC discharge and the mass selection is performed by adding a quenching gas which removes the unwanted excited or ionised species. A rather high pressure is used (of the order of 1 torr) which provides a good thermalisation of the ions before reaction.

In the ICR cell the ions are produced by a pulsed low energy electron beam (~ 50eV). The trapping potential is also pulsed in such a way that only the light coming from reactions of charged species is recorded. At the present stage there is no mass selection of the reactant ions.

The main differences with flowing afterglow are a lower pressure ($\sim 5 \ 10^{-4}$ to $\sim 10^{-2}$ torr), a controlled reaction time and the possibility in some cases to have a mass selection and to vary the ions kinetic energy up to about 2 eV.

III. - Charge exchange excitation

There is a general discussion on low energy charge exchange elsewhere in this book so that we shall only discuss here the informations obtained in luminescence experiments.

Table I gives the recombination energy for some projectile ions and the energy required to ionise a few diatomic and triatomic target molecules in different electronic states (the energies are given for $v' = 0$).

This table is very useful to determine the energy balance or energy defect ΔE of any charge exchange reaction in order to discuss the validity of Rapp and Francis resonance theory i.e. increasing cross sections with decreasing ΔE.

We shall now first give a brief survey of the results published for some of the molecules listed in table I. Then try to see if some general rule may be outlined.

1 - N_2

N_2^+ has certainly been the most extensively studied ion because of its aeronomical interest and also its ease of production and optical observation. Schematic potential curves of N_2^+ are given in Fig. 1 (Gilmore (1965)).

Table I

I^+ Projectile \ Product M^+	N_2^+	CO^+	HCl^+	HBr^+	CO_2^+	N_2O^+	eV
He^+	$C^2\Sigma_u^+$ 24.3						24.46
Ne^+							21.5
		$B^2\Sigma^+$ 19.6					
	$B^2\Sigma_u^+$ 18.84						
					$B^2\Sigma_u^+$ 18.08		
					$A^2\pi_u$ 17.23		
		$A^2\pi$ 16.4	$A^2\Sigma^+$ 16.26			$A^2\Sigma^+$ 16.4	
Ar^+	$X^2\Sigma_g^+$						15.76
				$A^2\Sigma^+$			15.6
N_2^+	15.6			15.31			
$Kr^+(1/2)$							14.67
$Kr^+(3/2)$		$X^2\Sigma^+$					14
H^+		14			$X^2\pi_g$ 13.68		13.5
						$X^2\pi_i$ 12.9	
			$X^2\pi_i$ 12.74				
Xe^+							12.13
				$X^2\pi_i$ 11.67			

Although there are many excited states of N_2 which may deexcite by radiation only the second negative :

$$C^2\Sigma_u^+ \longrightarrow X^2\Sigma_g^+$$

and the first negative :

$$B^2\Sigma_u^+ \longrightarrow X^2\Sigma_g^+$$

systems have been clearly identified in charge transfer reactions. However some additional lines have been observed some of which may be due to the Janin D'Incan system ($D^2\pi_g \longrightarrow A^2\pi_u$) (Holland, R. F. (1971)).

$N_2^+(C)$ has been produced by charge transfer with He^+, Ne^+ and Ar^+.

Reaction III. I : $He^+ + N_2 \longrightarrow N_2^+(C) + He$ provides an almost unique example of resonant charge transfer. Its cross section and vibrational level population has been determined in a very large energy range for the three isotopes $^{14}N_2 - {}^{15}N_2 - {}^{14}N^{15}N$. (Holland, R. F. (1971) ; Hughes, B. M. (1973) ; Govers, T. R. (1974); Van de Rumstraat, C. A. (1973)).

At energies higher than 10 KeV excitation cross sections $\sigma(c, v)$ are energy independent and identical to the electron impact excitation cross sections. Comparing $\sigma(c, v)$ with the values calculated for a Franck-Condon excitation leads to the conclusion that levels $v' \geqslant 3$ are strongly predissociated. As an example in $^{14}N_2$ the branching ratio A_{pred}/A_{rad} is 9 for $v' = 3$ and 18 for $v' = 4$ (Van de Rumstraat, C. A. (1973)). This means that experimental radiation cross sections $\sigma_{rad}(c, v')$ have to be corrected to obtain the true excitation cross sections $\sigma_{ex}(c, v')$.

At low He^+ kinetic energies there is a sharp increase in $v' = 3$ and 4 levels population indicating that resonant charge transfer is favored. Moreover it may be seen on Fig. I that increasing internuclear equilibrium distance of neutral N_2 (polarisation by He^+) would also increase the population of $N_2^+(C, v' < 5)$.

Correcting $\sigma_{rad}(C)$ for predissociation allows calculation of $N_2^+(C)$ contribution to N_2^+ (total) which is according to Van de Rumstraat C. A. (1973) about 80 % .

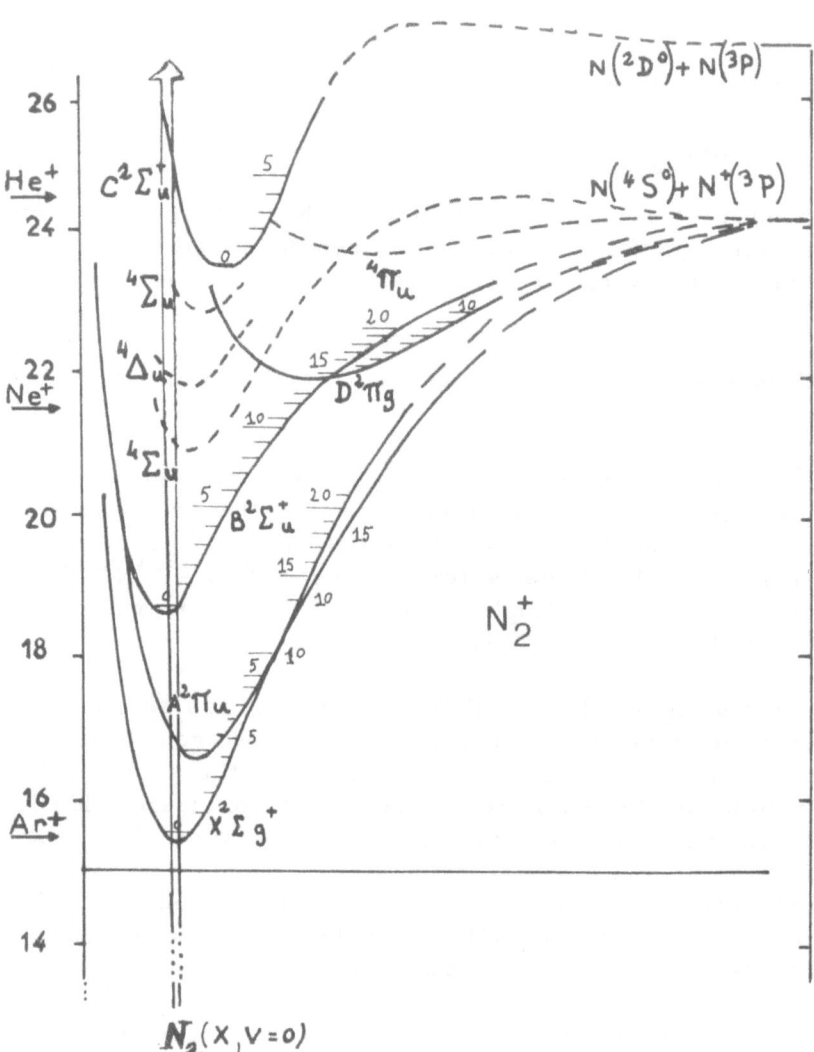

Figure 1 : Schematic potential curves taken from Gilmore

Isotope effect in the $C \longrightarrow X$ spectrum has been attributed to the decay mechanism of the corresponding C state and not to a difference in the excitation mechanism (Govers, T.R. (1974)). It may be worth noticing that with Ne^+ the high energy F.C limit is reached at lower velocities than with He^+ ($1.4 \ 10^7$ cm s^{-1}) (Van de Rumstraat, C.A. (1973)).

$N_2^+(B^2\Sigma_u^+)$ has been observed with a great number of different projectile ions : He^+, Ne^+, Ar^+, H_2^+, H^+, N^+ ... with energy defects ΔE ranging from - 5.34 eV to + 5.62 eV. With high energy ions there is a normal F.C. distribution of the vibrational levels population and the cross sections $\sigma_{tot}(B)$ as a function of ions velocity behaves according to the Massey adiabatic hypothesis.

Deviation to F.C. population are observed at low energy (Sheridan, J.R. (1965) ; Saban, G.H. (1972) ; Moore Jr, J.H. (1969) ; Fan, C.H. (1956) ; Brandt, D. (1973)). According to Doering (1969) this deviation appears at velocities $v \lessdot 10^8$ cm s^{-1} independently of the nature of the ion. Whereas high energy ions and electrons excites $N_2^+(B)$ mostly in its v' = 0 level, low energy ions excites very efficiently higher vibrational levels (Schmeltekopf, A.L. (1968) ; Moore Jr, J.H. (1969) ; Saban, G.H. (1972) ; Albritton, D.L. (1973) ; Brandt, D. (1973) ; Marx, R. (1974 b)). As a consequence the decrease observed in $\sigma(B, v' = 0)$ with decreasing energy down to 1 KeV does not mean that $\sigma_{tot}(B)$ decreases. In fact an increase of $\sigma(B, v' = 0)$ has been observed below 1 KeV. At our knowledge there is no data for the absolute value of $\sigma(B)$ in this energy range. An evaluation of $\sigma(B)/\sigma(A)$ for thermal He^+ is in progress (Marx, R. (1974 b)).

Production of $N_2^+(B)$ by charge transfer from He^+ and Ne^+ are both exothermic, however at thermal energies their behavior is very different : Ne^+ reacts only with vibrationaly excited N_2 (Albritton, D.L. (1973)) and populates exclusively the higher vibrational levels of $N_2^+(B)$ (v' = 11 to 20) producing the so called tail bands, while He^+ reacts with ground state N_2 and populates about equaly v' = 1 to 5 levels (Marx, R. (1974 b)). An excess kinetic energy is needed for Ne^+ and also Ar^+, to react with ground state N_2 and both ions produces mainly the tail bands (Brandt, D. (1973)). This behavior is qualitatively accounted for by Lipeles polarisation model (see section III. 2).

The mechanism of the two very exothermic reactions of Ne^+ and He^+ giving $N_2^+(B)$ has not been clearly established. For Ne^+,

population of the $^4\Delta_u$ state of N_2^+ followed by a radiationless transition to the near resonant high vibrational levels of the B state has
been proposed (Albritton, D. L. (1973)).

With thermal He^+ there is about 5. 5 eV excess energy to dissi-
pate and, as there seems to be no kinetic energy in the products
(Ferguson, E. E. (1974)) and no collisional deactivation of some
higher state (Marx, R. (1974 b)), a cascade process may also be
tentatively proposed. Apart the $C^2\Sigma_u^+$ state, He^+ may populate one
of the high lying quartet or doublet states of N_2^+ which then may
radiate to the B state giving some of the unidentified bands of the
spectrum. If the B state population is as low as indicated by our
preliminary results, even forbidden transitions may account for
its production.

To complete the description of charge exchange produced $N_2^+(B)$
we have to mention the abnormal excitation of high rotational levels
observed in the O - O band when low energy projectile ions are
used (Moore Jr, J.H. (1968) ; (1969)). The onset of abnormal
rotational excitation appears at about the same velocity as vibrational excitation and molecular ions seems to be more efficient
projectiles than atomic ions.

There is far less information on dissociative charge exchange
leading to excited products. N I lines have been observed with
He^+ ions of energy higher than 23 eV (Holland, R. F. (1971)). N II
lines appear with Ne^+ only above 100 eV but not with Ar^+ even at
high kinetic energies (Brandt, D. (1973)).

2 - CO

CO^+ is observed in most cases principaly in its $A^2\pi$ state (comet
tail system) (Utterback, N. G. (1965) ; Haugh, M. J. (1966) ; Lipeles,
M. (1969) ; Polyakova, G. N. (1970) ; Liu, C. (1970) ; Brandt, D.
(1973)). The first negative system coming from the $B^2\Sigma$ state is
always much less intense (Brandt, D. (1973) ; Simonis, J. (1974)).

As for N_2^+, in high energy collisions ($\geqslant 2$ KeV) cross sections
as a function of kinetic energy behave as expected on the basis of
Massey adiabatic hypothesis and there is a F. C. population of the
vibrational levels (Polyakova, G. N. (1970) ; Brandt, D. (1973)).

The effect of resonance is not clear. In some cases the resonance criterion operates and increasing σ are observed for decreasing $|\Delta E|$: with 1 KeV ions $\sigma(CO^+A)$ is higher for Ar^+ than for Kr^+ (Haugh, M. (1966)) and for Ar^+ than for Ne^+ (Brandt, D. (1973)). Moreover with 2.5 keV $Ar^+\sigma(A, v')$ increases with decreasing v' i.e. with decreasing $|\Delta E|$ (Haugh, M.J. (1972)).

The reverse is found on comparing He^+ and Ne^+ (Polyakova, G.N. (1970)) and also on comparing $\sigma(A^2\pi)$ and $\sigma(B^2\Sigma^+)$: even Ne^+ which is resonant with $CO^+(B^2\Sigma^+)$ populates mainly the $A^2\pi$ state ($\sigma(B^2\Sigma) < 1\% \, \sigma(A^2\pi)$). (Brandt, D. (1973)). Departure from simple behavior of $\sigma(v')$ as a function of energy is observed below about 1 keV but no systematic measurements have been made to find if this is an effect of velocity or of kinetic energy of the projectile ion. In the low energy range cross sections are much higher than predicted by Massey theory, they remain constant or increase with decreasing energy (Polyakova, G.N. (1970)). There is no data below 15-20 eV.

It is also worth noticing that using Ar^+ or N_2^+ which have about the same recombination energy (see table 1), cross sections are equal with 1 keV ions, whereas with 30 eV ions N_2^+ is 10 times more efficient than Ar^+ (Liu, C. (1970)).

Deviation of the vibrational level population of $CO^+(A^2\pi)$ from the Franck-Condon values has been first interpreted by Lipeles as due to a polarisation induced distortion of the ground vibrational state of the target molecule by the presence of an ion (Lipeles, M. (1969)).

Although the calculation of the distorted ground state potential curve is rather crude, there is a fairly good fit between calculated and experimental results both for CO and for N_2. This simple model accounts quite well for the opposite deviation to the F.C. values observed in the two molecules.

The same qualitative agreements is obtained for CO^+ in its A and B state. The internuclear equilibrium distance being 1.243 Å for the A state and 1.168 Å for the B state, an increase of the equilibrium distance of the neutral CO (ρ_o = 1.128 Å) due to polarisation would shift the vibrational level population toward low v' for $CO^+(A)$ and high v' for $CO^+(B)$ (Fig. 2). This has been observed recently with low energy Ar^+ and Ne^+ on CO (Simonis, J. (1974)).

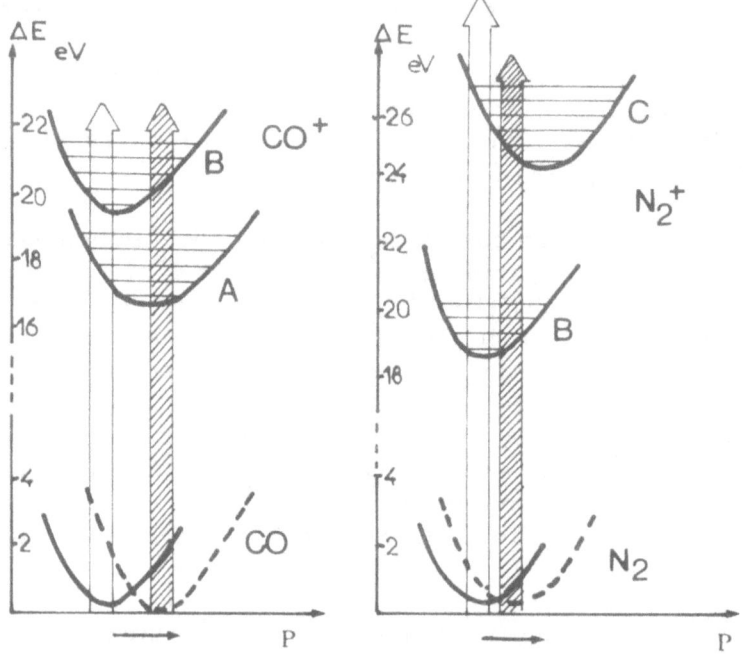

<u>Figure 2</u> : Lipeles polarisation model

F. C excitation from undistorted molecules

F. C excitation from polarisation distorted molecules

3 - HCl - HBr

Two papers have been published concerning the cross sections for the formation of the various vibrational levels of HCl^+ and HBr^+ in their $A^2\Sigma^+$ state produced by charge exchange from Ar^+ and Kr^+. One at constant collision energy (2500 eV) to investigate the cross sections variation as a function of $|\Delta E|$ (Haugh, M.J. (1972)), another varying the projectile ion velocity to test the validity of a simple resonance theory (Tomcho,L. (1972)).

With 2.5 keV, Ar^+ relative cross sections as a function of $|\Delta E|$ are not too far out of the curve calculated using Rapp and Francis theory. The reversal of the vibrational levels population for

$HBr(A^2\Sigma^+)$ when changing the projectile ion from Ar^+ to Kr^+ is also in qualitative agreement with the resonance argument.

At the highest velocities used for the projectile ions ($v \sim 10^7$ cm s^{-1}) relative cross sections seem to approach the values predicted from the F.C. factors.

In the velocity range investigated ($2 \cdot 10^6$ to $2 \cdot 10^7$ cm s^{-1}) relative cross sections, obtained by multiplying Rapp and Francis cross sections for near resonant charge transfer by the Franck-Condon factors, are in good agreement with the experimental values for (Kr^+, HBr) acceptable for (Ar^+, HCl) and very bad for (Ar^+, HBr).

This result is explained in terms of a competition between the effect of low ΔE (resonance) and large F.C. factors between the potential curves of the distorted molecule and the excited ion.

In low velocity collisions the ion-dipole interaction which tends to increase the equilibrium internuclear distance in the neutral HCl or Hbr molecule (Lipeles, M. (1969)) increases the overlap with the v' = 0 level of the excited ion. Resonance also favors v' = 0 for the (Ar^+, HCl) and (Kr^+, HBr) systems but not for (Ar^+, HBr).

However the authors conclude that this explanation is only qualitative and that one has probably to consider the molecular orbitals of the transient triatomic ions.

Production of excited HCl^+ and HBr^+ by collision with He^+ seems to be a two step process where He^+ reacts with HCl or HBr to form a reactive intermediate that subsequently reacts with a second molecule to produce the ion in its $A^2\Sigma$(v' = 0) state (Haugh, M.J. (1972)).

4 - CO_2

Electronically excited CO_2^+ is produced in its $A^2\pi_u$ and $B^2\Sigma_u$ state with H^+, He^+, Ne^+ (Haugh, M.J. (1974)) ; Coplan, M.A. (1973)), Ar^+, N_2^+ (Liu, C. (1966)). The ratio $\sigma(B)/\sigma(A)$ depends on the nature of the projectile ion and its kinetic energy (Haugh, M.J. (1974)). With He^+ this ratio is constant (~ 0.4) from 0.2 keV up to 10 keV and with Ne^+ it increases with increasing energy from 0.07 at 0.2 keV to 0.4 at 3 keV. Cross sections as a function of energy has only been studied for He^+ where

$\sigma(A)$ remains constant up to 5 KeV ($v \sim 5 \ 10^7$ cm s^{-1}) then starts increasing and for N_2^+ where $\sigma(A)$ increases by a factor of 10 between 0.025 keV ($\sim 10^6$ cm s^{-1}) and 1 keV ($\sim 10^7$ cm s^{-1}). These results do not show any correlation between cross sections and energy defect :

- B state is always less populated than A state as well with He$^+$ and Ne$^+$ ($\Delta E(B) \langle \Delta E(A)$) as with Ar$^+$ N$_2^+$ and H$^+$ ($\Delta E(B) \rangle \Delta E(A)$)

- σ_A measured for 0.9 - 1 KeV projectile ions ($v \sim 3 \ 10^7$ - $7 \ 10^7$ cm s^{-1}) have the following values :

$$He^+ : \quad \sigma = 0.35 \ \text{\AA}^2 \quad ; \ \Delta E = 7.28 \ eV$$
$$Ne^+ : \quad \sigma = 0.035 \ \text{\AA}^2 ; \ \Delta E = 4.26 \ eV$$
$$Ar^+ : \quad \sigma = 0.15 \ \text{\AA}^2 \quad ; \ \Delta E = -1.47 \ eV$$
$$N_2^+ : \quad \sigma = 1.2 \ \text{\AA}^2 \quad ; \ \Delta E = -1.63 \ eV$$

Deviation from F.C. population of the vibrational levels has been observed only with Ne$^+$ at energies below 1 keV (Haugh, M.J. (1974)). In general no deviation from F.C. population is observed for triatomic ions (CO_2^+, N_2O^+, CS_2^+) even if resonance would favor higher levels. It has also been pointed out (Haugh, M.J. (1974)) that for the reaction of He$^+$ and Ne$^+$ with CO_2, dissociative charge transfer to form $CO^+(B^2\Sigma^+)$ and $CO^+(A^2\pi)$ respectively is a nearly resonant process, however $\sigma(CO^+)/\sigma(CO_2^+)$ is always less than 1%.

5 - Common features of the low energy luminescent ion-molecule charge exchanges

As a conclusion of this review it is clearly impossible to outline any simple rule concerning charge exchange excitation in low and intermediate energy collisions even for diatomic molecules.

However there are a few common features which may be worth summarizing.

- In most cases the resonant charge exchange is not the predominant excitation channel except when there is a good vibrational overlap between the corresponding states of the ion and of the neutral molecule (high Franck-Condon factor). Otherwise vibrational overlap seems to be the dominant factor.

- Departure of the vibrational levels population from the values calculated using Franck-Condon factors are, at least qualitatively, correctly accounted for by Lipeles polarisation model (polarisation distorsion of the neutral prior to ionisation).

- Charge exchange cross sections stop decreasing and in some cases increase when the kinetic energy of the collision decreases, which indicates that the charge transfer mechanism probably changes also.

In this velocity range there must be a strong interaction between the colliding particles allowing polarisation induced distortion of the neutral by the ion or even formation of an intermediate complex state. In this case charge exchange would be very similar to ion molecule reaction and could be interpreted in the same way using correlation diagrams provided the molecular orbitals of the inter-mediate ion were known.

It has also to be mentioned that the theoretical calculations using a statistical phase space model which have been made to predict electronic, vibrational and rotational populations distribution of the ions produced in charge exchange reactions did not give very good results (Wolf, F.A. (1966) ; Saban, G.H. (1972) ; Moran, T.F. (1972)) especially for the (He^+, N_2) system.

IV. - <u>Ion molecule reactions</u>

There are, as far as I know, only two examples of true chemi-luminescent ion molecule reactions and they have both been obser-ved in Gottingen by Ottinger and coworkers (Brandt, D. (1973) ; Appell, J. (1974)

$$N^+ + NO \longrightarrow N_2^+(B\,^2\Sigma_u^+) + O(^3P)$$

$$C^+ + H_2 \longrightarrow CH^+(\tilde{A}\,^1\pi) + H\,^2(S)$$

Reaction of C^+ with NO giving CN has been shown to be a two step process, the chemiluminescent step beeing a neutral reaction : $C^* + NO \longrightarrow CN^* + O.$

Reaction $N^+ + N_2 \longrightarrow N_2^+(B) + N$ seems to proceed at least par-tly via an ion molecule reaction, i.e. a reaction where there is an N atom exchange (Brandt, D. (1974)).

These reactions have a similar behavior :

a) They occur only at low center of mass kinetic energy ($E_{CM} \leqslant$ 15 -20 eV) and the cross sections as a function of collision energy has a maximum at around 5-6 eV. This indicates that the interac-tion time must be long enough to allow nuclear rearrangments and long lived complex intermediates are possibly formed.

b) Threshold energies are of the order of 3-4 eV. So that slightly endothermic as well as exothermic pathways are possible depending on the existence of an activation barrier to reach the intermediate state.

The energy balance of reactions IV. 1 and 2 corresponding to ground state neutral and various possible states of the reactant ions are given below

	$\Delta E(eV)$		$\Delta E(eV)$
$N^+ (^3P)$	0.94	$C^+(^2P)$	3.4
(^1D)	- 0.94	$C^+(4P)$	- 1.9
(^1S)	- 3.1		

Orbital correlation diagrams may be used to interpret these reactions. To do this one needs to know not only the energy state of the reactants and products but also those of the complex inter-mediate state.

An approximate diagram has been proposed for reaction IV. 2 which leads to the conclusion that the electronic transition do not occur on an adiabatic potential hypersurface but at an avoided crossing (Appell, J. (1974)). C^+ is supposed to be in its $^2P°$ state because of the good fit between the corresponding reaction endo-thermicity and the threshold energy for the $CH^+(\tilde{A}\,^1\pi)$ emission.

Conclusion

The field of luminescent ion molecule reactions is in its very ° begining and it is difficult to guess its future developments.

The occurence of a given reaction is very hard to predict : energy considerations are of little use because one does not know how much translational energy may be converted into electronic energy and spin or symetry rules are not always obeyed. Even simple correlation diagrams cannot usually be constructed because the electronic states available for the reaction intermediate are not known. A great deal of experimental and theoretical work is needed to try many possible systems and to calculate the relevant energy states.

References

D. L. Albriton, Y. A. Bush, F. C. Fehsenfeld, E. E. Ferguson, T. R. Govers, M. McFarland and A. R. Schmeltekopf (1973). J. Chem. Phys. 58, 4036.

J. Appell, D. Brandt and Ch. Ottinger (1974). to be published.

D. Brandt, Ch. Ottinger and J. Simonis (1973a). Bericht 119/1973 Max Planck Institute Gottingen.

D. Brandt and Ch. Ottinger (1973b). Chem. Phys. Lett. 23, 257.

D. Brandt (1974). private communication.

M. A. Coplan and J. E. Mental (1972). J. Chem. Phys. 58, 4912.

C. Y. Fan (1956). Phys. Rev. 103, 1740.

E. E. Ferguson (1974). private communication.

T. R. Govers, F. C. Fehsenfeld, D. L. Albriton, P. G. Fournier and J. Fournier (1974). Chem. Phys. Lett. 26, 134.

F. R. Guilmore (1965). J. Quant. Spect. Rad. Transfer 5, 369.

M. J. Haugh, T. G. Slanger and K. D. Bayes (1966). J. Chem. Phys. 44, 837.

M. J. Haugh and K. D. Bayes (1970). Phys. Rev. A, 2, 1778.

M. J. Haugh (1972). J. Chem. Phys. 56, 4001.

M. J. Haugh and J. H. Birely (1974). J. Chem. Phys. 60, 264.

R. F. Holland and W. B. Maier, II (1971). J. Chem. Phys. 55, 1299.

B. M. Hughes, E. G. Jones and T. O. Tiernan (1973). VIII ICPEAC, 223.

M. Lipeles (1969). J. Chem. Phys. 51, 1252.

C. Liu and H. P. Broida (1970). Phys. Rev. A, 2, 1824.

J. H. Moore, Jr. and J. P. Doering (1968). Phys. Rev. 174, 178.

J. H. Moore, Jr. and J. P. Doering (1969a). Phys. Rev. 177, 218.

J. H. Moore, Jr. and J. P. Doering (1969b). Phys. Rev. 182, 176.

R. Marx, G. Mauclaire, M. Wallart and A. Deroulede (1974a). Advan. in Mass Spectrometry 6, 735.

R. Marx, G. Mauclaire, S. Fenistein and C. A. Van de Rumstraat (1974b). to be published.

T. F. Moran and D. C. Fullerton (1972). J. Chem. Phys. 56, 21.

G. N. Polyakova, V. F. Erko, A. V. Zats, Ya. M. Fogel' and G. D. Tolstolutskaya (1970). Zh ETF Pis. Red. II 562.

W. W. Robertson (1966). J. Chem. Phys. 44, 2456.

A. L. Schmeltekopf, E. E. Ferguson and F. C. Fehsenfeld (1968). J. Chem. Phys. 48, 2966.

J. R. Sheridan and K. C. Clark (1965). Phys. Rev. 140, A, 1033.

G. H. Saban and T. F. Moran (1972). J. Chem. Phys. 57, 895.

J. Simonis (1974). private communication.

L. Tomcho and M. J. Haugh (1972). J. Chem. Phys. 56, 6089.

N. G. Utterback and H. P. Broida (1965). Phys. Rev. Lett. 15, 608.

C. A. Van de Rumstraat (1973). Thesis, Amsterdam.

F. A. Wolf (1966). J. Chem. Phys. 44, 1619.

PHOTODISSOCIATION OF GAS-PHASE IONS

Robert C. Dunbar[*]

Chemistry Department, Case Western Reserve University

Cleveland, Ohio 44106

The study of photodissociation processes in gas-phase ions
dates back at least a decade, but recently there has been wide-
spread growth of interest in the chemistry and spectroscopy of
photodissociation, and in the photochemistry of complex ion-con-
taining systems under irradiation. (The photochemistry of solu-
tion-phase ions has had a parallel rise of interest [Cabell-Whiting,
1973], but thus far this work has had negligible interaction with
gas-phase work.) Substantial amounts of old and new results on ion
photodissociation were presented at the Biarritz Conference,
chiefly at the Photodissociation Panel; however, a comprehensive
overview of the field was not given. Since the field has never been
reviewed, it seems useful to give a somewhat broader picture of the
field here than was given at the Conference.

Among the diverse motivations behind the recent interest in
photodissociation of gas-phase ions, a few of the important ones
are these: first, the presence of a charge on one of the species
involved makes the study of ion photodissociation far easier than
the heroically difficult neutral-molecule photodissociation experiments [Busch, 1972; Diesen, 1971], and suggests small ions as much
more convenient systems for the detailed study of photodissociation
dynamics; second, the capability in photodissociation for depositing
a known amount of energy in an ion is attractive for unimolecular
decay studies; third, the wavelength dependence of photodissociation
provides in most cases the only feasible spectroscopic approach to
study of an isolated gas-phase ion; fourth, selective photodissociation is useful for producing state-selected ion populations; and
finally, such dissociation processes may be important in some at-

* Alfred P. Sloan Fellow, 1973-75

mospheric and interstellar environments. Photodissociation occurs
with large cross section at readily accessible wavelengths for such
a variety of ions, and is detectable by so many different techniques, that it is rather surprising that substantial interest has
developed in these processes only in the very recent past, but
enough has now been accomplished that a review of the field can
have some substance. The field will be surveyed with emphasis on
three aspects: instrumental techniques; photodissociation processes
in simple ions; and photodissociation and photochemistry in complex
ion-neutral systems. Of the two modes of photodecomposition of
negative ions, only photodissociation will be discussed, since the
spectroscopy and chemistry of electron photodetachment are quite
different from the spectroscopy and chemistry of photodissociation.

INSTRUMENTAL METHODS

    The instrumental approaches to studying photodissociation
divide naturally into two types, depending on whether the photon
beam is fired at a moving ion target (crossed-beam techniques) or a
stationary ion target (trapped-ion techniques).

    The first determined photodissociation study was that of Dunn
and his coworkers [Dunn, 1964; Busch, 1972], mainly on H_2^+. Wave-
length-selected light from a xenon arc was crossed with a slow beam
of mass-analyzed ions, and the exiting beam was electrostatically
mass-analyzed to separate product ions. For the simple systems
studied, high mass resolution was not needed, but sensitivity was
good, with broadband cross sections of the order of 10^{-20} cm^2 being
detectable.

    Durup and coworkers [Ozenne, 1972], in an experiment of widely
acknowledged elegance, crossed a fast H_2^+ beam with a pulsed polar-
ized laser beam and used magnetic sector analysis of the product ion
mass and momentum. For in-line polarization the product-ion momentum
splitting due to H_2^+ vibrational levels was readily resolved. While
this method is limited in resolution to the momentum resolution of
the mass spectrometer and does not take advantage of the high potential resolution of optical techniques, it nevertheless has interesting possibilities for studying the dynamics of the photodissociation
process. Van Asselt, Maas and Los [Van Asselt, 1973] have used a
similar technique, but a detailed description has not yet apparently
appeared.

    Vestal and Futrell [Vestal, 1974] have developed an instrument
using a high-pressure ion source, three tandem quadrupole mass
filters and particle-counting product detection. Following mass
analysis in the first quadrupole filter, the ion beam is crossed
with the light beam in the central quadrupole, and the products are

mass analyzed in the third quadrupole before detection. Sensitivity
to product-ion formation is high, and cross sections of the order of
2×10^{-20} cm^2 have been observed in the visible wavelength region.
An improved light source is expected to permit high resolution as
well.

Moseley, Bennett and Peterson [Moseley, 1974] have crossed an
intracavity argon-ion laser beam with the ion beam extracted from a
drift tube source to observe photodissociation of CO_3^-. The product
ions are mass analyzed with a quadrupole mass filter and detected.
The lack of primary beam mass analysis complicates the data analysis, but the high-pressure source capability makes this an instrument
of great promise for study of cluster ions and other high-pressure
species.

In a final crossed-beam technique, Ellefson, Denison and Weber
[Ellefson, 1972] focussed a tunable dye laser on a tightly-focussed
ion beam in the region between two quadrupole mass filters which
gave mass analysis of the primary and secondary ion beams. In this
approach a beam of ions, for instance CO_2^+, is irradiated at a
(variable) wavelength too long to dissociate ground-state ions, so
that only vibrationally excited ions dissociate. By sweeping the
wavelength it is possible in principle to map out the distribution
of vibrational states in the ions emerging from the ion source.
There is apparently not yet any detailed published description of
this interesting technique, or of results obtained from it.

There is less diversity of trapped-ion photodissociation experiments, which have, nonetheless, provided most of the known data
on ion photodissociation.

Apparently the only use of a quadrupole ion trap for this purpose was the work of Dehmelt and coworkers [Richardson, 1968], who
used selective photodissociation of H_2^+ in a quadrupole trap to
obtain a population of oriented H_2^+. An rf resonance technique
was used to determine the H_2^+ density, and substantial polarization
of the population was achieved.

The ion trapping capability of the ion cyclotron resonance
spectrometer has been extensively exploited for photodissociation
studies. (For recent surveys of the ion cyclotron resonance technique, see [Baldeschwieler, 1971; Beauchamp, 1971; Dunbar, 1974A].)
Imposition of a relatively high potential on the trapping plates of
an ICR cell creates a set of closed equipotentials in the cell along
which ions may drift indefinitely without leaving the cell. Our
group has exploited this effect in a standard ICR cell run with high
trapping potentials, and we have demonstrated the capability of
trapping ions for times in excess of five minutes. McIver [1973]
and others have used a trapped ion cell with closed ends in pulsed

ICR work. Beauchamp and his coworkers have developed an ingenious
use of a standard cell configuration in which ions are trapped and
photolyzed in the source region, and are subsequently drifted into
the analyzer for ICR detection.

The ions trapped in any of these configurations are irradiated
with a light source, and the decrease in primary ions and the
increase in photodissociation product ions are followed by ICR detection methods. Subsequent ion-molecule reactions of the photodissociation products may also be followed. The powerful capabilities of ICR double resonance methods for separating out the various
concurrent and consecutive processes occurring in a complex photochemical system are indispensable, and account for the attractive-
ness of ICR methods in studying complex photochemistry.

Xenon arc light sources with interference filters or monochromators have been used for most ICR work. Use of a flashlamp
was shown to be feasible but not very promising [Kramer, 1974].
Tunable dye lasers are now coming into use: we reported the first
such experiment [Dymerski, 1974] and Eyler [1974] has demonstrated
the advantage of putting the ICR cell intracavity.

The ICR instrumentation offers a means of examining the polarization dependence of photodissociation [Dunbar, 1973A]: taking advantage of the anisotropy of the ICR trap, which allows much easier
escape of fast photodissociation fragments from the cell along the
magnetic field direction, it is possible to determine at least
qualitatively the preferred direction of the optical plane of polarization relative to the molecular axes.

The ICR trapped-ion approach to studying photodissociation
makes possible a potentially valuable technique which might be
termed "selective depletion." If a population of ions of a given
mass is made up of more than one ion species having photodissociation absorption peaks at different wavelengths, then by strong
irradiation at an appropriate wavelength it is possible to select-
ively remove all of a chosen species from the ion population. This
concept has so far been used only in the application described
below with benzyl chloride [Fu, 1974], in which selective depletion
was used to demonstrate the presence of more than one ion species.
However, the ion population remaining after depletion of one species
is in principle available for ion-molecule reaction study or other
purposes, and such experiments will undoubtedly be reported soon.

PHOTODISSOCIATION OF SIMPLE IONS

For the purpose of our rather arbitrary classification, a
"simple" ion has no more than four or five atoms, and is simple

enough that one can with confidence characterize the low-lying
excited states from calculations, photoelectron spectra, intuition,
or other sources.

Dunn and coworkers [Busch, 1972] have extensively characterized the cross sections and vibrational structure in H_2^+. While a
high degree of theoretical sophistication seems necessary to obtain
quantitative agreement with the experiment, this system is probably
well understood and should present few new surprises.

Methyl chloride ion, CH_3Cl^+, has been examined several times
[Dunbar, 1971; Dunbar, 1973A; Vestal, 1974]. The peak near 3400Å
has not yet shown resolved structure; while the peak cross section
is not yet known with confidence, it is probably within a factor of
two of 4×10^{-18}. On energetic grounds the peak could correspond
to excitation to either of the known low-lying ^2E or $^2A_1$ states (or
to both), while the angular dependence of the photodissociation
(Dunbar, 1973A] shows that the transition moment is parallel to the
C-Cl bond and suggests predominant involvement of the ^2E excited
ion state. Vestal and Futrell [1974] find photodissociation at
longer wavelengths than any excited ion state could account for,
and suggest that direct excitation to the dissociated continuum may
occur. They also report a spectrum for CH_3Br^+ which shows evidence
of two resolved peaks near 3000Å and 4000Å.

Our early report [Dunbar, 1970] of photodissociation of N_2O^+
in the visible may have been in error, possibly due to a thermal
effect with the broadband light source then in use. This ion has a
significant photodissociation cross section in the ultraviolet
[Orth, 1974].

Dunn [1964] observed dissociation of N_2^+ with a cross section,
averaged over his xenon arc spectrum, of 1.7×10^{-20} cm^2. He also
set an upper limit of 10^{-20} cm^2 on the average cross section for H_3^+.

COMPLEX IONS - SPECTROSCOPIC ASPECTS

For all but a few very simple ions, there is little prospect of
obtaining useful gas-phase spectroscopic information by conventional
techniques, and photodissociation provides an indirect spectroscopic approach of very great value. The wavelength dependence of
the parent ion disappearance is commonly referred to as the photodissociation spectrum of the ion, and is a composite, with the
intensity at a given wavelength being determined both by the optical
absorption cross section and by the probability of subsequent dissociation of the excited ion.

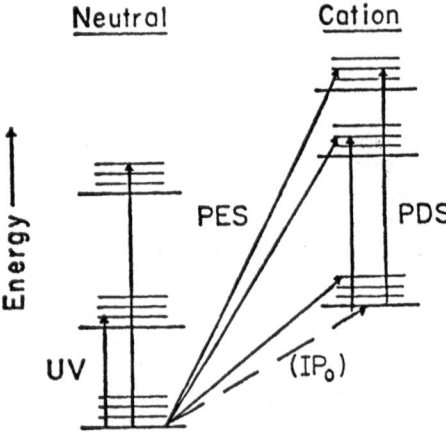

Fig. 1. Schematic comparison of molecular processes involved in
photodissociation and photoelectron spectroscopy.

     The most useful point of comparison in interpreting photodis-
sociation spectra has turned out to be the photoelectron spectrum
(PES spectrum) of the corresponding neutral molecule. As Fig. 1
indicates, the photoelectron spectrum gives peaks corresponding to
excitation from the neutral ground state to the ground or excited
states of the cation, while the photodissociation spectrum indicates excitations from the cation ground state to excited ion
states. Therefore the peak energies from these two techniques are
closely comparable, bearing in mind three points: (1) vibrational
effects (Franck-Condon factors) may differ for the two processes,
leading to peak shifts of tenths of an eV; (2) photoelectron
spectroscopy will not usually show excited cation states involving
electron excitation to a previously unoccupied orbital, while such
states may give photodissociation peaks; (3) photoelectron spectroscopy is much less constrained by selection rules and transition
probabilities and should generally show many more peaks than photodissociation spectroscopy.

     Turning to specific systems: several olefin cations (propene,
four isomeric butenes) were studied early [Kramer, 1973], and show
similar photodissociation spectra (Fig. 2 is typical): photodissociation shows a gradual onset in the visible, and then rises
sharply near 3300Å to give an intense UV peak. The first onset
probably corresponds to the lowest excited cation state from photoelectron spectroscopy, and the UV peak was tentatively proposed as
corresponding to a second cation excited state which appears in the
PES spectrum of four of the five ions studied. Of particular interest was the observation that the l-butene cation showed a dis-

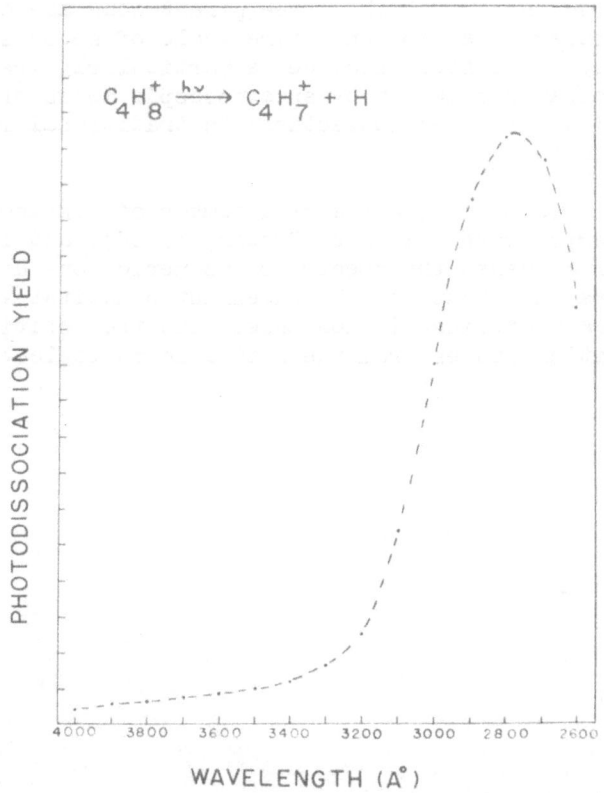

Fig. 2. The photodissociation spectrum for loss of H radical for trans-2-butene cation from 4000 to 2600 Å.

tinctly different spectrum from <u>cis</u>- and <u>trans</u>-2-butene, ruling out
complete conversion of these ions to a common structure even on a
time scale of seconds.

In a spectroscopic structure investigation of isomeric $C_7H_8^+$
cations [Dunbar, 1973B], the photodissociation spectra of the parent
ions obtained from toluene, cycloheptatriene and norbornadiene were
compared, as shown in Fig. 3. The three parent ions are obviously
different, and interconversion on a time scale of seconds is clearly not important. This study provides a particularly graphic illustration of the clean resolution by spectroscopic means of an ion-
structure question which was intractable to traditional mass spectrometric methods.

The photodissociation spectra of a number of alkylbenzenes in
the threshold region were compared [Dunbar, 1973C], and it was found
that in a number of cases the spectra of isomeric ions differed significantly. More important was the excellent quantitative agreement
found between the photodissociation onsets and the positions of the
first excited cation states from the available photoelectron

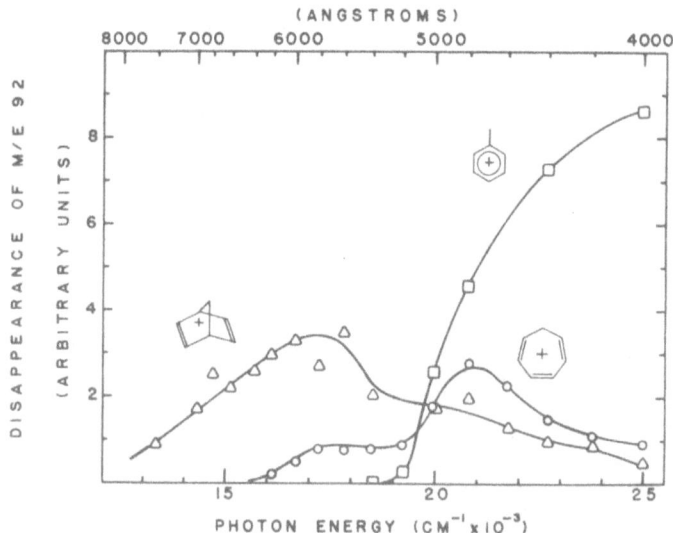

Fig. 3. Relative photodissociation rates of $C_7H_8^+$ ions as a
unction of wavelength for toluene (□), cycloheptatriene (o), and
norbornadiene (Δ). The three curves were normalized to a common
(arbitrary) vertical scale and may be compared directly.

spectra. Historically, these results served to give confidence both in the validity of the comparison with PES, and in the assumption (or hope) that the parent cations retained the neutral structure for the substituted benzenes.

An extensive series of substituted-benzene spectra in the visible and UV has recently been reported [Dymerski, 1974B], and compared with PES spectra. From these it is clear that the PES spectrum does indeed display many more peaks than the photodissociation spectrum. The photodissociation spectra are all qualitatively similar, showing a visible peak and a UV peak; the low energy peak is invariably in good correspondence with a low-lying PES peak. Comparison with molecular orbital calculations for toluene suggests that the low-lying photodissociation peak in toluene cation, and presumably in all the substituted benzenes, arises from an allowed π-π^* transition whose position is in good quantitative agreement with the molecular orbital prediction [Dunbar, 1974D]. The nature of the UV peak in these photodissociation spectra is uncertain, and may plausibly be argued to be either a second hole promotion π-π^* transition or the first electron-promotion transition.

The laser photodissociation spectrum of toluene cation near 4000Å [Eyler, 1974] does not differ significantly from the low resolution spectrum, but there is a preliminary indication of some fine structure. The cross section for this process at 4000Å, while not yet well established,[1] is probably within a factor of two of 5×10^{-18} cm^2.

The laser photodissociation spectrum of p-$C_6H_4CH_3I^+$ [Dymerski, 1974A] does not show evidence of vibrational structure, but the peak is narrowed considerably over what had been observed with low-resolution techniques.

Although the high dissociation endothermicity for benzene cation precludes observation of the interesting region of the first excited state around 4000Å, the spectrum has been observed at shorter wavelength [Beauchamp, 1974]. A peak is observed at 4.9 eV, with a shoulder at longer wavelength. $C_6D_6^+$ peaks in the same region, but the splitting of the peak shows a different pattern, and it more pronounced.

The spectrum of protonated benzene $C_6H_7^+$ [Beauchamp, 1974] provides the first direct comparison with a reliable solution

---

[1] The preliminary value which we reported at Biarritz was erroneously high due to a computational error. The value given here is in accord with our "revised preliminary" value and with Eyler's comparison with SH$^-$ photodetachment, but is clearly not very precise and is given only since there is not yet a definitive measurement.

absorption spectrum. The gas-phase photodissociation peak at 3250Å closely matches the solution spectrum.

The spectrum of parent ions obtained from benzyl chloride provides an interesting comparison with the chlorotoluenes [Fu, 1974]. The benzyl chloride case shows two peaks in the visible, one close to the chlorotoluene peak at 4400Å, and an additional longer wavelength peak near 6000Å. Selective depletion experiments with irradiation at either one of the two peak wavelengths in the benzyl chloride system showed the presence of two spectroscopically distinct $C_7H_7Cl^+$ species produced from benzyl chloride. It seems very likely that one of these ions is the unrearranged benzyl chloride parent ion, while the other is a rearranged ion of uncertain structure.

The $C_3F_5^+$ ion from perfluoropropene shows a sharp photodissociation peak at 2700Å [Freiser, 1974], which has been tentatively assigned as a π-π^* transition.

The photodissociation of negative ions has been investigated only for the CO_3^- system [Moseley, 1974] and for the transition-metal carbonyl anions. Photodissociation of CO_3^- was reported over the region 4500-5300Å. The largest cross section was about 1.2 x 10^{-18} cm^2, which was observed at both ends of the wavelength range, while there was a minimum of about 0.5 x 10^{-18} cm^2 near 5000Å. $CO_3^-\cdot H_2O$ had an approximately constant photodisappearance cross section of ∿1 x 10^{-18} cm^2 over the range 4800-5200Å.

The photodisappearance spectra of several transition metal carbonyl anions were obtained quite early [Dunbar, 1972]; however, only recently were the photochemical complications of these systems overcome sufficiently to be confident that, for Fe $(CO)_4^-$ at least, the process observed was in fact photodissociation, with spectrum as shown in Fig. 4 [Richardson, 1974]. An extensive comparison of the peak positions of the strong photodisappearance peaks for species $M(CO)_n^-$ has been made [Dunbar, 1974B]; it was concluded that these peaks are probably of charge-transfer type. The peaks shift consistently to higher energy either with an increase in atomic number of the metal or with an increase in n (Fig. 5), as expected for a charge-transfer process.

COMPLEX IONS - PHOTOCHEMICAL ASPECTS

Full elucidation of the photochemical processes in some of the systems discussed is not a simple undertaking because of the multitude of simultaneous dissociations and reactions which often take place in a trapped-ion experiment. This is both an asset and a liability of the trapped-ion approach, and the more straightforward

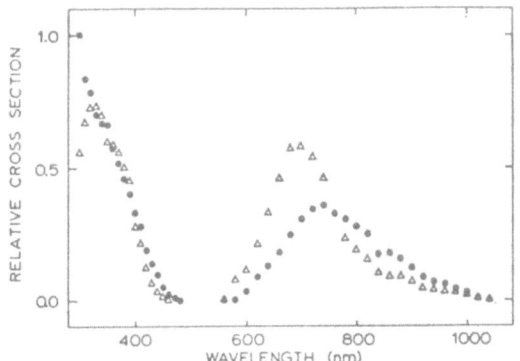

Fig. 4. Relative photodissociation cross sections. (●)Fe(CO)$_4^-$
disappearance: data below 550 nm, 9.6 nm band width (fwhm), average
of eight runs; data above 550 nm, 39.6 nm fwhm, seven runs, multiplied x 10. (Δ)Fe(CO)$_3^-$ appearance: data below 550 nm, 14.1 nm
fwhm, nine runs; data above 550 nm, 59.1 nm fwhm, four runs, multiplied x 10. The Fe(CO)$_3^-$ and Fe(CO)$_4^-$ curves are normalized
relatively so as to minimize the variance. All standard deviations
∿ ± 10%.

results from ion-beam technique should be valuable aids to understanding the "photochemical soup" produced in trapped-ion experiments.

The primary photodissociation products appear in most cases to
be the most stable products. In only a few cases does there appear
to be branching to more than one product: reported cases of multi-
ple-product formation include butene ions [Kramer, 1973] and possibly sec-butyl-benzene [Dunbar, 1973C]. Beauchamp [1974] has
noted the apparent generalization that the principal photodissociation product is also the most intense metastable mass-spectrometric
decomposition.

Ion-molecule reaction of the photodissociation product ion with
parent neutrals is common. The reactions generally follow expected
patterns and are not in themselves very interesting. Beauchamp
[1974] has noted the amusing photochemical cycle

$$C_3F_5^+ \xrightarrow{h\nu} CF_3^+ + C_2F_2$$
$$\uparrow \quad C_3F_6 \quad \downarrow$$
$$-CF_4$$

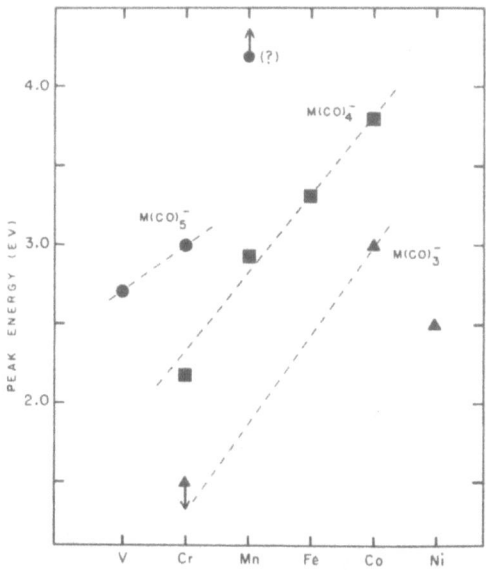

Fig. 5. Plot of optical absorption peak energy derived from threshold photodissociation curves. Tricarbonyl anions (▲), tetracar-
bonyl anions (■), and pentacarbonyl anions (●) are indicated, and
the dashed lines represent a guess as to the trend in each series.

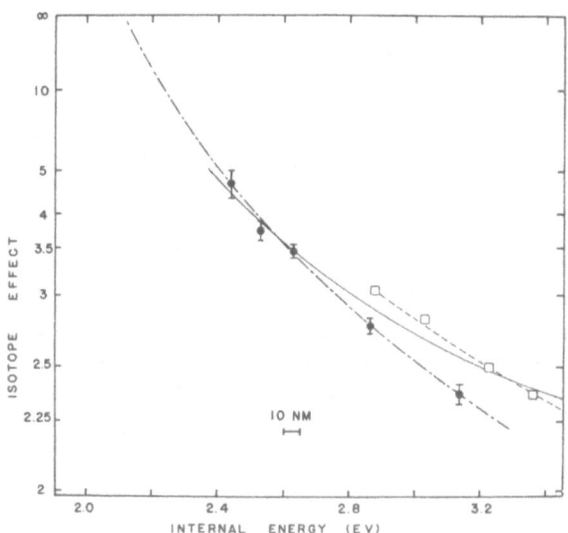

Fig. 6. Isotope effect for $C_6D_5CH_3$ vs. cation internal energy from
the ICR photodissociation experiments (—●———) compared with mass
spectrometric results (-- □ --) and quasi-equilibrium theory calculations (solid line).

which has the net effect of a photochemical conversion of C_3F_6 to CF_4 plus C_2F_2. The reactive processes can serve as a probe of the structure of the photodissociation product ion. This approach was used [Dunbar, 1974C] to study the product-ion structure from the photodissociation of toluene cation in the sequence

$$C_7H_8^+ \xrightarrow[-H]{h\nu} C_7H_7^+ \xrightarrow[-C_6H_6]{C_7H_8} C_8H_9^+ .$$

It was found that only a fraction of the photoproduced $C_7H_7^+$ was reactive, and that this fraction was wavelength dependent. These results were interpreted as showing competitive formation of tropylium and benzyl forms of $C_7H_7^+$.

The toluene cation is the only case for which careful isotopic labelling work has been carried out [Dunbar, 1973D]. Photodissociation of deuterium-labelled $C_7H_8^+$ to give $C_7H_7^+$ + H shows equal probability of losing a ring or methyl hydrogen. There is, however, a strong hydrogen-deuterium isotope effect shown in Fig. 6, which is in quantitative agreement with the isotope effect found for electron-impact fragmentation, and also with that calculated from quasi-equilibrium theory. As Fig. 6 shows, the photodissociation isotope effect may be extrapolated to give a rather precise value for the dissociation endothermicity, and this is an approach which seems likely to be valuable in thermochemical studies.

Acknowledgments: Support is gratefully acknowledged from the National Science Foundation and from the donors of the Petroleum Research Fund, administered by the American Chemical Society.

REFERENCES

1. N. Van Asselt, J. Maas and J. Los (1972). Private communication cited by J. Durup, 21st Annual Conference on Mass Spectrometry and Allied Topics, San Francisco, May 20-25.

2. J. D. Baldeschwieler and S. S. Woodgate (1971). Accts. Chem. Res. 4, 114.

3. J. L. Beauchamp (1971). Ann Rev. Phys. Chem. 22, 527.

4. J. L. Beauchamp (1974). Presented at the NATO Advanced Study Institute on Ion-Molecule Reactions, Biarritz, France, June 24-July 6.

5.   F. von Busch and G. H. Dunn (1972). Phys. Rev. A $\underline{5}$, 1726.

6.   G. E. Busch and K. R. Wilson (1972). J. Chem. Phys. $\underline{56}$, 3626, 3638, 3655.

7.   P. W. Cabell-Whiting and H. Hogeveen (1973). Advan. Phys. Org. Chem. $\underline{10}$, 129.

8.   R. W. Diesen, J. C. Wahr and S. E. Adler (1971). J. Chem. Phys. $\underline{55}$, 2812.

9.   R. C. Dunbar (1971). J. Am. Chem. Soc. $\underline{93}$, 4354.

10. R. C. Dunbar (1972). 163rd National Meeting of the American Chemical Society, Boston, Mass., April 9-14.

11. R. C. Dunbar and J. M. Kramer (1973A). J. Chem. Phys. $\underline{58}$, 1266.

12. R. C. Dunbar and E. Fu (1973B). J. Am. Chem. Soc. $\underline{95}$, 2716.

13. R. C. Dunbar (1973C). J. Am. Chem. Soc. $\underline{95}$, 6191.

14. R. C. Dunbar (1973D). J. Am. Chem. Soc. $\underline{95}$, 472 (1973).

15. R. C. Dunbar (1974A). "The Study of Gas-Phase Ions by Ion Cyclotron Resonance," in "Chemical Reactivity and Reaction Paths," (Ed. G. Klopman), Wiley-Interscience.

16. R. C. Dunbar and B. B. Hutchinson (1974B). J. Am. Chem. Soc. $\underline{96}$, 3816.

17. R. C. Dunbar (1974C). J. Am. Chem. Soc., in press.

18. R. C. Dunbar (1974D). Chem. Phys. Lett., submitted.

19. G. H. Dunn (1964). "Atomic Collision Processes," (Ed. M. R. C. McDowell). North-Holland Publishing, Amsterdam, p. 997.

20. P. P. Dymerski and R. C. Dunbar (1974A). Presented at the 22nd Annual Conference on Mass Spectrometry and Applied Topics, Philadelphia, May 19-24.

21. P. P. Dymerski, E. Fu and R. C. Dunbar (1974B). J. Am. Chem. Soc. $\underline{96}$, 4109.

22. R. E. Ellefson, A. B. Denison and J. H. Weber (1972). Presented at the 20th Annual Conference on Mass Spectrometry and Applied Topics, Dallas, June 4-9.

23. J. R. Eyler (1974). Presented at the NATO Advanced Study
     Institute on Ion-Molecule Reactions, Biarritz, France,
     June 24-July 6.

24. B. Freiser and J. L. Beauchamp, J. Am. Chem. Soc., in press.

25. E. Fu and R. C. Dunbar (1974). Presented at the 22nd Annual
     Conference on Mass Spectrometry and Allied Topics, Philadel-
     phia, May 19-24.

26. J. M. Kramer and R. C. Dunbar (1973). J. Chem. Phys. 59, 3092.

27. J. M. Kramer and R. C. Dunbar (1974). J. Chem. Phys. 60, 5122.

28. R. T. McIver, Jr. (1973). Rev. Sci. Instrum. 44, 1071

29. J. T. Moseley, R. A. Bennett and J. R. Peterson (1974). Chem.
     Phys. Lett. 26, 288.

30. R. Orth and R. C. Dunbar, to be published.

31. J. B. Ozenne, D. Pham and J. Durup (1972). Chem. Phys. Lett.
     17, 422.

32. C. B. Richardson, K. B. Jefferts and H. G. Dehmelt (1968).
     Phys. Rev., 2nd Series 165, 80.

33. J. H. Richardson, L. M. Stephenson and J. I. Brauman (1974).
     J. Am. Chem. Soc. 96, 3671 (1974).

34. M. L. Vestal and J. H. Futrell (1974A). Presented at the 22nd
     Annual Conference on Mass Spectrometry and Allied Topics,
     Philadelphia, May 19-24.

35. M. L. Vestal and J. H. Futrell (1974B), to be published.

THE DETERMINATION OF RATE CONSTANTS OF ION-MOLECULE REACTIONS.

SUMMARY OF THE PANEL DISCUSSION LED BY

G.G. Meisels

Department of Chemistry, University of Houston

Houston, Texas 77004

INTRODUCTION

The magnitude and the dependence on energy of the rate constants of ion-molecule reactions have been topics of considerable
interest both from an experimental and theoretical point of view.
They remain among the most important aspects of the study of ion-
molecule reactions with many facets as yet not known reliably nor
understood well (Henchman, 1972). This panel discussion was organized with the aim of elucidating, for each technique, the characteristics, problems, and limitations as seen by those acknowledged to
be leading practitioners of the method. In particular, participants
in the panel discussion were asked to address themselves to the
reliability of measured rate constants and to complications introduced by the possibility that a given technique may not be dealing
with ions whose translational and internal energy distributions are
adequately described by the Maxwell-Boltzman distributions defined
by the gas temperature. Moreover, the energy regime was limited to
the thermal and near-thermal (less than 1 eV) range. It is in this
energy range where most ion-molecule reactions of practical interest occur; yet most laboratory techniques of dealing with such reactions have limitations imposed on them as a result of unspecified
or unknown ion energy distributions.

When ions are not in thermal equilibrium with their surroundings, their translational and internal energies may well be different. The cross section σ for any given reaction must therefore
be expressed as a function of internal energy of the neutral E_n,
the internal energy of the ion E_i, and the relative translational
energy E_r (or relative center-of-mass velocity v_r) of the colliding
pair. It is usually assumed that the internal energies of the ion

595

and the neutral affect the outcome of the collision process indistinguishably so that one may set $E_I = E_n + E_i$ and σ becomes a function of E_I and v_r only. However, this has never been demonstrated quantitatively although differences may be existent when the reaction intermediate is very short-lived. For some exothermic reactions σ may be independent of E_I, and when ion-induced dipole forces predominate σ may then be approximated by the well-known Langevin $1/v_r$ variation. At near-thermal energies, however, absolute cross sections become difficult to measure, and the rate constant becomes the parameter whose knowledge is sought by choice because it is also the quantity usually most directly applicable to macroscopic or practical systems. Measured rate constants are, however, the convolution of the product of cross section and velocity with the appropriate distribution functions; if we do not retain the distinction of E_i and E_n in the excitation function (energy dependence by σ), the expression for the rate constant becomes formally

$$k(P_i,P_n,P_v) = \int_{v_r=0}^{\infty} \int_{E_I=0}^{\infty} \int_{E_i=0}^{E_I} \sigma(E_I,v_r) \cdot v_r \cdot P_i(E_i) \cdot P_n(E_I-E_i) \cdot P_v(v_r) \cdot dE_I dE_i dv_r \qquad [I]$$

where P_i is the probability distribution of internal energies of the ion, P_n that of the neutral, and P_v that of the relative velocities. $P_n(E_I-E_i)$ is the value of P_n at the energy E_I-E_i. A chemically conventional rate constant results when all three of these probabilities can be described by Maxwell-Boltzman distributions characterized by a single common temperature. In principle at least, when any one of these distributions does not fulfill this criterion as is commonly the case for ions, the distribution should be known and data be accurate enough to permit unfolding with subsequent refolding into the appropriate Maxwell-Boltzman distribution. Deviations from thermal equilibrium distributions are inherent and virtually unremovable limitations of those experimental techniques where the distributions are unknown or not measurable.

The remainder of this account is taken largely from written statements provided by the principal contributors; however, consistency and avoidance of redundancy required extensive modifications of some contributions. While the speakers are identified, responsibility for inaccuracies is entirely the author's.

1. LOW PRESSURE SINGLE SOURCE TECHNIQUES

The use of conventional or somewhat modified mass spectrometer ion sources with differential pumping sufficient to increase source pressure to the millitorr range has been extensive and has laid much of the foundation of what we know today about the general re-

activity of ions with neutral molecules, has provided a broad survey
of classes of reactions, and much qualitative information. The difficulties associated with quantitative measurements have been well
recognized for some time (Henchman, 1972), and as a consequence,
a series of modifications have been introduced which rely largely on
pulsing of the electron beam and various potentials in the source.
Professor Harrison (University of Toronto, Canada) discussed the
background and methodology of this approach.

Rate coefficients of ion-molecule reactions can be measured by
following ion concentrations either as a function of neutral molecule concentration at constant reaction time or as a function of reaction time at a constant neutral molecule concentration. Historically, pressure variation in a conventional (or, more recently,
a modified) continuous ionization mass spectrometer ion source was
the first technique used to estimate reaction rate constants. The
details of the method have been adequately reviewed elsewhere
(Franklin, 1972; Henchman, 1972). The method suffers from several
distinct disadvantages. First, the reactant ions are produced by
bombardment with electrons emitted from a heated filament and consequently $P_i(E_i)$ is not well-defined. Second, the reactant ions
are continuously accelerated by the ion extraction field and may
reach a final exit energy of several eV. Hence, the rate constants
measured are weighted averages over a range of internal and kinetic
energies.

The first attempt to measure reaction rates for ions with
thermal kinetic energies involved the zero-field pulsing technique
first introduced by Talroze and Frankevitch (Talroze and Frankevitch,
1959) and later used extensively by many others (for references, see
Franklin, 1972; Henchman, 1972). In this technique ions are produced in a field-free source by a short pulse of ionizing electrons.
After reaction for a defined length of time the reactant and product
ions are removed for mass analysis by a suitable voltage pulse applied to the repeller electrode, with the reaction rate constants
being determined by following relative ion concentrations as a function of reaction time. This technique also suffers from various
deficiencies. First, $P_i(E_i)$ is again not well defined. Second,
because of thermal kinetic energy the ions are lost from the sampling region at rates which are mass dependent. Thus to avoid ion
discrimination effects it is necessary to limit reaction times such
that the extent of ion loss is small (reaction times typically ≤ 1
μsec) and consequently the extent of reaction is small. Finally,
it has always been observed that the rate of ion loss is greater
than that anticipated for thermal energy ions indicating small residual fields; hence, $P_v(v_r)$ can be regarded as quasi-thermal only.
The technique and associated problems have been fully discussed by
Henchman (Henchman, 1972).

Because of the short reaction times available by the above technique and the problems associated with the pressure variation technique, considerable effort has been extended towards developing techniques in which the reaction time could be varied over a much wider range by using various forms of ion trapping. Those involving ICR instrumentation will be discussed below. Following earlier suggestions (Baker and Hasted, 1966; Bourne and Danby, 1968) Harrison and colleagues (Herod and Harrison, 1970) have developed a technique in which the ions are trapped in the negative space charge of an electron beam. The primary ions are produced by a short (ca. 1 μsec) pulse of energetic electrons and trapped for a variable length of time (up to 4 msec) in the negative space charge of a continuous electron beam of energy insufficient to cause ionization (ca. 4 eV). A known and variable time after primary ion formation the ions are removed for mass analysis by applying a suitable voltage pulse to the repeller electrode. The method still suffers from the disadvantage that $P_i(E_i)$ is ill defined. Moreover, the reactant ions do not have thermal kinetic energies although the average ion kinetic energies probably are quite low.

Dr. Ryan (CSIRO, Australia) discussed his calculation of the ion velocity distributions for the trapped ion technique. It is necessary to know the space charge induced potential at any point within the ion chamber to calculate the velocity distribution. The case of an electron beam of circular cross section flowing down an

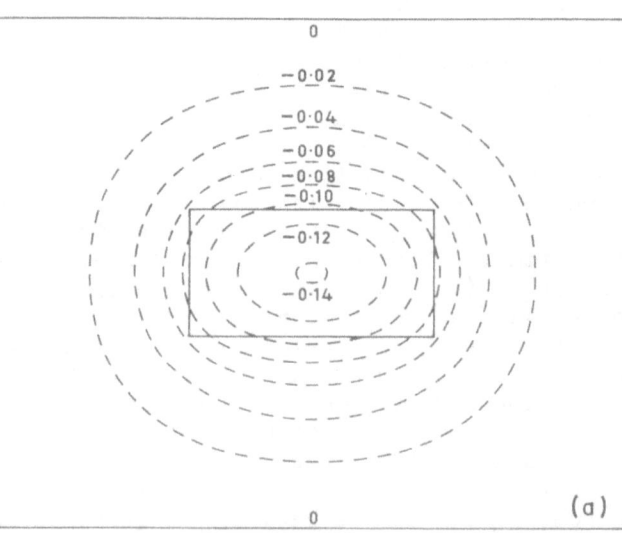

Figure 1. Potential distribution created in an ion chamber by the space charge of an electron beam. Beam dimensions 2 mm x 1 mm carrying 8 μamp at 4 eV. Chamber dimensions 5 mm x 4 mm.

infinitely strong magnetic field has been analyzed (Baker and Hasted, 1966) and a solution was obtained in closed form. Only the potential well within the electron beam was considered, but when an electron beam flows through a hollow conductor, it is necessary to calculate the field to the boundary of the conductor. Moreover in most ion chambers the electron beam is of rectangular rather than circular cross section.

Ryan considered the case of a 2 x 1 mm electron beam of 4 eV centered in a 5 x 4 mm chamber. This configuration is not amenable

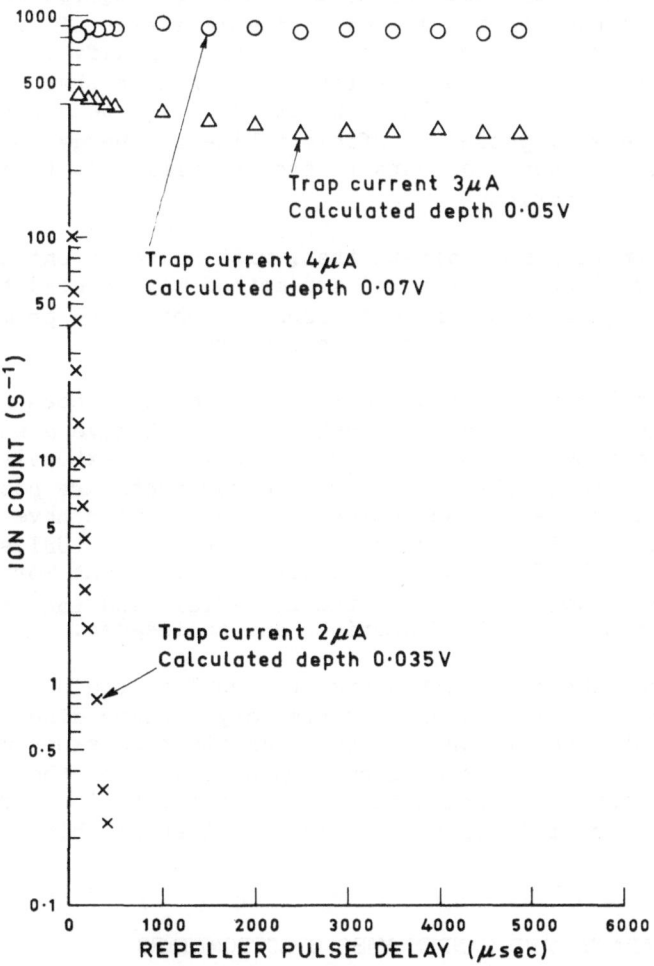

Figure 2.   Trapping characteristics for Ar^+ in pure argon as a function of trapping electron current.

to an analytic solution and it was necessary to solve Poisson's
equation numerically, both inside and outside the beam. Figure 1,
calculated for an 8 μA beam, shows that most of the well depth
lies outside the electron beam. Thus, while the total depth is
0.14 volts the maximum depth across the electron beam is only 0.06
volts. Ryan argues that, under these conditions, the maximum energy
a primary ion can derive from the trapping well is only 0.06 eV.

These computations are only approximate but may be tested by
observing the performance of the trap. Figure 2 shows the results
obtained for the trapping of Ar^+ in pure argon as the electron
current is varied. For an electron current of 2 μamp the trap depth
is computed to be 0.03 volts while at a source temperature of 340 K
argon ions will have an initial mean kinetic energy of 0.043 eV.
The effect of the space charge well will be to modify this distribution, but it is clear from the results at 2 μamp that even ions
formed in the center of the beam and hence unable to derive kinetic
energy from the well possess sufficient kinetic energy to escape.
At 4 μamp the well depth appears to become sufficient to give good
trapping characteristics.

These computations indicate that, with suitable choice of
operating conditions, the trapped ion method may be used to study
ion-molecule reactions under conditions for which the primary ions
have less than 0.1 eV of translational energy.

Professor Harrison noted that few systems have been studied by
the zero-field and ion-trapping techniques which have also been
studied by the flowing afterglow method; however, extensive comparisons with results obtained by ICR measurements are possible.
Table I presents a random selection of results which have been
studied for the most part, by the three techniques. Unless otherwise indicated, the ICR results are taken from recent work (Huntress and Pinizzotto, 1973) while the zero-field and ion-trapping
results are from the work of Harrison and colleagues.

In general the agreement among the results obtained by the
three techniques is reasonably satisfactory. Where discrepancies
exist it is not clear whether these originate from experimental
errors in one or more of the measurements or whether they reflect
the effect of different relative velocity distributions on the
measured rate coefficients as discussed earlier in this symposium
by Henchman.

2.   BEAM APPARATUS AND TANDEM MASS SPECTROMETERS

Beam experiments have been widely applied for the measurement
of cross sections and excitation functions for ion-neutral reactions
and much of the currently available data at near-thermal energies

Table I. Comparison of Measured Rate Coefficients

$$k \times 10^9 (cm^3 \; molec^{-1} \; sec^{-1})$$

Reaction	Zero Field (Flowing Afterglow)	ICR	Trapped Ion
$CH_4^+ + CH_4$	1.22	1.11	(1.20)
$CH_3^+ + CH_4$	-	0.89	1.15
$CD_4^+ + CD_4$	0.82	0.79	0.98
$CD_3^+ + CD_4$	-	0.76	1.08
$NH_3^+ + NH_3$	1.0	1.5 - 2.2	1.93
$NH_2^+ + NH_3$	0.65	2.31	1.95
$H_2O^+ + H_2O$	1.69	2.05	-
$OH^+ + H_2O$	~1.5	2.89	-
$H_2S^+ + H_2S$	0.77	0.59	0.72
$HS^+ + H_2S$	1.19	0.82	1.22
$(CH_3)_2O^+ + (CH_3)_2O$	1.93	-	1.58
$CH_3F^+ + CH_3F$	-	1.46*	1.87
$CH_4F^+ + CH_3F$	-	0.80*	0.90

*Marshall and Buttrill (1970)

and higher have been obtained using such techniques. These were discussed by T.O. Tiernan (Aerospace Research Laboratories, Wright-Patterson Air Force Base, Ohio).

Some of the obvious advantages common to beam methods include the capabilities for essentially unambiguous identification of re- actant and product ion species, low relative energy dispersion, and generally good intensities of product ions. Dr. Tiernan limited his discussion to applications of so-called "longitudinal" or "in- line" tandem mass spectrometers, such as that utilized in his lab- oratory. More detailed information on this apparatus, representative data, and a comprehensive discussion of the limitations and features of beam instrumentation in general have recently been presented elsewhere (Futrell and Tiernan, 1972; Henchman, 1972).

The ARL tandem mass spectrometer is of the "in-line" type, (that is, the product ion detector is directly in-line with the incident ion beam) and it is a beam-collision chamber device. Most beam experiments are restricted to the energy regime having a lower limit of a few tenths of an eV while the upper limit can range in- to the KeV regime if desired (merged beam methods are an exception).

The ARL tandem is capable of studying reactions at laboratory energies as low as 0.3 eV, and the upper limit is about 170 eV. The
chief advantage is the good definition and quite low energy dispersion (0.3 eV) over the entire accessible energy range; this
facilitates the determination of excitation functions.

    The quantity determined in experiments utilizing the ARL tan-
dem mass spectrometer is a phenomenological cross section which is
defined by the expression,

$$Q = I_s/(I_p n \ell) \hspace{5cm} [II]$$

where I_s is product ion current; I_p, the reactant ion current; n,
the neutral target number density; and ℓ, the reaction path length.
Applying the usual thin-target expression for the rate constant,
$k = Q v_i$ where v_i is the ion velocity, it is possible to obtain rate
constants via

$$k = (I_s/I_p)(v_i/n\ell) \hspace{4cm} [III]$$

The number density n of the neutral target can be accurately measured and controlled ($\pm 10\%$) using an MKS Baratron capacitance manometer. The collision chamber exit slit is large with respect to
the entrance slit and ions are refocused onto the collection slit
of the analyzing mass spectrometer after emerging from the collision
chamber. This improves the collection efficiency for product ions.
The collision chamber is very nearly field-free and well shielded
to eliminate field penetration and to ensure good energy definition
for the collision.

    The reactant ion beam current can be accurately measured at
the collision chamber using a removable Faraday cage and an elect-
rometer; the product ion beam is measured with an electron multiplier operated in the pulse counting mode. While the use of pulse
counting eliminates the need to apply corrections for multiplier
discrimination, the use of two different techniques to measure I_s
and I_p makes it difficult to assess I_s/I_p absolutely because the
measurements are difficult to interconvert. ℓ is known to a reasonable approximation from the dimensions of the collision chamber,
although streaming of gas through the entrance and exit slits effectively increases this length by an unknown amount. The method
actually applied, therefore, involves referencing of the measured
parameters for an unknown reaction to those for a "standard" reaction having a "known" rate, both reactions being studied under
identical conditions. In the case of positive ion processes, the
reaction typically used as a standard is the reaction of CH_4^+ ions
with CH_4 giving CH_5^+, for which the rate constant (1.2 x 10^{-9}cc/
molecule sec) is relatively insensitive to collision energies up to

about 0.5 eV (Clow and Futrell, 1970); therefore measurements made in the tandem at 0.3 eV will not be seriously in error by using this reference reaction. There are, however, limitations and problems with such methods, such as

a) The measured cross section is the average of $\sigma(E_I, v_r)$ over the translational energy spread which characterizes the re-axtants and is not thermal. When the dependence on v_r is appreciable between 0 and 0.6 eV, measured values of the rate constants may differ substantially from that which applies for a thermal distribution.

b) E_i is difficult to select. Some success can be achieved by operating the ion source at lower electron energies, but in general this is limited to controlling electronic excitation. Vibrational and rotational excitation levels are generally unspecified and unknown. The first stage of the instrument has recently been fitted with a high pressure ion source which will permit the generation of reactant ions which are themselves the products of higher order ion-neutral reactions (clustered species and large condensates, for example). This source will also promote collisional relaxation of some of the reactant ion internal energy.

c) Large corrections for the thermal motion of the neutral target molecules are necessary in some cases in order to accurately specify the collision energy. This results from so-called Doppler broadening (See "Reactions of Negative Ions" elsewhere in this volume).

d) Perhaps the most severe problem is that the collection efficiency of the tandem mass spectrometer (and of beam instruments in general) cannot be specified. The collection efficiency will depend upon many factors including the mechanism and dynamics of the collision, the collision chamber and collector slit dimensions (which determine the appropriate emergence and collection angles), instrument transmission, and so forth. Obviously with the ARL tandem mass spectrometer which has a field-free collision chamber, it is expected that collection of forward scattered products will be maximized. The ARL tandem has a fixed collector stage and collection over the entire range of scattering angles is not possible. Experiments with extraction fields have established that indeed only a portion of the product ions emerging from the collision chamber are collected, in spite of the efficiency improvements noted above. A crude attempt has been made to estimate the angular divergence of the product beam as it exits from the collision chamber by using a lateral field across two split focusing electrodes immediately following the chamber. The product ion beam can thus be swept across the collector slit and an angular profile obtained. Such experiments have been conducted for a variety of reactions which are expected to exhibit different mechanisms. The results indicate that at the low energy at which these measurements are made ($\sim$0.3 eV, lab.),

essentially all product ions have about the same emergence angle
profiles. This suggests that under these conditions, the energy
distribution of the product ions is essentially the same regardless of
the mechanism of their formation and so the detection efficiencies
should not be markedly dissimilar (even some thermal products will
have velocities and trajectories which permit their detection).
Obviously, these methods do not provide a complete answer to the
questions regarding collection efficiency.

One may properly ask how the accuracy of any given technique
for measuring rate constants can be assessed. Some indication can
be obtained by comparing values derived by a given method with data
reported using other instrumentation. This is not strictly valid because, as has been noted at this conference by several speakers, the
energy distribution functions which characterize the reactant species
certainly differ between various techniques. Comparisons of rate
data obtained for various negative ion-molecule reactions using the
tandem mass spectrometer with data for the same processes from other
methods are shown in the chapter on Reactions of Negative Ions.
Similar comparisons for positive ion processes have been reported
earlier (Futrell and Tiernan, 1972). In general, the agreement is
reasonably good and appears to indicate that the energy definition
specified for the ARL tandem mass spectrometer is correct. This
is supported by measurements of excitation functions.

It seems probable that the uncertainties inherent in all beam
techniques currently utilized for rate constant measurements limit
the absolute accuracy of reported rate constants to $\pm 100\%$ at best;
this probably exceeds differences arising from detailed disparities
in the energy distribution functions.

3. FLOWING AFTERGLOW TECHNIQUES

Dr. F.C. Fehsenfeld (NOAA Environmental Research Laboratories,
Boulder, Colorado) presented a detailed summary of the flowing
afterglow techniques developed at the NOAA laboratories; this was
followed by a lively and extensive discussion. The basic flowing
afterglow apparatus consists of four sections (Fehsenfeld, 1974):

a) The Flow Development Section. A buffer gas is introduced
into a stainless steel reaction tube which has a radius of 4 cm
and length of 1 meter. Pressure ranges typically from 0.1 torr
to 2 torr with an average flow velocity between 10^3 and 10^4 cm/sec.
The point of admission of the buffer gas is followed by a section
of tube to allow development of the gas flow, which should be
almost laminar with no more than a few percent slip at the wall due
to molecular flow.

b) <u>Reactant Ion Production</u>. An electron gun operating at tube pressure or above is located at the beginning of this region, and excites and ionizes the gas as it flows past it. For ions other than those of the buffer gas, a suitable source gas, at a concentration typically less than 0.2%, is added just before or just after the electron gun, so that ions are created either directly by inter- action of the energetic electrons emitted by the electron gun or by secondary reactions with the ions, excited species and electrons downstream from the electron gun. In this fashion it is possible to "tailor make" a wide variety of reactant ions and excited states by subsequently adding various source gases and passing through several reaction sequences. This is one of the most powerful as- pects of this technique.

c) <u>Reaction Region</u>. Reactant neutrals are added through an inlet downstream from the region where reagent ions are produced, at a point where the reactions to form reagent ions have gone to completion before the reactant neutral is added. A sufficient in- terval also assures that ions, electrons and long-lived reactant neutrals have been thermalized to the buffer gas (wall) temperature before they are mixed with the reactant neutrals. In order to perform the measurements as a function of temperature between 80°K and 900°K the afterglow tube is equipped with electric heaters and cooling coils which are mounted in copper heat sinks directly in contact with the tube, and permit temperature stability and uni- formity of 1% or better above room temperature and 5% or better below room temperature. The gases are introduced in such a fashion that they are in thermal equilibrium with the walls of the flow tube before they enter the tube.

d) <u>Ion Sampling and Mass Analysis</u>. The reaction region is ter- minated by the sampling orifice of the mass spectrometer. The ions are sampled, mass analyzed, and counted. The sampling orifice is located in the center of the end of a right-angle cone which is usually truncated. The distance between the front of the nose cone and the end of the flow tube is about equal to the flow tube diameter. Because of this positioning of the sampling port, it is expected that the flowing ionized gas near the front of the nose cose will chara- cterize the central portion of the steady state flowing ionized gas just upstream from the end of the flow tube. The flowing gas is smoothly separated by the conical surface of the nose cone and subsequently exhausted into the pump. If a stagnant sampling layer existed before the blunted face, the ions and electrons might be trapped in this "dead gas" causing a significant increase in the reaction time. The NOAA observations indicate that this does not occur. Results were identical to those obtained using blunt nose cones when measurements were made with a conical nose cone tapered to a sharp knife edge at the orifice.

The mass spectrometer sampling orifice is operated at or near
ground potential. This is necessary if weakly bound cluster ions are
to be studied. Tests indicate that weakly bound cluster ions such
as $NO^+(H_2O)_3$ can be detectably broken up with draw-in potentials
in excess of 1 volt. The pressure in the chamber following the
sampling orifice is about 10^{-4} torr or less. When detecting weakly
bound clusters the pressure in this chamber is kept below 10^{-4}
torr in order to avoid collisional breakup.

Aside from breakup, sampling problems may also arise from distortion of the extraction fields between the mass filter and the
nose cone caused by surface charges on the inside surface of the
nose cone and on ion-exposed surfaces of the mass filter. This can
be minimized by periodic electrocleaning followed by a distilled
water rinse; in a heated system cleaning is supplemented by baking.
Reliable sampling, especially when oxygen atoms were in the system,
could only be obtained when the sampling orifice was made of molybdenum or carbon. In general practical considerations dictate the
use of molybdenum. The ions are directed into the mass filter by a
bias voltage ranging from about 1 to 30 V, applied between the quadrupole rails and the nose cone. The advantages of more complicated
ion optics are more than offset by the surface charging problems
and reduced pumping speeds which additional electrodes would produce.
It is extremely important that the sampling efficiency be independent of the reactant ion density so that no change in overall efficiency occurs when the reactant gas is added. For most cases this
may be tested when no further reactions of the product ions occur,
when recombination and diffusion of the product ions are taken into
account, and when the mass discrimination of the mass spectrometer
is corrected; the product ion signal should then equal the initial
reactant ion signal. For associative detachment reactions involving
negative ions, non-reacting ions are produced and monitored for constancy during experiments.

The variation of reactant and product ion signals with reactant
concentration leads quantitatively to the rate constants. An alternative is the variation of reaction length at a fixed reactant flow.
It is necessary to solve the transport equation

$$v(r) \frac{[A^+]}{Z} = (D_A+)[\frac{1}{r} \frac{\partial}{\partial r} r \frac{\partial [A^+]}{\partial r} + \frac{\partial^2 [A^+]}{\partial Z^2}] - \begin{cases} 0 & Z<0 \\ k[A^+][B] & Z\geq 0 \end{cases} \quad [IV]$$

to obtain rate constants from flowing afterglow data. Here $[A^+]$
is the reactant ion concentration, D_A^+ the (ambipolar) diffusion
coefficient for A^+, k is the binary reaction rate constant between
A^+ and B and [B] is the reactant neutral concentration. v(r) is
the radial velocity distribution. For laminar (viscous) flow v(r)
is parabolic, i.e.

$$v(r) = 2v_o[1 - (r/a)^2] \qquad\qquad\qquad [V]$$

where v_o is the average flow velocity and a is the radius of the
tube. The transport equation [IV] is solved using a finite difference method (Ferguson, Fehsenfeld and Schmeltekopf, 1969) which
is quite general and flexible. It can equally well handle a mixture
of reactant gas producing several concurrent reactions, ion-ion or
ion-electron recombination, a neutral reactant gas which is destroyed at the wall or by reactions in the gas phase, and complicated
diffusion coefficients which are functions of position. However, if
these effects become important the accuracy of the solutions one
obtains is limited by the accuracy with which all fixed parameters
are known.

Rather than solving the transport equation to match each run of
data, Fehsenfeld et al. use a family of solutions which assume only
one reaction. It depends on D_A^+, and therefore on the identity of
the reactant ion, the buffer gas, the nature of the other charged
particles and the gas pressure, and on [B]. Over a wide range of
experimental conditions these solutions are not strongly dependent
on the assumed experimental parameters, and the rate constant may
be expressed in the form

$$k = \frac{\alpha \Pi a^2 v_o}{L} \frac{\Delta \ln[A^+]}{\Delta Q_B} \qquad\qquad\qquad [VI]$$

where L is the reaction length, and ΔQ_B is the change in rate of
addition of the reactant gas which causes the change in the loga-
rithm of the detected ion signal, $\Delta \ln[A^+]$. The constant of pro-
portionality, α, is a complicated function of experimental conditions. In the NOAA apparatus α has a value between 1.4 and 1.6 and
is constant to within 5% over the range of conditions used in the
experiment (Ferguson, Fehsenfeld, and Schmeltekopf, 1969). When
α is established the rate constant is computed from the slope of
the logarithmic decline of the reactant signal as a function of
reactant neutral addition.

An independent test has been devised to judge the correctness
with which hydrodynamics and diffusion have been treated in the
present model (Fehsenfeld, 1974). The analysis was used to calculate the behavior of a diffusing ion sheet in the afterglow tube,
and was compared to an experiment in which a thin sheet of ions was
generated at z = 0 and t = 0 by a pulsed beam of electrons incident perpendicular to the axis of the tube. The pulse width was
characteristically 100 μsec. For ions other than those of helium
a suitable source gas was added with the flowing buffer. Thus,
the diffusive properties of various ions, negative as well as posi-

tive, were tested in helium. The diffusing ion sheet was carried
downstream by the flowing gas. The detector was gated with a width
of 100 μsec; thus, an ion arrival-time spectrum could be recorded.
Agreement between experimental and calculated arrival-time spectra
was excellent for all gases and flow conditions.

The flowing afterglow results are not complicated by diffusion effects because reaction rates are usually determined under
conditions where diffusion and reaction are independent of each
other. Consequently, a maximum error of ±10% can be ascribed to
rate constants carefully measured by this technique. Since reagent
ions are permitted to undergo a relatively large number of collisions with a buffer gas before encountering reactant gas, P_i, P_v,
and P_n should be Maxwell-Boltzman distributions for the same temperature. The flowing afterglow is therefore currently the principal technique not subject to uncertainties in this regard.

4. DRIFT TUBES AND HIGH PRESSURE (CHEMICAL IONIZATION) SOURCES

These devices are grouped together because they employ an
arrangement in which an electric field is normally, but not necessarily, applied in the reaction region. They were originally a
tool used almost exclusively by physicists (McDaniel, et al., 1970;
Heimerl, Johnsen and Biondi, 1969), but the development of chemical
ionization as an important analytical tool (Field, 1968) led to the
use of high pressure sources for rate constant measurements (Field,
1969; Vredenberg, Wojcik and Futrell, 1971). Reliable rate constant measurements using such sources became available when residence time measurements were introduced (Chang, Sroka and Meisels,
1971; Sroka, Chang and Meisels, 1972). With both drift tubes and
high pressure sources, ion motion is largely dictated by applied
fields when these are present; the former technique differs largely
through the employment of uniform fields and well defined experimental conditions, while in the latter ions usually drift through a
range of field strengths (Chang, Sroka and Meisels, 1973). The
physical models are therefore very similar, and because of the
greater simplicity of conditions in drift tubes, discussion by
Dr. Johnsen (University of Pittsburgh) was confined to this technique, which has been used extensively to study ion-molecule reaction rates in the energy range from thermal energies to several
eV (McFarland, et al., 1973; Johnsen and Biondi, 1973; Hasted,
1974). This is the result of experimental difficulties in raising
flowing afterglow apparatus to temperatures above 900°K ($\overline{E_r}$ =
0.12 eV), and the inaccuracies of beam techniques at energies below
about 2 or 3 eV. While drift tube instrumentation and associated
techniques have been refined considerably over the last few years,
caution has always been expressed that rate constants measured
were obtained where $P_i(E_i)$ is a non-Maxwellian velocity distribution.
It is this point which shall be discussed here.

The standard practice so far has been to show the measured rate coefficient as a function of the parameter E/N or E/P, i.e. the ratio of the applied electric field to the gas density or pressure. In addition, the ions' mean energy has usually been computed from the measured drift velocity by means of a theoretical relationship (Wannier, 1953) which states that the ions' mean energy, E_i, in the laboratory frame of reference is related to the drift velocity via

$$\overline{E_i} = 3kT/2 + (m+M)\, v_d^2/2 \qquad\qquad\qquad [VII]$$

where m and M are the mass of the ion and of the neutral, respectively and v_d is the ions' drift velocity. This formula, while derived under the assumption of a constant mean free time between ion-neutral collisions (i.e. $\sigma \, \alpha \, 1/v_r$) has nevertheless been applied to other experimental conditions. Skullerud has recently shown that Wannier's equation agrees within about 10% with results of Monte Carlo calculations even if drastically different cross section models are used (Skullerud, 1972). It thus appears that no major errors have been caused by the somewhat uncritical application of Wannier's formula.

The same relationship has also been used to assign so-called "effective" or "equivalent" temperatures to drift tube data, to permit comparison of data with results obtained in flowing afterglow systems. While this approach obviously disregards completely the differences in the velocity distributions in the two experiments, the agreement has been surprisingly good. This is obviously trivial for reactions where $\sigma \, \alpha \, 1/v_r$) since in this case the rate constant will be energy independent of $P_v(v_r)$. In the case of reaction cross sections falling off faster than $1/v_r$ most of the contribution to the rate constant will come from the low energy part of the velocity distribution and the high energy tail will contribute very little. It is the high energy tail, however, which is most likely to deviate most from Maxwellian distributions.

More pronounced differences between drift tube and thermal measurements are to be expected where rate constants increase with energy, but few good data are available for such systems. The assignment of effective temperatures without consideration of velocity distributions is frequently useful but still not more than a first approximation; for a more rigorous analysis of drift tube rate coefficients it is necessary to obtain the velocity distribution of the drifting ions and to infer the velocity dependence of the reaction cross section from the measured rate constants. The effort required is considerable. The effect of the applied field on the rate constant k(E/N) can be understood from equation I.

The right hand side of I depends on E/N principally through

the dependence of $P_v(v_r)$ on E/N if internal modes are not excited.
If k(E/N) is measured and P_v is known, the cross section $\sigma(v_r)$ can
be obtained by deconvolution. This procedure is not trivial but
can be handled adequately by assuming a functional dependence of σ
on v_r, executing the integral numerically and adjusting certain
parameters in the assumed cross section until the best fit to the
measured rate constants is obtained. Demands on the accuracy of the
experimental data and the E/N range are considerable if reliable
cross sections are to be inferred. Woo and Wong have discussed the
application of such deconvolution procedures to drift tube data on
endothermic reactions (Woo and Wong, 1971). A more difficult problem is that of obtaining the distribution function $P_v(v_r)$ which is
composed of the ion velocity distribution, $P(v_i)$, and that of the
neutral reactant molecules. The latter is simply a Maxwell distribution and thus known. The ion velocity distribution can be
estimated at least qualitatively as follows.

A drift tube may be viewed as a cylindrical reaction cell with
an electric field present. The motion of the ions through the
drift tube is controlled by the strength of the electric field and
the effect of collisions with the gas molecules contained in the
tube. The path of an individual ion may be visualized as consisting
of a very large number of parabolic segments, each segment beginning
and ending with a collision which changes the direction and magnitude of the ion's velocity. The average of the velocity components
in the field direction is the drift velocity, while the average of
the components perpendicular to the field is zero. The corresponding distribution of the absolute velocities (regardless of direction)
is the speed distribution.

Several theoretical methods are available to obtain the distribution function, $P(v_i)$. Early approaches using a two-temperature
representation give qualitative insights but cannot be expected to
work well for mean ion energies in excess of a few tenths of an eV
(Hong and Woo, 1972; Chang, Meisels, and Taylor, 1973). A very
powerful technique is that of performing a computer simulation of
the path of an ion through an idealized drift tube. The only essential input information required is a realistic scattering model
for the individual ion-neutral collisions which has to be derived
from the relevant ion-neutral interaction potentials. All random
quantities are simulated by random number generators. In addition
to information on velocity distributions these "Monte Carlo" techniques yield values for ion mobilities and for lateral and longitudinal diffusion coefficients. Comparison with experimental data
can be used to test and improve the assumed scattering model. The
first such Monte Carlo calculations were performed for the case of
a constant mean free time between collisions (Wannier, 1953).
Making use of the considerable improvements in computer technology,
similar calculations have been performed for different scattering
models (Skullerud, 1972). Lin, Bardsley, and Biondi are presently

developing a program for a variety of scattering models with proper inclusion of the neutral particles' thermal motion (Biondi, 1974).

It appears that Monte Carlo calculations of this type may be the most promising approach to the problem since they circumvent the rather formidable task of solving the Boltzman transport equation in the general case. Under certain assumptions, however, the Boltzman equation can be treated successfully and a number of authors have used expansion methods successfully to obtain distribution functions and other parameters of interest (Wannier, 1953; Moruzzi and Harrison, 1974).

Several authors have attempted experimental determinations of the ion velocity distributions (Kihara, 1953; Hasted, 1974). A retarding potential analysis performed on the ions effusing from the drift tube through a sampling orifice seems to be the obvious approach. However, a number of interfering effects such as contact potentials, field inhomogeneities near the sampling orifice and gas density variations near the orifice, are difficult to take into account. While the measured distribution functions are qualitatively in harmony with expectations, it has been noticed that the mean ion energy inferred often exceeds that expected from Wannier's formula. It is not clear, at present, if this disagreement is significant or an experimental artifact.

Monte Carlo calculations currently appear to offer the best route to quasi-experimental determination of drift properties and ion velocity distributions. While they may thus remove uncertainties associated with $P_v(v_r)$ they do not necessarily resolve the question to what extent the collision process leads to conversion of collision energy to internal energy; $P_i(E_i)$ therefore remains uncertain but is probably nearly equivalent to a Maxwell-Boltzman distribution defined by the gas temperature and only shifted to higher energies by small amounts. Drift tubes can therefore solve a useful purpose in providing accurate rate constant measurements as a function of relative velocities in an energy regime which is not easily accessible otherwise. Almost all drift tubes and high pressure sources can also be operated in a field-free mode so that these techniques can also be employed to measure thermal rate constants.

5. ION CYCLOTRON RESONANCE

A large number of rate constants have been measured by this technique within the last few years. In most instances, these have been in good agreement with thermal rate constants measured by other techniques, such as the flowing afterglow. As noted earlier, agreement is not significant when the rate constant is independent of velocity, that is, when σ varies as $1/v_r$. Since ICR experiments

employ trapping fields, one may ask to what extent these affect the
velocity distribution of reactant ions. Professor Bowers (University of California at Santa Barbara, California) summarized some
of the recent work he and his colleagues have undertaken to gain
insight into this problem (Su, Chesnavich, and Bowers, 1974).

When ions are initially formed in an ICR cell, the velocity
distribution is Maxwellian at any point along the electron beam
(z direction) as a result of the Maxwellian distribution of the
neutral precursor target gas. These newly formed ions then begin
to oscillate back and forth in the trapping potential well (z direction). They also drift from the source to the analyzer region,
but the drift potential distorts the velocity distribution only to
an immeasurably small extent and can be ignored ($v_D = E/B \ll \bar{v}_{th}$
where $\bar{v}_{th}$ is the average thermal velocity). The ion kinetic energy
distribution can then be calculated by combining the thermal energy of the ion with the energy associated with the oscillating motion in the trapping potential. The frequency of the oscillating
ion depends on the trapping voltage (Beauchamp and Armstrong, 1969)
while its amplitude depends on its initial velocity and position.
The probability that an ion has a velocity v_z in the z direction is
then

$$P(v_z) = \int_{ICR\ cell} P_T(v_z)_{z_0} P(z_0)dz \qquad\qquad [VIII]$$

where $P(z_0)$ is the probability that the ion will oscillate with
amplitude z_0 and $P_T(v_z)_{z_0}$ is the probability that an ion has velocity
v_z in the z direction for a given amplitude, z_0. The functions
$P_T(v_z)_{z_0}$ and $P(z_0)$ can be calculated for any trapping voltage. When
studying ion-molecule reactions, we are interested in the relative
velocity distribution $P_V(v_r)$. This function can be evaluated by
convoluting the ion velocity distribution with the neutral molecule
velocity distribution (thermal) as follows. The probability of a
relative velocity v_{rz} in the z direction between the ion and the
neutral is

$$P(v_{rz}) = \int P(|v_z|) P_{Nz}(v_z - v_{rz})dv_z \qquad\qquad [IX]$$

where $P_{Nz}(v_z - v_{rz})$ is the velocity distribution of the neutral in
the z direction (which is the one-dimensional thermal distribution
function). Equation IX is integrated over all available v_z in
the ICR cell. The three dimensional ion-molecule relative velocity
distribution is then calculated by convoluting $P(v_{rz})$ with the two
dimensional thermal energy distribution function, P_{xy}

$$P(v_r^2)d(v_r^2) = \int_0^{v^2} P_{xy}(v^2 - v_{rz}^2)P(v_{rz}^2)d(v_{rz}^2)d(v^2) \qquad\qquad [X]$$

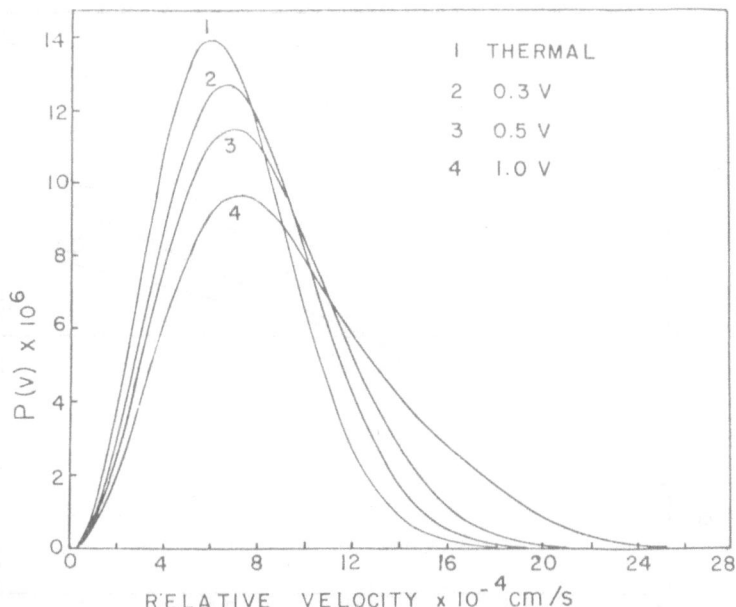

Figure 3. Comparison of ion-molecule relative velocity distribution functions in a drift mode ICR cell associated with 0.3, 0.5 and 1.0 volt trapping potentials and with the thermal velocity distribution function for the system $N_2^+ + N_2$.

By transforming v^2 to v in Equation X, one obtains the three dimensional ion-molecule relative velocity distribution in the ICR drift cell prior to any collisions.

Figure 3 compares the ion-molecule relative velocity for 0.3, 0.5 and 1.0 volt trapping potentials with the thermal velocity distribution for the system $N_2 + N_2^+$. As trapping voltage increases, the zero collision relative velocity distribution in the ICR cell deviates increasingly from the thermal distribution.

An analysis of the effect of this change on average rate constant is facilitated by defining the velocity dependence of the cross section by $\sigma(v_r) \alpha v_r^n$. Figure 4 shows the deviation of k from the thermal energy value as a function of n for different trapping potentials. At n = -0.5, and 1 volt trapping voltage, k_{trap}/k_{therm} is ~1.17 or a deviation of 17%. Figure 5 shows the relative thermal energy macroscopic rate constant as a function of

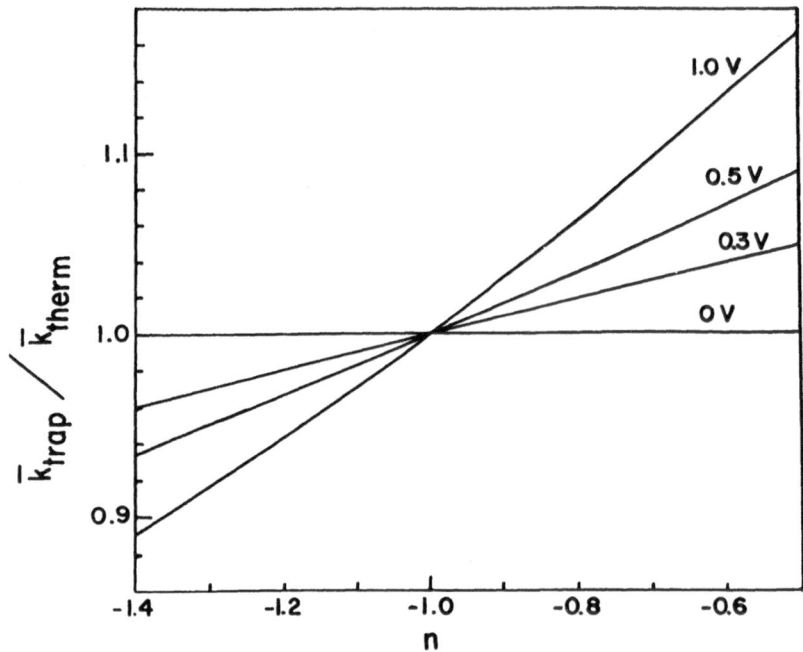

Figure 4. Ratio of macroscopic ion-molecule reaction rate constants
associated with various trapping potentials in a drift mode ICR cell,
k_{trap}, to the thermal velocity macroscopic rate constant, k_{therm},
as a function of n for the system $N_2^+ + N_2$.

n. For a typical ICR experiment, the trapping potential is ca.
0.5 volts. From the calculations presented here, the measured
rate constants should not deviate more than a few percent from those
at thermal energy defined by the gas temperature when n lies between
-1.2 and -0.8. Comparison of ICR rate constants with thermal flowing afterglow data appear to support this suggestion (See for
example Laudenslager, Huntress, and Bowers, 1974).

An experimental approach to determine the effect of the trapping well potential on rate constants measured by the ICR trapped
ion method (McMahon and Beauchamp, 1972) was presented by Dr.
Huntress (Jet Propulsion Laboratory, Pasadena, California).

Extensive measurements of the dependence of rate constants on
trapping potential were made for the reactions

$$H_2S^+ + H_2S \rightarrow H_3S^+ + HS \tag{1}$$

and

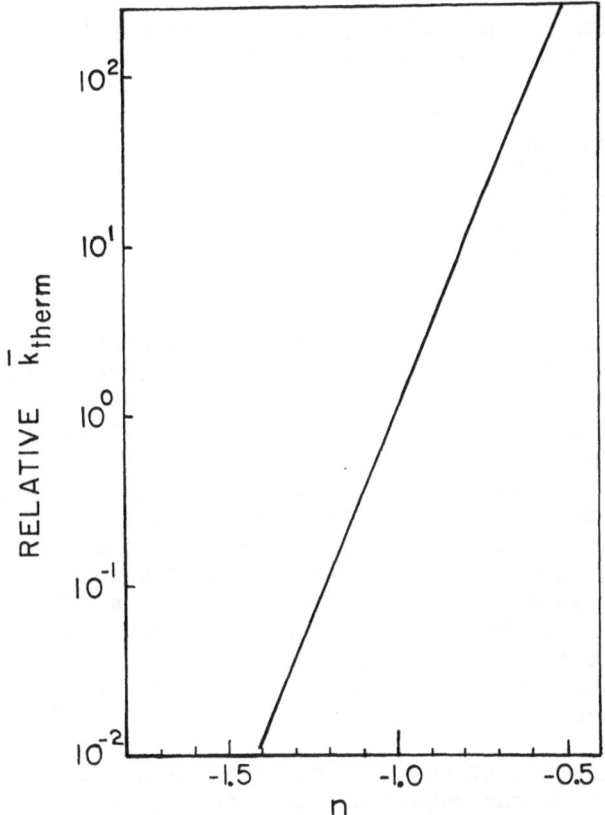

Figure 5. Relative thermal velocity macroscopic reaction rate constant as a function of n for the system $N_2^+ + N_2$.

$$Kr^+ + CO \rightarrow CO^+ + Kr \tag{2}$$

The rate constant for reaction (1) decreases rapidly with increasing ion kinetic energy, while that for reaction (2) increases rapidly with increasing ion kinetic energy. Thus σ does not vary as $1/v_r$ for either.

The rate constant for reactions (1) and (2) did not show any dependence on the trapping voltage used to store the ions from V = 0.50 to 3.0 volts. However, the rate constants did change significantly when rf heating at the cyclotron frequency of the reactant ion was applied shortly following the ionization pulse. The kinetic energy of the ion was calculated from the amplitude and duration of the applied rf pulse (Huntress, Mosesman, and Elleman, 1971).

Using reaction (1) as an example, the rate constant decreased from 6×10^{-10} with no applied rf pulse to a value of 3×10^{-10} with an rf pulse applied to heat the ions to a mean energy of about 1.5 eV. The trapping voltage in these experiments was 0.50 volts. However, with no applied rf, the measured rate constant remained constant at 6×10^{-10} with applied trapping potentials from 0.50 to 3.0 volts. Similar results were obtained on reaction (2), and suggest that energetic ions formed near to the trapping plates by the electron beam pulse are not as efficiently trapped in the ICR cell as are ions with very low kinetic energy formed near the center of the trapping well.

At high pressures, epithermal ions can apparently be thermalized by non-reactive collisions, and can be subsequently trapped before loss from the well. This phenomenon has been detected by observing the storage characteristics of the Ar^+ ion in argon as a function of pressure. At very low pressures (5×10^{-7} torr) where only one or two collisions can occur during the storage time (100 msec), the intensity of the $Ar+$ ion is constant with time. At higher pressures (1×10^{-5} torr), where many collisions can occur during the storage time, the Ar^+ intensity increases at first at short times (~ 10 msec) before slowly decaying as a consequence of ion diffusion losses. The increase in Ar^+ intensity at short times may in part result from the trapping of additional ions via collisional deactivation of kinetically energetic Ar^+ ions by the resonant charge transfer reaction.

The discrimination of the ICR ion trap against ions of higher kinetic energy may be the result of the fringing fields at the longitudinal ends of the source region. The trapping fields at these points collapse to small values; nearly zero at the end-plate, and a small value (the analyzer trapping voltage) at the boundary between the source and analyzer regions of the ICR cell. The trapping voltage in the analyzer region is usually around 0.3 volts. Energetic ions may therefore escape from the cell at the end-points of their oscillations in the longitudinal direction. Selective retention of near-thermal energy ions in the ICR trap is consistent with the absence of any noticeable effect of trapping voltage on measured rate constants, and the close agreement with thermal energy flowing afterglow values for a large number of rate constants.

One may conclude from these arguments that the ICR technique permits the measurement of rate constants very nearly at thermal energies provided that the trapping potential is maintained at a minimum. Particularly when the velocity dependence of the cross section is not too different from the Langevin form, where the rate constant is obviously insensitive to the form of the distribution, errors will be small. However, measured rate constants for endothermic processes with a strong energy dependence may tend to be larger than those which should be observed if all species were at thermal energy.

SUMMARY

The five techniques summarized and discussed by the panel clearly all have assets. In the low energy regime, the flowing afterglow is a versatile technique with well-defined energy distributions and permissive of good accuracy in rate constant measurements. Drift tubes operated in the field-free mode are not quite as versatile but have similar virtues otherwise. While energy distributions for other techniques are not as clearly defined, their ability to penetrate into the higher energy regime and rapidity make them equally valuable. Considerable progress in developing knowledge of velocity distributions applicable to these methods is now being made and should make these techniques even more powerful than they are now.

ACKNOWLEDGEMENTS

The author is indebted to the United States Atomic Energy Commission and to the Robert A. Welch Foundation for support of this work.

REFERENCES

Baker, F.A. and Hasted, J.B. (1966) Phil. Trans. Roy. Soc. London A261, 33.

Bourne, A.J. and Danby, C.J. (1969) J. Sci. Instr. 2, 155.

Chang, C., Meisels, G.G. and Taylor, J.A. (1973) Int. J. Mass Spectrom. Ion Phys. 12, 411.

Chang, C., Sroka, G. and Meisels, G.G. (1971) J. Chem. Phys. 55, 5154.

Chang, C., Sroka, G. and Meisels, G.G. (1973) Int. J. Mass Spectrom. Ion Phys. 11, 367.

Clow, R.P. and Futrell, J.H. (1970) Int. J. Mass Spectrom. Ion Phys. 4, 165.

Fehsenfeld, F.C. (1974) Int. J. Mass Spectrom. Ion Phys., (in press).

Ferguson, E.E., Fehsenfeld, F.C. and Schmeltekopf, A.L. (1969) Adv. in Atomic and Molec. Phys. 5, 1.

Field, F.H. (1968) Accts. of Chem. Res. 1, 42.

Field, F.H. (1969) J. Amer. Chem. Soc. 91, 2827.

Franklin, J.L. (1972) Positive Ion-Molecule Reaction Studies in a Single Electron-Impact Source, in "Ion-Molecule Reactions," J.L. Franklin, Editor, Plenum Press, New York.

Futrell, J.H. and Tiernan, T.O. (1972) Tandem Mass Spectrometric
    Studies of Ion-Molecule Reactions, in "Ion Molecule Reactions,"
    J.L. Franklin, Editor, Plenum Press, New York.
Hasted, J.B. (1974) elsewhere in this volume
Henis, J.M.S. (1972) Ion Cyclotron Resonance Spectrometry, in "Ion-
Molecule Reactions," J.L. Franklin, Editor, Plenum Press, New
    York.
Heimerl, J., Johnsen, R. and Biondi, M. (1969) J. Chem. Phys. $\underline{51}$,
    5041.
Henchman, M. (1972) Rate Constants and Cross Sections, in "Ion-
    Molecule Reactions," J.L. Franklin, Editor, Plenum Press, New
    York.
Herod, A.A. and Harrison, A.G. (1970) Int. J. Mass Spectrom. Ion
    Phys. $\underline{4}$, 415.
Hong, S.P. and Woo, S.B. (1972) J. Chem. Phys. $\underline{57}$, 2597.
Huntress, W.$\underline{T}$. and Pinizzotto, R.F. (1973) J. Chem. Phys. $\underline{59}$,
    4742.
Johnsen, R. and Biondi, M. (1973) J. Chem. Phys. $\underline{59}$, 3504.
Kihara, T. (1953) Rev. Mod. Phys. $\underline{25}$, 844.
Marshall, A.G. and Buttrill, S.E. (1970) J. Chem. Phys. $\underline{52}$, 2752.
McDaniel, E.W., Cermak, V., Dalgarno, A., Ferguson, E.E. and
    Friedman, L. (1970) "Ion-Molecule Reactions," Wiley-Interscience,
    New York.
McFarland, M., Albritton, D.L., Fehsenfeld, F.C., Ferguson, E.E.,
    and Schmeltekopf, A.L. (1973) J. Chem. Phys. $\underline{59}$, 6610, 6620, and
    6629.
Moruzzi, J.L., and Harrison, L. (1974) Int. J. Mass Spectrom. Ion
    Phys. $\underline{13}$, 163.
Skullerud, H.R. (1972) J. Phys. $\underline{B6}$, 728.
Sroka, G., Chang, C., and Meisels, G.G. (1972) J. Am. Chem. Soc.
    $\underline{95}$, 1052.
Talroze, V.L. and Frankevich, E.L. (1959) Izv. Acad. Nauk SSSR, Otd.
    Khim. Nauk $\underline{7}$, 1351.
Vredenberg, S., Wojcik, L. and Futrell, J.H. (1971) J. Phys. Chem.
    $\underline{75}$, 590.
Wannier, G.H. (1953) Bell Syst. Tech. J. $\underline{32}$, 170.

CHARGE TRANSFER

SUMMARY OF THE PANEL DISCUSSION LED BY:

J. Durup

Collisions Ioniques, Universite de Paris-Sud

91405 - ORSAY, France

The present discussion will be concerned only with charge transfer at low energies, more precisely at energies from thermal to a few eV, where ion-molecule reactions are also taking place.

Three topics will be discussed, viz:

1) long-distance (resonant and near-resonant) charge transfer;
2) charge transfer through multiple crossings of potential surfaces;
3) short-distance effects, including intermediate complex formation, internal energy dependence of charge transfer cross sections, etc.

This introduction intends to recall some basic knowledge about these topics.

RESONANT AND NEAR-RESONANT CHARGE TRANSFER

I first wish to emphasize the conceptual difference between resonant and near-resonant charge transfer from the quantum-mechanical viewpoint.

In the simple example of the resonant charge transfer of ground-state He^+ on He, the quasi-molecular system will behave as a coherent superposition of the $X^2\Sigma_u^+$ and $A^2\Sigma_g^+$ states of He_2^+. The electronic wave function of the incoming system may be expressed as $[\Psi_g \exp(-iEt/\hbar) + \Psi_u \exp(-iEt/\hbar)]/\sqrt{2}$. The g and u states are uncoupled during the collision and thus behave independently, but they experience phase shifts different from each other so that the electronic wave function of the outgoing system will be (except for a common phase term)

$$[\cos \omega \ (\Psi_g + \Psi_u) + \sin \omega \ (\Psi_g + \Psi_u)]/\sqrt{2},$$

where the first term gives the elastic scattering amplitude and the second term the charge transfer amplitude, and where in the time-dependent classical-trajectory formalism

$$\omega = (1/2 \ \hbar) \int (E_g - E_u) \ dt.$$

Here E_g and E_u are the potential energies of the g and u states, respectively. They depend on time t through the time dependence of the internuclear distance, R, along the trajectory.

Thus, at large impact parameters, where $E_g - E_u$ is small all along the trajectory, ω, and therefore the charge transfer probability $P = \sin^2 \omega$, will be small. For decreasing impact parameters, P increases up to the point where ω is of the order of magnitude of $\pi/2$. Thereafter, i.e. for all smaller impact parameters, P will oscillate between 0 and 1 about the mean value 0.5. The critical impact parameter R_c under which P is thus essentially equal to 0.5 depends for a given collision velocity on the magnitude of the splitting between the u and g states, which in turn is related to the transparency of the barrier the "active" electron has to tunnel through in order to jump from one He atom to the other one.

From this picture it is clear that the charge transfer cross section (which will be a little bit larger than $0.5 \ \pi R_c{}^2$) is not directly connected with the Langevin cross section, except in the case where the orbiting radius R_0 happens to be larger than R_c, so that the charge transfer probability will rise abruptly from a small value up to 0.5 when the impact parameter reaches (by decreasing values) R_0.

Now let us turn to near-resonant charge transfer $A^+ + B \rightarrow A + B^+$ where we shall assume that the potential curves of the relevant states undergo an avoided crossing at some large internuclear distance. If the charge exchange is energetically allowed, it is clear that it will take place if the system undergoes one (and only one) transition at the avoided crossing, either on its way in or on its way out. Classically a maximum charge transfer probability $P = 0.5$ will be reached if the transition probability at the avoided crossing (which may be estimated in a good approximation from Landau-Zener formula) is 0.5. On the other hand, P will be very small if the transition probability is either close to 0 or close to 1. The actual quantum-mechanical charge exchange probability will of course oscillate about the value 0.5 in some range of impact parameters (at a given collision velocity) and decrease to 0 as well for larger as for smaller impact parameters. If the splitting between the diabatic states at the avoided crossing is

large (i.e., if the tunneling probability at this point is large)
the charge transfer process will be favored by a somewhat large
collision velocity; conversely, if the interaction between the
diabatic states is small, the charge transfer process will occur
with least difficulty at low radial relative velocities.

Again there is no direct connection between the charge transfer cross section and the Langevin cross section.

It is expected from the above description and usually verified
that the cross sections for near-resonant charge transfer will be
smaller than for resonant charge transfer.

Up to now I have considered very simple systems with widely
spaced states. In the case of <u>polyatomic</u> species with a large
density of vibronic states, the occurrence of <u>near-resonant</u> charge
transfer (with the smallest possible energy difference between
reactants and products) is evidenced by the observation that in
the process

$$X^+ + M \rightarrow M^+ + X$$

the largest charge transfer probability occurs when M^+ is formed
in a state with an internal energy equal to the recombination energy of X^+. Thus, various authors, in particular Bowers and coll.,
found that for a given M the most efficient X^+ projectiles are
those whose recombination energy equals the internal energies of M^+
corresponding to the most intense peaks in the photoelectron spectrum of M, i.e., to the largest Franck-Condon factors between the
ground state of M and the accessible electronic states of M^+.
Similar results were obtained by Munson using a given X^+ and various M.

This near-resonant principle has often been verified by these
and other techniques (ICR line shapes; kinetic energy distribution
of fragments; breakdown patterns; emission spectroscopy of M^+,
etc.). But sometimes a slight shift between expected and observed
most probable transitions indicates a perturbation of the Franck-
Condon factors connecting M and M^+. This is often referred to in
terms of the so-called Lipeles mechanism (cf Lipeles 1969), according to which the Franck-Condon factors have to be taken between
modified potential curves of M and M^+, due to the perturbing electric field of X^+.

However, some confusion will arise if one uses the same concept in the very different cases of high-energy collisions and
thermal-energy collisions. In the former case, more specifically
when the collision time is much shorter than a vibrational period,
the search for a perturbed vibrational wave function during the
collision would have no real significance; one may rather consider

that the initial distribution of internuclear distances is preserved
after the collision but that one or a few of the nuclei experience
a momentum change similar to the deflection of the X^+ projectile,
giving rise to an apparent distortion of the Franck-Condon factors.
In the latter case, in contrast, the perturbation of the Franck-
Condon factors is expected to appear only under thermal-energy conditions and not at higher energies. Here there is time for a re-
adjustment of the vibrational wave function of M perturbed by X^+
and one may well speak of Franck-Condon factors connecting the
perturbed states of M and M^+.

CHARGE TRANSFER THROUGH MULTIPLE CROSSINGS

No special comment.

SHORT-DISTANCE EFFECTS

Under this heading we shall consider cases where an extensive
relaxation of the excess energy takes place in the process leading
to charge transfer.

Direct evidence for more or less <u>long-lived intermediate com-
plexes</u> can be brought only from contour maps of the product velocity distribution, such as those J. Futrell will show from his work
with Z. Herman.

Evidences of other kinds are more ambiguous. A sharp dependence of the cross section with respect to the vibrational energy
of a reactant may as well be due to curve crossing effects as shown
in the case of ion-molecule reactions (Tully and Preston, 1971;
Chapman and Preston, 1974). Peculiar shapes of the translational
energy dependence of the charge transfer cross section may be due
to various effects, in particular the occurrence of a minimum may
be due to the superposition of two processes. A resonance-type
behavior of the cross section, which would be indicative of the
formation of an intermediate complex in an endothermal process,
has never been observed to my knowledge. Finally we shall see
that some phenomena are as yet not clear even in the simple $He^+ + N_2$
system, where the fate of the excess energy in the process leading
to the $\widetilde{B} \to \widetilde{X}$ emission is still unknown.

W. T. Huntress, Jr.: COMMENTS ON FRANCK-CONDON FACTORS IN THERMAL
        ENERGY CHARGE TRANSFER REACTIONS

Charge transfer reactions at thermal kinetic energies are
unusual among ion-molecule processes because of the apparent specificity of the reaction towards the details of energy resonance and
Franck-Condon factors. Exothermic charge transfer reactions appear
to require a close matching of energy levels between reactants and
products together with a large Franck-Condon factor for coupling

of the product ion with the ground state neutral. If these conditions are not met, then the reaction is generally quite slow even
if highly exothermic.

A great deal of the work on energy resonance and Franck-Condon
factors in ion-molecule reactions at thermal energies has come from
Michael Bowers at the University of California at Santa Barbara
(M. T. Bowers and D. D. Elleman, 1972) and from our own laboratories (J. B. Laudenslager et al., 1974).

In our laboratory, the rate constants for several dozen reactions of rare gas ions with diatomic molecules were measured using
the ICR trapped ion technique, and the results were compared to
correlations of the rare gas ion recombination energies with states
of the product ions given in photoelectron spectra.

If k_L is the Langevin rate constant, the measured rate constant
when divided by k_L gives an indication of the reaction "efficiency",
or the approximate probability of reaction per collision. For Ne^+,
Ar^+, and Kr^+ there are no states visible in the photoelectron spectrum of O_2 which are energy resonant with the recombination energy
of the ion, and these reactions are observed to be slow, or k/k_L
< 0.1. These latter reactions are clearly much slower than the
reactions of He^+ and Xe^+ with O_2, for which the recombination energy of the ion falls underneath a rotation-vibrational envelope
of a state which is observed in the photoelectron spectrum.

The low rate constants observed for the reactions of Ne^+, Ar^+,
and Kr^+ ions with O_2 may correspond to charge transfer in complex
formation as Eldon Ferguson has proposed to explain the difference
in reactivities between H^+ and O^+ ions in charge transfer reactions
with NO. If charge transfer can take place during long-lived encounters where adiabatic rather than vertical transitions are more
appropriate, then Franck-Condon factors are less likely to be important and correlations between near resonant states are more
likely to control the probability for charge transfer in such encounters. There are two other alternative mechanisms for charge
transfer reactions where energy resonance and Franck-Condon factors
are not favorable. The reaction may occur with a very small probability at large impact parameters due to finite probabilities for
curve crossings, or the reaction may occur with high probability
at only very small impact parameters where the energy levels and
Franck-Condon factors are highly distorted due to close interaction.
Both of these latter mechanisms yield small rate constants.

There are several pitfalls in using the photoelectron spectrum
to obtain the states and Franck-Condon factors for ions. First,
the photoelectron spectrum displays only those ionic states which
are connected to the ground molecular state by an <u>optical</u> transition. There may be other states available to the electron transfer

process which are forbidden by optical transitions. Second, auto-
ionizing transitions do not appear in photoelectron spectra; and
third, repulsive states do not appear in photoelectron spectra.

With regard to dissociative vs. non-dissociative charge transfer, the thermodynamic availability of a dissociative pathway may
not necessarily result in a large rate constant. For example, exothermic dissociative pathways are available for charge transfer
reactions of Ne^+ ions with O_2, and CO, but these reactions have
very small rate constants. The He^+-O_2 reaction yields a large
percentage of O^+ ions, which is probably the result of charge
transfer into the $^2\Pi_u$ state observed in the photoelectron spectrum
of O_2 near the recombination energy of He^+. The diffuse nature of
this state in the photoelectron spectrum indicates that it is most
likely predissociated to O^+ and O. For dissociative charge transfer, a repulsive state of the product ion must cross the Franck-
Condon region in the vicinity of the recombination energy of the
reactant ion. This requirement provides an explanation for the
failure to observe a large constant for the highly exothermic
charge transfer reaction between He^+ ions and H_2. Examination of
the potential energy curves of H_2^+ shows that in the region of the
recombination energy of He^+ there are no bound states of H_2^+, and
the repulsive potential curves cross this region at much larger
internuclear distance than exists in the ground state of H_2.
Biondi and Johnsen have considered this point in more detail in
their work on the He^+-H_2 reaction.

I also wish to report our observations on two charge transfer
reactions which are accidentally near resonant. These are

$$Kr^+(^2P_{3/2}) + CO \rightleftarrows CO^+(X^2\Sigma^+, v=0) + Kr$$

and

$$Xe^+(^2P_{3/2}) + O_2 \rightleftarrows O_2^+(X^2\Pi g, v=0) + Xe .$$

Reaction (1) is 0.014 eV endothermic in the forward direction.
This is well within thermal energies, 0.04 eV, so that the reaction
is observed to proceed in the forward direction at thermal ion kinetic energies. Reaction (2) is exothermic in the forward direction by 0.07 eV.

Both the forward and reverse reaction were observed in ICR
trapped ion experiments for these reactions. In order to prevent
back reaction, and to obtain both forward and reverse rate constants, ICR experiments were performed in which the decay of one
or the other of the Kr^+, CO^+, Xe^+, or O_2^+ ions is observed in the
absence of the product ion. This is achieved by continuous rf
cyclotron ejection of the product ion during observation of the
decay of the reactant ion.

The rate constants measured for these accidentally resonant reactions are all significantly less than the capture rate constants predicted by the Langevin theory: Kr^+-CO, 7.1×10^{-10}; CO^+-Kr, 8.0×10^{-10}; Xe^+-O_2, 5.8×10^{-10}; and O_2^+-Xe, 9.2×10^{-10} cm^3/sec.

Manfred A. Biondi and Rainer Johnsen: COMMENTS ON NON-RESONANT CHARGE TRANSFER BETWEEN SIMPLE ION-NEUTRAL SYSTEMS

Results of our drift tube/mass spectrometer measurements of several charge transfer reactions between simple atomic ion-neutral atom or diatomic molecule systems are examined to clarify, if possible, the parameters which determine the probability of charge transfer at thermal energies. Earlier studies of symmetric, resonant charge transfer processes for noble gas systems (Chanin and Biondi, 1957; Sheldon, 1962), e.g.

$$He^+ + He \rightarrow He + He^+ , \tag{1}$$

reveal large thermal energy charge transfer coefficients, $k_{res} \gtrsim 10^{-9}$ cm^3/sec and cross sections which vary slowly with energy,

$$Q_{res} = A - B \ln \varepsilon_i , \tag{2}$$

where A and B are constants and ε_i is the ion-atom relative energy. These results are in excellent agreement with theory (Holstein, 1952; Bates, D. R. and McCarroll, R., 1962).

By way of contrast, exoergic non-resonant charge transfer processes vary from rapid (Johnsen et al., 1973), i.e.

$$He^+ + Hg \rightarrow He + Hg^{+*} , \tag{3}$$

$k(He^+, Hg) = (1.6 \pm 0.3) \times 10^{-9}$ cm^3/sec at 300 K; to moderately slow, $k \sim 10^{-11}$ cm^3/sec for (Ar^+, Hg) and (Kr^+, Hg), to immeasurably slow, $k < 10^{-12}$ to 10^{-13} cm^3/sec, for (Ne^+, Hg), (Xe^+, Hg), (He^+, Ne) and (He^+, Ar).

In cases where a neutral molecule is involved and dissociative charge transfer occurs, the rates at thermal energy vary from rapid, $k \approx 1 \times 10^{-9}$ cm^3/sec for

$$He^+ + N_2 \nearrow He + N^+ + N \tag{4a}$$
$$\searrow He + N_2^+ \tag{4b}$$

and similarly for $He^+ + O_2$ to very slow, $k = (1.1 \pm 0.1) \times 10^{-13}$ cm^3/sec for

$$He^+ + H_2 \nearrow He + H^+ + H \tag{5a}$$
$$\rightarrow He + H_2^+ \tag{5b}$$
$$\searrow HeH^+ + H \tag{5c}$$

In reaction (4) the dissociative branch (a) occurs roughly half the time, while in reaction (5) it occurs over 80% of the time (Heimerl, et al., 1969; Johnsen and Biondi, 1974).

The ideas of Turner-Smith et al. (1973) have been used to calculate the reaction rate in terms of transitions from the initial state $X^+ + Y$ to the final state $X + Y^{+*}$ at a pseudo-crossing ("avoided" crossing) of the potential curves of the system. Landau-Zener formalism, an empirical matrix element H_{12}, the known polarizabilities of X and of Y and the (exoergic) energy difference $\Delta\varepsilon(\infty)$ between initial and final states at infinite separation are used to estimate the reaction rate. The theory predicts large rate coefficients only for small (but non-zero) values of $\Delta\varepsilon(\infty)$. To achieve such energy differences the product ion must often be formed in an excited or dissociated state.

This Turner-Smith theory is extremely sensitive to the values chosen for H_{12} - the uncertainties in the empirical values permit many orders of magnitude variation in the predicted rate coefficients. Further, only outer pseudo-crossings between states of the same symmetry are considered, and so smaller contributions from inner crossings and real curve crossings of states of different symmetry are neglected - these contributions may account for the small rate coefficients (10^{-12} - 10^{-11} cm^3/sec) noted in cases where the theory predicts an extremely small value.

In the case of thermal energy charge transfer involving diatomic molecules it appears that, in addition to the exoergic, near-resonant energy requirement, one must also consider the Franck-Condon overlap between the initial and final molecular states. For $He^+ + N_2$ and $He^+ + O_2$ there is a good overlap between N_2 and O_2 molecules in their ground vibrational state and both an excited and a dissociated state of the N_2^+ and O_2^+ ions at energy resonance [i.e., small $\Delta\varepsilon(\infty)$] for the reaction. In the case of $He^+ + H_2$, at energy resonance there is very poor overlap between the Franck-Condon region of the v=0 state of H_2 and the two available states of H_2^+. The small rate coefficient for the dissociative process probably results from near-cancellation of the overlap as a result of the rapid oscillations in the H_2^+ (σ_g) wave function at internuclear separations within the Franck-Condon region of the ground state (gaussian) H_2 wave function.

Thus, it appears that in simple atomic ion-atom or diatomic molecule systems a necessary (but by no means sufficient) condition for rapid thermal energy charge transfer is exoergic near-resonance. Further, for molecules, available states at near-energy resonance must occur within the Franck-Condon overlap region.

J. H. Futrell: COMMENTS ON A CROSSED MOLECULAR BEAM STUDY OF CHARGE
TRANSFER OF Ar^+ WITH NO

I wish to report briefly on what I believe is the first study
of charge transfer reactions at low relative energy utilizing a
crossed-beam ion-neutral collision apparatus. This work was carried out in collaboration with Z. Herman, A. Yensha, V. Pacak and
J. Heyrovsky at the Czechoslovak Academy of Sciences. Characteristics of the specific apparatus have been presented in an earlier
lecture (Futrell, 1974) and a more detailed description has been
published elsewhere (Birkinshaw, 1973). The present experiments
have applied these techniques in an investigation of the charge
exchange reaction of Ar^+ with NO at a relative kinetic energy of
0.65 eV.

This system appeared especially favorable for a crossed-beam
study in view of the results obtained in injected-ion drift tube
studies by Kobayashi (Kobayashi, 1974) and by Birkinshaw and Hasted
(Birkinshaw, 1971). These experiments have shown that several rare
gas-diatomic molecules exhibit a minimum in the energy dependence
of the charge transfer cross-section. With NO as the reactant
neutral, the cross section exhibited an $E^{-1/2}$ dependence at low
energy, consistent with the hypothesis that orbiting collisions
governed by the long-range ion-induced-dipole force between the
reactants dominates the reaction. At higher relative energy the
cross section increases in a manner described at least qualitatively
by a rectilinear trajectory model (Rapp and Francis, 1962). Of the
several systems studied, Ar^+/NO had the largest cross-section
(about four tenths of the Langevin collision frequency limit) and
also the highest energy minimum (1.3 eV relative kinetic energy).
Moreover, a recent photoelectron spectroscopy study (Edquist, et
al., 1969) provided precise information on energy levels and
Franck-Condon factors for population of vibrational levels of the
various electronic states of NO^+.

A contour diagram in Cartesian velocity space giving the
angular distribution in the collision plane and velocity relative
to the center of mass of NO^+ for a collision energy of 0.65 eV
reveals two general features. First, the region of maximum intensity at $\chi = 180°$ shows that an important mechanism is electron
transfer without momentum exchange, such that the product recoils
with the velocity and direction of the initially neutral NO beam.
Second, there is substantial intensity at $\chi = 0°$ (forward scattered)
and at all angles; this process falls within the "resonant charge
transfer" circle and is moderately inelastic, particularly at
moderate CM angles.

A further analysis of the center-of-mass angular distribution
of products leads to the rough estimate that the relative importance of the two processes "complex"/"electron exchange" $\simeq 10$ at

0.65 eV This result is entirely consistent with the injected-ion
drift tube result if we assume the $E^{-1/2}$ dependence is evidence for
the "complex" mechanism for charge transfer. Extrapolation of
Kobayashi's curves (Kobayashi, 1974) as two overlapping straight
line segments suggests that 90-100 percent of the reaction should
proceed via this mechanism at 0.65 eV.

The data may be further analyzed in terms of energy conversion,
from translational to internal energy, by examining the translational energy distribution of products (relative to reactants).
This led to the interesting result that the distributions matches
very well that predicted by the Franck-Condon factors for the $^3\Sigma^+$
first excited state of NO^+ (Edquist, 1969), assuming our ion beam
contains both $^2P_{3/2}$ and $^2P_{1/2}$ states of Ar^+ in the ratio of their
statistical weights. This state is nearly resonant with the recombination energy of Ar^+. At a relative energy of 0.65 eV only the
$^1\Sigma^+$ ground state and $^3\Sigma^+$ first excited state are energetically
accessible. Franck-Condon factors for populating very high vibrational levels of the ground state of the ion to achieve energy
matching with the recombination energies of argon are very small.

It is at first surprising that the inelastic, momentum-exchange
"complex" mechanism for charge transfer appears to obey relative
Franck-Condon factors for the unperturbed molecule, since the lifetime of the complex ($\geq \tau_{rotation}$ of complex) provides ample opportunity for nuclear motion. In this connection, it is useful to
consider the structure of the molecule NOCl, which is isoelectronic
with the postulated $ArNO^+$ complex. In this molecule the O-N bond
distance is 1.16 Å, the N-Cl bond distance is 1.95 Å, and the O-N-Cl
angle is 116° (Cotton, 1972 a). The bond distance for ground state
NO is 1.15 Å (Cotton, 1972 b), practically the same as in the triatomic molecule. Thus, if we postulate that the structure and
interatomic distances of the intermediate complex are similar to
those for the isoelectronic triatomic molecule it is not unreasonable to postulate that electronic transfer should follow approximately the Franck-Condon factors for the isolated NO molecule for
this particular case.

In conclusion we have obtained in a crossed beam study of the
detailed kinematics of a charge transfer reaction direct evidence
for the transition from a (predominately) complex mechanism at low
relative energy to an electron stripping mechanism at higher energy.
Moreover, the analysis of translational-to-internal energy states
of products. This "translational spectroscopy" aspect of the interpretation of crossed-beam measurements will doubtless become increasingly important in future work with the improved resolution
now available for these experiments.

T. O. Tiernan: COMMENTS ON ISOTOPIC SCRAMBLING IN THE PRODUCTS OF
CHARGE TRANSFER PROCESSES

In connection with the investigations of endoergic negative-
ion charge-transfer processes which were discussed earlier in the
session on "The Reactions of Negative Ions", it is appropriate to
present here some data on isotopic scrambling in the products
of these reactions. In beam experiments, where the mass of the
incident ion can be selected, it is possible to use isotopically
labelled species which are distinct from those in the neutral
target molecule. Mass analysis of the product ion then gives an
indication of the extent to which the incident ion has been incorporated into the product and provides some indication of the
mechanism of the process. These techniques have been widely
applied, (Futrell and Tiernan, 1972). Obviously, the conclusions
which can be drawn from such experiments regarding the mechanism
are not as definitive as detailed scattering data such as that
which has been described at this meeting. Isotopic data really
provide little or no information on the lifetimes of intermediate
collision complexes, and in some cases scrambling apparently
occurs even at energies where velocity distribution measurements
indicate a direct mechanism. Durup has recently discussed this
situation, (Durup, 1974).

It appears however, that a marked change in the extent of
isotopic scrambling in the products of ion-neutral processes is
indicative of a change in mechanism. This may be illustrated by
the results shown in Table I, which reports the experimentally
measured isotopic product distribution from the reactions of $^{79}Br^-$
and $^{81}Br^-$ with Br_2 molecules having the natural isotopic distribution (more detailed data has been published in Hughes, Lifshitz,
and Tiernan, 1973). The data are presented at two energies, at
1.9eV, which is very near the threshold for the reaction, and at
10.5eV, which is well above threshold and in the region where the
reaction cross section attains a maximum. The observed distributions for these two energies are obviously quite different.

It can be seen that at the higher energy, where the cross
section for electron transfer from Br^- to Br_2 is essentially independent of energy, the isotopic distribution of the Br_2^- product
ion approaches that which is characteristic of natural bromine,
(that is, the relative ratio of $^{79}Br^{79}Br/^{79}Br^{81}Br/^{81}Br^{81}Br$ in
natural bromine is 0.255/0.499/0.244.) Moreover, at these energies, the isotopic distribution of Br_2^- products is virtually
identical from the reactions of both $^{79}Br^-$ and $^{81}Br^-$, indicating

TABLE I. Isotope distributions in products from endothermic
charge transfer reactions of Br^- with Br_2

Mechanism	Impacting ion	Calculated product ion Distributions (%)		
		$79,79Br_2^-$	$79,81Br_2^-$	$81,81Br_2^-$
I. $Br^- \rightarrow Br_2^-$ (long-range electron transfer)	$^{79}Br^-$	25.5	49.9	24.4
	$^{81}Br^-$	25.5	49.9	24.4
II. $Br^- \ldots Br \rightarrow Br$ (atom abstraction)	$^{79}Br^-$	50.6	49.4	0.0
	$^{81}Br^-$	0.0	50.6	49.4
III. $Br \ldots \overset{\ominus}{Br} \ldots$ (linear complex)	$^{79}Br^-$	38.1	49.7	12.2
	$^{81}Br^-$	12.8	50.2	37.0
IV. (cyclic complex)	$^{79}Br^-$	42.2	49.6	8.2
	$^{81}Br^-$	8.5	50.4	41.7

Reaction	Ion energy (eV)	Impacting ion	Experimental product ion distributions (%)		
			$79,79Br_2^-$	$79,81Br_2^-$	$81,81Br_2^-$
$Br^- + Br_2 \rightarrow Br_2^- + Br$	1.9	$^{79}Br^-$	38.0	50.5	11.9
	1.9	$^{81}Br^-$	12.1	50.5	37.4
	10.5	$^{79}Br^-$	25.6	50.2	24.5
	10.5	$^{81}Br^-$	25.2	50.4	24.5

that the impacting ion is not incorporated into the products. At
translational energies nearer threshold, however, the Br_2^- products
exhibit an isotopic distribution which is quite different from
the normal bromine distribution, and obviously some isotopic
scrambling has occurred. Also, at lower energies, the isotopic
distributions of the product ions are significantly different
for the cases where $^{79}Br^-$ and $^{81}Br^-$ are the impacting ions.

Also shown in Table I are the expected isotopic distributions
calculated on the basis of several possible mechanisms. One
such mechanism is long-range electron transfer which involves no
bonding between the incident negative ion and the neutral halogen

molecule, but which results in formation of the negative molecular ion. In this case, the isotopic distributions of the product Br_2^- ions should be just that which corresponds to the natural isotopic abundances of the halogen molecules. Another possibility is an atom abstraction mechanism, in which the product molecular ion is formed by the creation of a new bond between the incident ion and one atom of the target molecule, with simultaneous rupture of the bond between the two atoms of the target molecule. In this case, it is assumed that there is equal probability of abstracting either atom of the target molecule, but the incident ion is always incorporated in the product ion. A third possibility is a mechanism which involves formation of a linear intermediate $[Br_3]^-$ complex, in which the available collision energy is distributed equally among the vibrational degrees of freedom. Here, it is assumed that it is equally probable that either of the two "partial" bonds of the complex will be broken to yield the product molecular ion. A somewhat similar mechanism for the charge transfer process would entail a cyclic intermediate complex, having three "partial" bonds. Again, the collision energy is assumed to be randomized and loss of any one of the three atoms to give the product ion is considered to be equally likely. The isotopic product distributions which are calculated for these four reaction models for the reactions of $^{79}Br^-$ and $^{81}Br^-$ with natural bromine molecules are given in Table I. It is evident, from these data that the isotopic distributions of the Br_2^- products, at translational energies in the threshold region, are in striking agreement with the distributions calculated for a reaction mechanism which involves a linear triatomic intermediate complex. At translational energies several volts above threshold however, the experimental product distributions are quite consistent with the isotopic distributions calculated for a long range electron transfer mechanism, where no isotopic scrambling occurs. More detailed data suggests that there is a gradual transition from the complex mechanism to the long range mechanism as the energy increases above threshold. Probably, more than one mechanism is operative in the intermediate energy regime. It is difficult to visualize an endoergic reaction of this type which does not involve an intimate complex, however. In any event, this data strongly suggests that an intermediate $[Br_3]^-$ collision complex is indeed formed in the energy region near threshold. Since this complex was never actually detected in our experiments, (at collision chamber pressures up to 30μ, the maximum used), its lifetime must be shorter than 20-30μsec, which is approximately the transit time for this species from the collision chamber to the detector in our tandem instrument.

Linear symmetric structures of the polyhalide ions which correspond to the intermediate complexes discussed above were predicted from molecular orbital calculations, (Pimentel, 1951),

which showed that the bonding in these ions could be described
in terms of "p" orbitals without involving the higher "d" orbitals.
Later measurements of the nuclear quadrupole coupling constants
in the polyhalide ions confirmed this interpretation of the bonding, (Yamasaki and Cornwell, 1959). The infrared and Raman
spectra of the polyhalide ions determined by Person et al.,
(Person et al., 1962), are also consistent with this structural
picture. The symmetric structure which our isotopic data
suggests for the $[Br_3]^-$ intermediate complex therefore appears to
be quite reasonable.

REFERENCES

Bates, D. R. and McCarroll, R. (1962). Adv. in Phys. (Phil. Mag.
    Suppl.) 11, 39.
Birkinshaw, K. and Hasted, J. B. (1971). J. Phys. B4, 1711.
Birkinshaw, K. and Herman, Z. (1973). Ber. Bunsenges. Phys. Chem.
    77, 566.
Bowers, M. T. and Elleman, D. D. (1972). Chem. Phys. Lett. 16, 486.
Chanin, L. M. and Biondi, M. A. (1957). Phys. Rev. 106, 473.
Chapman, S. and Preston, R. K. (1974). J. Chem. Phys. 60, 650.
Cotton, F. A. and Wilkinson, G. (1972). Advanced Inorganic Chem-
    istry, p. 356, 365. Wiley-Interscience, New York.
Durup, J. (1974). Recent progress in the understanding of the
    mechanisms of ion-molecule reactions. Advances in Mass Spec-
    trometry (Ed. A. R. West). Applied Science Publishers, Ltd.,
    Barking, Essex.
Edquist, O., Lindholm, E., Selin, L. E., Sjogren, H. and Asbrink,
    L. (1969). Arkiv for Fysik, 40, 439.
Futrell, J. H. (1975). These Proceedings.
Futrell, J. H. and Tiernan, T. O. (1972). Tandem mass spectrometric
    studies of ion-molecule reactions. Ion-Molecule Reactions (Ed.
    J. L. Franklin). Plenum Press, New York.
Gauglhofer, J. and Kevan, J. (1972). Chem. Phys. Lett. 16, 492.
Heimerl, J., Johnsen, R. and Biondi, M. A. (1969). J. Chem. Phys.
    51, 5041.
Holstein, T. D. (1952). J. Phys. Chem. 56, 832.
Hughes, B. M., Lifshitz, C. and Tiernan, T. O. (1973). J. Chem.
    Phys. 59, 3162.
Johnsen, R., Leu, M. T., and Biondi, M. A. (1973). Phys. Rev. A8,
    1808.
Johnsen, R. and Biondi, M. A. (1974). J. Chem. Phys.
Kobayashi, N. (1974). J. Phys. Soc. of Japan 36, 259.
Laudenslager, J. B., Huntress, W. T. and Bowers, M. T. (1974). J.
    Chem. Phys.
Lipeles, M. (1969). J. Chem. Phys. 51, 1252.

Person, W. B., Anderson, G. R., Fordemwalt, J. N., Stammreich, H.,
  and Foneris, R. (1962). J. Chem. Phys. 35, 908.
Pimentel, G. C. (1951). J. Chem. Phys. 19, 446.
Rapp, D. and Francis, W. E. (1962). J. Chem. Phys. 37, 2631.
Sheldon, J. W. (1962). Phys. Rev. Lett. 8, 64.
Tully, J. C. and Preston, R. K. (1971). J. Chem. Phys. 55, 562.
Turner-Smith, A. R., Green, J. M. and Webb, C. E. (1973). J. Phys.
  B6, 114.
Yamasaki, R. S. and Cornwell, C. D. (1959). J. Chem. Phys. 30, 1265.

PLANETARY IONOSPHERES

SUMMARY OF THE PANEL DISCUSSION LED BY:

Rocco S. Narcisi

Air Force Cambridge Research Laboratories

Hanscom AFB, MA   01730

This paper is intended to provide a very brief introduction to some current concepts of planetary atmospheres for the purpose of considering planetary ion molecule reactions. Model calculations of neutral and ionic species concentrations for the upper atmospheres of Venus, Mars and the major planets (Jupiter, Saturn, Uranus and Neptune) are discussed. The models presented were somewhat arbitrarily chosen and are not to be construed as being more valid than others that are available. Indeed, it is to be emphasized that much of this material is highly speculative owing to a lack of detailed measurements in planetary environments and to gaps in our knowledge of basic atomic and molecular processes.

Ion-molecule reactions are indispensable to the study of planetary ionospheres. Since many hundreds of reactions are involved, the subject is far too extensive to be handled properly in a panel session; rather, only selected topics are treated. Reactions important in the earth's ionosphere are also discussed here but are covered in more detail by E.E. Ferguson.

VENUS

Several attempts have been made to model the Venus atmosphere, e.g. Kumar and Hunten (1974) and references therein. Fig. 1 depicts the neutral atmosphere model of Venus derived by Kumar and Hunten (1974). The composition in the lower atmosphere is assumed to be 98% CO_2, 1% CO, 1% O, 0.1% N, 2×10^{-4}% H_2 and 2×10^{-2}% He. The exospheric temperature is 350K and the total particle number density at 125 km is 10^{13} cm^{-3}. The calculation includes an eddy diffusion coefficient of 10^8 cm^2 s^{-1} with an assumed turbopause

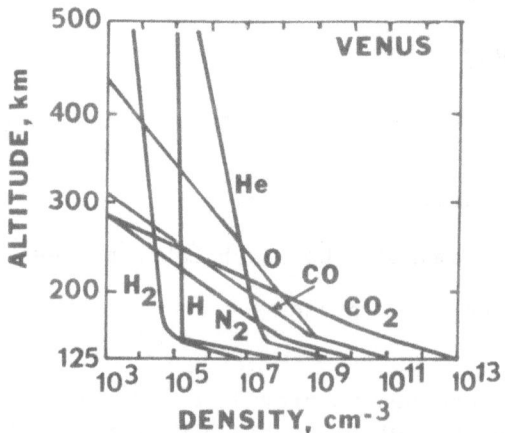

Fig. 1.   Neutral atmosphere model of Venus from Kumar and Hunten
(1974).

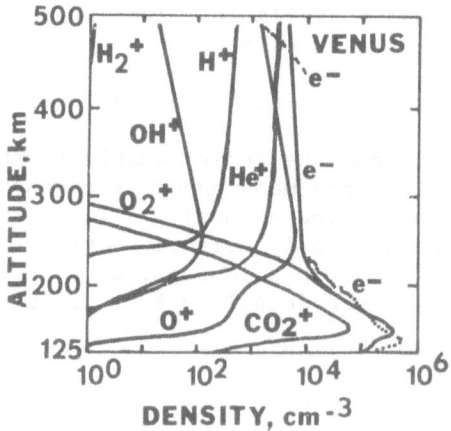

Fig. 2.   Ionosphere model of Venus from Kumar and Hunten (1974)
compared with Mariner 5 electron density measurements
(points and dashes).

height at 145 km above which mixing ceases and the constituents distribute diffusively according to their molecular weights. Recent ultraviolet observations of Venus from Mariner 10 (Broadfoot et al., 1974) have indicated the additional presence of A, Ne and C as well as He, H, O, CO and He^+ in the atmosphere.

Once the neutral species distributions are available, the ionosphere may be determined. Ionization rates are calculated using the scaled solar flux, ionization cross sections, etc. for such reactions as:

$$h\nu + CO_2 \begin{cases} \longrightarrow CO_2^+ + e \\ \longrightarrow O^+ + CO + e \\ \longrightarrow CO^+ + O + e \end{cases}$$

which are important in the lower ionosphere. These primary ions then participate in various ion molecule reactions:

$$O^+ + CO_2 \rightarrow O_2^+ + CO$$
$$CO_2^+ + O \rightarrow O^+ + CO_2$$
$$\rightarrow O_2^+ + CO$$
$$CO^+ + CO_2 \rightarrow CO + CO_2^+$$

which eventually produce O_2^+ as the major ion. Photochemical equilibrium prevails below 200 km where chemistry is faster than diffusion. At higher altitudes the reverse is true so that the ion diffusion equations must be included. Finally, the ions are removed, typically and most efficiently by dissociative electron-ion recombination. Fig. 2 gives the Kumar and Hunten (1974) Venus ionosphere model for daytime conditions using the neutral atmosphere in Fig. 1. The electron density profile deduced from Mariner 5 data (Kliore et al., 1967) is also shown in Fig. 1 for comparison. The plasmapause, or the sudden drop in electron density near 500 km shown by the dashed curve in Fig. 2 (Fjeldbo and Eshleman, 1969), is not predicted by the model and is presumably caused by the solar wind sweeping away the ions at high altitudes. More recent measurements of the Venus dayside and nightside electron densities from Mariner 10 (Howard et al., 1974) are presented in Fig. 3. The nighttime Venus ionosphere may well be maintained, at least in part, by long-lived meteoric atomic-ions (Fe^+, Mg^+, Si^+, etc.) that are deposited by ablating meteoroids. These species are measured in the earth's ionosphere where similar layered structure and plasma densities as in Fig. 3 are often observed. Measurements of meteoric species in the earth's ionosphere have precipitated laboratory

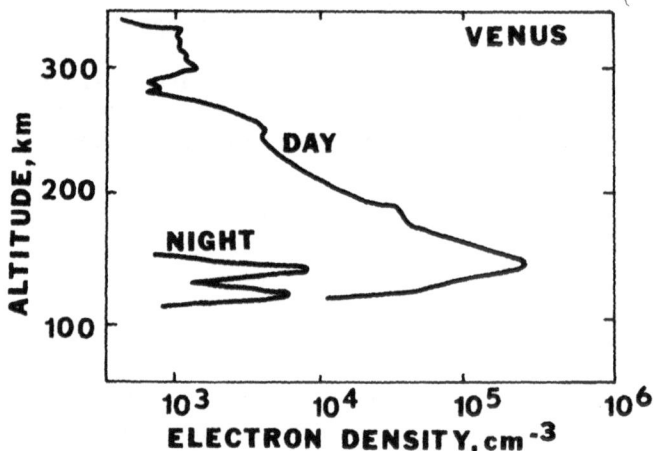

Fig. 3.   Daytime and nighttime electron density profiles of Venus
          from Mariner 10 measurements (Howard et al., 1974).

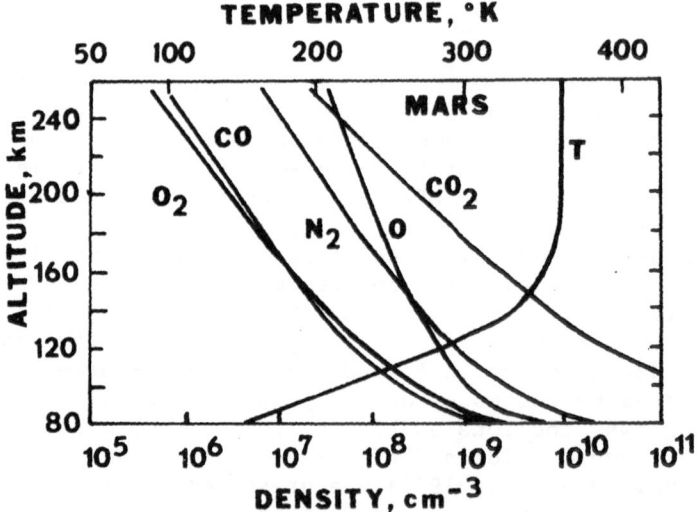

Fig. 4.   Neutral atmosphere model of Mars from McConnell (1973).

studies of numerous reactions of meteor type ions, e.g.
Ferguson and Fehsenfeld (1968).

MARS

Mars has an upper atmosphere that is strikingly similar to
that of Venus. A model of the neutral atmosphere of Mars
(McConnell, 1973; McElroy and McConnell, 1971) which is consistent
with the available data is presented in Fig. 4. Our knowledge of
the martian upper atmosphere (>90 km) comes from the Mariner 6 and
7 UV spectrometer measurements (Barth et al., 1971) and more recently from Mariner 9 (Barth et al., 1972). All of the species in
Fig. 4 have been experimentally identified except N_2 for which a
5% upper limit has been set. H and C were also observed by the
Mariner probes but are not included in Fig. 4. O_2 and CO_2 were
detected by ground-based measurements. The ionospheric chemistry
of Mars is essentially the same as for Venus; the principal processes for the lower portion of the ionosphere were described earlier. Fig. 5 shows computed ionospheric profiles for the two major
species, O_2^+ and CO_2^+, using the neutral atmosphere in Fig. 4
(McConnell, 1973). In order to obtain the good agreement between
the measured electron density from Mariner 7 (Rasool and Stewart,
1971) and the model computation, seen in Fig. 5, the solar flux
was arbitrarily increased by a factor of 2.7! The only ion constituents definitely identified in the martian ionosphere by the
Mariner UV spectrometers are CO_2^+ and CO^+.

MAJOR PLANETS

The upper atmospheres of the four major planets are known to
contain H_2 and CH_4. Atomic hydrogen is produced by photodissociation of H_2, and He is probably an abundant constituent also. The
Jovian atmosphere is composed of H_2, He and H with smaller amounts
of CH_4 and NH_3. Photochemistry of CH_4 apparently produces other
hydrocarbons: C_2H_4, C_2H_2, CH_3 and C_2H_6 (Strobel, 1973). Although
still questionable, it is believed that the upper atmospheres of
the major planets are approximately isothermal and very cold; some
estimates: Jupiter 150K, Saturn 88K, Uranus 64K, and Neptune 50K
(McElroy, 1973). However, analysis of occultation measurements
by Veverka, Wasserman and Sagan (1974) suggests that Neptune's
upper atmospere is not isothermal but is 135K at the level of 10^{15}
cm^{-3} and above this height possesses a small positive temperature
gradient. In any case ion-molecule rate constants need to be measured at very low temperatures. In this chapter the Jovian atmosphere is well summarized by Huntress and some Jovian ionospheric
processes are also discussed by Biondi and Johnsen. Here we shall
demonstrate the uncertainties in the model calculations and the
drastic effects caused by erroneously estimating rate constants
and omitting important reactions.

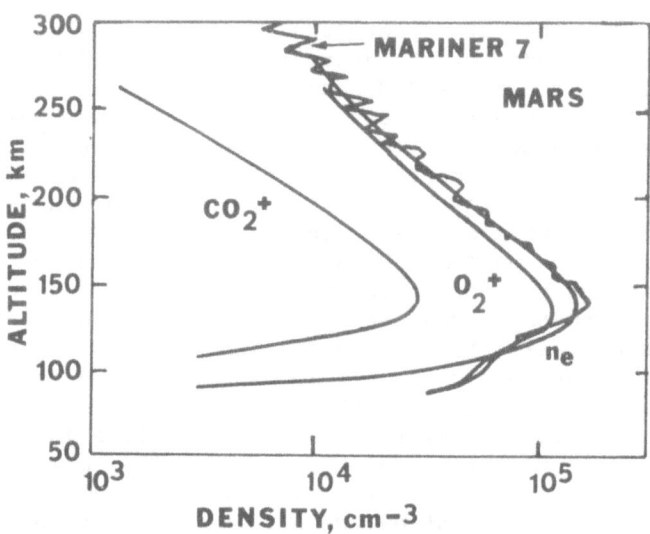

Fig. 5.   Lower ionosphere model of Mars (McConnell, 1973) compared
with the electron density measurements from Mariner 7
(Rasool and Stewart, 1971).

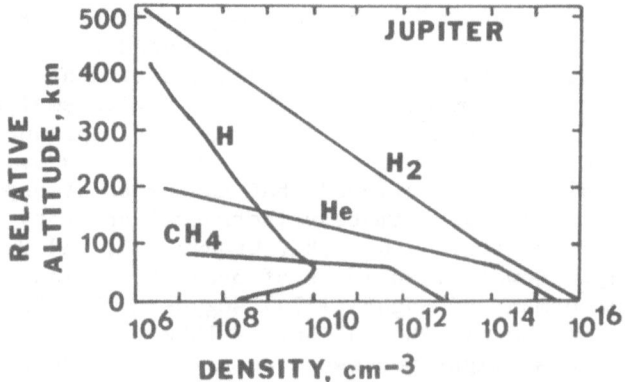

Fig. 6.   Neutral atmosphere model of Jupiter from McElroy (1973).
The height scale is arbitrary.

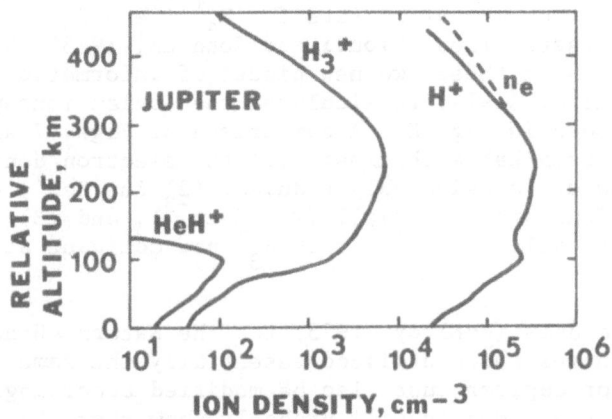

Fig. 7. Ionosphere model of Jupiter from McElroy (1973). The
height scale is arbitrary.

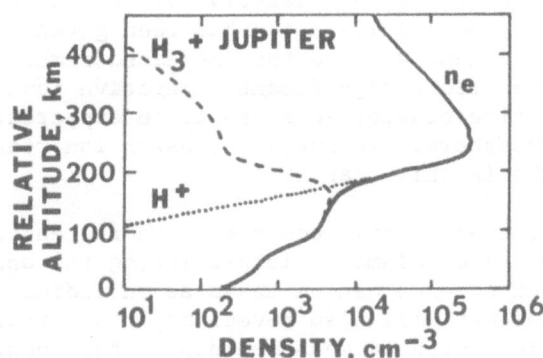

Fig. 8. Model for Jupiter's ionosphere from Atreya, Donahue and
McElroy (1974). Zero altitude is taken as the level where
the neutral atmospheric density is 10^{16} cm^{-3}.

Figs. 6 and 7 present McElroy's (1973) neutral and ionospheric models for Jupiter assuming an eddy diffusion coefficient of 10^5 cm^2 s^{-1} and 150K exospheric temperature. Shortly after McElroy's calculations, the very fast reaction (not included in his model), HeH$^+$ + H$_2$ → H$_3^+$ + He was measured (Theard and Huntress, 1974) and the electron-ion recombination rate for H$_3^+$ + e → H$_2$ + H was found to be 40 times faster (Leu, Biondi and Johnsen, 1973) than the value assumed. With these two new pieces of information, Atreya, Donahue and McElroy (1974) recalculated the Jovian ionosphere and the result is seen in Fig. 8. A comparison of Figs. 7 and 8 clearly reveals the remarkable changes: (1) the electron densities at lower altitudes are considerably reduced, (2) the HeH$^+$, He$^+$ and H$_2^+$ concentrations are very small (<< 10 cm^{-3}), and (3) the H$^+$/H$_3^+$ ratio is dramatically different with H$_3^+$ now dominant at the lower heights.

Previous models (McElroy, 1973) for the Saturn, Uranus and Neptune ionospheres which utilized essentially the same ionospheric chemistry as for Jupiter must also be modified accordingly. A recalculation of these models will probably show that the upper ionospheres of Saturn, Uranus and Neptune are similar to the new Jupiter model (Fig. 8) with Neptune's electron density somewhat reduced.

CONCLUDING REMARKS

Many details have been necessarily omitted in this brief overview. For example, no consideration has been given to the lower ionospheres of the planets where the increased number densities make three-body reactions significant. Negative ions and both positive and negative cluster ions are to be expected, as in the earth's lower ionosphere. (Planetary cluster ion reactions are discussed by Burke in this chapter.)

In the past, theoretical models of both the earth and martian ionospheres have failed dismally in predicting the observations. If this earlier experience can be taken as an indicator, future planetary measurements will also reveal major departures from the models. Meanwhile continued measurements of pertinent ion-molecule reaction rates will provide the data necessary to update the models.

W. T. Huntress, Jr.: COMMENTS CONCERNING THE IONOSPHERE OF JUPITER

In the ionosphere of Jupiter - that uppermost extremity of the atmosphere where solar photoionization produces a thin plasma - the neutral species of major importance are H$_2$, He, and H atoms.

In mixtures of H_2, He, and H, the major ionization processes are

$$h\nu + H_2 \longrightarrow H_2^+ + e^- \tag{1a}$$

$$\longrightarrow H^+ + H + e^- \cdot \tag{1b}$$

$$h\nu + He \longrightarrow He^+ + e^- \tag{2}$$

$$h\nu + H \longrightarrow H^+ + e^- \tag{3}$$

Secondary ionization can also occur during the thermalization of hot electrons produced by the solar photoionization process. The subsequent ion-molecule reactions which can take place following ionization are:

$$H_2^+ + H_2 \longrightarrow H_3^+ + H \tag{4}$$

$$+ He \longrightarrow HeH^+ + H \tag{5}$$

$$+ H \longrightarrow H^+ + H_2 \tag{6}$$

$$H^+ + H_2 \longrightarrow H_2^+ + H \tag{7}$$

$$+ He \longrightarrow \text{no reaction} \tag{8}$$

$$+ H \longrightarrow H^+ + H \tag{9}$$

$$He^+ + H_2 \longrightarrow H^+ + H + He \tag{10a}$$

$$\longrightarrow H_2^+ + He \tag{10b}$$

$$\longrightarrow HeH^+ + H \tag{10c}$$

$$+ He \longrightarrow He^+ + He \tag{11}$$

$$+ H \longrightarrow H^+ + He \tag{12}$$

The rate constant for reaction (4) is well-known (Theard and Huntress, 1974, and refs. therein). Reaction (5) can occur only for H_2^+ ions with excess vibrational energy, but the production of H_2^+ ions in vibrationally excited states by photoionization, and the rate constant for reaction (5) as a function of H_2^+ vibrational energy have both been extensively studied (Chupka, Russell and Refaey, 1968), (Chupka and Russell, 1968). Reaction (6) has not been experimentally examined, but this reaction is probably important only at extreme heights in the ionosphere where the H atom density becomes comparable to the H_2 density. Reaction (7) can occur only for kinetically excited protons produced by the dissociative photoionization of H_2 above $\sim$ 30 eV photon energy. These

kinetically excited protons are also rapidly thermalized by the
resonant charge transfer process, reaction (9), and by collisions
with H_2 and He (see Monahan, et al., 1974).

One of the biggest questions in Jovian ionospheric modeling
to date has been the rate constant and product distribution for
reaction (10). Just recently, Johnsen and Biondi (these proceedings) have shown that reaction (10) proceeds at a rate of
1×10^{-13} cm^3/sec at 300°K and that at least 80% of the products,
if not all the products, are protons. For the atomic ions H^+ and
He^+, any bimolecular rate constant greater than about 10^{-17} cm^3/
sec, which converts these ions into diatomic or polyatomic ions,
is a fast reaction in the Jovian ionosphere (Hunten, 1969). This
circumstance arises because of the long lifetime of atomic ions
with respect to recombinations with electrons. The rate constants
for radiative recombination of atomic ions with electrons are very
small, $\sim 6 \times 10^{-12}$ cm^3/sec (Bates and Dalgarno, 1962)

$$H^+ + e^- \longrightarrow H + h\nu \tag{13}$$

$$He^+ + e^- \longrightarrow He + h\nu \tag{14}$$

Reaction (10) is therefore the major loss mechanism for He^+ ions in
spite of the small rate constant, and is also a significant source
of kinetically excited protons in the Jovian ionosphere. Reaction
(12) is an extremely non-resonant, thermal-energy charge transfer
reaction and probably has a rate constant even smaller than 10^{-17}
cm^3/sec.

The terminal ions produced by ion-molecule reactions in the
upper ionosphere are therefore H_3^+ and H^+. The HeH^+ ions
produced by reaction (5) are converted to H_3^+ by the fast reaction
(Theard and Huntress, 1974)

$$HeH^+ + H_2 \longrightarrow H_3^+ + He \tag{15}$$

The loss of H_3^+ ions is controlled by dissociative recombination
with electrons

$$H_3^+ + e^- \longrightarrow H + (H + H) \tag{16}$$

Although the rate constant for reaction (16) is well-known (Leu,
Biondi and Johnsen, 1973), the product distribution is not and
could either be three hydrogen atoms or one hydrogen atom and one
H_2 molecule. The loss rate for thermal energy protons in the
Jovian ionosphere is very slow since there are no bimolecular ion-
molecule reactions available. As a consequence, protons are the
major ionic component throughout most of the ionosphere (except

at the lowest altitudes) and the electron density remains quite high ($\sim 10^5$ cm^{-3}).

Slow processes which might possibly compete with radiative recombination of protons with electrons in the Jovian ionosphere include radiative association

$$H^+ + H_2 \longrightarrow H_3^+ + h\nu \tag{17}$$

collisional association

$$H^+ + 2H_2 \longrightarrow H_3^+ + H_2 \tag{18}$$

bimolecular reaction with methane

$$H^+ + CH_4 \longrightarrow CH_4^+ + H \tag{19a}$$

$$CH_3^+ + H_2 \tag{19b}$$

and bimolecular reaction with vibrationally excited H$_2$ molecules

$$H^+ + H_2{}^* \longrightarrow H_2^+ + H \tag{20}$$

Reaction (18) can occur in the lower part of the ionosphere where the atmospheric density exceeds about 10^{11}/cm^3. This reaction converts protons to H$_3$ ions followed by rapid loss of ionization via reaction (16). If methane reaches to sufficient heights in the ionosphere, reaction (19) can occur in competition with reaction (18). An upper limit for the rate constant of reaction (17) can be obtained from the known rate constant for reaction (18), which yields a lower limit on the lifetime towards dissociation of the (H$_3$)$^+$ intermediate complex. For vibrationally excited ions, the lifetime towards radiation is on the order of 1 msec, and $k_{18} = 3.2 \times 10^{-29}$ cm^6/sec so that $k_{17} \sim 10^{-19}$ cm^3/sec. This value is too small for reaction (17) to be a significant loss mechanism for protons in the ionosphere. Reaction (20) is the only other possible bimolecular process which might compete with radiative recombination, but this reaction has not yet been studied in the laboratory and the vibrational distribution in H$_2$ molecules in the Jovian ionosphere is presently unknown.

In spite of the progress in the laboratory over the past several years on ionic processes in reducing atmospheres, the present laboratory data we now have is still inadequate to model accurately the Jovian ionosphere. There are challenges yet left to the aeronomer as well as the experimentalist. For the former, the question of the amount of helium in the atmosphere still remains, as well as the location of the turbopause in the upper atmosphere and an explanation of the strong ionic layering effect seen by the

radio occultation experiment on Pioneer 10. For the experimentalist, a list of the type of laboratory data required would include
the following:

1) Rate constant for reaction (6) vs. H_2^+ vibrational energy. This
   reaction may be important at extreme heights in the ionosphere.
2) Product and energy distribution for the reaction $H_3^+ + e^-$. This
   reaction is important for H atom production and energy distribution in the neutral ionosphere.
3) Rate constant for the reaction $H^+ + H_2^V$ vs. vibrational temperature of the H_2 molecule. This reaction may be an important
   proton loss mechanism.
4) Electron attachment reactions in H_2-He-CH_4 mixtures. These
   reactions may be important for the formation of negative ions
   in the lower ionosphere.
5) Rate constants and product distributions for the reactions of
   protons with simple hydrocarbons. These reactions may be important for proton loss, and for the production of hydrocarbon
   ions at the lower boundaries of the ionosphere.
6) Rate constants and product distributions for the reactions of
   hydrocarbon ions with H atoms. The H atom concentration remains significant at low altitudes because of the photodissociation of H_2.
7) Three-body reactions of H^+ and H_3^+ ions in mixtures of H_2, He,
   and CH_4.
8) Reactions of metal ions with H_2 and with electrons. Metallic
   ions may be responsible for the layering in the ionosphere
   observed by Pioneer 10. Metal atoms would result from the
   impact of meteorites on the atmosphere or by transport from
   the ionosphere of Io.

   The whole question of whether methane can reach to ionospheric
heights is very important, since in this circumstance the chemistry
in the ionosphere becomes much more rich. Strong mixing of the
lower atmosphere would also result in some methane photoionization
as well (Prinn, 1970) and it has recently been suggested that methyl
radicals formed by photodissociation of CH_4 may be readily ionized
at subionospheric heights because of their low ionization potential
(S. Prasad, private communication). In any of these circumstances,
synthesis of new hydrocarbon species in the Jovian ionosphere can
occur by ionic processes (Huntress, 1974)

Manfred A. Biondi and Rainer Johnsen: COMMENTS ON   ION-MOLECULE
REACTIONS IMPORTANT IN THE JOVIAN IONOSPHERE

Since the primary ions of the Jovian upper ionosphere, H_2^+,
H^+ and He^+ are slow to recombine with electrons it is important to
determine the rates at which they, as well as the intermediate ion
HeH^+, are converted to H_3^+ which rapidly is removed by dissociative
recombination with electrons. [The recombination coefficient between H_3^+ ions and electrons at $150°K$ has been determined to be
large, $\alpha(H_3^+) = (3.4 \pm 0.4) \times 10^{-7}$ cm^3/sec (Leu, et al, 1973).]
Consequently we have used our drift tube-mass spectrometer apparatus to study both the fast and slow reactions of H^+, He^+, and HeH^+
ions with H_2.

The rate coefficient for the three-body reaction

$$H^+ + 2H_2 \rightarrow H_3^+ + H_2 \tag{1}$$

has been determined (Johnsen and Biondi, 1974) to be k =
$(3.1 \pm 0.1) \times 10^{-29}$ cm^6/sec at $300°K$, in good agreement with the
results of other investigators (Graham, et al, 1973). In the
higher regions of the Jovian ionosphere, where the neutral concentrations are small, this three-body reaction will be very slow.

The intermediate ion HeH^+ is found to convert rapidly
(Johnsen and Biondi, 1974) to H_3^+ at $300°K$ via the two-body,
proton transfer reaction

$$HeH^+ + H_2 \rightarrow He + H_3^+ \tag{2}$$

with $k \gtrsim 10^{-9}$ cm^3/sec, in agreement with other results (Theard and
Huntress, 1974) and is of comparable speed to the analagous
$H_2^+ + H_2 \rightarrow H_3^+ + H$ reaction (Theard and Huntress, 1974). Thus,
paths from H^+, HeH^+ and H_2^+ to the rapidly recombining ion H_3^+
are defined.

Previously, the rate coefficient for the reaction of He^+ ions
with H_2 at low energies ($\sim 300°K$) had been found to be too small to
measure. We find (Johnsen and Biondi, 1974) a total rate coefficient $k_T = (1.1 \pm 0.1) \times 10^{-13}$ cm^3/sec at $300°K$ for the reaction,

$$
\begin{array}{ll}
& \rightarrow He + H^+ + H \tag{3a} \\
He^+ + H_2 & \rightarrow He + H_2^+ \tag{3b} \\
& \rightarrow HeH^+ + H , \tag{3c}
\end{array}
$$

which increases to $\sim 2.8 \times 10^{-13}$ cm^3/sec at ~ 0.18 eV ion mean
energy. The corresponding reaction with the isotope D_2 yields a
total coefficient $k_T = (5 \pm 1) \times 10^{-14}$ cm^3/sec at $300°K$.

The principal branch of reaction (3) is dissociative charge transfer to form H^+, reaction (3a), with the molecular-ion-forming branches (3b) and (3c) accounting for $\leq 20\%$ of the total rate coefficient. Thus atomic ions (H^+ and/or He^+) have long lives in the Jovian upper ionosphere.

Reasons for the slow rate ($k \sim 10^{-13}$ cm^3/sec) of the dissociative charge transfer process, reaction (3a), are discussed in a contribution to the Panel on Charge Transfer elsewhere in these Proceedings.

Rudolf R. Burke: COMMENTS ON CLUSTER IONS IN PLANETARY IONO-
  SPHERES AND THE INTERSTELLAR MEDIUM

Cluster ions are important in planetary atmospheres for two reasons. First, cluster ions recombine dissociatively with electrons at rates typically one to two orders of magnitude faster than small polyatomic ions (Bardsley and Biondi, 1970), resulting in an appreciable reduction of the electron density at the levels where cluster ions are formed. Second, the formation of cluster ions is a significant process in ion-induced heteromolecular nucleation, which typically operates at far lower supersaturations than does homogeneous nucleation (Castleman, 1974).

The $H^+(H_2O)_n$ cluster ions in the Earth's mesosphere have been discussed by Arnold (1975) and Ferguson (1975). Here should be added that Witt (1969) has suggested that these cluster ions might provide nuclei for the growth of the noctilucent clouds.

Under conditions which approximate those of the Martian ionosphere, traces of CO and O_2 are incorporated in ion clusters via ion-molecule reactions initiated by the CO_2^+ ion. Sieck et al. (1973) observed the following reactions:

$$CO_2^+ + 2\ CO_2 \rightarrow CO_2^+(CO_2) + CO_2$$

$$CO_2^+(CO_2) + CO \rightarrow [(CO_2)(CO)]^+ + CO_2$$

$$[(CO_2)(CO)]^+ + CO \rightarrow (CO)_2^+ + CO_2$$

$$(CO)_2^+ + 2\ CO_2 \rightarrow [(CO)_2\ (CO_2)]^+ + CO_2$$

$$(CO)_2^+ + O_2 \rightarrow O_2^+ + 2\ CO$$

$$O_2^+ + 2 \; CO_2 \;\rightarrow\; O_2^+ \; (CO_2) + CO_2$$

$$O_2^+ \; (CO_2) + 2 \; CO_2 \;\rightarrow\; O_2^+ \; (CO_2)_2 + CO_2$$

In the presence of H_2O Sieck (1974; personal communication) observes:

$$CO_2^+ \; (CO_2) + H_2O \;\rightarrow\; CO_2^+ \; (H_2O) + CO_2$$

$$CO_2^+ \; (H_2O) + H_2O \;\rightarrow\; H^+(H_2O)(OH) + CO_2$$

$$H^+ \; (H_2O) \; (OH) + H_2O \;\rightarrow\; H^+ \; (H_2O)_2 + OH$$

The latter scheme has been proposed by Aikin (1972) as the initiating sequence for $NH_4Cl(H_2O)_n$ formation on Venus, at a time when it was suggested that the Venus ultraviolet haze layer might be composed of NH_4Cl particles. Coffey and Mohnen (1972) have shown that $H^+(H_2O)_n$ ions can catalyze $NH_4Cl(H_2O)_n$ formation from NH_3 and HCl by the chain:

$$H^+(H_2O)_{n+1} + NH_3 \;\rightarrow\; H^+(NH_3)(H_2O)_n + H_2O$$

$$H^+(NH_3)(H_2O)_n + HCl \;\rightarrow\; NH_4Cl(H_2O)_m + H^+(H_2O)_{n-m}$$

On the outer planets with their low exospheric temperatures, cluster ions, in particular $H^+(H_2)_n$ ions, are probably prominent at certain altitudes. To construct model Jovian ionospheres Atreya et al. (1974) include the reaction:

$$H^+ + 2 \; H_2 \;\rightarrow\; H_3^+ + H_2 \tag{1}$$

and Capone and Prasad (1973) even add:

$$H_3^+ + 2 \; H_2 \;\rightarrow\; H_5^+ + H_2 \tag{2}$$

These authors use room temperature rate constant data, and do not consider the temperature dependence of the rate constants for these reactions. In analogy with the data obtained by Payzant et al. (1973) for the reactions

$$O_2^+ + 2 \; O_2 \;\rightarrow\; O_2^+ \; (O_2) + O_2 \qquad k_3 \propto T^{-3.2} \tag{3}$$

$$O_2^+(O_2) + 2 \; O_2 \;\rightarrow\; O_2^+(O_2)_2 + O_2 \qquad k_4 \propto T^{-5.1} \tag{4}$$

one expects $k_1 \propto T^{-2}$ and $k_2 \propto T^{-4}$, i.e., rate constants which increase rather quickly with decreasing temperature. This indicates that $H^+(H_2)_n$ clusters may influence the electron densities on the outer planets to a larger degree than surmised at present. Inclusion of CH_4 and NH_3 and their photochemical reaction products (Strobel, 1973a and 1973b) in the ion chemistry could change the picture

dramatically, of course. The possible presence of sodium in the
atmosphere of Jupiter (Atreya et al., (1974) would be even more
perturbing.

The formation of most interstellar molecules can be accounted
for by ion-molecule reactions (Dalgarno, 1975). Two basic problems
subsist however, the formation of complex molecules and the lack
of condensation of the molecules onto the grains. Cluster ions
may provide an answer to these problems. Thus far, cluster ions
have not been suggested to occur in the interstellar medium, because cluster ion formation is usually a three-body process which
requires densities much higher than those prevailing in the interstellar medium. Nevertheless, cluster ion formation may become a
two-body process with radiative stabilization of the intermediate
cluster complex. Consider the formation of H_5^+ via the following
elementary steps:

$$H_3^+ + H_2 \xrightarrow{k_a} H_5^{+*} \qquad \text{(association)}$$

$$H_5^{+*} \xrightarrow{k_d = 1/\tau_d} H_3^+ + H_2 \qquad \text{(dissociation)}$$

$$H_5^{+*} \xrightarrow{k_r = 1/\tau_r} H_5^+ + h\nu \ (IR) \qquad \text{(radiation)}$$

The rate of H_5^+ formation is:

$$\frac{d(H_5^+)}{dt} = k_a (H_3^+)(H_2) \ \frac{\tau_d}{\tau_d + \tau_r}$$

Since k_a may be expected to have the magnitude of a Langevin orbiting rate constant, it is the ratio of the dissociative lifetime
τ_d of the excited complex to its radiative lifetime τ_r which determines whether H_5^{+*} formation is a fast or a slow process. A
simple estimation of τ_d for H_5^{+*} based on RRK theory shows that
below 50K τ_d becomes as long as 10^{-3} sec, which has to be compared
with τ_r which is also on the order of 10^{-3} sec. Formation of H_5^+
via the overall process

$$H_3^+ + H_2 \longrightarrow H_5^+ + h\nu(IR)$$

has therefore to be considered seriously in the study of the interstellar medium.

Higher clusters are expected to be formed even more easily
than H_5^+, and one might hypothesize that the formation chain con-
tinues until it is interrupted by dissociative recombination with
electrons. The formation of the complex stellar molecules could
then be explained by reactive incorporation of C, O, N and other

atoms into the $H^+(H_2)_n$ clusters, or by clustering around these elements in their ionic form. The molecules which constitute the cluster ions are brought into the gas phase upon dissociative recombination, which provides the energy necessary for the evaporation. Grains could be either large cluster ions or neutral remnants of recombined cluster ions, bound to become ionized again and to recombine with electrons, eventually releasing the molecules occluded in the grains. The dynamic cycle or ionization - clustering - nucleation - recombination - evaporation would thus remove the two remaining basic problems of interstellar molecule formation.

E. E. Ferguson: COMMENTS ON SOME REQUIRED RATE CONSTANTS FOR
  PLANETARY IONOSPHERIC REACTIONS

There are reactions which need to be measured for planetary ionosphere modelers. Among these are three reactions with atomic oxygen of interest for the ionosphere of Venus:

$$H_2^+ + O \rightarrow OH^+ + H \tag{1}$$

$$H_3^+ + O \rightarrow OH^+ + H_2 \tag{2}$$

$$He^+ + O \rightarrow O^+ + He \tag{3}$$

The difficulty of measuring rate constants for ions reacting with O is an obvious problem and reaction (3) will be especially difficult as it is almost certainly quite slow.

In a similar vein, the reaction

$$H_2^+ + H \rightarrow H^+ + H_2 \tag{4}$$

has been assumed to have a rate constant $\sim 1 \times 10^{-10}$ cm^3 sec^{-1} by Atreya, Donahue and McElroy (1974) in an analysis of the Jupiter ionosphere and this of course should be measured.

In addition to unknown reaction rate constants, there are several branching reactions which have been measured but for which only crude branching ratios are available. These include

$$He^+ + N_2 \rightarrow N^+ + N + He \tag{5}$$

$$\rightarrow N_2^+ + He$$

$$He^+ + CO_2 \rightarrow O^+ + CO + He \tag{6}$$

$$\rightarrow CO^+ + O + He$$

$$CO_2^+ + H \rightarrow COH^+ + O \tag{7}$$

$$\rightarrow H^+ + CO_2$$

as well as others.

Further, there have been no thermal energy measurements of a number of important charge-transfer reactions of doubly charged ions, e.g.

$$O^{++} + He \rightarrow He^+ + O^+ \tag{8}$$

$$O^{++} + O_2 \rightarrow O^+ + O_2^+ \tag{9}$$

$$O^{++} + O \rightarrow O^+ + O^+ \tag{10}$$

which are of interest in the earth's upper ionosphere where O^{++} has been observed (as has N^{++}) and for which the chemical loss rates are completely unknown.

The needs of planetary atmospheric modelers certainly press the capabilities of laboratory experimentalists at the present time and will probably do so for some time.

REFERENCES

Aikin, A. C. (1972). Nature Phys. Sci. 235, 10.

Atreya, S. K., Donahue, T. M., and McElroy, M. B. (1974). Science 184, 154.

Bardsley, J. N. and Biondi, M. A. (1970). Advances in Atomic and Molecular Physics, 6 (Eds. D. R. Bates and I. Esterman), p. 1, Academic, New York.

Barth, C. A., Hurd, C. W., Pearce, J. B., Kelly, K. K., Anderson, G. P. and Stewart, A. I. (1971). J. Geophys. Res. 76, 2213.

Barth, D. A., Hurd, C. W., Stewart, A. I. and Lane, A. L. (1972). Science 175, 309.

Bates, D. R. and Dalgarno, A. (1962). Atomic and Molecular Processes, Academic Press, New York.

Broadfoot, A. L., Kumar, S., Belton, M. J. S. and McElroy, M. B. (1974). Science 183, 1315.

Capone, L. A. and Prasad, S. S. (1973). Icarus 20, 200.

Castleman, A. W. (1974). Space Sci. Rev. 15, 547.

Chupka, W. A. and Russell, M. (1968). J. Chem. Phys. 49, 5426.

Chupka, W. A., Russell, M. and Refaey, K. (1968). J. Chem. Phys. 48, 1518.

Coffey, P. and Mohnen, V. A. (1972). Bull. Am. Phys. Soc. 17,392.

Dalgarno, A. (1975). These Proceedings.

Ferguson, E. E. (1975). These Proceedings.

Ferguson, E. E. and Fehsenfeld, F. C. (1968). J. Geophys. Res. 73, 707.

Fjeldbo, G. and Eshleman, V. R. (1969). Radio Sci. 4, 879.

Graham, E., James, D. R., Keever, W. C., Gatland, I. R., Albritton, D. L., and McDaniel, E. W. (1973). J. Chem. Phys. 59, 4648.

Howard, H. T., Tyler, G. L., Fjeldbo, G., Kliore, A. J., Levy, G. S., Brunn, D. L., Dickinson, R., Edelson, R. E., Martin, W. L., Postal, R. B., Seidel, B., Sesplaukis, T. T., Shirley, D. L., Stelzried, C. T., Sweetnam, D. N., Zygielbaum, A. I., Esposito, P. B., Anderson, J. D., Shapiro, I. I., and Reasenberg, R. D. (1974). Science 183, 1297.

Hunten, D. M. (1969). J. Atm. Sci. 26, 826.

Huntress, W. T., Jr. (1974). Planet. Space Sci., in press.

Johnsen, R. and Biondi, M. A. (1974). J. Chem. Phys.

Kliore, A., Levy, G. S., Cain, D. L., Fjeldbo, G. and Rasool, S. I. (1967). Science 158, 1683.

Kumar, S. and Hunten, D. M. (1974). J. Geophys. Res. 79, 2529.

Leu, M. T., Biondi, M. A. and Johnsen, R. (1973). Phys. Rev. A8, 413.

McConnell, J. C. (1973). Physics and Chemistry of Upper Atmospheres (Ed. B. M. McCormac), p. 309. D. Reidel Publishing Co., Dordrecht-Holland.

McElroy, M. B. (1973). Space Sci. Rev. 14, 460.

McElroy, M. B. and McConnell, J. C. (1971). J. Atmospheric Sci. 28, 879.

Monahan, K. M., Huntress, W. T.,Jr., Lane, A. L., Ajello, J. M., Burke, T. E., LeBreton, P. and Williamson, A. (1974). Planet. Space Sci. 23, 143.

Payzant, J. D., Cunningham, A. J. and Kebarle, P. (1973). J. Chem. Phys. 59, 5615.

Prinn, R. G. (1970). Icarus 13, 424.

Rasool, S. I. and Stewart, R. W. (1971). J. Atmospheric Sci. 28, 869.

Sieck, L. W., Gorden, R. and Ausloos, P. (1973). Planet. Space S Sci. 21, 2039.

Strobel, D. F. (1973a). J. Atmos. Sci. 30, 489.

Strobel, D. F. (1973b). J. Atmos. Sci. 30, 1205.

Strobel, D. F. (1973c). Physics and Chemistry of Upper Atmospheres (Ed. B. M. McCormac), p. 345. D. Reidel Publishing Co., Dordrecht-Holland.

Theard, L. P. and Huntress, W. T. Jr., (1974). J. Chem. Phys. 60, 2840.

Veverka, J., Wasserman, L. and Sagan, C. (1974). Astrophys. J. 189, 569.

Witt, G. (1969). Space Res. IX (Eds. K. S. W. Champion, P. A. Smith, R. L. Smith-Rose(, p. 157. North-Holland Publishing Co., Amsterdam.

NEW INSTRUMENTATION FOR THE INVESTIGATION OF ION-MOLECULE REACTIONS

SUMMARY OF THE PANEL DISCUSSION LED BY

Jean H. Futrell

University of Utah

Department of Chemistry, Salt Lake City, Utah 84112

All aspects of ion-molecule reactions, including detailed descriptions of instrumentation available for these investigations, has been reviewed relatively recently (Franklin, 1972). For this reason the panel on instrumentation emphasized the description of new developments since this review was published. Instrumentation and techniques which were discussed in other sessions of this Institute are not included in the present discussion.

New instrumental developments were classified for purposes of discussion as single ion source chamber experiments, beam methods, ion cyclotron resonance, flowing afterglow and drift tube measurements, and tandem mass spectrometers. I am indebted to Drs. Keith Ryan, Bill Chupka, A. Ding, D. Brandt, John Eyler, Rudy Burke, Eldon Ferguson, Veronica Bierbaum, Dave Smith, Gerard Mauclaire, Rainer Johnsen and John Todd, and to Professors Sidney Buttrill, Marvin Vestal, Mel Comissarow, Jean Durup, and Walter Koski for their contributions.

SINGLE ION CHAMBER EXPERIMENTS

Trapped Ion Source

Dr. K.R. Ryan of the National Measurement Laboratory, CSIRO (Australia), reviewed recent work with pulsed, single ion chamber, trapped ion source measurements. This technique was suggested by Baker and Hasted (Baker, 1966) and developed by Harrison and his coworkers (Herod, 1970).

In this method positive ions are produced and then held in the space charge well generated by the passage of a beam of low energy electrons through the ion source. During this storage period reactive collisions may occur between the trapped ions and the molecular species flowing through the source. Typically the neutral particle pressure is only 10^{-4} Torr, but because reaction times of 10^{-2} sec are readily achieved, it is a simple matter to observe the reactions of both primary and secondary ions.

Figure 1 gives a simplified representation of the pulse sequence required in the experiment. Ions are produced for 1 μsec by raising the energy of the electrons to 30 eV. For the remainder of the cycle the electron energy is reduced to 4 eV which, for a current density of 2 μA mm^{-2}, is sufficient to trap positive ions but in most systems will be low enough to ensure no further ionization occurs. Reactions between ions and neutrals then proceed until the trap is destroyed by applying a withdrawal pulse to the repeller, thereby causing the ions to move into the mass analyzer.

The trapped ion technique has at least two attractive features: (1) Reaction rate coefficients may be measured under conditions for which the translational energy of the reactive ion is significantly less than 0.1 eV, (2)

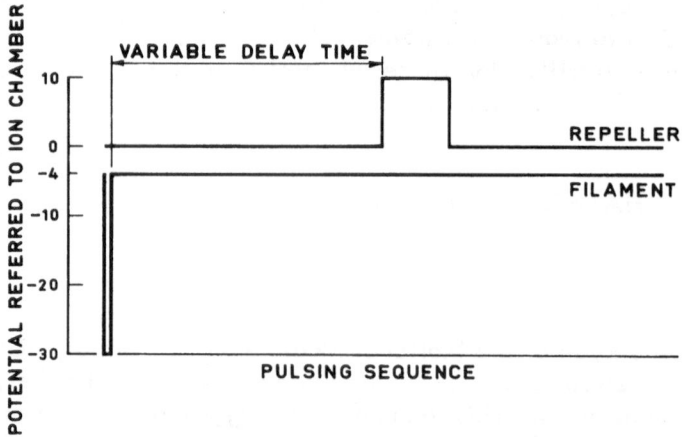

Figure 1. Simplified representation of pulse sequence applied to ion source for ion trapping.

the technique may be incorporated into any conventional ion source to give it chemical ionization capabilities at low pressure and low cost.

Setting up and successful operation of the trapped ion technique require considerable care, particularly with regard to electric fields around the ionization chamber where even slight field penetration into the chamber can destroy completely the performance. In the CSIRO instrument this field-free condition is established by optimizing the trapping characteristics of a non-reactive ion by means of the offset potential on the repeller electrode. Further, it is essential to regulate pulse amplitudes and offset voltages to within 5×10^{-3} volts per 24 hours. Schematic circuit diagrams, a description of their functions, and typical operating parameters have recently been published (Ryan, 1973).

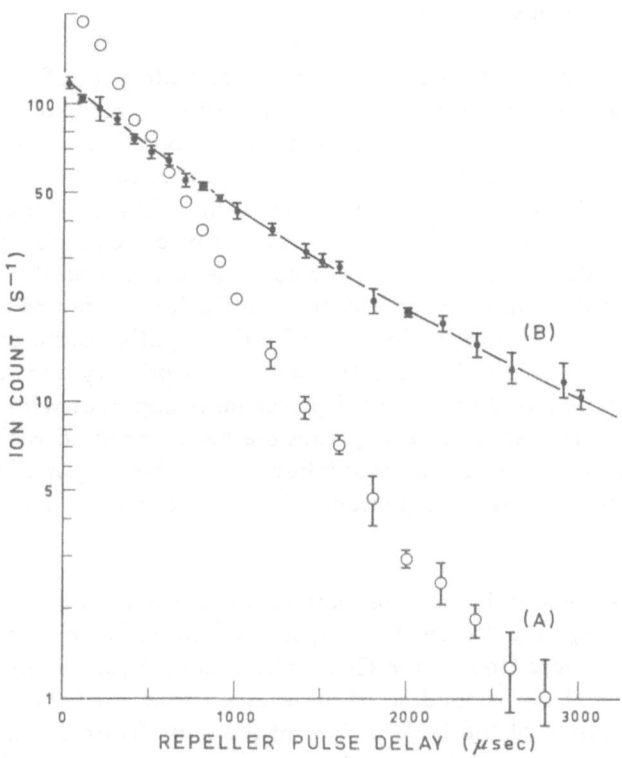

Figure 2. Illustration of simultaneous electron transfer reactions of CO_2^+ with methane to produce CH_4^+ in competition with proton transfer from CH_4^+ to CO_2.

Figure 2 gives an example of the potential of this technique. Shown there are CH_4^+ decay characteristics in a 1:1 (points A) and a 4:1 (points B) $CO_2:CH_4$ mixture. The CH_4^+ signal does not decay in a simple exponential function because of its simultaneous production via charge transfer. Thus the time dependence of the CH_4^+ is controlled by the three reactions

$$CO_2^+ + CH_4 \rightarrow CO_2H^+ + CO_2 \qquad\qquad R(1)$$

$$CO_2^+ + CH_4 \rightarrow CH_4^+ + CO_2 \qquad\qquad R(2)$$

$$CH_4^+ + CO_2 \rightarrow CO_2H^+ + CH_3. \qquad\qquad R(3)$$

The smooth curve through the points (B) represents a fit to the kinetic equation governing the removal of CH_4^+. From this it is a simple matter to deduce the rate coefficients for all three reactions.

Photoionization Mass Spectrometry

Dr. W.A. Chupka, Argonne National Laboratory (U.S.A.) reviewed the principal features of photoionization measurements (Chupka, 1972). Professor Sidney Buttrill of the University of Minnesota then reviewed a modification of this general technique recently introduced at the Jet Propulsion Laboratories, Pasadena, California (LeBreton, 1974). The principal change is that the hydrogen light source is pulsed at a frequency of 350 Hz by means of a Cober Model 605P high power pulse generator and the amplified output pulses from the Channeltron electron multiplier are fed to a multi-channel analyzer. To maximize ion intensity the repeller plate of the ion source is shaped to provide electrostatic focusing of primary and secondary ions onto the exit slit, and the 14 mm light beam is coplaner with the axis of the ion lens of the mass spectrometer. Ions are mass analyzed with an Extra-nuclear Model 324-9 quadrupole mass filter, while the output of the dispersed light source is monitored using a photomultiplier tube coated with sodium salicylate.

The master time pulse for the experiment is provided by the Cober pulse generator. This trigger pulse generates a scan of the multi-channel analyzer. Whenever an ion is detected by the Channeltron multiplier, a single count is added to the channel corresponding to the elapsed time from the beginning of the lamp power pulse until the ion's arrival at the mass filter detector. In a typical experiment carried out with toluene at a pressure of 6×10^{-5} Torr in the ion source, approximately one ion is detected for every ten lamp flashes. Accumulation of the arrival time distribution for a mass selected ion typically will require $2-5 \times 10^5$ light flashes.

BEAM EXPERIMENTS

Reactive Scattering

The importance of crossed beam experiments for understanding detailed reaction kinematics has already been reviewed at this conference. Two examples of new experimental apparatus have recently been placed in operation and will be described briefly. One of these, located at the Heyrovsky Institute of Physical Chemistry and Electrochemistry in Prague, is an improved version of the first true crossed-beam ion-neutral spectrometer developed earlier (Herman, 1969). It has been described briefly elsewhere (Herman, 1973). In this apparatus ions generated by electron impact are extracted and mass analyzed at about 150 electron volts by a 4 cm. radius, 90° permanent magnet and decelerated to the desired energy by means of a modified Lindholm-type lens. Useable ion beams at energies from 0.5 eV upwards are readily obtained. The ion beam is crossed perpendicularly by a jet of thermal energy target molecules created by a multi-channel capillary array containing about 60 channels per mm^2.

The two beam sources are mounted at right angles on the scattering chamber lid, which can be rotated about the scattering center between -15° and $+90^\circ$. The neutral beam is mechanically chopped at 166 cps and lock-in detection of the ion product signal is employed. Reaction and product ions pass from the shielded scattering zone into the detector slit and undergo energy analysis by a retarding potential analyzer. They are then refocused, accelerated into the detector mass spectrometer--a 15 cm radius, 60° magnetic deflection instrument--and are finally detected and counted by means of a 17 stage electron multiplier. Through careful shielding and avoidance of stray potentials it is possible to detect thermal velocity ions with apparently little discrimination. A Fabritek Model 1010 signal averager is used to reduce the signal noise in generating the experimental stopping potential curve of product ions. Integration times of an hour or less are required to obtain a set of data for a single laboratory scattering angle.

The angular spread of the reactant ion beam in the Prague apparatus is about 1° full width at half maximum (FWHM); the energy spread of the ion beam is about 0.2 to 0.3 eV FWHM. The thermal energy reactant neutral beam has an angular spread of about 20° FWHM and the effective size of the scattering volume is $0.04 \times 0.1 \times 0.15$ cm^3. The laboratory angular resolution as defined by the detection slit of the energy and mass analyzers is one degree in the horizontal plane and three degrees in the vertical plane. Some typical results obtained with this apparatus are presented in the panel on charge transfer.

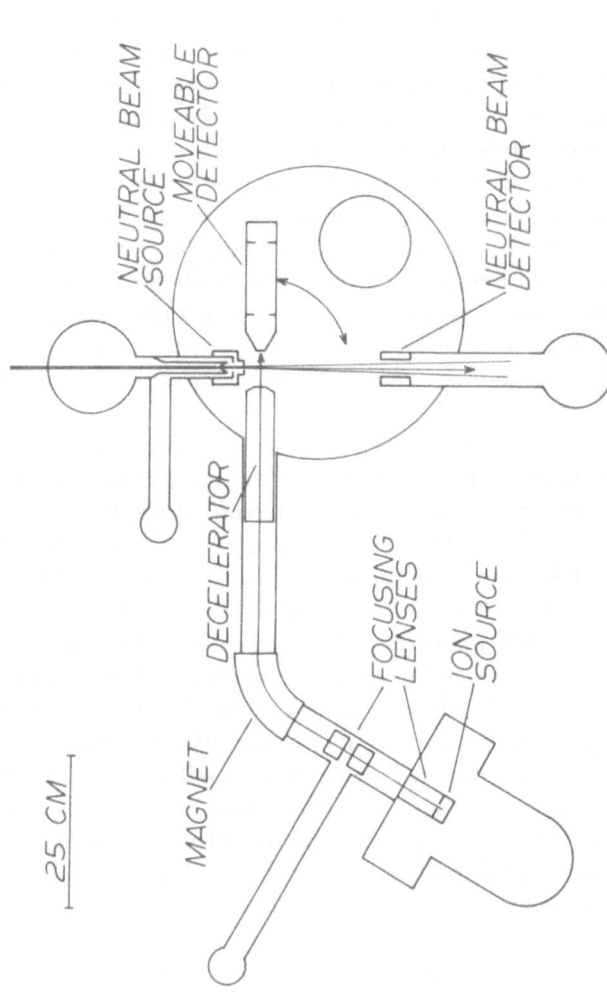

Figure 3. Schematic illustration of ion-neutral crossed-beam apparatus at the University of Utah. Primary ions are generated by a differentially-pumped, high pressure (1-20 Torr) ion source, decelerated to 0.2-30 eV, crossed by a supersonic nozzle neutral beam, and detected by a quadrupole mass filter and energy analyzer which may be rotated through nearly a full octant of a sphere.

A new crossed beam apparatus with several novel features has recently been placed in operation at the University of Utah (U.S.A.); it is shown schematically in Figure 3. In this apparatus a high pressure (1-20 Torr) chemical ionization source is used which, by virtue of the large number of collisions ($\sim 10^6$) experienced by the average ion before exiting, ensures that the ions are vibrationally relaxed. The primary mass analyzer is a 12.5-cm radius, 60 degree sector magnet whose flight tube is electrically isolated to allow the collision region to be at ground potential. With this arrangement, the final nominal ion energy is the same as the ion source voltage relative to ground. The ions are decelerated to ground potential by a decelerating lens which consists of a stack of 42 identical 1.25-cm aperture, plates, spaced 0.5 cm apart by ceramic spacers and connected to an exponentially decreasing resistive voltage divider. Typical ion intensities through a 3-mm diameter aperture located at the collision center remain very nearly constant at about 10^{-9} A down to 0.2 eV before falling off; energy spreads (FWHM) are about 0.3 eV.

The neutral beam source is a differentially pumped free-jet similar to one described in the literature (Parson, 1972). This supersonic nozzle source gives total intensities of approximately 10^{16} molecules sec^{-1} with a beam diameter of 4 mm at the collision center and an angular divergence of 4 degrees.

The secondary ion analyzer consists of a series of collimating apertures, a retarding filter lens energy analyzer, a quadrupole mass filter, and a secondary electron multiplier. With the 3-mm diameter collimating apertures presently used, the analyzer acceptance angle is 4 degrees. The resolutions of the energy analyzer and mass filter are approximately 10 percent and 50, respectively. The detector is mounted so that it may be rotated about an axis through an angle of 70 degrees relative to the neutral beam while remaining perpendicular to the ion beam. Thus the detector may be rotated through nearly a full octant of a sphere. This detector is interchangeable with a differential energy analyzer quadrupole mass filter assembly which can be mounted to coincide with the collision plane. With both detectors, time-averaged, phase-sensitive pulse counting is used to obtain the energy spectra of the mass resolved product ions which are then processed off-line on a Digital Equipment Corporation PDP 11/20 computer.

Merged Beam Apparatus

Professor W.R. Gentry of the University of Minnesota (U.S.A.) has recently placed in operation a merged beam apparatus which is an improved

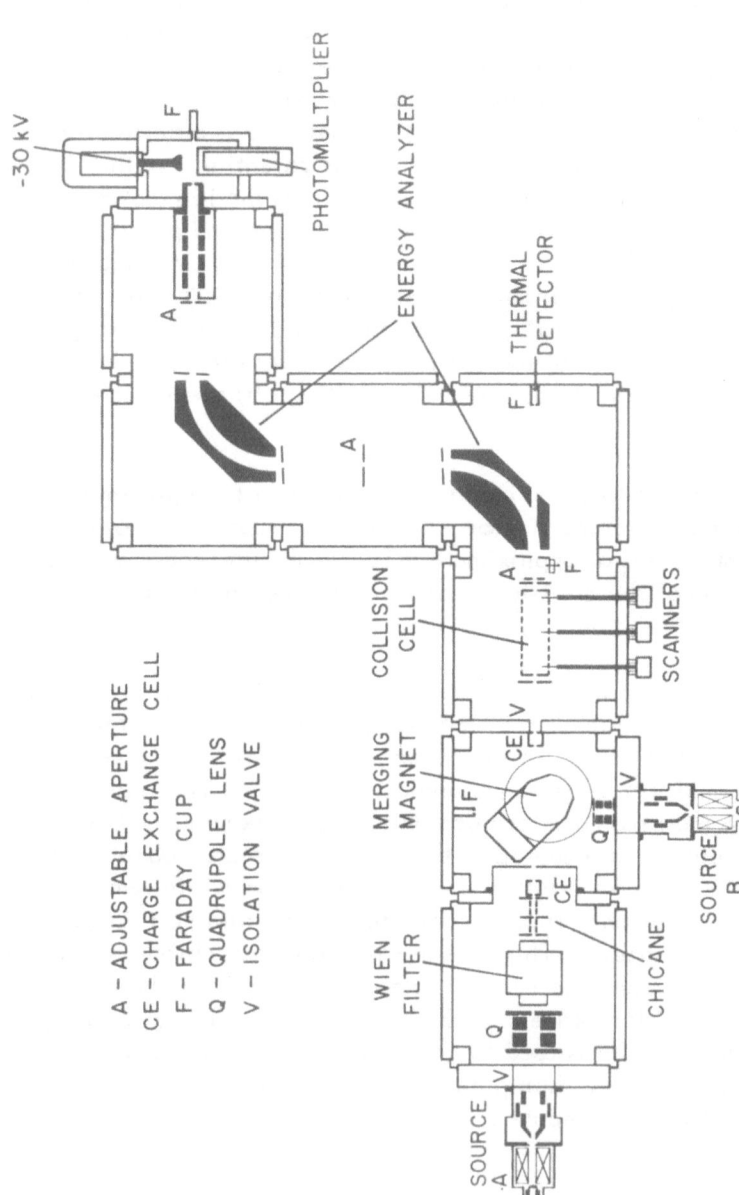

A – ADJUSTABLE APERTURE
CE – CHARGE EXCHANGE CELL
F – FARADAY CUP
Q – QUADRUPOLE LENS
V – ISOLATION VALVE

Figure 4.   Schematic illustration of merging beam apparatus at the University of
Minnesota. Ions from source A are mass-selected by a Wien filter, steered through
a chicane mago, neutralized in a change exchange chamber, and merged with an ion beam
from source B. Final reactant relative energy is adjusted by the potential on the
collision cell, and, the products are identified by their velocities as measured by
twin sperical energy analyzers.

version of the instrument and technique originally described by Trujillo, Neynaber, and Rothe (Trujillo, 1966). This particular beam technique is distinguished by (1) the enormous range of relative kinetic energy which is experimentally accessible, (2) favorable kinematics for the collection of all products, for accurate measurement of total reaction cross sections, and (3) the ability to study reactions of exotic neutral species. These advantages are achieved by sacrificing information on the velocity components of scattered products and loss of precise control over the internal states of the neutral reactants. So far, the apparatus described here has been used to study reactions in the systems $H_2^+ + H_2$, $D_2^+ + HD$ and $D_2^+ + N_2$ (Gentry, 1974).

A diagram of the apparatus, with most features drawn approximately to scale, is shown in Figure 4. Both reactant beams originate as ion beams from electron-bombardment or high-frequency discharge sources and are extracted at the desired laboratory kinetic energy (up to 10 keV). An einzel lens focuses each beam after extraction, and quadrupole doublets are used for stigmatic correction and beam steering. The ion beam from source A is mass-analyzed by a Wien filter, passed through a chicane to remove all accompanying high velocity neutrals and neutralized by charge transfer. The ion beam from source B is mass selected and merged with the A beam in a 90° sector β-focusing magnet. After merging, the beams pass through an isolation valve and collimating apertures into a collision cell, where the ions may be accelerated or retarded relative to the neutrals.

The relative velocity at which the beams are merged is ordinarily chosen to be either sufficiently large or small that the signal arising from products formed outside the collision cell is negligible. A relatively small potential applied to the center element is then used to establish the desired relative kinetic energy within the collision cell without significantly altering the ion beam focusing. The collision cell is 14.6 cm in length, and corrections to the absolute total cross sections for end effects usually amount to ~3 percent. Because only the collision cell voltage must be changed to vary the relative kinetic energy of collision, the energy dependence of the reaction cross-section can be measured much more easily than in instruments which require retuning of the primary ion beam at each energy.

Inside the collision cell are three beam scanners with a spatial resolution of 0.05 mm, used for the measurement of the beam intensity profiles. Besides permitting the direct determination of the three-dimensional beam overlap integral, which is necessary for obtaining absolute reaction cross sections, the scanners are used in focusing the reactant beams so as to minimize the transverse component of relative velocity. The absolute ion beam intensity is measured at one of several Faraday cups. The neutral beam intensity can be determined by either secondary electron emission or the power transferred by the beam to a thermistor bolometer. Relative kinetic energies of collision of ~0.005 eV have been achieved with the apparatus.

Since the reactants and products have approximately the same laboratory velocity in a merged beam experiment, they can be separated according to mass by energy analysis. The energy analyzer consists of two 15.24 cm radius, 90° spherical electrostatic sectors in tandem, the first having its object point at the center of the collision cell. Several aperture sizes are selectable to give resolutions of 0.04 percent to 1 percent, without prior retardation. The two-stage analysis system is extremely helpful in eliminating noise arising from collisions of the primary beams with background gas and various surfaces. The high energy resolution in both the primary beams and the product analyzer has enabled the Gentry group to detect asymmetry in the forward-backward product angular distribution ratio at initial relative kinetic energies as low as 0.010 eV. The detector is a scintillation counter of the Daly type (Daly, 1960), with an intrinsic noise of about 5 counts per minute and essentially unit efficiency as estimated from the pulse height distribution.

The performance of the apparatus depends highly on the system being studied. Ion and neutral beam intensities in the collision cell are typically equivalent to 10^{-6} amp and 10^{-7} amp, respectively, for the major constituents of the beams before mass analysis. The resolution in initial relative kinetic energy typically varies from about 0.005 eV at a nominal zero kinetic energy up to about 0.10 eV at an energy of 10 eV. In the systems studied so far typical product signals have ranged from 1 to 100 counts/sec while the noise from all sources has been typically less than 1 count/sec. The precision of an absolute cross section determination is about 5 percent. Although all sources of known systematic error have been accounted for, the absolute accuracy may be somewhat lower.

Photodissociation Mass Spectrometer

An ion-photon crossed beam apparatus constructed at the University of Utah (U.S.A.) is shown schematically in Figure 5. It consists of three quadrupole mass filters arranged in tandem, with an ion source at one end and an electron multiplier detector at the other. Ions produced by electron impact in the source are accelerated into the first quadrupole, which is set to transmit the reactant ion of interest. The ions are decelerated to ca. 0.3 eV as they enter the second quadrupole, which is excited with RF only and acts as a strong focusing lens. The third quadrupole is set to transmit the product of interest. Ions are detected by a Bendix Model 4700 electron multiplier and counted using an SSR Model 1110 counting system. The light beam from a

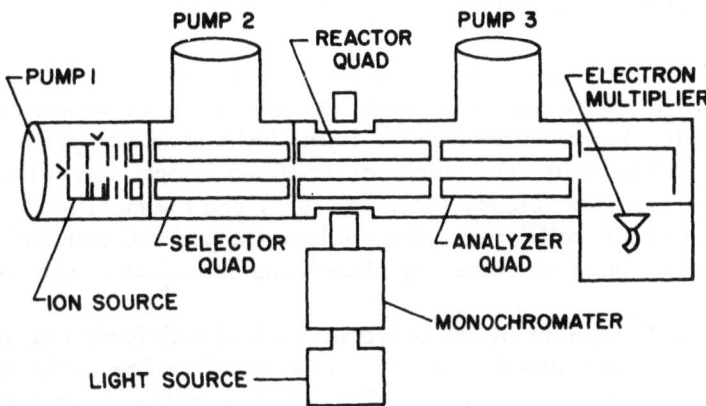

Figure 5. Schematic illustration of photodissociation mass spectrometer at the University of Utah.

Bausch and Lomb high intensity monochromater equipped with a 200 watt super pressure Hg light source and a mechanical chopper intersects the ion beam near the center of the second quadrupole. An alternate light source now being constructed replaces the DC line source with a continuum pulsed light source.

The light passes into the vacuum chamber through an optical grade sapphire window, passes between the quadrupole rods, intersects the ion beam, passes out of the vacuum chamber through a second window and is detected by a radiometer. Differential pumping of the source, primary analyzer, and reaction region is employed. The pressure in the reaction region is typically 2×10^{-8} Torr.

Primary ion beam intensities are of the order of 3×10^{-11} amperes. The principal background signal results from collision-induced dissociations which may amount to as much as 10^{-6} of the primary ion beam. Light fluxes with the DC source are from 2 to 20 milliwatts/cm^2 at the local maxima of the spectrum produced by the Hg source. The intersection volume of the beams is 1 cm long in the direction of ion travel and 0.3 tall in the transverse direction. Photodissociation processes with cross sections of 10^{-20} cm^2 may be studied at 20 Å resolution with the present apparatus, while the improved S/N of the pulsed source should increase the sensitivity by about a hundredfold.

ION CYCLOTRON RESONANCE

The general technique of ion cyclotron resonance and instrumentation for carrying out these measurements were well covered in other sessions of the Institute. Two ICR topics were discussed in the instrumentation panel. Dr. John Eyler of the National Bureau of Standards described trapped ion cell ion cyclotron resonance mass spectroscopy (McIver, 1973) and the modifications necessary to a standard ICR cell for carrying out both trapped and conventional ICR in a single spectrometer developed by Miasek and Beauchamp (Miasek, 1973).

Professor Comisarow of the University of British Columbia (Canada) described the quite new development of Fourier transform ion cyclotron resonance mass spectroscopy (Comisarow, 1974). This technique is an analogous development to Fourier transform magnetic resonance and infrared spectroscopy which share with ICR the common feature that a very broad frequency band width characterizes the spectrum. For this reason conventional ICR detection is achieved by scanning slowly across the spectrum with a detector having a narrow frequency window and a long time constant to reduce rapid noise fluctuations. Pulsed Fourier techniques reduce by orders of magnitude the time required to obtain and absorption spectrum for such a situation. Consequently FT-ICR mass spectrometry promises broad band detection essential to observation of a wide mass range without the long observation period inherent to conventional slow sweep operation.

In FT-ICR, excitation of all ions present is accomplished by applying a burst of rf for several microseconds centered at frequency appropriate to the mass range of interest. The excitation of ion motion and its subsequent decay yield rf interference patterns which result from the characteristic resonance frequencies of the ions in the ICR cell. These are amplified by a broad band rf amplifier and mixed with the output of a second oscillator. The difference frequency is extracted using a 1.5 kHz low pass filter and the resulting signal is digitized using a signal averager to give the digitized time-domain response. Data points obtained by the signal averager are recorded on magnetic tape and Fourier transform analysis is carried out off-line by a digital computer.

The FT-ICR technique can readily be extended to such applications as automated signal averaging, enhancement of mass resolution or S/N ratio, ICR double resonance, and ion molecule reaction kinetics and equilibrium measurements. The major advantages of the technique may be appreciated by noting that no information is lost in this technique. All signals are measured simultaneously and stored; the use of a digital computer then permits the experimenter to manipulate the data in the manner most appropriate to that particular aspect of an ICR experiment which is being investigated.

FLOWING AFTERGLOW AND DRIFT TUBE TECHNIQUES

Principles and operating characteristics of the flowing afterglow apparatus--which appears to be the method of choice for the investigation of thermal rate constants for ion molecule reactions of most species of interest-- was discussed by Dr. Eldon Ferguson of the National Oceanographic and Atmospheric Administration (U.S.A.). This well-known technique has been discussed in detail elsewhere (Ferguson, 1972). A relatively new adaptation of this basic apparatus (McFarland, 1973) is the addition of a drift tube to the flowing afterglow as illustrated schematically in Figure 6. This flow-drift apparatus may be used to study rate constants of ion molecule reactions as a function of mean ion kinetic energy from 0.05 to about 5 eV center-of-mass. The same experiment yields information on ion mobility in the buffer gas and ion reaction rate constants with added reactant gases .

Ions are produced by electron impact at the upstream end of the tube in the ion production section of Figure 6. This section is a unipotential flow tube which operates at a selected elevated voltage with respect to ground. Reactant ions are generated by electron impact or by secondary chemical reactions in the flow section, and chemical reactions which produce reactant ions are allowed to go to completion   in the production section. The concentration of source gas is generally held below 0.2 percent of the buffer gas concentra- tion, thereby insuring that ion drift velocity is determined by the buffer gas alone. Ions then transit a separation section where they are subjected to an electric field which accelerates negative charges in one direction and positive

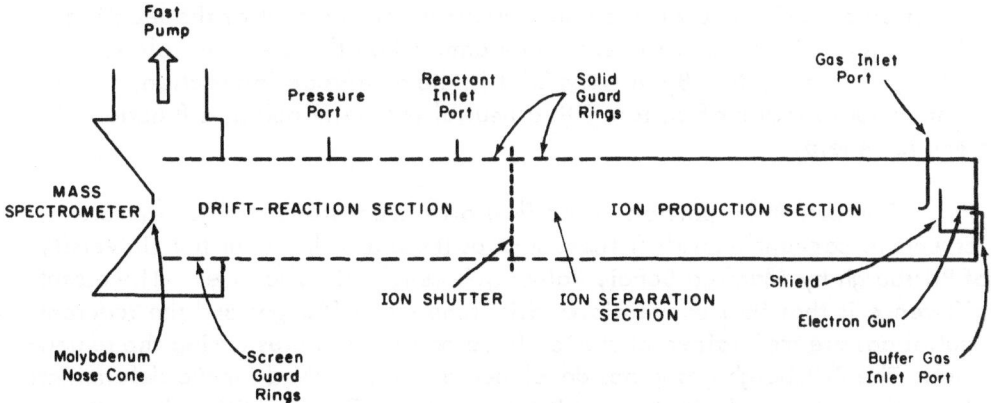

Figure 6. Schematic illustration of NOAA flow-drift apparatus. The ion production section is a conventional flowing afterglow which generates ions for the drift cell reaction section in which a uni- form electric field is imposed on reactant ions.

charges in the other. This ion filter establishes that ions of only one charge reach the ion shutter.

Measurement of drift velocity requires narrow pulses to be injected into the drift and reaction section at specified times. In the present apparatus the ion shutter is constructed from two 100 line per inch tungsten grids spaced 0.3 mm apart. In the open mode these grids are electrically connected to form an open gate through which ions pass. When operated in pulsed mode, the upstream grid is normally biased at +3 (or -3) eV with respect to the down-stream grid, so that a field of 23 volts per centimeter prevents positive (negative) ions from entering the drift-reaction region. An ion pulse of +3.8 V (-3.8 V for negative ions) applied to the upstream grid releases ions into the drift section.

The drift reaction section is a 55.1 cm flow tube in which there is an uniform electric field directed along the axis. This field is maintained by 70 stainless steel guard rings 0.747 cm thick, 7.90 cm inside diameter, and 10 cm outside diameter. Separating each pair of rings is a viton "O" ring for a vacuum seal and a 0.018 cm Mylar spacer for electrical insulation.

In the pulsed mode, ion pulses drift and flow through the drift-reaction section and are sampled, mass-analyzed, and counted according to arrival time by a multi-channel analyzer. The arrival time spectrum of a single mass species yields the drift velocity, from which the mobility at the applied electric field gradient may be computed. In the open mode the ions pass continuously through the shutter. Neutral reactant gas is added at a predetermined rate from an electrically insulated port. The decrease in signal of reactant ion as a function of added neutral reactant permits a measurement of the reaction rate constant at the mean ion energy determined by the electric field to buffer gas ratio E/N. By varying E/N in the drift-reaction section, the rate constant for reaction of an ion with a neutral is determined as a function of mean ion energy.

The drift tube portion of the flow-drift apparatus is similar in most respects to conventional drift tubes such as those developed at the University of Pittsburgh by Manfred Bondi, Rainer Johnsen, and colleagues. The essential difference is that in a conventional drift tube the buffer gas and the reactant neutral gas are maintained at static, fixed partial pressures during the experiment. The Pittsburgh group has developed a novel method for the determination of reaction rate coefficients in drift tube studies. This "additional residence time" method (Heimerl, 1970), colloquially termed the "ping-pong" method, measures the time decay of pulses of parent ions in the presence of the reactant gas for different residence times, while keeping the parent ion drift velocity constant. This is achieved by rapidly reversing the drift field for a particular time delay, then restoring the drift field to its original electrical configuration. After the original drift field has been restored, parent ions transverse the rest of the distance to the entrance apperture of a quadrupole mass spectrometer where they are sampled and mass analyzed.

The number of ions remaining in a cycle during which the total residence time has been varied for a specified time (by twice reversing the field direction and direction of travel) is then compared to the number of reactant ions remaining in the reference cycle in which the field remains constant and ions transverse the drift tube without interruption. The two cycles, with and without interruption of drift velocity and direction, are stored in separate sections of a multi-channel analyzer. Ions observed over a large number of cycles are coherently summed to provide satisfactory statistics in data analysis. The resulting signals correspond to two different residence times at the same mean energy in the drift and reaction region.

The ratio of the ion signals reflects the loss of ions during the additional residence time, Δt, which includes all ion loss processes--mainly lateral diffusion and reaction. The diffusion loss must be determined in a separate experiment, usually by repeating the experiment without the reactant gas present. In the case of very slow reactions it is necessary to resort to a theoretical analysis of lateral diffusion losses (Johnsen, 1973).

TANDEM MASS SPECTROMETERS

Tandem ICR Spectrometer

The tandem mass spectrometer technique has proved invaluable for the elucidation of reaction of ion molecule reaction mechanisms (Futrell, 1972). Several versions of the basic apparatus have been developed recently for addressing additional details of ion molecule reactions. At the University of Utah we have developed a tandem instrument which utilizes ion cyclotron resonance as the detection method for product ions (Smith, 1974). This apparatus is shown schematically in Figure 7.

The first stage of the tandem spectrometer consists of a 180° (radius 5.7 cm) magnetic deflection mass spectrometer which injects a mass analyzed beam of reactant ions into an ion cyclotron resonance mass spectrometer. The entire assembly fits between the poles of a 12 inch Varian electromagnet.

For low pressure and/or high repeller field, the ion source functions as a conventional electron impact mass spectrometer source. Raising the

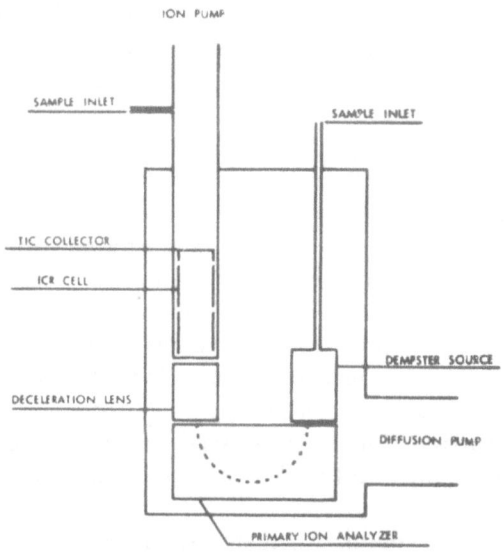

Figure 7. Schematic illustration of tandem ion cyclotron resonance
mass spectrometer at the University of Utah. Ions are mass analyzed
by 180° mass spectrometer and injected into conventional drift ICR
cell for reaction and analysis. The auxiliary ion collector is used
to mass analyze primary ions generated in the electron impact ion
source.

pressure to as much as ca. 100 microns and lowering the repeller voltage
permits the ions formed in the electron beam to undergo up to 50 collisions
with neutrals prior to extraction. Reaction collisions generate a variety of
reactant ions, such as CH_5^+ and H_3^+, which are of interest for elucidating
sequential ion molecule reaction schemes. In addition, energy transfer processes and collisional stabilization may be studied when these ions are allowed
to undergo a specified number of unreactive collisions at higher pressure.

Ions leaving the source are accelerated by a simple, three element
lens system to a few keV. A first order focus of the ion beam is achieved for
those ions executing a deflection angle of 180°. Those ions with a radius of
curvature of 5.7 cm pass through a deceleration lens and into the ICR cell
which is nominally at ground potential. Dempster geometry permits symmetry
in design of the lens system and provides a mass-resolved ion beam which has
the same nominal energy as ions exiting the ion source. After passage through
the final deceleration element, ions are transmitted with high efficiency along
an equipotential line of the ICR cell. These special characteristics permit us
to achieve ion beams of very low energy, typically less than 0.1 eV, using a
simple mass analyzer and deceleration lens system. The ICR cell functions as
both a reaction chamber and an ICR mass spectrometer. Because no slits separ-

ate the collision chamber and the final mass analyzer, product ion detection
is independent of reaction kinematics.

Guided Beam Apparatus

Dr. E. Teloy of the University of Freiburg (Ger) has developed a guided
beam tandem mass spectrometer (Teloy, 1974) shown in Figure 8. The distinctive feature of this apparatus is the use of inhomogeneous oscillatory
electric fields to store and guide slow ions. By using such fields, it is possible
to devise arrangements with a 2 or 3 dimensional minimum of the effective
potential. This in turn makes it possible to achieve (1) ground state primary
ions with a narrow distribution of kinetic energies, (2) increased ion intensity
at low collision energies, and (3) close to unit collection and detection probability for primary and secondary ions, independent of mass, energy and
scattering angle.

In the storage ion source, ions are formed by electron impact and
stored for milliseconds by means of an inhomogeneous rf-field. This field is
produced by a stack of 10 parallel, equally spaced molybdenum plates, in
which U-shaped holes are cut, and 2 solid end plates. The rf voltage (20 to
150 V rms at 10 MHz) is connected alternatively to these plates. The resulting
effective potential is flat near the middle of the channel and has steep repulsive walls at the borders of the enclosed volume. Ions are allowed to effuse
through an outlet opening in the rf walls. Energy-relaxing collisions with the
source gas during the long storage time generally leads to a narrow energy distribution of predominately ground state ions.

The combined mass and velocity filter serves to separate the desired
ion species from the other ones that may have been produced in the ion source.
The filter consists of two rf quadrupoles arranged in series, similar to conventional mass filters; however, in the present apparatus mass selection is not
effected by stability or instability of the ion trajectories. Rather each quadrupole represents a lens with a mass-dependent focal length. Ions with correct
mass ratio and velocity are focused through a small tube, 1 mm D x 8 mm, at
the exit, which serves as a collimator for the injection into the adjacent
octopole and is the only leak between the two separately pumped vacuum
chambers. The collimating entrance tube permits maximum angles of $10°$
with respect to the axis.

The primary ions are accelerated or decelerated to the desired kinetic
energy into the reaction chamber. An rf octopole is used to guide the primary
ions through the scattering chamber, to collect the secondary ions and to guide
them to the mass spectrometer together with unreacted primary ions. It consists

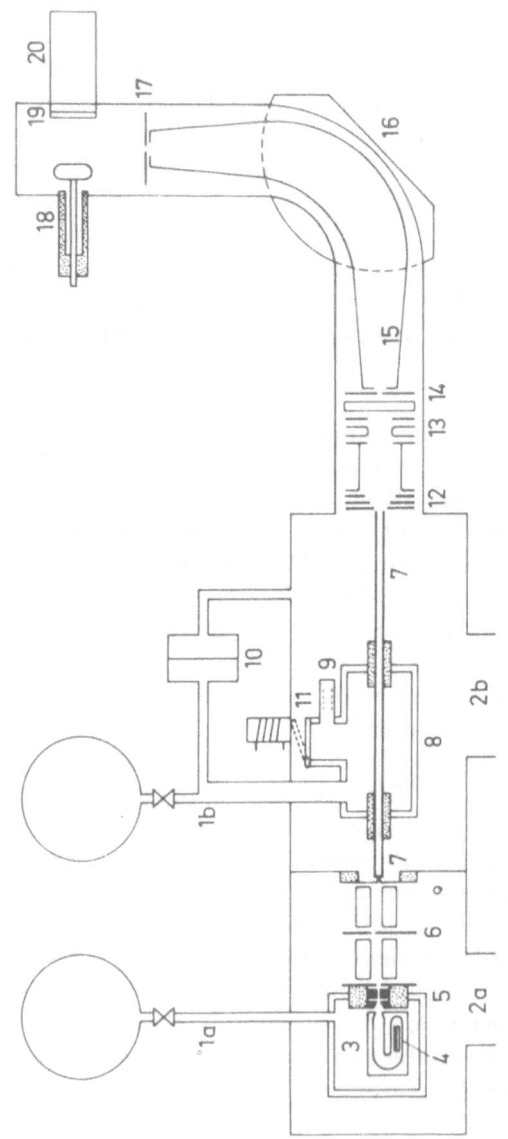

Figure 8. Simplified scheme of the guided beam apparatus. 1a, b gas inlets for ion source and scattering chamber; 2a, b pumping ports, 3 storage ion source, 4 cathode and electron beam; 5 electrodes for shaping and pulsing of the ion beam; 6 rf-mass and velocity filter (8 cm long); 7 rf octopole; 8 scattering chamber (effective length 9.1 cm); 9 ionization gauge; 10 MKS Baratron capacitance manometer; 11 discharge valve for the scattering chamber, operated by an electromagnet; 12, 13 acceleration and shaping lenses; 14 entrance slit of the mass spectrometer (0.15 x 1.4 cm); 15 metal flight tube (-3 kV); 16 poles of the 90° magnet (radius 12.5 cm); 17 exit slit (0.3 x 1.4 cm); 18 voltage feed through and converter (-30 kV); 19 scintillator; 20 photomultiplier.

of 8 parallel cylindrical rods which are equally spaced in an octagonal array.
Neighboring poles are connected to opposite phases of the rf voltage. The
effective potential is proportional to the sixth power of the distance from the
central axis. The longitudinal velocity of the ions is not influenced by the rf
field, their motion being similar to series of specular reflections in a cylindrical
tube. Rf voltages range from 50 to 200 V rms at 10 to 40 MHz.

During the interaction with the oscillatory field, an ion's kinetic
energy in the transverse direction is momentarily changed; however, it has
been shown that the influence on the distribution of collision energies is
negligible.

The average kinetic energy of the primary ions and its approximate
distribution are determined in two independent ways: (1) By measuring the
time-of-flight through the octopole and (2) by retarding potential analysis.
Typical energy spreads are from 0.06 eV FWHM (Ne^+, Ar^+) to 0.2 eV (H^+,
N^+). Thus, in most cases the distribution of relative collision energies is
essentially determined by the thermal motion of the scattering gas, and the
primary ion energy spread can be neglected.

Secondary ions, which are formed in the scattering chamber, (1 to 5 $\times$
10^{-4} Torr) will generally be directed forward, resulting from the center-of-
mass motion. Backward moving ions will--in most cases--be reflected at the
injection end of the octopole by the potential of the collimating tube and the
end-plate of the mass-filter. At low primary energies, however, this potential
may be too low and pulsed mode operation is useful. In this mode a burst of
ions is injected into the octopole, and the potential of the mass filter is raised
immediately after each pulse so that all ions are reflected into the product
analyzer mass spectrometer. This is a comparatively large magnetic mass
spectrometer, with wide slits, utilizing a scintillation detector. Ion transmission
is very high and it is estimated from indirect evidence that at least 95 percent
of all ions of a given mass generate a counting pulse.

Variable Angle Tandem Mass Spectrometer

A tandem mass spectrometer with velocity analysis of both reactant and
product beams and angular analysis of products has been fabricated by Walter
Koski and colleagues at the John Hopkins University (U.S.A.) A schematic
diagram of the spectrometer is shown in Figure 9. It consists of two quadrupole
mass filters and two hemisperical electrostatic analyzers in tandem connected
by a scattering chamber. Cylindrical lenses are used throughout.

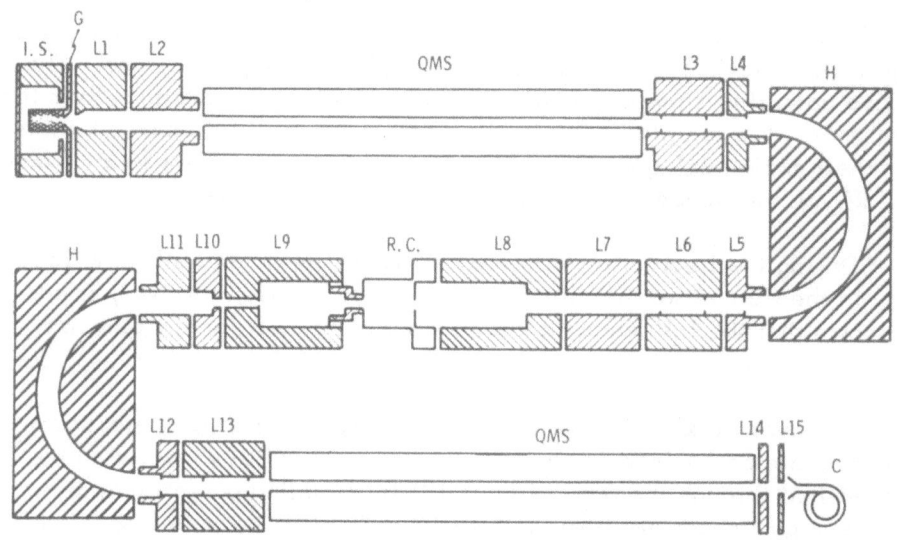

Figure 9. Schematic illustration of rotatable tandem mass spectrometer a Johns Hopkins University. Ions generated by electron impact in the ion source (IS) are mass selected by a quadrupole mass
spectrometer (QMS) and energy analyzed by a hemispherical analyzer,
decelerated to final reactant energy, and injected into the reaction
chamber (RC). Product ions are accelerated, energy analyzed, and
mass analyzed by an identical optical system and counted by a Channeltron electron multiplier.

The ion source (IS) is a high pressure (2 Torr) source; a conventional
Nier source is also available. Ions are extracted and focused by the lens
system L_1 and L_2. This lens system produces an image of the ions leaving the
source at the entrance plane of the quadrupole mass filter (QMS) 12.7 cm.
long. The 180° electrostatic hemispherical analyzer is of the type used
extensively in electron scattering instruments and has been described in the
literature (Simpson, 1964; Kuyatt, 1967). The reaction chamber is made of
concentric cyclinders and is at ground potential. The rotation of the concentric
cylinders provides a detector scanning angle in excess of ±45° relative to the
ion beam. Gas pressure is monitored by a capacitance manometer.

The detector assembly is geometrically identical to the ion gun. However, energy analysis of the product ions must be accomplished with a constant FWHM of the energy distribution passed by the hemispherical analyzer, which requires analysis at some constant value of the ion energy. Since product ions are scattered with a large energy range the ion optics must bring all ions into focus at the entrance plane of the hemisphere at some preset energy. This is accomplished by the lens pair L_9 and L_{10}. This lens has the property that over large changes in voltage ratios of the lens, the position of the principal plane changes very little. Consequently, the position of the image formed at the second principal plane of the lens changes very slightly regardless of the voltage ratio on the lens. L_9 is maintained at laboratory ground with the reaction chamber, and the remaining lens elements $L_{10} - L_{13}$ may be biased appropriately with respect to each other to obtain the proper FWHM.

The entire detector assembly is mounted on a movable plate whose axis of rotation passes through the center of the reaction chamber perpendicular to the ion beam. A worm gear with a 360:1 ratio produces the necessary mechanical motion, and a direct drive is brought through the vacuum wall.

With 1 mm orifices in L_3, L_6 and the reaction chamber, an H_2O^+ projectile ion beam had the angular and energy distributions of $\leqslant 1.0°$ FWHM and 100 meV FWHM, respectively. Doubling the area of these orifices gave an angular spread of $1.5°$ FWHM and energy spread of 175 meV FWHM for a 10 eV F^+ beam.

Ion Luminescence Spectrometer

Professor Christoph Ottinger and colleagues have developed a tandem mass spectrometer at the Max Planck Institut in Gottingen which couples an optical spectrometer to the collision chamber of a tandem mass spectrometer. The apparatus is shown schematically in Figure 10. The ions are produced in a plasma source (Colutron) accelerated to 1 K eV, and mass selected by a 5 cm radius, 16° magnetic sector mass spectrometer. The neutral reactant gas is introduced into a collision chamber at a pressure from 10^{-4} to 10^{-2} Torr. Before entering the collision chamber, ions can be decelerated down as low as 2 eV. Light emission from the collision region is observed using quartz optics and a fast grating spectro photo meter. A cooled photomultiplier detector and dc detector photon counting are used; in the photon counting mode a portion of the spectrum is scanned repetitively and pulses are stored in a multi-channel analyzer.

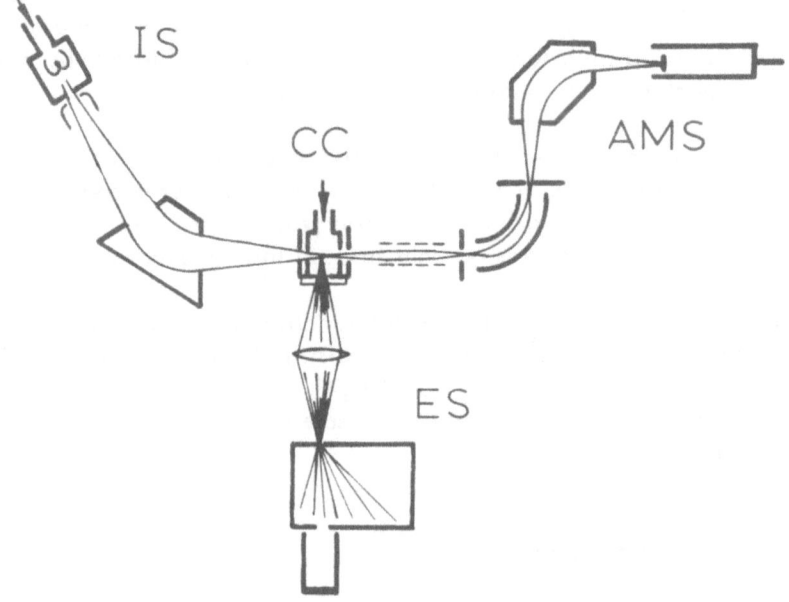

Figure 10. Schematic illustration of ion luminescence spectrometer
at Gottingen. Ions generated in a plasma source are mass analyzed
and injected into a collision chamber (CC). Light emission is
monitored by an optical spectrophotometer (ES) and product ions are
identified by a second stage mass spectrometer.

Also illustrated in the figure is an ion optical system for measuring the ion products from ion-neutral interaction. After re-acceleration to about 1 K eV, ions are focused onto the entrance slit of a 90° electric sector energy analyzer. To this will shortly be added a 90° magnetic sector mass spectrometer as indicated in the figure. The ion source chamber, mass selector, collision chamber, and energy analyzer and mass spectrometer are differentially pumped using 600 liter/sec mercury diffusion pumps.

Some statistical operating parameters are the following: The primary ion intensity is about 10^{-6} A at 1 keV and 10^{-8} A at 2 eV with an optical resolution of 10. This results in photon counting rates of about one thousand counts/sec for a favorable optical transition, well within the realm of dc measurement. With faint emissions, the counting rate is comparable to the dark current of about one count/sec. This requires signal averaging for about 15 hours if the spectrum is to be scanned from 1,000 to 2,000 A. Energy spread in the ion beam is about 1.5 eV FWHM.

Thus far the apparatus at Gottingen has already been used to observe emission of excited states formed by ion excitation, ion molecule reaction, charge exchange, and dissociative charge transfer. Both chemical specificity and the effects of ion translational energy have been demonstrated.

References

Baker, F.A. and Hasted, J.B. (1966). Phil Trans. Roy. Soc. London, A261, 33.

Chupka, W.A. (1972). "Ion Molecule Reactions" (Ed. J.L. Franklin) Plenum Press, New York.

Comisarow, M.B. and Marshall, A.G. (1974). Chem. Phys. Lett. 25, 282.

Daly, N.R. (1960) Rev. Sci. Instr. 31, 264.

Ferguson, E.E. (1972). "Ion Molecule Reactions" (Ed. J.L. Franklin) Plenum Press, New York.

Franklin, J.L. (1972). "Ion Molecule Reactions" (Ed. J.L. Franklin) Plenum Press, New York.

Futrell, J.H. and Tiernan, T.O. (1972) "Ion Molecule Reactions" (Ed. J.L. Franklin) Plenum Press, New York.

Gentry, W.R., et. al. (1974). To be Published.

Heimerl, J., Johnsen, R., and Biondi, M.A. (1970) J. Chem. Phys. 52, 5080.

Herman, Z., Kerstetter, J., Rose, T., and Wolfgang, R. (1969). Rev. Sci. Instrum. 40, 538.

Herman, Z. and Birkinshaw, K. (1973). Ber. Bunsenger. Physik. Chem. 77, 566.

Herod, A.A. and Harrison, A.G. (1970). Int. J. Mass. Spectrom. Ion Phys. 4, 415.

Johnsen, R. and Biondi, M.A. (1973). J. Chem. Phys. 59, 3504.

Kuyatt, C.E. and Simpson, J.A. (1967). Rev. Sci. Instr. 38, 103.

LeBreton, P., Williamson, A.D., Huntress, W.T., and Beauchamp, J.L. (1974). J. Chem. Phys. In Press.

MacFarland, M., Albritton, D.L., Fehsenfeld, F.C., Ferguson, E.E., and Schmeltekopt, A.C. (1973). J. Chem. Phys. 59, 6610.

McIver, R.T., Jr. (1970). Rev. Sci. Instr. 41, 1953.

Miasek, P.G. and Beauchamp, J.L. (1974). Submitted to Rev. Sci. Instr.

Parson, J.M. and Lee, Y.T. (1972). J. Chem. Phys. 56, 4558.

Ryan, K.R. and Graham, I.G. (1973). J. Chem. Phys. 59, 4260.

Simpson, J.A. (1964). Rev. Sci. Instr. 35, 1698.

Smith, D.L. and Futrell, J.H. (1974). Int. J. Mass Spectron. and Ion Phys. 14, 171.

Teloy, E. and Gerlich, D. (1974). Chem. Phys. 4, 417.

Trujillo, S.M., Neynaber, R.H., and Rothe, E.W. (1966). Rev. Sci. Instrum. 37, 1655.

PARTICIPANTS

N. G. ADAMS, University of Birmingham, Birmingham B15 2TT, United
  Kingdom.
M. A. ALMOSTER-FERREIRA, Universidade de Lisboa, Lisboa 1, Portugal.
J. APPELL, Universite de Paris-Sud, 91405 Orsay, France.
F. ARNOLD, Max-Planck-Institut, Heidelberg, West Germany.
P. AUSLOOS, National Bureau of Standards, Washington, D. C. 20234,
  USA.
T. BAER, University of North Carolina, Chapel Hill, North Carolina
  27514, USA.
A. BARASSIN, U. E. R. Sciences Orleans, 45045 Orleans Cedex, France.
J. L. BEAUCHAMP, California Institute of Technology, Pasadena,
  California 91109, USA.
V. M. BIERBAUM, University of Pittsburgh, Pittsburgh, Pennsylvania,
  15260, USA.
M. A. BIONDI, University of Pittsburgh, Pittsburgh, Pennsylvania
  15260, USA.
K. BIRKINSHAW, University College London, London WCIE 6BT, United
  Kingdom.
D. BOHME, York University, Downsview, Ontario, Canada.
L. BOUBY, Universite de Paris-Sud, 91405 Orsay, France.
M. T. BOWERS, University of California, Santa Barbara, California
  92706, USA.
D. BRANDT, Max-Planck-Institut, Gottingen, West Germany.
R. R. BURKE, GRI CNRS, 45045 Orleans Cedex, France.
S. E. BUTTRILL, JR., University of Minnesota, Minneapolis, Minnesota
  55455, USA.
F. CACACE, Universita di Roma, 00100 Rome, Italy.
W. A. CHUPKA, Argonne National Laboratory, Argonne, Illinois 60439,
  USA.
M. COMISAROW, University of British Columbia, Vancouver, B. C.,
  V6T 1W5, Canada.
D. C. CONWAY, Texas A and M University, College Station, Texas 77843,
  USA.
A. DALGARNO, Harvard University, Cambridge, Massachusetts 02138, USA.
N. R. DALY, AWRE, Aldermaston, Reading RG7 4RX, Berks, United Kingdom.
J. DANON, Universite de Paris-Sud, 91405 Orsay, France.

J. J. DeCORPO, U. S. Naval Research Laboratory, Washington, D. C.,
    20375, USA.
R. DERAI, Universite de Paris-Sud, 91405 Orsay, France.
A. DING, Hahn-Meitner-Institute, 1 Berlin 39, West Germany.
H. DISPERT, Hahn-Meitner-Institute, 1 Berlin 39, West Germany.
I. DUCHOVNY, The Weizmann Institute of Science, Rehovot, Israel.
R. C. DUNBAR, Case Western Reserve University, Cleveland, Ohio 44106,
    USA.
J. DURUP, Universite de Paris-Sud, 91405 Orsay, France.
M. DURUP, Universite de Paris-Sud, 91405 Orsay, France.
I. DZIDIC, Baylor College of Medicine, Houston, Texas 77025, USA.
J. R. EYLER, University of Florida, Gainesville, Florida 32601, USA.
F. C. FEHSENFELD, NOAA Environmental Research Laboratories, Boulder,
    Colorado 80302, USA.
S. FENISTEIN, Universite de Paris-Sud, 91405 Orsay, France.
E. E. FERGUSON, Aeronomy Laboratory, NOAA, Boulder, Colorado 80302,
    USA.
A. J. FERRER-CORREIA, University of Warwick, Coventry EV4 7AL,
    United Kingdom.
J. L. FRANKLIN, Rice University, Houston, Texas 77001, USA.
J. H. FUTRELL, University of Utah, Salt Lake City, Utah 84112, USA.
A. GIARDINI-GUIDONI, Istituto Chimico dell Universita, Rome, Italy.
M. L. GROSS, University of Nebraska, Lincoln, Nebraska 68508, USA.
E. GRUNWALD, Brandeis University, Waltham, Massachusetts 02154, USA.
A. G. HARRISON, University of Toronto, Toronto, Ontario M5S 1A1,
    Canada.
J. B. HASTED, Birkbeck College, London W. C. 1, United Kingdom.
L. HELLNER, Equipe de Recherche du CNRS, Paris 5^e, France.
M. J. HENCHMAN, Brandeis University, Waltham, Massachusetts 02154,
    USA.
A. HENGLEIN, Hahn-Meitner-Institut, 1 Berlin 39, West Germany.
J. A. HERMAN, Universite Laval, Quebec, Quebec, Canada.
K. HIRAOKA, University of Alberta, Edmonton, Alberta, Canada.
P. M. HIERL, Kansas University, Lawrence, Kansas 66045, USA.
M. L. HUERTAS, Universite Paul Sabatier, 31077 Toulouse Cedex, France.
D. F. HUNT, University of Virginia, Charlottesville, Virginia 22901,
    USA.
W. T. HUNTRESS, JR., Jet Propulsion Laboratory, Pasadena, California
    91103, USA.
D. HYATT, Polytechnic of North London, Holloway, London N7 8DB,
    United Kingdom.
K. R. JENNINGS, University of Warwick, Coventry CV4 7AL, United
    Kingdom.
R. JOHNSEN, University of Pittsburgh, Pittsburgh, Pennsylvania
    15260, USA.
J. KARLAU, Hahn-Meitner-Institut, D1 Berlin 39, West Germany.
Z. KARPAS, The Weizmann Institute of Science, Rehovot, Israel.
J. J. KAUFMAN, Johns Hopkins University, Baltimore, Maryland 21218,
    USA.
P. KEBARLE, University of Alberta, Edmonton, Alberta, Canada.

P. F. KNEWSTUBB, Cambridge University, Cambridge, United Kingdom.
W. S. KOSKI, Johns Hopkins University, Baltimore, Maryland 21218, USA.
P. J. KUNTZ, Hahn-Meitner-Institut, 1 Berlin 39, West Germany.
K. LACMANN, Hahn-Meitner-Institut, 1 Berlin 39, West Germany.
F. W. LAMPE, Hahn-Meitner-Institut, 1 Berlin 39, West Germany.
Y. LECOAT, Universite de Paris-Sud, 91405 Orsay, France.
S. G. LIAS, National Bureau of Standards, Washington, D. C. 20234,
    USA.
B. H. MAHAN, University of California, Berkeley, California 94720,
    USA.
R. MARX, Universite de Paris-Sud, 91405 Orsay, France.
G. MAUCLAIRE, Universite de Paris-Sud, 91405 Orsay, France.
T. M. MAYER, Hahn-Meitner-Institute, 1 Berlin 39, West Germany.
G. G. MEISELS, University of Houston, Houston, Texas 77004, USA.
M. MENTZONI, University of Oslo, Oslo 3, Norway.
P. MICHAUD, CNRS-CRCCHT, 45045 Orleans Cedex, France.
T. MOCHIZUKI, Hahn-Meitner-Institut, 1 Berlin 39, West Germany.
B. MUNSON, University of Delaware, Newark, Delaware 19711, USA.
R. S. NARCISI, Air Force Cambridge Research Laboratories, Hanscom
    Field, Bedford, Massachusetts 01730, USA.
V. NESTLER, Max-Planck-Institut, Mainz, West Germany.
N. M. M. NIBBERING, University of Amsterdam, Amsterdam, The Netherlands.
J. NIEDZIELSKI, Warsaw University, 02-089 Warsaw, Poland.
J. E. PARKER, Heriot-Watt University, Riccarton, Midlothian, Scotland,
    United Kingdom.
D. PERNER, Kernforschungsanlage Julich, Julich, West Germany.
R. F. PORTER, Cornell University, Ithaca, New York 14850, USA.
C. PROFOUS, Universitat Frankfurt/Main, 6000 Frankfurt am Main 1,
    West Germany.
R. E. REBBERT, National Bureau of Standards, Washington, D. C. 20234,
    USA.
A. RICARD, Universite de Paris-Sud, 91405 Orsay, France.
D. P. RIDGE, University of Delaware, Newark, Delaware 19711, USA.
A. RINTISCH, Hahn-Meiter-Institut, 1 Berlin 39, West Germany.
K. R. RYAN, CSIRO National Standards Laboratory, Chippendale, N. S. W.
    2008, Australia.
A. A. SCALA, Worcester Polytechnic Institute, Worcester, Massachusetts
    01609, USA.
L. W. SIECK, National Bureau of Standards, Washington, D. C. 20234,
    USA.
J. SIMONIS, Max-Planck-Institut, D34 Gottingen, West Germany.
D. SMITH, Universite de Paris-Sud, 91405 Orsay, France.
J. J. SOLOMON, The Rockefeller University, New York, New York 10021
    USA.
T. SUGDEN, Shell Research Laboratories, Thornton, United Kingdom.
I. SZABO, Chemical Center, S-220 07 LUND 7, Sweden.
P. W. TIEDEMANN, Universidade de Sao Paulo, Sao Paulo, Brazil.
T. O. TIERNAN, Aerospace Research Laboratories, Wright-Patterson
    AFB, Ohio 45433, USA.
J. F. J. TODD, University of Kent, Canterbury, Kent, United Kingdom.

E. TSCHUIKOW-ROUX, University of Calgary, Calgary, Alberta, Canada.
W.J. VAN DER HART, Gorleeus Laboratories, Leiden, The Netherlands.
M.L. VESTAL, Salt Lake City, Utah 84103, USA.
J. WEINER, Dartmouth College, Hanover, New Hampshire 03755, USA.
J. WEISE, Hahn-Meitner-Institut, 1 Berlin 39, West Germany.
F.A. WOLF, Universite de Paris-Sud, 91405 Orsay, France.